Auditing
Integrated Concepts
and Procedures

Auditing
Integrated Concepts and Procedures

Donald H. Taylor, Ph.D., C.P.A.
University of Arkansas, Fayetteville

G. William Glezen, M.B.A., C.P.A.
Former partner of Arthur Andersen & Co.
University of Arkansas, Fayetteville

John Wiley & Sons, Inc.
New York   Chichester   Brisbane   Toronto

Library of Congress Cataloging in Publication Data:

Taylor, Donald H., 1933–
Auditing.

1. Auditing. 2. Auditing—Data processing.
3. Sampling (Statistics) I. Glezen, G. William,
joint author. II. Title.

HF5667.T295   657'.45   78-16264
ISBN 0-471-84654-6

Printed in the United States of America

10 9 8 7 6 5 4 3 2 1

To Rosetta, Michael, and Timothy
and
Sylvia, Paul and John,
and Addie Bell

About the Authors

Donald H. Taylor is an Associate Professor of Accounting at the University of Arkansas. He received a B.S. in accounting from Louisiana Tech University and an M.B.A. and Ph.D. from Louisiana State University. He is a CPA in several states.

He has had approximately four years of public accounting experience with national and local CPA firms. During his academic career, he has taught basic and graduate auditing courses. Much of his practical and academic experience is reflected in this text. He is the author of *Mathematical and Statistical Techniques for Accountants,* published by Prentice-Hall. Also, he is the coauthor of the case study in auditing that is available with this book.

G. William Glezen is a part-time instructor in accounting at the University of Arkansas. A native of Texas, he received a B.B.A. in accounting from Texas A & M University and an M.B.A. from the University of Arkansas. He is a CPA in several states and a member of a number of professional accounting organizations.

Many of the concepts, examples, and illustrations in this book are the result of his seventeen years of experience in public accounting with Arthur Andersen & Co. During this period, which included seven years as a partner, he participated at all levels in, and supervised, audit engagements of both large and small enterprises in a broad range of industries. He is also the coauthor of the case study in auditing that is available with this book.

Preface

In considering the format of this book, we agreed on several objectives. First, we wanted the book to retain some of the character of the texts that, until a few years ago, were traditional in the auditing course. However, we believed that the overall approach to the book should emphasize the environment in which auditors perform their tasks. We believed our combination of experience would enable us to accomplish this objective.

Second, we believed that the various topics covered in auditing books should be integrated better to overcome the inclination of readers and users to regard some of these topics as "out of context" with the mainstream of study. Particularly, we had in mind the newer areas of statistical sampling and EDP which, until a short time ago, were treated in a semi-appendix manner within the framework of text outlines.

Third, we wanted to create enough versatility to enable users with different backgrounds and teaching approaches to be "comfortable" with the book. Auditing is not structured as firmly as introductory and intermediate accounting. The course can be, and is, taught in a variety of ways. For example, some instructors prefer to teach the course as a type of continuation of intermediate and advanced accounting, with stress on audit procedures and working-paper techniques. Other instructors put special emphasis on the legal and ethical framework of the auditors' environment, preferring to leave audit procedures and working-paper analysis to public accounting (if the student chooses that field).

Fourth, we wished to develop a textbook package that would minimize the extra readings and material needed by auditing instructors. In view of the ever-changeable nature of the subject, it is not practicable to attempt a book that can serve as a self-contained unit. There will always be a need for auditing students to read journal articles, official pronouncements, and other outside material. One of our objectives, however, was to write a text that would reduce considerably the amount of supplementary material needed (such as extra books on EDP and statistical sampling).

Fifth, we wanted to provide material that would allow students *some* exposure to the procedural aspects of auditing without changing the basic structure of the book.

We believe this book accomplishes our original objectives. It has been class-tested thoroughly, and the material has been taught in different ways. One of us, who has taught for several years, used a conceptual approach. The other, who has had extensive public accounting experience, used a procedural approach.

Integration has been accomplished in several ways. (1) Various aspects of EDP and statistical sampling are discussed in relation to the appropriate auditing functional areas rather than out of context. (2) Legal cases and provisions of the AICPA *Code of Professional Ethics* are summarized at the beginning of the book and are discussed throughout the book as they relate to the auditing areas of internal control, evidence-gathering, and reports. (3) Securities and Exchange Commission influences and rulings are discussed throughout, instead of in one section or chapter only. (4) Certain system descriptions and numerical illustrations are carried over from one chapter to the next. For example, a type of test of compliance is shown in the introductory chapter on internal control. The same illustration is used in the chapter on estimation sampling for attributes (which is a special type of sample design and sample evaluation for testing compliance with internal control).

Versatility has been achieved by designing several potential routes through the book.

1. If the instructor wishes to teach a conceptual approach to auditing, he or she should spend considerable time on Chapters 1–3 and then cover Chapters 4, 8, 9, 15, 16, and 17. Some of the other chapters could be touched on briefly or assigned as outside reading.

2. If the user also desires to include procedures, he or she can spend less time on Chapters 1–3 and cover Chapters 4–9 and 15–17. These chapters provide information on statistical sampling, EDP, and working-paper techniques but omit other discussions of these three topic areas.

3. If the user wishes to emphasize both the conceptual and the procedural aspects of the course, all the chapters should be covered in sequence.

4. If a procedural approach is desired, all chapters could be covered, but less time spent on Chapters 1–3, 4, 8, 9, and 15–17. The course would be concentrated on Chapters 5, 6, 7, and 10–14.

Suggested time schedules for the different routes follow. The class periods represent 50-minute sessions on a semester basis. For classes taught on a quarterly basis, appropriate adjustments should be made.

Topic	Chapter	A Conceptual Approach	Coverage of Some Procedures	An Equal Emphasis on Concepts and Procedures	A Procedural Approach
Introduction	1	2 periods	2 periods	1 period	1 period
Ethics	2	5 periods	4 periods	3 periods	2 periods
Legal liability	3	6 periods	4 periods	3 periods	2 periods
Planning	4	2 periods	2 periods	2 periods	2 periods
Internal control—general	5	1 period	2 periods	4 periods	4 periods
Attribute sampling	6	1 period	2 periods	2 periods	3 periods
EDP controls	7	1 period	3 periods	4 periods	4 periods
Internal control reports	8	2 periods	1 period	1 period	1 period
Evidence-gathering—general	9	6 periods	4 periods	2 periods	1 period
Audit procedures	10–12	1 period	4 periods	6 periods	9 periods
Variable sampling	13	1 period	3 periods	3 periods	5 periods
EDP generalized audit programs	14	1 period	1 period	2 periods	2 periods
Reports—general	15	3 periods	2 periods	2 periods	1 period
Reports—modified opinions	16	5 periods	4 periods	4 periods	3 periods
Special reports	17	5 periods	4 periods	3 periods	2 periods
		42 periods*	42 periods*	42 periods*	42 periods*

*Excluding time for exams, etc.

Although the text was written to be followed in sequence, the user can cover certain chapters differently without affecting the continuity. Here are some examples.

1. Chapter 3 (legal environment) can be covered prior to Chapter 2 (ethical environment).

2. Chapter 15, which covers the general nature of audit reports, can be introduced after Chapter 3 in order to expose students to the output of the audit process early in the course.

3. The chapters on statistical sampling (6 and 13), EDP (7 and 14), and internal control and evidence-gathering (5 and 9–12) can be covered concurrently. Some users believe there is "reinforcement" value in discussing these technical areas together.

4. The chapters on variable sampling (13) and EDP generalized audit programs (14) can be covered after the chapter on general evidence-gathering (9). Then the three chapters on detailed audit procedures (10–12) can be discussed.

A comprehensive seven-part case study on internal control and evidence-gathering is available with this text. Each of the seven sections can be used after Chapters 5, 6, 7, 8, 10, 13, and 14, respectively. The case study includes evaluations of internal control systems, statistical and nonstatistical tests of compliance, use of EDP test data, a letter of internal control, a simulation of accounts receivable confirmation work, variable sampling exercises, and use of a computer program to extract data from hypothetical client files. Insofar as we know, no other auditing case study now in use is similar to this one. Despite the comprehensive nature of the case study, each section can be used alone if it is used with the appropriate chapter in the book. We urge users to try at least some sections of the case study. We believe that it is an interesting and practical simulation.

End-of-chapter material is ample, regardless of which teaching approach is taken.

1. Approximately 600 questions are taken from the reading material in the chapter. The sequence of these questions follows the narrative and provides a convenient mechanism for review.

2. More than 300 objective questions are taken from CPA examinations. These questions range over the many areas covered in the semi-annual CPA examinations. They should provide a good source of study material, as well as topics for class discussion. Many of the questions can be assigned as daily homework. The Instructor's Manual includes the authors' opinions of why the answers to the objective questions are correct.

3. There are more than 200 discussion questions, cases, and problems. Many of them are taken from actual audit situations experienced by the authors, and have been class-tested with students. We believe that the learning process will be maximized by covering these questions during the last class period of each chapter. The Instructor's Manual includes student responses to the class-tested discussion and case questions.

Included among the end-of-chapter materials are statistical sampling problems in Chapters 6 and 13. We suggest the use of some of these problems when the applicable material is covered. The ability to derive definite quantitative results is a good criterion for assessing the degree of understanding attained by the students.

There are other special features of the book.

1. Chapters 9–12 provide a special type of coverage of evidence-gathering and audit procedures. Basically, Chapter 9 presents a conceptual approach to the subject with *some* working-paper illustrations. Chapters 10–12 are procedurally oriented and contain numerous examples of working papers (which are interrelated) and accompanying explanations. The format of these working papers is exactly like the format of those used in public practice. Only numbers and names are changed. An added feature is the explanation of applicable accounting principles and audit objectives that accompanies each set of working papers. The user may cover any combination of the three chapters.

2. The appendix to Chapter 13 is a primer on estimation sampling and tests of hypotheses. The material can be used for review by students who are "rusty" on statistical sampling. It has been our experience that the gap can be very wide between learning such material in a basic statistics course and actually applying it in auditing. We hope this primer can serve to close the gap before coverage of Chapter 13.

3. A new method of planning and conducting an audit is discussed in the text. The method, called Systems Evaluation Approach (SEA), is currently used in public practice to relate internal control evaluation to substantive audit tests. The textbook user can study the entire system in sequence by reading appendix I to the text. References to SEA are contained in Chapters 4, 5, and 9.

4. The text follows closely the AICPA official pronouncements on auditing and accounting. There are more than 150 references to Volumes 1, 2, and 3 of AICPA *Professional Standards,* and the discussions of these pronouncements are integrated with other material throughout the book. The user can refer easily to the full text of the pronouncements.

5. Where applicable, we have included footnotes to some discussions referring the user to articles that explore the issues in greater detail. The same articles are included in the reference list at the end of the chapters.

6. Learning objectives, listed at the beginning of each chapter, should assist the instructor in conceptualizing learning outcomes. The learning objectives also outline important concepts to students before the chapter is studied, and serve as standards against which the instructor can evaluate the degree to which the desired level of learning was achieved.

The book is divided into five major sections. Section I (Chapters 1–3) is an overview of the organizational, ethical, and legal environment in which auditors operate. Section II consists of Chapter 4, on audit planning. Section III (Chapters 5–8) contains discussions and examples of various aspects of internal control. Chapter 5 is a broad overview of the nature of internal control. Chapters 6–8 cover statistical sampling for tests of compliance, characteristics of EDP control systems, and letters of internal control.

Section IV (Chapters 9–14) concerns evidence-gathering. Chapter 9 is similar in structure to Chapter 5 in that it describes the general nature of evidence-gathering.

On the basis of the conceptual material in Chapter 9, Chapters 10–12 cover audit procedures and working-paper techniques, Chapter 13 covers statistical sampling for evidence-gathering, and Chapter 14 discusses computer-generated audit program concepts.

Section V (Chapters 15–17) covers reports. Chapter 15 is a discussion of reports in general, and Chapters 16–17 cover modified opinions and special reports. The treatment of auditors' reports in a three-chapter sequence is another unique feature of this text. We believe that this increasingly important subject matter deserves considerable attention in an auditing book.

The major theme of the text is integration. The concepts of auditing are integrated with practical examples. Auditing objectives are integrated with auditing procedures. EDP and statistical sampling are integrated with internal control and evidence-gathering. Legal issues and ethical considerations are integrated with auditing theory and practice. AICPA pronouncements are integrated with topics in each chapter of the book. We believe that this text conveys a firm understanding of the role of auditing in our society and how auditing draws on many disciplines to accomplish its task.

We express our appreciation to the American Institute of Certified Public Accountants for permission to quote or reproduce material from (1) *Statements on Auditing Standards,* (2) the *AICPA Code of Professional Ethics,* (3) questions from uniform CPA examinations, (4) tables from AICPA statistical sampling volumes, and (5) procedural flow charts from AICPA booklets on internal control.

We thank Arthur Young & Co. and Coopers & Lybrand for permission to illustrate some of their EDP and statistical sampling applications in auditing. Ernst & Ernst was kind enough to allow use of several of their case studies as discussion and problem material. We are grateful to Peat, Marwick, Mitchell & Co. for permission to reproduce and discuss portions of their Systems Evaluation Approach.

Professor Paul Hooper, of Tulane University, reviewed the chapter on the auditor's legal environment. Dr. Gale Sullenberger, Associate Dean of the College of Business Administration and Professor in the Department of Data Processing and Quantitative Analysis at the University of Arkansas, reviewed the chapters on statistical sampling. Robert Kelley, who is in charge of the Administrative Services Division of the New Orleans office of Arthur Andersen & Co., reviewed the chapters on EDP. We are thankful to each of these people for important suggestions.

The manuscript was reviewed by many individuals, all of whom contributed useful suggestions. These include Professors Thomas Hubbard of the University of Nebraska at Lincoln, Anthony Krzystofik of the University of Massachusetts, Dan Guy of Texas Tech University, Hugh D. Grove of the University of Denver, Ronald Huefner of the State University of New York at Buffalo, and Joseph McHugh of Boston College.

Various parts of the manuscript were typed by Mrs. Jeanine Springer, Mrs. Mary McClain, Mrs. Sherry Rowe, and Mrs. Joan Brookman. We express to them our sincere appreciation.

Finally, our appreciation goes to our editors, John Crain, Donald Ford, Chuck Pendergast, and Jean Varven for furnishing valuable aid in the writing and production of the book.

Donald H. Taylor
G. William Glezen

Contents

Chapter 1

The Audit Function

Learning Objectives *After reading and studying the material in this chapter, one should*

Understand the function of auditing and the role that it plays in our society.

Know how the auditor communicates with the users of his work.

Understand the standards of auditing for independent auditors, company auditors, and governmental auditors.

Know the professional and standard-setting organizations that directly influence the audit function.

Of all the subjects taught in the various accounting curricula, auditing is one that can be regarded as truly interdisciplinary. During a semester's study, business law, ethics, accounting theory, statistical sampling, and EDP are all referred to and, in one form or another, are integrated into the course material.

This interdisciplinary approach to the topic of auditing is not followed merely to pull together material taught at earlier points in the business administration curriculum (although there is some justification for performing this synthesis). The integrated teaching techniques are an expression of the fact that practicing auditors, whether public, industrial, or governmental, actually are called upon to use the entire range of business knowledge. Whether individuals are engaged in auditing functions *or* are the preparers of audited data, they need to understand the purposes, techniques, and limitations of this subject matter.

In addition, the work of auditors is in the public spotlight. The financial statement opinions of public accountants serve as a basis for securities trading and credit extensions. The reports issued by governmental auditors contribute to various forms of legislative scrutiny. In modern-day society, it is imperative that many types of financial representations be subjected to an audit. Management, stockholders, credit institutions, regulatory agencies, and legislative and executive branches of federal, state, and local governments require such audits.

Many years ago the word "auditor" conjured an image of an individual wearing a green eye shade and sitting on a high stool. This image is no longer accurate. The modern auditor must be a talented businessman or woman who has the ability to make vital decisions on many important issues, and the courage and depth of character to stand by personal convictions. The auditing function offers individuals an opportunity that is rare in other fields of endeavor; that is, the opportunity, on almost a daily basis, to be responsible for making decisions and judgments as to what is right and what is wrong, and to stand by those decisions and judgments regardless of the pressures that may be brought to bear. This opportunity has attracted many outstanding individuals into the field of auditing and has helped to retain them.

Broad and Narrow Definitions of Auditing

Auditing can be defined broadly as any function that involves the scrutiny of someone else's representations. In a narrower sense, a financial audit can be described as a systematic examination of financial statements, records, and related operations to determine adherence to generally accepted accounting principles, management policies, or stated requirements.[1] One can refer to statements and records as financial representations.

The Reasons For Audits

The main reason for conducting financial audits is that many people who are affected by financial representations or who make financial decisions are not in a position to know whether the information at their disposal is fair and reliable. The following sections contain examples.

The Needs of the Present or Potential Investor

Assume that you are considering an investment in the securities of a company. You are able to make a reasonably intelligent analysis of financial statements, but you do not have sufficient assurance that the figures are fair representations because you were not privy to the accounting system that produced them. One potential problem is that you cannot be certain of what may have been omitted, such as an important contingent liability. How can you tell whether the published information is fair and not misleading?

It is the duty of the independent auditors to provide the potential investor with an unbiased expert *opinion* as to whether the statements are fairly

[1]Robert E. Schlosser, "The Field of Auditing," *Handbook for Auditors,* 1971, p. 1–4.

presented. They are the only group that can perform this function because (1) they have the knowledge and training, (2) they are allowed to examine the necessary records and gather sufficient evidence, and (3) they are independent of management.

In the latter part of the nineteenth century and the early part of the twentieth century, the owners and managers of a company knew each other. In many cases they were the same people. Today, widely dispersed ownership of business has created a class of absentee stockholders who have little, if any, contact with management's operations. Some person or organization must perform an audit function for them. To the extent that their decisions are dependent on management's figures, present and potential investors must place a certain amount of trust in the auditors. They tend to regard the public accountants as guardians of the integrity of the financial statements.

The Needs of Management

Presidents, vice-presidents, department managers, and others use financial data produced by the firm's accounting system in making decisions on such questions as pricing, budgeting, and plant expansion, among others. To provide a reasonable certainty that this system produces reliable figures, operates efficiently, and follows company policy, auditors must perform various checks of the operations. This function, commonly referred to as *internal auditing,* is performed by employees of the firm rather than by independent parties. In many cases, however, internal auditors report directly to a high-echelon executive. Though this organizational arrangement cannot provide internal auditors with the same measure of autonomy as independent auditors, it does increase the probability that internal audit recommendations will be acted upon by top management.

Although there is some overlap, the responsibilities of the internal auditor differ from those of the outside or independent auditor. The former is concerned with many financial and operational aspects of the company; the latter is more concerned with the overall fairness of the firm's financial statements. Though the work of the internal auditor cannot be a substitute for the work of the independent auditor, the internal audit procedures performed must be evaluated by the independent auditor and may affect the nature, timing, and extent of his work.

The Needs of Governmental Officials

In the public sector of our society, many uses are made of the auditing function. As accounting students should know, net income for financial statement purposes is the starting point for determining federal, state, and local income taxes. How, for example, can the Internal Revenue Service assure itself that taxpayers' profit figures are valid bases for the purpose of determining tax liability? The answer, of course, is that periodic audits are made of tax returns. This function is performed on a sample basis because it is impossible for several thousand revenue agents to check millions of sets of records.

Another illustration is the need of Congress for some verification of financial data. Each year, billions of dollars are spent on various social, economic, and military programs. Congress created the General Accounting Office (GAO) and gave it responsibility to act as a type of financial "watchdog" over the executive agencies that spend these funds. Each year the GAO issues financial reports directly to the legislative branch. Many of them are reported in the newspapers, especially when overspending has taken place. Often the duties of the GAO go beyond financial audits, i.e., the agency may ascertain whether governmental agencies have complied with applicable laws and regulations.

The Department of Defense and other agencies of the executive branch award a substantial number of contracts to private industrial firms, such as the one the Defense Department gave to Lockheed to build cargo airplanes. Some of the contracts call for the government agency to pay for the "cost" of the goods or services plus a stipulated fee. To determine that the contractor's costs have been charged to the contract in accordance with regulations, a government agency audits the company's records. Occasionally the findings are newsworthy, e.g., Lockheed's cost overruns several years ago.

One may be inclined to wonder whether there are many actions in the financial world that do not require an audit. The answer is "not many." Whenever one group must make critical decisions based on the financial representations of another group, an audit is needed.

The Role of Public Accounting Firms[2]

Although auditing is performed by a number of organizations and for a variety of purposes, the focus here is on the organizational structure of *CPA firms* because most of the text is devoted to financial auditing.

CPA firms are professional organizations that may take the form of proprietorships, partnerships (the most common) or, in some states, corporations. These firms render a variety of accounting services to the public. Among these are auditing, tax, and management services. The extent to which each of these services is rendered within a given firm depends on its clients' needs.

The Types of Services

1. *Auditing*—In many CPA firms this is the dominant service in terms of both time and revenue. In some cases the audit services are performed for a purpose other than reporting on the fairness of financial statements. For example, the CPA may be asked to audit the records of institutions that receive Medicare funds from the federal government. Political candidates may request an audit of their campaign finances (in keeping with the spirit of the times). Labor unions may need to have their pension funds checked. The CPA is a logical choice to perform all these tasks. In general, however, most audits are conducted by CPAs

---

[2]For a detailed description of the activities of large CPA firms, see T. A. Wise, "The Auditors Have Arrived," *Fortune,* November–December, 1960.

for the purpose of attesting to the fairness of financial statments, and the audit is made because management, investors, creditors, or governmental units ask for one.

2. *Tax Services*—Most accounting students probably are well aware of the demand for tax return preparation because of the large body of tax laws passed by federal, state, and local governments. This service appears to be a "natural" one for public accountants because the methods for determining financial statement net income are basically the same as those for arriving at taxable net income. In addition, many CPAs engage in tax planning and act on behalf of their clients in matters of tax litigation. Because the latter services often require legal training, it is not unusual for an individual to acquire both a CPA certificate and a law degree.

3. *Management Services*—This area is known by several names, such as management consulting or administrative services. They all refer to the newest, and in some cases the fastest-growing, function performed by current public accountants. One way to define management services might be to say that it is practically everything done by the large public accounting firm other than auditing and tax services. Some examples are a computer systems study, a revision of the management accounting system, or the installation of budgeting techniques. Management services may be rendered in addition to the auditing service or as the only function. If the CPA firm renders management services *and* conducts audits, an obvious question arises. How can a CPA firm conduct an independent audit for a company for which it performs other services, i.e., helps set up parts of the accounting system? This is by no means a settled question. At present, most practicing CPAs believe that management services are compatible if advice is given and no decisions are made for the client. This matter is discussed further in Chapter 2.

4. *Bookkeeping Services*—Many CPA firms, particularly smaller ones, perform bookkeeping or "write-up" work for their clients. For some clients, the services may be confined to such things as ledger account postings. Other clients ask CPAs to prepare financial statements for internal managerial use and/or for submission to creditors. These statements have not been audited; they are merely prepared from the client's records, with particular attention given to proper classifications and disclosures. Generally, such services are referred to as preparation of "unaudited financial statements." They are discussed at length in Chapter 17.

The Sizes of CPA Firms

Small firms normally have one office and conduct their business within the immediate area of a town or city. The ownership is held by one or a few CPAs and generally the clients require substantial bookkeeping services and tax preparation. These small firms also render management advisory services, although this type of service usually is not structured in a formal way. Their audit clientele is somewhat limited because most companies that offer securities to the public are audited by large firms. However, local banks sometimes require audited financial statements and much of this work is performed by the local CPAs.

Firms that have offices and clients in a few cities are referred to as regional. Usually one or more of the owners live in the areas served by the firm. Tax preparation and bookkeeping represent a smaller percentage of the total business than do the same services in a local company. Auditing comprises a larger share of the work, and in some cases formalized management or consulting services are provided.

Companies that have offices in the major cities are called national firms; many operate in other countries. The eight largest public accounting firms in the country, all of which have international operations, commonly are referred to as the "Big Eight." These firms are (in alphabetical order) Arthur Andersen & Co., Arthur Young & Co., Coopers & Lybrand, Deloitte Haskins & Sells, Ernst & Ernst, Peat, Marwick, Mitchell & Co., Price Waterhouse & Co., and Touche Ross & Co. All of these firms have annual revenues in the hundreds of millions of dollars. National firms have a considerably larger number of owners than do local or regional firms. National firms offer auditing, tax service, management services, and some bookkeeping. However, auditing accounts for a large percentage of the services because most large business enterprises are clients of these national firms.

The hierarchy of a public accounting firm is not significantly different from that of other organizations. General staff personnel (sometimes called juniors) serve under the direct guidance of experienced members (seniors) and have little or no supervisory responsibility. Seniors generally direct one engagement at a time, such as the audit of a particular client. Supervisors and managers have responsibility for several engagements that run concurrently. Sometimes their duties extend to an entire activity, such as all company audits. Partners (or stockholders if the firm is a corporation) are at the top of the hierarchy and represent the ownership. They sign the audit reports, tax returns, and other documents. In many large organizations one individual has primary responsibility for an office. This person generally is referred to as a managing partner. The entire firm also has a managing partner who acts as the major representative or spokesman.

The Modern-Day Audit of Business Enterprises

The most common type of audit performed by CPAs is an examination of financial statements of business enterprises. Its purpose is to provide an *opinion* of the *fairness* of the statements, in conformity with generally accepted accounting *principles*.[3] The following is an example of such a report.

---

[3]*AICPA Professional Standards,* Volume 1, Commerce Clearing House, Inc., Chicago, AU Section 110.01 *(SAS No. 1).*

To the Shareholders and Board of Directors
of X Company:

We have examined the Statement of Financial Position of X Company and consolidated subsidiaries as of January 31, 19X9 and 19X8, and the related Statements of Income, Shareholders' Equity and Changes in Financial Position for the years then ended. Our examination was made in accordance with generally accepted auditing standards, and accordingly included such tests of the accounting records and such other auditing procedures as we considered necessary in the circumstances.

In our opinion, the aforementioned financial statements present fairly the financial position of X Company and consolidated subsidiaries at January 31, 19X9 and 19X8, and the results of their operations and changes in their financial position for the years then ended, in conformity with generally accepted accounting principles applied on a consistent basis.

Jones & Jones, CPA
April 4, 19X9

It should be pointed out that the formulation of the financial statements is done by individuals who work for management. This process typically is referred to as the accounting function. It is in part a process of *compiling* figures by using the system designed and supervised by company personnel. The accountants who perform this function work for the company that publishes the statements.

In contrast, an audit is a process of performing certain mechanical and analytical procedures on the figures that back up the statements. Included are such steps as observation, inspection, confirmation, comparison, analysis, computation, and inquiry. On the basis of the evidence gathered by applying these audit procedures, a report is issued on the financial statements.

In the following sections we will discuss the meaning and significance of (1) auditing standards and procedures, (2) opinion, (3) fairly, and (4) generally accepted accounting principles.

Auditing Standards and Procedures

For almost forty years an AICPA group called the Committee on Auditing Procedure had the authority to make pronouncements on auditing guidelines. The first of these publications, entitled *Extensions of Auditing Procedure,* was issued in 1939. It became the first of fifty-four pronouncements entitled *Statements on Auditing Procedure* (known as SAPs). In 1948 the Committee issued a booklet entitled *Generally Accepted Auditing Standards—Their Significance and Scope.* The nine standards listed in the publication were divided into three categories of three each. Later a tenth standard was added. (The profession generally makes a distinction between standards and procedures. Standards are broad guidelines and procedures are detailed acts.)

In 1962, *Statement on Auditing Procedure No. 33* codified all the previous pronouncements on standards and procedures issued by the Committee.

In 1973, another codification of the *Statements on Auditing Procedure* was made. It is called *Statement on Auditing Standards No. 1 (SAS No. 1)* and includes edited portions of all fifty-four previous pronouncements. The Committee on Auditing Procedure changed its name to Auditing Standards Executive Committee (AudSec). Since the issuance of *SAS No. 1,* additional pronouncements have been promulgated by this committee. A codification of statements on auditing standards has been published by Commerce Clearing House, Inc. The book is entitled *AICPA Professional Standards, Volume 1.*

The auditing standards adopted by AICPA are explained briefly and listed hereafter.[4] Many of them are self-evident, even to non-accountants. Some of them, however, need considerable explanation. The explanations are given in the following chapters.

The first three are referred to as *general standards.* They provide summary guidelines on the traits and qualifications that auditors should possess. Auditors should have adequate technical training, be independent in mental attitude, and exercise due professional care. Of these three, the standard on independence is the most adaptable to definite guidelines. A considerable part of Chapter 2 is devoted to a discussion of auditors' independence.

The second three, which are called *field standards,* concern the conduct of the audit. Work is to be adequately planned and assistants properly supervised. The company's financial control system is to be evaluated so that auditors can know how much reliance to place on the system and how much testing to do. Enough evidence is to be gathered to provide auditors a reasonable basis for an opinion. Within these broad standards, a wide variety of procedures may be employed. The second standard of field work is discussed extensively in Chapters 5, 6, and 7. The third standard of field work is elaborated upon in Chapters 9–14.

The last four are *reporting standards.* They describe the coverage of the auditor's report. Taken together, they require that the report state the manner in which the financial statements are presented, contain a definite expression of opinion on the statements, and express a clear-cut statement of the degree of responsibility that the auditor is taking. The standards of reporting are illustrated at length in Chapters 15 through 17.

General Standards

1. The examination is to be performed by a person or persons having adequate technical training and proficiency as an auditor.

2. In all matters relating to the assignment, an independence in mental attitude is to be maintained by the auditor or auditors.

3. Due professional care is to be exercised in the performance of the examination and the preparation of the report.

[4]AU Section 150.02 *(SAS No. 1).*

Qualifications of the Auditor

Most independent audits are performed by a person or a group of people with the designation of "Certified Public Accountant" or "CPA." It is not necessary to hold this title to perform bookkeeping services or to prepare income tax returns (in contrast to the requirement of an M.D. degree to practice medicine). However, CPAs enjoy the highest reputation for competence of any of the various groups that engage in these activities. In an attempt to insure a high quality of service, many states require that accountants be CPAs before they perform independent audits.

CPA is a designation granted by the states, which have their own rules as to what criteria must be satisfied. For this reason there is no way for an accountant to acquire a national CPA certificate. Generally, the states' rules have three common components.

1. *An Educational Requirement*—Many states already require a college degree with the equivalent of a major in accounting, and other states are moving in this direction. This trend is indicative of the evolutionary change in the accounting profession during the past several years. Like other professions, accounting had, in earlier times, a "conventional wisdom" that put primary emphasis on the "apprenticeship" approach to public practice. Little or no formal education was required of one who wished to become a CPA. But as the body of knowledge increased, it became more and more apparent that formal education was a necessary supplement to on-the-job training. The major emphasis today is on a college prerequisite.

2. *An Experience Requirement*—The type and length of experience required depend on the state's explicit or implicit philosophy on granting the CPA certificate. Traditionally, issuing the certificate has served to give someone permission to practice as a CPA. Thus, experience in public practice was a highly desirable, if not a necessary, prerequisite. Today, beliefs are mixed about the question of whether the CPA certificate should be associated automatically with practice experience. The fact that some states allow industrial work and/or teaching as substitutes seems to indicate that they view acquiring the certificate as a type of academic achievement.

3. *A Testing Requirement*—On this point there is unanimity among the states. A licensed CPA must have passed a written examination that tests his ability in several accounting areas. A uniform examination now is administered by the American Institute of Certified Public Accountants, and the results are passed on to the respective states.

Quality Control of the Audit Practice

A CPA's most valuable possession is his reputation for integrity and the performance of quality work. If a CPA firm has only a few partners and employees, direct and almost daily control can be exercised over the auditing practice. As CPA firms become larger, this control becomes more difficult and policies and procedures must be established to assure quality work. AU Section 160 *(SAS No. 4)*, which is applicable to firms of all sizes, covers many factors that a CPA must consider in evaluating the quality controls of his firm. Elements of quality control include policies and procedures to assure (1) independence, (2) the assigning of qualified personnel to engagements, (3) consultation with others to the extent required, (4) reasonable supervision, hiring, training, and advancement of qualified personnel, (5) acceptability and reliability of clients, and (6) review and continuous monitoring of all quality-control procedures. Some CPA firms engage other CPA firms to review and publicly report on their quality-control policies and procedures.

Standards of Field Work

1. The work is to be adequately planned, and assistants, if any, are to be properly supervised.

2. There is to be a proper study and evaluation of the existing internal control as a basis for reliance thereon and for the determination of the resultant extent of the tests to which auditing procedures are to be restricted.

3. Sufficient competent evidential matter is to be obtained through inspection, observation, inquiries, and confirmations to afford a reasonable basis for an opinion regarding the financial statements under examination.

The standards of field work are more specific than the general standards. They are given to auditors as guidance in gathering evidence to support an opinion. Basically, this evidence comes from two sources, (1) the reliance that the auditor places on the client's controls (the system of authorizations, accountability, checks and balances, etc. that any company must have to insure reliable records and efficient operations), and (2) the evidence obtained from testing the financial statement balances and the transactions that support these balances.

The extent to which auditors must gather evidence of the second type listed above depends on the reliance they place on the client's controls. This is why the second standard of field work specifies that the auditor's study and evaluation of internal control determine the extent of audit tests.

The third standard of field work underlines the importance of evidence-gathering to support the auditor's opinion of the financial statements. Without proper evidence, the auditor would have no basis to decide on the proper opinion to render and would be committing a dishonest act if he gave one. To provide additional guidance in this regard, the third standard specifies the *general* evidence-gathering techniques an auditor should follow, i.e., confirmation, observation, etc.

Audit Working Papers—The Supporting Evidence for an Opinion

Because the purpose of today's audit is to render an opinion on fairness, the auditor must accumulate written documentation that he has followed the standards of field work and gathered sufficient evidence to support the opinion. Traditionally, this documentation is in the form of audit "working papers."

Audit working papers should show:

1. Data sufficient to demonstrate that the statements were in agreement with the client's books.

2. That the standards of field work were observed.

3. How exceptions or unusual matters were treated.

4. Appropriate comments by the auditor indicating his conclusions concerning significant aspects of the audit.[5]

Within this framework, the auditor tries to design and prepare the working papers in such a manner that any knowledgeable reader has a fairly clear view of the evidence gathered and the conclusions derived. The format of a working paper depends on the related phase of the audit. For example, working papers on cash might include a tally of the contents of the petty cash drawer and/or a bank reconciliation. Working papers on accounts receivable probably would include letters from customers indicating that they agree or disagree with the year-end dollar balance of the customers' accounts receivable as shown on the company records. Such letters are considered good evidence because people independent of the client's organization verify the accuracy of certain of the client's accounting records used to prepare the financial statements. Chapters 9–12 contain numerous illustrations of working papers and descriptions of their use in an audit.

Standards of Reporting

1. The report shall state whether the financial statements are presented in accordance with generally accepted principles of accounting.

2. The report shall state whether such principles have been consistently observed in the current period in relation to the preceding period.

3. Informative disclosures in the financial statements are to be regarded as reasonably adequate unless otherwise stated in the report.

4. The report shall either contain an expression of opinion regarding the financial statements, taken as a whole, or an assertion to the effect

[5]AU Section 338.05 *(SAS No. 1)*.

that an opinion cannot be expressed. When an overall opinion cannot be expressed, the reasons therefore should be stated. In all cases where an auditor's name is associated with financial statements, the report should contain a clear-cut indication of the character of the auditor's examination, if any, and the degree of responsibility he is taking.

Unqualified Opinions

The two-paragraph report shown earlier in the chapter contains an unqualified opinion. The auditors stated that they followed generally accepted auditing standards in the conduct of the examination and that in their opinion the financial statements are fairly stated in accordance with generally accepted accounting principles.

However, conditions may prevent the auditors from completely following generally accepted auditing procedures and/or the auditors may find something during the course of the examination that prevents them from reporting that the financial statements are fair in all respects. In such a case, one of three types of opinion other than unqualified is issued. The other three types of opinions that may be used are (1) a qualified opinion, (2) a disclaimer, or (3) an adverse opinion.

Qualified Opinions

Basically there are three general reasons why auditors might issue a qualified opinion. First of all, circumstances might prevent them from performing all the audit procedures necessary to follow generally accepted auditing standards. For example, the auditors may be unable to observe the count of inventory quantities conducted by the client. If inventory represents a significant portion of the balance sheet total (as it often does), the auditors may have to issue a qualified opinion because of the inadequate scope of their examination.

Also, there may be an uncertainty about the potential impact of a future event on the financial statements. A good example is a pending lawsuit against the client. In such a case, the auditors may have to issue a qualified opinion, not because of inadequate audit procedures but because of the inability to foresee the effect of the uncertainty on the financial statements.

During the course of the audit, it might be revealed that certain accounting techniques followed by the client are not in accordance with generally accepted accounting principles, or that principles are not followed consistently from year to year, or that all proper informative disclosures have not been made in the financial statements. In other words, the auditors have conducted the examination in accordance with generally accepted auditing standards and have *found* omissions and/or discrepancies that require a qualified opinion.

Disclaimers

In some cases the scope of an audit might be so inadequate that auditors will not render any opinion on the financial statements. In other situations the uncertainty might have such a serious potential impact on the financial statements that auditors would refuse to give an opinion. The applicable report in these circumstances is one that *disclaims* an opinion on the financial statements and gives the reasons for so doing. A disclaimer in an auditor's report can have a serious impact on readers' views of the accompanying financial statements. This type of report is rendered only if the auditors are convinced that the inadequate scope or the uncertainty is too serious to warrant a qualified opinion.

Adverse Opinions

If, as the result of audit *findings,* the auditor concludes that the financial statements taken as a whole are not fairly stated in accordance with generally accepted accounting principles, an *adverse* opinion is appropriate. For example, such a conclusion might be formed because the client carries a significant amount of its fixed assets at appraisal value rather than cost.

An adverse opinion, like a disclaimer of opinion, can have a serious effect on the views of readers of the accompanying financial statements. Such an opinion would be issued by the auditors only if they believe that the deviation from generally accepted accounting principles is too serious to warrant a qualified opinion.

A FLOW CHART OF AUDITORS' OPINIONS

1. If the scope of the audit is adequate,

2. If no significant uncertainty exists,

3. If generally accepted accounting principles have been followed and applied consistently, and there is adequate disclosure,

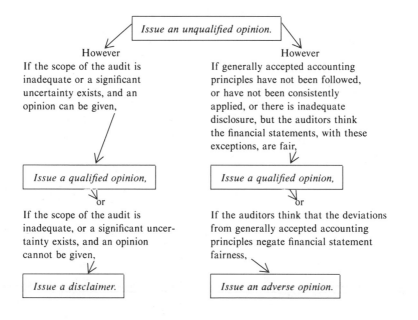

A Guarantee or an Opinion

The question of whether an auditor is issuing a guarantee or an opinion on the financial statements he examines needs some explanation. The word *"opinion"* is particularly important. In the early days of the accounting profession in this country, the reports issued by auditing firms seemed to indicate that a type of accuracy guarantee was being made to management and the ownership. Such words as "certify" and "correct" were used. An evolutionary process has convinced the accounting profession that indication of a *guarantee* is both unwise and improper. An audit of a commercial institution is largely a process of *test-checking* the underlying evidence supporting transactions and balances. Consider, for example, a large inventory figure on the financial statements. There is no practical way that a group of auditors could examine every document supporting that figure or assure themselves that all the quantities of inventory had been counted, so the auditors gather a sample of evidence as a basis for their opinion.

Unfortunately, the distinction between a guarantee and an opinion often escapes uninitiated users of financial statements. From their standpoint, it may be perfectly natural to assume that an audit of financial statements by public accountants leaves no question of accuracy. The auditing profession must continue to educate those who use their audit reports.

Fair or Correct

In the audit report, why do the auditors prefer to use the term "fair" rather than the term "correct" to describe the financial statements? The reason is simply that there is no such thing as a correct set of financial statements. A very wide range of net income figures can be produced by applying different inventory methods, different depreciation methods, and so on.

At the present time several accounting and reporting options are acceptable. It is the auditor's responsibility to determine that one of the approved methods has been employed in the financial statements being examined and that it is appropriate in the circumstances.

The concept of fairness, however, is broader than the application of different inventory or depreciation methods. An attempt to define "fairness" within the context of the auditor's report is included in AU Section 411 and is discussed fully in Chapter 16. Here is a summary of the AU Section 411 *(SAS No. 5)* definitions.

The auditor's opinion that financial statements present fairly an entity's financial position, results of operations, and changes in financial position in conformity with generally accepted accounting principles should be based on his judgment as to whether (a) the accounting principles selected and applied have general acceptance, (b) the accounting principles are appropriate in the circumstances, (c) the financial statements, including the related notes, are informative of matters that may affect their use, understanding, and interpretation, (d) the information presented in the financial statements is classified and summarized in a reasonable manner, that is, neither too detailed nor too condensed, and (e) the financial statements reflect the underlying events and transactions in a manner that presents the financial position, results of operations, and changes in financial position stated within a range of acceptable limits, that is, limits that are reasonable and practicable to attain in financial statements.

Generally Accepted Accounting Principles

One way to define generally accepted accounting principles is to say that they represent a series of recording and reporting methods that are suggested in the pronouncements of leading accounting organizations or have gained widespread acceptance in practice. Many examples are covered in basic accounting courses (cost, matching, etc.).

From the 1930s until the early 1970s, the responsibility for issuing pronouncements on accounting principles rested with two successive committees of the AICPA. The first, called the Committee on Accounting Procedure, operated until 1959. It issued fifty-one pronouncements, each referred to as an *Accounting Research Bulletin.* These bulletins covered such diverse topics as inventory pricing, depreciation methods, and business combinations.

The Committee had no authority to enforce its opinions on public accountants. However, the pronouncements carried substantial weight because they represented the collective thoughts of some of the best minds in the profession. For example, *Bulletin No. 44* contained a conclusion that deferred income taxes should be recorded when declining-balance depreciation was used for tax purposes. If auditors observed that one of their clients was not following this method, they might ask the client to make an adjustment to his records in order to conform to the practice recommended in *Bulletin No. 44.*

Many of the recommendations contained in the *Accounting Research Bulletins* have been modified or superseded. A substantial number of them still remain in force, however, and until they are changed or modified by future pronouncements of a designated authority, they will be considered a part of generally accepted accounting principles.

In 1959 the AICPA abolished the Committee on Accounting Procedure and established in its place the Accounting Principles Board (APB). Like its predecessor, the Board had no authority to compel public accountants to adhere to its pronouncements, although generally this adherence occurred voluntarily. However, in 1964 the AICPA did give the opinions of the Board more impact by specifying that material departures from the pronouncements should be disclosed either in the auditor's report or in a separate note.

During its existence, the APB issued thirty-one pronouncements covering such areas as the investment credit, intangible assets, and disclosure of accounting policies. The topical emphasis was similar to that of the *Accounting Research Bulletins,* i.e., most pronouncements covered narrow subject areas.

In 1973 the APB was dissolved, and the function of formulating pronouncements on generally accepted accounting principles was transferred to an organization outside the AICPA. But the *APB Opinions* now carry the same status as the *Accounting Research Bulletins.* Until they are superseded, they constitute generally accepted accounting principles.

Also, in 1973 both sets of pronouncements were given elevated status when the AICPA amended its *Code of Professional Ethics.* A new provision of the Code states that no member of the AICPA should express an opinion that statements are presented in accordance with generally accepted accounting principles if these statements contain a departure from a principle promulgated by the body designated by the AICPA Council, unless the member can demonstrate that owing to unusual circumstances the financial statements would otherwise have been misleading. The Council has designated *Accounting Research Bulletins* and *APB Opinions* as statements of accounting principles until these documents are superseded. In other words, auditors who belong to the AICPA may find themselves in technical violation of the *Code of Professional Ethics* if they do not adhere to the designated pronouncements. This situation is discussed further in Chapters 2 and 16.

The Auditor's Responsibility for Errors and Irregularities

There is considerable misunderstanding about the purpose of an audit of financial statements. The following boundaries of responsibility for detecting errors (unintentional mistakes) and irregularities (intentional distortions of financial statements, including fraud) are set out in official auditing pronouncements.

The auditor must plan his audit to search for material errors and irregularities, and his report implicitly indicates that the financial statements are not materially misstated as a result of errors or irregularities.

If the auditor has reason to believe that material errors or irregularities exist in financial statements he is examining, he must discuss the matter at a level of management above the one involved (and, ultimately, the board of directors, if it is not resolved at a lower level) and attempt to obtain evidence to make such a determination. If errors or irregularities are found to exist which the management refuses to correct, he should disclose this fact in his report on the financial statements and consider withdrawing as auditor of the company.

If the auditor has reason to believe that immaterial errors or irregularities exist, he should refer the matter to management at least one level above the one involved *and* consider the effect on the company's system of internal control of the role of the personnel involved.

The pronouncement goes on to state that because of the inherent limitations of the audit process:

> the subsequent discovery that errors or irregularities existed during the period covered by the independent auditor's examination does not, in itself, indicate inadequate performance on his part. The auditor is not an insurer or guarantor; if his examination was made in accordance with generally accepted auditing standards, he has fulfilled his professional responsibility.[6]

[6]AU Section 327.13 *(SAS No. 16).*

To illustrate, assume a situation in which the client is deliberately overstating accounts receivable and sales by the use of fictitious customer accounts. If the auditor is lax in gathering evidence on the authenticity of the accounts, the failure to detect this fraud might be the direct result of the failure to follow generally accepted auditing standards. It is possible, though unlikely, that the auditors could use acceptable procedures (examination of documents, confirmations, etc.) and still not detect the fraud because of massive collusion by several key employees. In this case, the auditors would attempt to use as a defense the fact that generally accepted auditing standards were followed.

The Auditor's Responsibility for Illegal Acts

In addition to the possibility of errors or irregularities, management might be guilty of illegal acts. Such acts do not necessarily result in intentional mistakes in or intentional distortions of the financial statements. But if they exist and are discovered, the acts could be a source of concern to the auditor. As discussed in the official pronouncements of the accounting profession, illegal acts refer to illegal political contributions, bribes, and other violations of laws and regulations.

Historically, many auditors have taken the position that illegal acts by management should be of no concern to them unless the fairness of the company's financial statements is affected. The revelations of illegal acts made in recent years and shifting public attitudes have caused many members of the accounting profession to adopt different views on this matter. As a result of these changing attitudes, a pronouncement was issued that specified the action auditors should take if they discover that illegal acts have been performed by a client.

The pronouncement contains a recommendation that the auditor report the illegal acts to a high enough level of management to insure appropriate action (but not to parties other than the client). Also, the auditor should consider changing his opinion of the fairness of the financial statements if he believes that illegal acts that are significant in amount have not been properly accounted for or disclosed in these financial statements. However, even if the auditor does not believe that the illegal act has a significant effect on the financial statements, he should consider withdrawing from the engagement if management does not give appropriate consideration to the act.[7]

Regardless of the official pronouncement, some members of the accounting profession believe that the detection and reporting of illegal acts by client officials is beyond the scope of the auditor's function. They maintain that in many cases the determination of an illegal act is beyond the auditor's competence.

Another viewpoint is that the existence of illegal acts, unattended to by management when such acts are discovered, might affect the auditor's ability to

[7]AU Section 328 *(SAS No. 17)*.

rely on management's financial statement representations. Such reliance is necessary to perform most audits.

Organizations of the Accounting Profession

The State-Level Regulatory Board

Because each state grants its own CPA certificate, the public agency that regulates accounting practice within a state generally is called a *State Board of Public Accountancy* (or a similar title). Although its functions differ from state to state, its typical duties and responsibilities include: (1) providing for the granting of certificates and for the registration of certified public accountants, (2) providing for examinations, and (3) suspending or revoking certificates.

In addition, many state boards have issued a *code of ethics* specifying the behavioral patterns to which holders of a CPA certificate should adhere.

The State-Level Voluntary Organizations

In addition to a regulatory agency, there are voluntary organizations of CPAs within each state, generally called *state societies*. Their major purpose is to improve professional standards and to enhance the reputation of professional accounting within the state.

State societies also issue a code of ethics that are enforceable only against members. Because membership is voluntary, this code is not as effective as that of a state board. An individual is not required to belong to a state society in order to practice as a CPA.

The National-Level AICPA

One of the most influential accounting organizations in the United States is the American Institute of Certified Public Accountants (AICPA). Its voluntary memberships include CPAs from all of the states and the various territories. On a national level, the AICPA performs many of the functions that the state societies perform on a local level. In addition, various committees have issued bulletins on accounting principles and auditing standards. Over the years these pronouncements have formed the basis of "generally accepted accounting principles" and "generally accepted auditing standards."

The AICPA has a code of ethics that has evolved over the years. Because association with the AICPA is voluntary, a person may practice as a CPA without joining this organization. Many of its major provisions, however, have been adopted by state societies and state boards. As a result, practicing CPAs operate under one or more of the codes of ethics of a state board, a state society, or the AICPA. Chapter 2 contains extensive discussion of auditors' ethical considerations.

The Financial Accounting Standards Board

Although the *Accounting Research Bulletins* and the *APB Opinions* gained wide acceptance among AICPA members and others in the business community, some criticism was levied against the committees that formulated these pronouncements. The Committee on Accounting Procedure and the Accounting Principles Board were part of the AICPA and were composed, for the most part, of owners of public accounting firms who worked on a part-time volunteer basis, without meaningful research assistance, and who were subjected to pressures from clients, government agencies, and others.

To overcome these criticisms, the accounting profession took the lead in creating the Financial Accounting Standards Board (FASB). It was hoped that this organization could be more effective in the formulation of accounting principles. The FASB was given the status of an independent organization. Its members consist of *former* owners of CPA firms as well as other business and professional people. To avoid conflicts of interest, all of the members are required to divest themselves of any affiliation with their former organizations and to devote full-time effort to the FASB. The Board is not affiliated with the AICPA.

The business climate in which the FASB is operating is different from that of either the Committee on Accounting Procedure or the Accounting Principles Board. Today, investor discontent with the credibility of financial statements has spawned a record number of lawsuits directed at companies for what is alleged to be faulty reporting in their published financial statements. This situation has caused governmental authorities and courts to take an active role in influencing the formulation and application of accounting principles.

The organization's pronouncements, entitled *FASB Statements,* supersede (where applicable) pronouncements of the Committee on Accounting Research and the Accounting Principles Board. (Pronouncements of all three of these organizations are contained in a book entitled *AICPA Professional Standards, Volumes 3 and 4,* published by Commerce Clearing House, Inc., Chicago.)

The Securities and Exchange Commission

One of the organizations that is taking an active role in influencing the formulation and application of accounting principles is the Securities and Exchange Commission (SEC). In 1934, Congress created this federal agency to regulate the distribution of securities to the public and the trading of securities on open exchanges. This action was taken partly in response to public pressure that arose because of a financial statement "credibility gap." During the 1920s, lax financial reporting practices resulted in overstated profits and overstated stock prices. Many investors lost substantial sums of money on such stocks. By 1934, Congress realized that the consumer needed protection against fraudulent representations of this type.

One concern of the SEC is that investors have adequate financial information about firms that sell securities to the public. Therefore, regulations have been written requiring such firms to file reports with the agency. Most of these reports are public information and can be secured by anyone. Many companies now include this information in the published financial statements issued to stockholders.

The SEC has statutory authority to determine accounting principles for use in published financial statements of companies that sell securities to the public. Its pronouncements on accounting principles are distributed in documents called *Accounting Series Releases*. Until a few years ago, the SEC generally did not exercise this authority, because its traditional position was that the private sector of the profession should set standards for itself. However, with public pressure growing and creditor-investor lawsuits increasing, the SEC has taken a more active role in matters of full disclosure and fair representation. It stated in *Accounting Series Release No. 150* that accounting principles issued by the FASB would be considered as having substantial authoritative support, while those contrary to FASB promulgations would be considered to have no such support. The SEC activity has led to a certain amount of conflict with the private sector of the profession, e.g., several accounting firms and the AICPA formally objected to a provision of *Accounting Series Release No. 177* requiring the approval of any accounting change by a company's independent public accountant. They maintained there was no basis for choosing between equally acceptable alternative methods. The belief among some accountants is that the SEC eventually may take a dominant role in pronouncements on accounting principles. More is said about the SEC's role in Chapter 3. Pronouncements of the SEC are discussed in Chapters 2, 9, and 17.

The Institute of Internal Auditors

Most audits performed by CPA firms are done for the purpose of attesting to the fairness of financial statements. However, another important type of auditing is carried out by employees within a firm. This type is referred to as internal auditing. The Institute of Internal Auditors in its booklet entitled *Statement of Responsibilities of the Internal Auditor* has described this function in the following way.

> Internal auditing is an independent appraisal activity within an organization for the review of operations as a service to management. It is a managerial control, which functions by measuring and evaluating the effectiveness of other controls.

This definition implies a broad scope of activities ranging from financial auditing procedures (similar to that performed by CPA firms) to review of operations. The following objective and scope of internal auditing taken from the booklet referred to above provides additional evidence as to the broadness of the function.

> The objective of internal auditing is to assist all members of management in the effective discharge of their responsibilities, by furnishing them with analyses, appraisals, recommendations, and pertinent comments concerning the activities

reviewed. The internal auditor is concerned with any phase of business activity wherein he can be of service to management. This involves going beyond the accounting and financial records to obtain a full understanding of the operations under review. The attainment of this overall objective involves such activities as:

Review and appraising the soundness, adequacy, and application of accounting, financial, and other operating controls, and promoting effective control at reasonable cost.

Ascertaining the extent of compliance with established policies, plans, and procedures.

Ascertaining the extent to which company assets are accounted for and safeguarded from losses of all kinds.

Ascertaining the reliability of management data developed within the organization.

Appraising the quality of performance in carrying out assigned responsibilities.

Recommending operating improvements.

The independent auditor is able to make an objective appraisal of financial statement fairness because he is not an employee of the firm that prepares the statements. Although the same amount of overt independence cannot be provided the internal auditor, steps can be and are taken by management to allow these auditors to operate with objectivity and to insure that their reports and recommendations will be acted upon by the proper authorities. The proper level of management might be an audit committee of the board of directors or the company's chief executive officer.

In addition to providing a valuable service to management, internal auditors can aid the independent auditors by improving the financial operations of their company. Because the independent auditors' opinion on fairness might provide a company with a better basis for attracting investment funds, the internal auditors are able to provide an additional service indirectly.

More is said in Chapter 5 about the relationships between the functions of the independent and internal auditors.

The General Accounting Office

As pointed out in an earlier part of the chapter, the General Accounting Office (GAO) is under the administrative supervision of the United States Congress and has broad responsibility for auditing various executive branches of the United States government. The GAO has had this responsibility since 1921, when the GAO was established as a result of an administrative change transferring the federal government audit function from the executive to the legislative branch. Since that time, the scope of GAO audits has expanded from examination of financial transactions to a range of activities that cover many areas of Executive Department programs. Two significant duties are (1) ascertaining that Executive Department spending programs are carried out in accord with the intent of Congress, and (2) determining that contract payments to private companies are made in compliance with federal regulations.

The audit responsibilities of the GAO do not include rendering an opinion on the fairness of financial statements of private companies that use such an opinion to obtain or retain investment funds. However, many of the audit standards used by the GAO are similar to the standards issued by the AICPA and discussed in an earlier section of the chapter.

Such similarity between the GAO and AICPA auditing standards implies similar approaches to certain audit work and similar reasons for conducting certain audits, namely ascertaining the fairness of financial reports. However, GAO reports are issued on the financial statements of government agencies. Most reports issued by CPA firms are on financial statements of private companies.

In summary, it is easy to see that the audit responsibilities of the GAO differ from those of independent CPAs or internal auditors. Nevertheless, all the aforementioned audit functions in the chapter have several things in common. These are (1) adequate technical training, (2) good planning and supervision, (3) a certain measure of independence, (4) sound judgment in conducting the audit, and (5) the ability to provide an adequate and understandable report to the appropriate parties.

Commentary

Auditing is not a sterile subject. Its value to the functioning of economic society in the United States has been firmly established, and the demand for auditing services continues to increase. At the same time, its objectives and techniques are being challenged by various groups that have a financial interest in the quality of this service. Accounting students need to understand (1) the ethical and legal environment in which auditors function, (2) the responsibilities that an auditor assumes when engagements are undertaken, (3) the techniques that are used to satisfy the audit objectives, and (4) the nature and limitations of the auditor's output.

It is the purpose of the succeeding chapters to satisfy these needs.

Chapter 1 References

Bevis, Herman W., "The CPA's Attest Function in Modern Society," *The Journal of Accountancy,* February 1962, pp. 28–35.

Carey, John L., "What is the Professional Practice of Accounting?" *The Accounting Review,* January 1968, pp. 1–9.

Catlett, George R., "Relationship of Auditing Standards to Detection of Fraud," *Contemporary Auditing Problems—Proceedings of the 1974 Arthur Andersen University of Kansas Symposium on Auditing Problems,* University of Kansas, Lawrence, Kansas, pp. 47–56.

Kohler, Eric L., "Fairness," *The Journal of Accountancy,* December 1967, pp. 58–60.

Willingham, John J., "Discussant's Response to Relationship of Auditing Standards to Detection of Fraud," *Contemporary Auditing Problems—Proceedings of the 1974 Arthur Andersen University of Kansas Symposium on Auditing Problems,* University of Kansas, Lawrence, Kansas, pp. 57–62.

Wise, T. A., "The Auditors Have Arrived," *Fortune,* November–December 1960.

Appendix to Chapter 1

Glossary of Auditing Terms

Throughout the text, a number of important auditing terms will be used. Definitions of some of these terms are provided here to help the reader better understand the discussions and illustrations that follow. Definitions taken from numerous *Statements on Auditing Standards (SAS)* are identified as such.

ACCOUNTING CONTROLS—the plan of organization and procedures that relate to the safeguarding of assets and the reliability of the accounting records. *AU Section 320.10 (SAS No. 1)*

ACCOUNTING ESTIMATES—estimations made within accounting principles. Service lives and salvage values of fixed assets are examples.

ADMINISTRATIVE CONTROLS—the plan or organization and procedures that are concerned mainly with operational efficiency and adherence to managerial policies. *AU Section 320.10 (SAS No. 1)*

APPLICATION CONTROLS—specific tasks performed by EDP that provide reasonable assurance that the recording, processing, and reporting of data are properly performed. *AU Section 321.08 (SAS No. 3)*

AUDITING PROCEDURES—acts to be performed such as reviewing, inspecting, confirming, etc. *AU Section 150.01 (SAS No. 1)*

AUDITING STANDARDS—describes the quality of the performance of auditing procedures and the objectives to be attained by the use of the procedures. *AU Section 150.01 (SAS No. 1)*

AUDITOR'S INDEPENDENCE—impartial and without bias. Fair to all users of the financial statements. *AU Section 220.02 (SAS No. 1)*

BUSINESS APPROACH TO AUDITING—using knowledge of industry operations, business and economic trends, and other business-related information to aid in determining the scope of audit procedures or the general conduct and approach to the audit.

CONTINGENCY—an existing condition or situation involving uncertainty as to possible gain or loss to a company that will be resolved when future events occur or fail to occur. *AU Section 337B.01 (SAS No. 12)*

CORROBORATING EVIDENCE—documentary material (checks, invoices, contracts, minutes, etc.) *AU Section 330.05 (SAS No. 1)*

DUAL-PURPOSE TESTS—combination tests of compliance and substantive tests. *AU Section 320B.37 (SAS No. 1)*

ERRORS—unintentional mistakes in financial statements, including mathematical or clerical mistakes, mistakes in the application of accounting principles, and oversight or misinterpretation of facts existing at the time financial statements were prepared. *AU Section 327.02 (SAS No. 16)*

EXPLANATORY, MIDDLE, OR SEPARATE PARAGRAPH—a paragraph usually placed between the scope and opinion paragraph of an audit report. It can be used for clarification or to explain the reason for the type of opinion expressed by the auditor.

FIELD WORK—the actual conduct of the audit (performance of audit procedures) at the client's business establishment, where records can be examined, questions can be asked of the client, etc.

GENERALLY ACCEPTED ACCOUNTING PRINCIPLES—the conventions, rules, and procedures necessary to define accepted accounting practice at a particular time. *AU Section 411.02 (SAS No. 5)*

GENERAL PUBLIC—present and prospective investors, creditors, customers, and others interested in particular organizations. *AU Section 640.02 (SAS No. 1)*

INTERIM FINANCIAL INFORMATION—complete financial statements or summarized financial data issued at intervals other than a year (monthly, quarterly, etc.). *AU Section 720.04 (SAS No. 10)*

INTERIM WORK—audit procedures conducted prior to the end of the client's fiscal year under examination. Generally consists of tests of the records and procedures to determine the extent of reliance on them. *AU Section 310.05 (SAS No. 1)*

INVENTORY OBSERVATIONS—auditor's act of observing the client count of physical inventory at year-end or at another appropriate date.

IRREGULARITIES—intentional distortions of financial statements, such as management fraud, or misappropriations of assets, referred to as defalcations. *AU Section 327.03 (SAS No. 16)*

LIMITED REVIEW—application of auditing procedures on a limited basis rather than a full examination in accordance with generally accepted auditing standards. The term is used in connection with interim financial information.

LONG-FORM REPORTS—auditor's reports that include, in addition to the basic financial statements, such items as financial statement details, statistical data, explanatory comments, etc. *AU Section 610.01 (SAS No. 1)*

MANAGEMENT—directors, officers, and others who perform managerial functions. *AU Section 640.02 (SAS No. 1)*

MATERIALITY—often used to describe items in the financial statements that are relatively important and/or items in which the possibility of serious error is greater. The dollar size of the financial statement figure is one factor. An item is considered material if it would cause financial statements readers to change their investment opinion. A definition included in Securities and Exchange Commission regulation is "those matters about which an average prudent investor ought reasonably to be informed."

MODIFIED AUDITOR'S OPINION—an opinion other than that found in the standard short-form report.

NEGATIVE ASSURANCE—wording in an audit report that "tempers" the qualification or disclaimer by using an expression such as "However, nothing came to our attention which would indicate that these amounts are not fairly presented." Not generally an acceptable form of reporting. *AU Section 518.01 (SAS No. 1)*

NEGATIVE ACCOUNTS RECEIVABLE CONFIRMATION REQUEST—the client's customer is asked by the client to respond directly to the auditor only if he disagrees with the balance on the client's records. *AU Section 331.04 (SAS No. 1)*

OTHER INFORMATION—information that appears in annual reports or other documents other than audited financial statements and the auditor's report. *AU Section 550.01 (SAS No. 8)*

POSITIVE ACCOUNTS RECEIVABLE CONFIRMATION REQUESTS—the client's customer is asked by the client to respond directly to the auditor whether or not he is in agreement with the balance on the client's records. *AU Section 331.04 (SAS No. 1)*

PRECISION (NON-TECHNICAL DEFINITION)—the range or limits within which the sample result is expected to be accurate. *AU Section 320B.09 (SAS No. 1)*

PRECISION (TECHNICAL DEFINITION)—a range of values, plus and minus, around the sample result. *AU Section 320B.09 (SAS No. 1)*

PREDECESSOR AUDITOR—an auditor who has resigned or whose services have been terminated by the client. *AU Section 315.01 (SAS No. 7)*

PRINCIPAL AUDITOR—the auditor who is considered to have performed the major portion of the examination as evidenced by such things as the materiality of the portion he examined and his knowledge of the overall financial statements. *AU Section 543.02 (SAS No. 1)*

REGULATORY AGENCIES—governmental and other agencies, such as stock exchanges, that exercise regulatory, supervisory, or other public administrative functions. *AU Section 640.02 (SAS No. 1)*

RELATED PARTIES—the company, its affiliates, principal owners, management and immediate members of their families, entities accounted for by the equity method, or any other party that might significantly influence or be significantly influenced by the company. *AU Section 335.02 (SAS No. 6)*

RELIABILITY OR CONFIDENCE LEVEL (NON-TECHNICAL DEFINITION)—the mathematical probability of achieving the degree of precision accuracy. *AU Section 320B.09 (SAS No. 1)*

RELIABILITY OR CONFIDENCE LEVEL (TECHNICAL DEFINITION)—the proportion of precision ranges from all possible similar samples of the same size that would include the actual population value. *AU Section 320B.09 (SAS No. 1)*

REVIEW OF INTERNAL CONTROL—obtaining information about the organization and prescribed procedures. *AU Section 320.51 (SAS No. 1)*

SPECIAL REPORTS OR SPECIAL-PURPOSE REPORTS—financial statements prepared on a basis other than in accordance with generally accepted accounting principles, or special elements of a financial statement, or audited financial

statements that comply with contractual agreements or regulatory requirements, or financial information prepared in such a manner that requires a
prescribed form of auditor's report. *AU Section 621.01 (SAS No. 14)*

SPECIALIST—a person or firm possessing special skill or knowledge in a particular field other than accounting or auditing. Examples include attorneys,
engineers, and geologists. *AU Section 336.01 (SAS No. 11)*

SUBSEQUENT EVENTS—events or transactions that occur between the balance
sheet date and the auditor's report date, that have a material effect on the
financial statements, and that require adjustment or disclosure in the
financial statements. *AU Section 560.01 (SAS No. 1)*

SUBSEQUENT PERIOD—the time period between the balance sheet date and the
auditor's report date. Auditors are responsible for certain auditing procedures during this period. *AU Section 560.10 (SAS No. 1)*

SUBSTANTIVE TESTS—tests of details of transactions and balances, and analytical
review of significant ratios and trends. *AU Section 320.70 (SAS No. 1)*

SUCCESSOR AUDITOR—an auditor who has accepted or has been invited to make a
proposal for an engagement. *AU Section 315.01 (SAS No. 7)*

TESTS OF COMPLIANCE—tests designed to provide reasonable assurance that
accounting control procedures are being applied as prescribed. *AU Section
320.55 (SAS No. 1)*

THIRD PARTIES—creditors, customers, and other parties with which a particular
enterprise has a relationship. Excluded are directors, officers, and employees of that enterprise.

UNAUDITED FINANCIAL STATEMENTS—financial statements to which the CPA
either has not applied any auditing procedures or has not applied enough
auditing procedures to permit expression of an opinion. *AU Section 516.02
(SAS No. 1)*

UNDERLYING ACCOUNTING DATA—books of original entry, journals, ledgers,
accounting manuals, etc. and informal and memorandum records, work
sheets, supporting cost allocations, computations, etc. *AU Section 330.04
(SAS No. 1)*

WORKING PAPERS—the records kept by the independent auditor of the procedures followed, tests performed, information obtained, and conclusions
reached pertinent to an audit. *AU Section 338.03 (SAS No. 1)*

Chapter 1
Questions Taken from the Chapter

1–1.   Give a broad definition and a narrow definition of auditing.

1–2.   What is the difference between accounting and auditing?

1–3.   Name three groups that have a need for audits. How does the need differ among these groups?

1–4.   Why do present and potential investors have more need of an audit today than they did sixty years ago?

1–5.   What is the difference between the relationship of the internal auditor to a company and the relationship of an external auditor to the same company?

1–6.   Why do members of the Internal Revenue Service conduct audits?

1–7.   What is the major responsibility of the General Accounting Office? To whom does it report?

1–8.   Why do agencies of the executive branch of the federal government conduct audits?

1–9.   What is the purpose of a modern-day audit of profit-making enterprises?

1–10.   Why do auditors issue an opinion rather than a guarantee?

1–11.   Why do auditors avoid the term "correct" in their reports?

1–12.   Define "generally accepted accounting principles."

1–13.   Under what conditions might an auditor be held responsible for fraud detection?

1–14.   What steps should be taken by the auditor if he suspects that there is material fraud affecting his opinion? What if the potential fraud is not material?

1–15.   What are working papers?

1–16.   Name four essential elements of good working papers.

1–17.   Name the three common requirements of most states for becoming a CPA.

1–18.   Distinguish among:
    a.   A state board of public accountancy,
    b.   A state society of certified public accountants, and
    c.   The American Institute of Certified Public Accountants.

1–19.   Briefly distinguish between auditing standards and auditing procedures.

1–20.   Name or indicate the main theme of the three general standards, the three field standards, and the four reporting standards.

1–21.   What was the function of the Committee on Accounting Procedure and the Accounting Principles Board?

1–22.   What is the difference between the structure of the Financial Accounting Standards Board (FASB) and that of the former Accounting Principles Board (APB)?

1–23.   What is the purpose of the Securities and Exchange Commission (SEC)? In what way does its role differ from that of the AICPA?

1–24.   Name three types of services performed by CPA firms.

1–25.   Briefly distinguish among the characteristics of local, regional, and national CPA firms.

1–26.   This question is designed to test your ability to distinguish the differences among the structures and responsibilities of the various organizations discussed in the chapter. Beside each numbered description, list the letter of the appropriate organization(s) taken from the following list (an organization may be listed more than once, and more than one organization may be listed beside each number).
    a.   State boards of public accountancy
    b.   State societies of CPAs
    c.   American Institute of CPAs (AICPA)
    d.   Committee on Accounting Procedure
    e.   Accounting Principles Board (APB)
    f.   Financial Accounting Standards Board (FASB)

    g. General Accounting Office (GAO)

    h. Securities and Exchange Commission (SEC)

        (1) A part of the legislative branch of the federal government.

        (2) Issued fifty-one bulletins concerning accounting principles.

        (3) Regulates the practice of public accountancy within states.

        (4) Independent private organization designated to issue pronouncements on accounting principles.

        (5) A part of the executive branch of the federal government.

        (6) Membership in the organization may be obtained by CPAs who meet the qualifications. Membership is optional.

        (7) From 1959 to 1973, issued thirty-one pronouncements on accounting principles.

        (8) Has the authority to revoke a CPA's license to practice public accountancy within a given state.

        (9) Companies that trade securities to the public on open exchanges are required to file periodic reports with this organization.

     (10) Performs audits of various executive agencies of the federal government.

     (11) CPAs that belong to this organization cannot be associated with a public accounting firm.

     (12) Concerned with financial statement disclosure of companies with securities traded on open exchanges.

     (13) Issues codes of professional ethics.

1–27.   Name six elements of quality control as contained in AU Section 160 *(SAS No. 4)*.

1–28.   Define internal auditing according to *Statement of Responsibilities of the Internal Auditor.*

1–29.   Name three activities of the internal auditor as described in *Statement of Responsibilities of the Internal Auditor.*

1–30.   Name two significant duties of the General Accounting Office.

1–31.   Name the four types of auditor's opinions.

1–32.   Name the three general reasons for a qualified opinion.

1–33.   Name the two reasons for a disclaimer.

1–34.   Name the reason for an adverse opinion.

1–35.   Briefly define each of the terms listed in the Appendix.

Chapter 1
Objective Questions Taken from CPA Examinations

1–36.   When is the auditor responsible for detecting fraud?
    a.   When the fraud did not result from collusion.
    b.   When third parties are likely to rely on the client's financial statements.
    c.   When the client's system of internal control is judged by the auditor to be inadequate.
    d.   When the application of generally accepted auditing standards would have uncovered the fraud.

1–37.   When the auditor's regular examination leading to an opinion on financial statements discloses specific circumstances that make him suspect that fraud may exist and he concludes that the results of such fraud, if any, could *not* be so material as to affect his opinion, he should
    a.   Make a note in his working papers of the possibility of a fraud of immaterial amount so as to pursue the matter next year.
    b.   Reach an understanding with the client as to whether the auditor or the client, subject to the auditor's review, is to make the investigation necessary to determine whether fraud has occurred and, if so, the amount thereof.
    c.   Refer the matter to the appropriate representatives of the client with the recommendation that it be pursued to a conclusion.
    d.   Immediately extend his audit procedures to determine if fraud has occurred and, if so, the amount thereof.

1–38.   What is the independent auditor's responsibility prior to completion of field work when he believes that a material fraud may have occurred?
    a.   Notify the appropriate law enforcement authority.
    b.   Investigate the persons involved, the nature of the fraud, and the amounts involved.
    c.   Reach an understanding with the appropriate client representatives as to the desired nature and extent of subsequent audit work.

   d.   Continue to perform normal audit procedures and write the audit report in such a way as to disclose adequately the suspicions of material fraud.

1–39.   Which of the following criteria is unique to the auditor's attest function?
   a.   General competence.
   b.   Familiarity with the particular industry of which his client is a part.
   c.   Due professional care.
   d.   Independence.

1–40.   What is the general character of the three generally accepted auditing standards classified as general standards?
   a.   Criteria for competence, independence, and professional care of individuals performing the audit.
   b.   Criteria for the content of the financial statements and related footnote disclosures.
   c.   Criteria for the content of the auditor's report on financial statements and related footnote disclosures.
   d.   The requirements for the planning of the audit and supervision of assistants, if any.

1–41.   The primary responsibility for the adequacy of disclosure in the financial statements and footnotes rests with the
   a.   Partner assigned to the engagement.
   b.   Auditor in charge of field work.
   c.   Staffman who drafts the statements and footnotes.
   d.   Client.

1–42.   Auditing standards differ from audit procedures in that procedures relate to
   a.   Measures of performance.
   b.   Audit principles.
   c.   Acts to be performed.
   d.   Audit judgments.

1–43.   Independent auditing can *best* be described as
   a.   A branch of accounting.
   b.   A discipline which attests to the results of accounting and other functional operations and data.
   c.   A professional activity that measures and communicates financial and business data.
   d.   A regulatory function that prevents the issuance of improper financial information.

1–44.   An independent audit aids in the communication of economic data because the audit

a. Confirms the accuracy of management's financial representations.
b. Lends credibility to the financial statements.
c. Guarantees that financial data are fairly presented.
d. Assures the readers of financial statements that any fraudulent activity has been corrected.

1–45. An investor is reading the financial statements of Sundby Corporation and observes that the statements are accompanied by an unqualified auditor's report. From this the investor may conclude that
a. Any disputes over significant accounting issues have been settled to the auditor's satisfaction.
b. The auditor is satisfied that Sundby is operationally efficient.
c. The auditor has ascertained that Sundby's financial statements have been prepared accurately.
d. Informative disclosures in the financial statements but not necessarily in the footnotes are to be regarded as reasonably adequate.

1–46. In order to achieve effective quality control, a firm of independent auditors should establish policies and procedures for
a. Determining the minimum procedures necessary for unaudited financial statements.
b. Setting the scope of audit work.
c. Deciding whether to accept or continue a client.
d. Setting the scope of internal control study and evaluation.

1–47. A CPA is most likely to refer to one or more of the three general auditing standards in determining
a. The nature of the CPA's report qualification.
b. The scope of the CPA's auditing procedures.
c. Requirements for the review of internal control.
d. Whether the CPA should undertake an audit engagement.

1–48. In pursuing its quality-control objectives with respect to assigning personnel to engagements, a CPA firm may use policies and procedures such as
a. Rotating employees from assignment to assignment on a random basis to aid in the staff training effort.
b. Requiring timely identification of the staffing requirements of specific engagements so that enough qualified personnel can be made available.
c. Allowing staff to select the assignments of their choice to promote better client relationships.
d. Assigning a number of employees to each engagement in excess of the number required so as *not* to overburden the staff and interfere with the quality of the audit work performed.

1-49.   The primary reason why a CPA firm establishes quality-control policies and procedures for professional development of staff accountants is to
   a.   Comply with the continuing educational requirements imposed by various states for all staff accountants in CPA firms.
   b.   Establish, in fact as well as in appearance, that staff accountants are increasing their knowledge of accounting and auditing matters.
   c.   Provide a forum for staff accountants to exchange their experiences and views concerning firm policies and procedures.
   d.   Provide reasonable assurance that staff personnel will have the knowledge required to enable them to fulfill responsibilities.

1-50.   The accuracy of information included in footnotes that accompany the audited financial statements of a company whose shares are traded on a stock exchange is the primary responsibility of
   a.   The stock exchange officials.
   b.   The independent auditor.
   c.   The company's management.
   d.   The Securities and Exchange Commission.

1-51.   In pursuing its quality-control objectives with respect to acceptance of a client, a CPA firm is *not* likely to
   a.   Make inquiries of the proposed client's legal counsel.
   b.   Review financial statements of the proposed client.
   c.   Make inquiries of previous auditors.
   d.   Review the personnel practices of the proposed client.

1-52.   The independent auditor lends credibility to client financial statements by
   a.   Stating in the auditor's management letter that the examination was made in accordance with generally accepted auditing standards.
   b.   Maintaining a clear-cut distinction between management's representations and the auditor's representations.
   c.   Attaching an auditor's opinion to the client's financial statements.
   d.   Testifying under oath about client financial information.

1-53.   Which of the following *best* describes the reason why an independent auditor reports on financial statements?
   a.   A management fraud may exist and it is more likely to be detected by independent auditors.
   b.   Different interests may exist between the company preparing the statements and the persons using the statements.
   c.   A misstatement of account balances may exist and is generally corrected as the result of the independent auditor's work.
   d.   A poorly designed internal control system may be in existence.

1–54.　The independent audit is important to readers of financial statements because it
a.　Determines the future stewardship of the management of the company whose financial statements are audited.
b.　Measures and communicates financial and business data included in financial statements.
c.　Involves the objective examination of and reporting on management-prepared statements.
d.　Reports on the accuracy of all information in the financial statements.

1–55.　The securities of Donley Corporation are listed on a regional stock exchange and registered with the Securities and Exchange Commission. The management of Donley engages a CPA to perform an independent audit of Donley's financial statements. The primary objective of this audit is to provide assurance to the
a.　Regional stock exchange.
b.　Investors in Donley securities.
c.　Securities and Exchange Commission.
d.　Board of Directors of Donley Corporation.

Chapter 1
Discussion/Case Questions

1–56.　The following two statements are representative of attitudes and opinions sometimes encountered by CPAs in their professional practices:
a.　Today's audit consists of test-checking. This is dangerous because test-checking depends upon the auditor's judgment, which may be defective. An audit can be relied upon only if every transaction is verified.
b.　It is important to read the footnotes to financial statements, even though they often are presented in technical language and are incomprehensible. The auditor may reduce his exposure to third-party liability by stating something in the footnotes that contradicts completely what he has presented in the balance sheet or income statement.

Required:
Evaluate each of the statements above and indicate areas of agreement or disagreement with the statements.

(AICPA adapted)

1–57.　Odom Co. is a medium-sized producer of fishing rods with approximately $5 million of annual sales. The Company is wholly owned by Mike Odom, who serves as its president and is active in its day-to-day operations. The Company at present has no long-term debt. Mr. Odom has no interest in expanding his company's operations in the future, because to do so would reduce the amount of time he is able to devote to his favorite pastime, fishing.

Some of Mr. Odom's friends who operate businesses of the same size as Odom Co. have annual audits performed of their company financial statements. Should Mr. Odom have the financial statements of Odom Co. audited by an independent CPA? If so, why? If not, why not?

1–58.   Mr. Larry Wilson, president of Wilson Co., which operates hamburger restaurants in Houston, New Orleans, Memphis, and Dallas, believes that his company may have reached the size where an internal audit function should be established. Certain accounting records are maintained at each restaurant, and others are maintained centrally. He has estimated the cost of establishing and operating an internal audit department, and would like to compare these costs with the benefits he could anticipate from such an internal audit function. Because Mr. Wilson has a sales background, he asks you, a recent accounting graduate, to outline for him in general the benefits he could expect from a continuing internal audit of Wilson Co.

Discuss the general benefits Mr. Wilson could expect.

1–59.   Timmon's Bakery has just completed its first year of operations and it must choose the generally accepted accounting principles it will use in the preparation of its financial statements. Depending on the accounting principles that are selected, either of the two following statements of income could result.

	A	B
Sales	$10,000,000	$10,000,000
Cost of Sales	8,000,000	8,500,000
Gross Profit	$ 2,000,000	$ 1,500,000
Selling Expenses	1,000,000	1,000,000
General and Administrative Expense	500,000	600,000
Net Income (Loss) Before Income Taxes	$ 500,000	$ (100,000)
Provision for Income Taxes	250,000	—
Net Income (Loss)	$ 250,000	$ (100,000)

The different income statements result from the following accounting methods:

	A	B
Inventories	First-in, First-out Method	Last-in, First-out Method
Depreciation	Straight-line Method	Declining-balance Method
Pension Cost	Prior Service Cost Amortized Over 40 Years	Prior Service Cost Amortized Over 10 Years

a. Which financial statement fairly reflects the results of operations of Timmon's Bakery for the year?
b. How can an auditor express an opinion as to the fairness of financial statements of such varying amounts?

1–60. Large international public accounting firms operate in all 50 states of the United States and in many foreign countries, so only a small percentage of their total operations is regulated by any one regulatory agency (state boards of accountancy, foreign regulatory agencies, etc.).

Is there any effective regulation of the total firm? If so, how is it accomplished? If not, how could it be accomplished?

1–61. At the time the Securities and Exchange Commission was established, the possibility of the formation of a corps of government auditors to audit the financial statements of all publicly held companies was discussed and discarded. Recently, this proposal has been revived. List what you consider to be the advantages and disadvantages of such a system and your conclusion regarding it.

1–62. A member of the business community makes the following statement to an associate who is a CPA. "The auditing standards for your profession are too general and provide no real guidelines. In fact, it would be very difficult to tell whether an audit is substandard or not." Reply to this comment by indicating specific acts or omissions in an audit engagement that would result in violations of generally accepted auditing standards. Name the auditing standard that is violated.

1–63. Read the following audit report and indicate places where it appears that generally accepted auditing standards have not been met. Name the auditing standard violated.

As treasurer of ABC Corporation, I was asked to make an examination of the records of that company and render an opinion on the fairness of the financial statements. The internal control is adequate and I gathered as much evidence as I considered necessary in view of the fact that as treasurer I compiled the statements.

In my opinion the financial statements are fair and correct and reflect the system of accounting traditional for this company.

1–64.   Read the following descriptions and indicate why you consider each description to be (1) compilation of financial statements (accounting), or (2) examination of financial statements (auditing).
   a.   Posting from sales invoices to the accounts receivable subsidiary ledger.
   b.   Inquiring about the status of doubtful accounts.
   c.   Sending confirmation letters to accounts receivable customers.
   d.   Preparing footnotes for the financial statements.
   e.   Tracing data from the general ledger to purchase invoices.
   f.   Recording invoice data as additions to fixed assets.

1–65.   A business associate of a CPA asks why the opinion paragraph of the auditor's standard short-form report is so "guarded" and full of "equivocal" statements. It seems to the business associate that the paragraph says hardly anything. Reply to this criticism.

1–66.   After reading a "qualified" opinion on an audit report, someone asks a CPA what it means. He is confused as to why such an opinion would be issued, rather than one indicating whether or not the statements are fair. Describe the meaning of a qualified opinion and the reasons why qualified opinions would be issued.

Chapter 2

The Auditor's Ethical Environment

Learning Objectives *After reading and studying the material in this chapter, one should*

Understand the ethical environment in which a CPA operates.

Know the major provisions of the AICPA *Code of Professional Ethics.*

Understand the role of management services and its relationship to auditing.

Realize the unique independence of CPAs and the difference between independence in fact and independence in appearance.

Most groups that refer to themselves as professionals have a code of ethics. Although the individual provisions may differ, all codes generally have one thing in common, a requirement that the members maintain a standard of conduct "higher" than that called for by the law. The code of the accounting profession is no exception.

In its relationship with clients and third parties, the accounting profession is unique in that auditors are expected to remain independent of their clients at the same time that they are serving their client's needs. Therefore, the major objective of this chapter is to explain the ethical environment in which auditors perform their unique function. Because many of the provisions of the AICPA's Code of Ethics have been adopted by state boards and state societies, its provisions are used as the model. The focus of this chapter is not only on the prohibitions embodied in the Rules of Conduct, but also on the concepts of professional ethics that are given positive emphasis in the Code.

The Structure of the AICPA Code of Professional Ethics

The AICPA *Code of Professional Ethics* is designed to encourage high moral behavior on the part of members. The public expects this type of behavior from persons in a respected profession, and a formal code acts as a guideline by providing answers to the complex questions that confront practicing accountants. It also provides a positive image, a message to the public that CPAs are willing and are expected to act with integrity.

The real enforcement of a profession's code of ethics lies in the willingness of the members of that profession to obey the provisions of their code. For these restrictions to be fully effective, however, there should be a means of disciplining persons who violate the articles of the code. The bylaws of the AICPA provide authority for the enforcement of the Code. If a member is charged with misconduct, a trial board conducts a hearing and, as a result, the member(s) may be suspended or expelled. A CPA may practice accountancy without maintaining membership in the AICPA, as pointed out in Chapter 1, but expulsion from this organization could damage his reputation.[1]

A committee of professional ethics has issued interpretations of the Rules of Conduct based on questions that arise among practitioners regarding correct conduct in specific situations. These interpretations have been formally compiled and published as a part of the Code.

The Code itself is divided into three parts. Part I, Concepts of Professional Ethics, consists of non-enforceable suggestions on proper behavior for members. This section is very general and stresses the positive aspects of the code. Part II, Rules of Conduct, consists of enforceable prohibitions, and Part III contains the formal interpretations.[2]

The Rules of Conduct generally apply only to members in public practice (industrialists, government employees, and academicians also belong to the AICPA). And, with some exceptions, the Rules apply to all public accounting activities, such as tax practice and management services engagements.

Both the Concepts of Professional Ethics and the Rules of Conduct have the following major divisions:

1. Independence.

2. Technical standards.

3. Relationship with clients.

4. Relationship with colleagues.

5. Other.

[1] Disciplinary action by a state board of accountancy, however, could result in the revocation of an individual's CPA certificate.

[2] The concepts, Rules of Conduct, and interpretations have been compiled into a book entitled *AICPA Professional Standards, Volume 2,* which is published by Commerce Clearing House, Inc., Chicago.

Independence is such a necessary characteristic of the auditor's function that it logically warrants a separate discussion. Of equal importance are the ability and willingness of auditors to comply with the technical standards of the profession (generally accepted auditing standards and generally accepted accounting principles).

Though the desirability of good relationships with clients and colleagues is self-evident, certain complicated situations need to be explained, for example, confidentiality between auditors and clients and encroachment on a colleague's practice.

Independence

No ethical concept is more important to the CPA's service and the image he projects than independence. Unlike members of the legal profession, auditors cannot act solely as advocates of their clients' interest, even though clients' fees provide their source of income. Over the last sixty years, the role of the auditor has evolved to that of a type of financial guardian for third-party groups. And auditors are constantly reminded of this role by the courts, the SEC, and the third parties themselves.

Independence in Fact and in Appearance

There are two aspects of auditors' independence, (1) independence of mind (independence of mental attitude), and (2) the image projected to the public (appearance of independence). For example, there is little likelihood that the auditors of a large industrial firm would conduct a substandard examination in order to profit from a small amount of stock ownership in the company. But this ownership of stock by auditors might give the public the wrong impression of the relationship between the auditors and their clients. Therefore, the AICPA decided several years ago to adopt a rule that prohibits *any* direct ownership of the client's stock by the auditor during the period of the audit engagement or at the time he or she expresses an opinion.

History of the AICPA Rule on Independence

The accounting profession has not always had a strict rule on independence. Until 1933, it was common practice for public accountants to own stock and/or hold an office or directorship in companies they audited. Although the concept of independence was supported by other provisions of the Code, there was no specific rule. Apparently, it was considered redundant.

However, in 1933, the Securities Act was passed. Subsequent regulations based on this act prohibited an external auditor from: (1) acting as an officer or director of the company he audited, or (2) owning any substantial financial interest in the company he audited. The passage of this act and the creation of the SEC in 1934 were steps in a process that ultimately resulted in adoption of an

AICPA rule. But the present rule was established only after a gradual shift had occurred in the prevailing attitude of accountants.

As an illustration, in 1941, the AICPA adopted a rule that prohibited a member from giving opinions on statements of a company financed in any part by public distribution of securities in which he had a substantial financial interest or was committed to acquire such an interest. Nothing was said about an external auditor serving as an officer or director. Only *disclosure* was required if the auditor held a substantial financial interest in a company whose statements were used as basis for credit.

But in 1950 the SEC changed its regulations by taking out the word "substantial." This meant that no auditor of a company that listed its securities on an exchange (and thus came under SEC jurisdiction) could own *any* stock of the client.

A few years later, the AICPA amended its own rule by inserting a provision that prohibits: (1) *any* direct or *material* indirect financial interest, or (2) a position as a promoter, underwriter, voting trustee, director, officer, or key employee. The effect of this change was to create a rule on independence which is similar to SEC regulations.[3]

Contrast Between the AICPA and the SEC Rules on Independence

The similarity between the AICPA and the SEC rules on independence is shown in the following narrative.

Part of ET Section 101.01
(Rule 101) of the AICPA Code of Professional Ethics

A member or a firm of which he is a partner or shareholder shall not express an opinion on financial statements of an enterprise unless he and his firm are independent with respect to such enterprise. Independence will be considered to be impaired if, for example:

A. During the period of his professional engagement, or at the time of expressing his opinion, he or his firm
   1. Had or was committed to acquire any direct or material indirect financial interest in the enterprise; or . . .

B. During the period covered by the financial statements, during the period of the professional engagement or at the time of expressing an opinion, he or his firm
   1. Was connected with the enterprise as a promotor, underwriter or voting trustee, a director or officer or in any capacity equivalent to that of a member of management or of an employee. . . .

---

[3]It should be pointed out that the AICPA formalized a rule that already was being followed by many CPA firms.

Part of Rule 2-01 (b)
of Regulation S-X—Issued by the SEC

The Commission will not recognize any certified public accountant or public accountant as independent who is not in fact independent . . . with respect to any person or any of its parents, its subsidiaries, or other affiliates
(1) in which, during the period of his professional engagement to examine the financial statements being reported on or at the date of his report, he or his firm or a member thereof had, or was committed to acquire, any direct financial interest or any material indirect financial interest; or (2) who which, during the period of his professional engagement to examine the financial statements being reported on, at the date of his report or during the period covered by the financial statements, he or his firm or a member thereof was connected as a promoter, underwriter, voting trustee, director, officer or employee . . .

For the purposes of rule 2–01 (b), the term "member" means all partners in the firm and all professional employees participating in the audit or located in an office of the firm participating in a significant portion of the audit.

In addition, the AICPA Code contains provisions prohibiting (1) closely held business investments that are material, (2) loans, and (3) positions as executor or administrator of an estate or trust when any of the foregoing are connected with the client. With the exception of these additional rules, the AICPA Code and the SEC regulations are very similar. This similarity is no coincidence, because auditors of companies that sell stock to the public must conform to SEC stipulations.

There is one apparent difference between the two rules. The text of the AICPA Code prohibits *partners* or *shareholders* from owning stock of a client, whereas the SEC regulations extend this prohibition to include *employees* who work on the audit or are located in any office that participates in a substantial part of the engagement. However, an ethics ruling by the AICPA professional ethics division's executive committee places a similar restriction on employees of CPA firms. For all practical purposes, the AICPA and SEC rules are very similar. As a practical matter, several national CPA firms have a standing rule against client stock ownership by *any* employee in *any* office, regardless of whether or not that office participates in the audit. Obviously, the image of independence is of major importance to members of the accounting profession.

The Meaning of Indirect Financial Interest

By way of additional explanation, the AICPA's rule states that no material *indirect* financial interest in the client may be held. Although "material" has never been defined clearly, there seems to be a general understanding about the term "indirect." It refers to (1) the holding of mutual funds (the auditor holds stock in a company that in turn owns the client's stock as one of its securities), and (2) investment in a client by an AICPA member's "close kin."[4]

The Audit of a CPA's Own Bookkeeping

Another interpretation of the rule on independence concerns the question of auditing a client for which the CPA has performed bookkeeping services. In light of the stringent rules on stock ownership, loans, and directorships, one may be surprised to learn that the AICPA considers auditing and bookkeeping to be compatible if certain conditions are met. According to ET Section 101.04 (Interpretation 101-3) the conditions are (paraphrased):

1. The member must not have any conflict of interest that would impair his objectivity.

2. The client must know enough about his activities and financial condition to be able to accept responsibility for the statements.

3. The member must not assume the role of employee or manager, i.e., have custody of assets or exercise authority on behalf of the client.

4. The auditor must still make sufficient tests and conduct the audit in accordance with proper standards.

In contrast, the SEC is emphatic about this particular relationship. According to SEC regulations, auditors lose their independence if they keep books for their client, and in most cases the SEC will not accept auditors' reports in such situations. This prohibition includes such services as making and/or posting journal entries and performing other related clerical functions.[5]

Why is there a conflict between the AICPA and the SEC, particularly in view of the relative uniformity of their respective rules on stock ownership? The answer may be that small CPA firms (which normally do not audit companies offering stock to the public) find it difficult to separate auditing and bookkeeping. Many small clients do not have the necessary clerical staff to keep complete sets of records; yet they may need an audit. To date, the AICPA (but not the SEC) has deemed it unnecessary to require such a strict rule on auditing and bookkeeping. Generally, the consensus is that independence in fact can be retained if auditors perform only perfunctory bookkeeping functions without assuming an employee's role.

---

[4]Interpretation 101-4 lists the following: nondependent children, brothers and sisters, grandparents, parents, parents-in-law, and the respective spouses of the foregoing.

[5]Technically, the auditors may propose that the client make certain adjusting journal entries to their books.

Auditors' Independence and Management Services[6]

Another issue related to independence needs to be explored—the question of whether CPAs lose their independence when they perform management services for their audit clients. As pointed out in Chapter 1, management services are a special type of accounting service in which the CPA furnishes aid and advice on some aspect of a company's operations. Examples are installation of a computer system, budgeting techniques, and inventory level problems. Unlike an audit, management services do not include an opinion on the fairness of financial statements.

There is a difference of opinion among accounting firms as to the range of management services that should be offered. Some believe that such services should be limited to those functions that have some financial base or relationship, such as the installation of a responsibility accounting system or a computerized billing system. Others feel that this range is too narrow and that management services should include merger assistance, actuarial services, and executive recruiting. Of the foregoing examples, it would seem that executive recruiting has the most potential for damage to independence in appearance, if not in fact. The public might find it difficult to believe that an accounting firm could recommend that an individual be hired as a controller, and later, during an audit, be completely objective in its review of the controller's work. The public might believe that the accounting firm would rather overlook certain errors during an audit than suffer the embarrassment of having it known that it recommended a less than competent individual.

The AICPA takes the position that no impairment of independence occurs when management services are confined to *advice* rather than management *decisions*. The following comments on the subject are pertinent.

> *When providing management advisory services, the independent accounting firm must, as in all areas of practice, give particular consideration to independence as set forth in the Code of Professional Ethics, particularly in Rule 101 ... Although the appearance of independence is not required in the case of management advisory services and tax practice, a CPA is encouraged to avoid the proscribed relationships with clients regardless of the type of services being rendered. In any event, the CPA, in all types of engagements, should refuse to subordinate his professional judgement to others and should express his conclusions honestly and objectively.*[7]

In MS Section 410.01 of the same publication, the Management Advisory Services Executive Committee describes the proper relationship between audit

---

[6]For more information on this subject, see D. R. Carmichael and R. J. Swieringa, "The Compatibility of Auditing Independence and Management Services—An Identification of Issues," *The Accounting Review*, October 1968, pp. 697–705.

[7]MS Section 410.09 of *AICPA Professional Standards, Volume 1 (Statement on Management Advisory Services No. 1)*, issued by the Management Advisory Services Executive Committee of the AICPA.

services and management services when both are performed by the same CPA.

> *When the services to a client also include expression of an opinion on the fairness of financial statements, the matter of role has special significance, since it also relates to the independence of the accounting firm. The accounting firm's role is to provide advice and technical assistance and to avoid making management decisions or taking positions that might impair the firm's objectivity.*

Some members of the accounting profession do not believe that these statements fully answer the difficult questions. For example, how can CPAs truly conduct an independent audit of a computer system when they have helped the client design that system? At the present time, the general consensus of the Management Advisory Services Executive Committee is that it can be done if a CPA insists on a proper arrangement of roles.

MS Section 430.28 *(Statement on Management Advisory Services No. 3)* contains these additional comments:

> *The compelling need for independence in that attest function is such, however, that these principles must be carefully followed in actual practice. For example, in some cases management may indicate a willingness to abdicate its role as decision maker and the member will need to insist on a proper arrangement of roles. Failing in this, he may need to decline the engagement.*

If members of the AICPA who undertake both an auditing and a management services engagement with the same client follow the foregoing advice, the problem will be alleviated. However, the question of these dual roles will continue to be debated for years.

Commentary on Independence

A high level of integrity is absolutely essential to the auditor. Without it, there is no independence, and without independence, the audit function would be meaningless. Though it is difficult to define and impossible to measure, integrity certainly involves rightness, sincerity, and honesty. Few auditors would accept a cash payoff to express an unqualified opinion on financial statements known to be in error. This decision would not be a difficult test of an individual's integrity. The difficult tests involve choices that are subtler and less clear. One example would test your resistance to a client's pressure to approve the use of an acceptable accounting principle when both you and the client believe that other auditors might approve it, but you believe that another acceptable principle is preferable. Another example would test the willingness of the auditor to lose a major client because he believes that additional audit procedures and higher fees are necessary, even though others might reduce their work rather than lose the client. Such situations provide the true test of integrity. Situations of these types are faced by auditors constantly, and they must be dealt with in a forthright manner. Chapter 1 discusses opportunities for an auditor to take a stand for what he thinks is proper. It is hoped that the meaning of that discussion is beginning to become more apparent.

Technical Standards

History of the Rules on Technical Standards

It is almost axiomatic that members of a profession should be expected to adhere to its technical standards. Failure to do so would result in a loss of public confidence. For forty years, members of the AICPA have been expected to uphold auditing standards and to require client compliance with accounting principles. (Both auditing standards and accounting principles were formalized into pronouncements during the 1930s.)

However, before restatement of the *Code of Professional Ethics* in 1973, it was difficult to enforce these technical standards because the previous Code made no mention of specific AICPA pronouncements that produced these standards. The former rule specified:

> *That a member in expressing an opinion on financial statements may be held guilty of a discreditable act if . . . he fails to direct attention to any material departure from generally accepted accounting principles or to disclose any material omission of generally accepted auditing procedures applicable in the circumstances.*

But there was no mention in the Code of any specific AICPA pronouncement or of any body of literature that comprised generally accepted accounting principles or generally accepted auditing standards. For example, if a member gave an unqualified opinion on financial statements that contained a material departure from a pronouncement of the Accounting Principles Board (e.g., amortization of goodwill), the member might not be disciplined for violation of the Code.

In 1973, this situation changed.[8] The revised *Code of Professional Ethics* of the AICPA contains two new rules that are specific about the authoritative source of the terms "generally accepted accounting principles" and "generally accepted auditing standards."

---

[8]For an explanation of why the *Code of Professional Ethics* was revised in 1973, see Thomas G. Higgins and Wallace E. Olson, "Restating the Ethics Code: A Decision for the Times," *The Journal of Accountancy,* March 1972, pp. 33–39.

The Technical Rule on Auditing Standards[9]

ET Section 202.01 (Rule 202) of the *Code of Professional Ethics* states:

> *A member shall not permit his name to be associated with financial statements in such a manner as to imply that he is acting as an independent public accountant unless he has complied with the applicable generally accepted auditing standards promulgated by the Institute. Statements on Auditing Standards issued by the Institute's Auditing Standards Executive Committee are, for purposes of this rule, considered to be interpretations of the generally accepted auditing standards, and departures from such statements must be justified by those who do not follow them.*

Statements on Auditing Standards and all future statements will serve as the official interpretation referred to in the foregoing rule.[10]

An illustration of the effect of this rule is provided by the opinion paragraph of the short-form auditor's report shown in Chapter 1.

> *In our opinion, the aforementioned consolidated financial statements present fairly the financial position of X Corporation and consolidated subsidiaries at January 31, 19X9 and 19X8, and the results of their operations and changes in their financial position for the years then ended, in conformity with generally accepted accounting principles* applied on a consistent basis. *(Roman added.)*

The second standard of reporting (contained in *SAS No. 1*) requires that the auditor state in the report whether the accounting principles have been consistently observed in the current period in relation to the preceding period. If the auditors had omitted the phrase "applied on a consistent basis" from the foregoing report, they might have violated Rule 202 of the Code. This is not to say that a similar omission by the auditors might not have resulted in a violation of the pre-1973 Code. But the current rule is more explicit and states that departures from AICPA pronouncements designated as auditing standards must be justified.

---

[9]For purposes of this chapter, the term "technical standards" refers to generally accepted auditing standards and generally accepted accounting principles, both of which are discussed in Chapter 1.

[10]All the numbered *Statements on Auditing Standards* have been compiled into a book published by Commerce Clearing House, Inc. entitled *AICPA Professional Standards, Volume 1.*

Technical Rule on Accounting Principles

ET Section 203.01 (Rule No. 203) of the *Code of Professional Ethics* states:

> *A member shall not express an opinion that financial statements are presented in conformity with generally accepted accounting principles if such statements contain any departure from an accounting principle promulgated by the body designated by Council to establish such principles which has a material effect on the statements taken as a whole, unless the member can demonstrate that due to unusual circumstances the financial statements would otherwise have been misleading. In such cases his report must describe the departure, the approximate effects thereof, if practicable, and the reasons why compliance with the principle would result in a misleading statement.*

The Council of the AICPA has designated *Accounting Research Bulletins* (ARB) and *Accounting Principles Board Opinions* (APB) as accounting principles until superseded by *Financial Accounting Standards Board Statements* (all three of these pronouncements are discussed in Chapter 1). This rule, then, has the practical effect of requiring auditors who are members of the AICPA to ascertain carefully whether clients' statements conform to these designated pronouncements. If the statements do not conform to these pronouncements, the burden of proof is on the auditor to show that the exception is justified.

As an example, AC Section 4211.12–.13 of *AICPA Professional Standards, Volume 3 (FASB Statement No. 2)* contains the following accounting requirements:

> *All research and development costs encompassed by this statement shall be charged to expense when incurred. Disclosure shall be made in the financial statements of the total research and development costs charged to expense in each period for which an income statement is presented.*

Because *FASB Statement No. 2* is a designated accounting principle, auditors are obligated to ask for client conformity with this method before rendering an unqualified opinion on the financial statements. But are there conditions that justify a departure from the FASB Statement? Yes; if, owing to unusual circumstances, the financial statements would have been misleading if the client had conformed to this method, the auditors do not have to require such conformity.

An interpretation of Rule 203 contains two examples of conditions that justify exceptions: (1) new legislation, and (2) a new form of business transaction. The interpretation also specifies two circumstances in which departures are not justifiable: (1) an unusual degree of materiality, and (2) conflicting industry practices. Thus, auditors cannot use the fact that research and development costs are excessively large, or that the companies within the industry have different recording practices as a reason for giving an unqualified opinion on statements containing a material amount of deferred research and development costs.

General Technical Standards

In an attempt to establish general standards applicable to all areas of practice, the AICPA has two additional rules.

ET Section 201.01 (Rule 201) states that:

A member shall comply with general standards of (a) professional competence, (b) due professional care, (c) planning and supervision, (d) sufficient relevant data, and (e) forecasts.

ET Section 204.01 (Rule 204) states that:

A member shall comply with other technical standards designated by the AICPA Council.

Specifically, part (e) of Rule 201 of the Code states that:

A member shall not permit his name to be used in conjunction with any forecast of future transactions in a manner which may lead to the belief that the member vouches for the achievability of the forecast.

This specific part of the rule on general technical standards originates in the time-honored tradition that auditors render opinions on *historical* statements rather than future events. In this "age of litigation" it is unlikely that members of the profession will change this tradition in the near future.

But a careful reading of this rule reveals that it does not completely prohibit assisting the client in preparing forecasts of future transactions. It does prohibit rendering this assistance in such a manner that readers of the report gain the impression that the CPA predicts the accuracy of the forecast. To do so would place the CPA in an undesirable position.

For example, assume that a company prepared a report forecasting specific figures for sales and net income. One can easily see that to lend the prestige of a CPA's name to such a report, without proper explanation, could give readers a misleading impression of its accuracy. In addition, one can visualize the possible legal consequences to the CPA if readers made monetary decisions based on a forecast (with a CPA's opinion attached) and the actual results were substantially different.

What type of assistance, then, can be given? A paraphrased interpretation of Rule 204 states: This rule does not prohibit a member from preparing or assisting in the preparation of forecasts of the results of future transactions.

It is presumed that data of this type may be used by parties other than the client. Therefore, the member should disclose:

1. The sources of information,

2. The major assumptions made in preparation of the statements and analyses,

3. The character of the work, and

4. The degree of responsibility that the member is taking.

This interpretation places the CPA in the role of a projector rather than that of a forecaster. One possible application might be assistance in the preparation of pro forma financial statements in which the client is contemplating a large bond or stock sale. He wishes to project an income statement, a balance sheet, and a statement of changes in financial position for the next year based on the assumptions that the bonds or stocks are sold and that certain other events occur. The CPA should state the client's assumptions and the nature of the client's information, and indicate that he is not vouching for the accuracy of the forecasts. By following these suggestions, he should be able to avoid some of the pitfalls of being associated with such statements.

Despite the suggested safeguards, some accounting firms believe that any work they perform in connection with forecasts has little substance, and they refuse to be associated publicly with them.

Relations with Clients

Confidentiality

Though the concept of independence is unique to a CPA's services, the notion of confidentiality is not. Members of all professions are expected to use discretion in disclosing information that they acquire from clients. Lawyers should not reveal the results of their confidential conversations with civil or criminal clients. Medical doctors are obligated to keep secret any sensitive facts gathered from their patients. Also, certain members of the clergy are sworn to secrecy about facts learned during the course of their duties.

A CPA, like members of these other professions, acquires confidential information during the course of audits, tax return preparations, and management services engagements. If this confidence were to be broken, the CPA's credibility would drop and clients would no longer be willing to allow him access to the records that are necessary to carry out engagements. Consider, for example, the damage that could result if a CPA revealed details of a firm's payroll or a plan to acquire property for plant expansion.

The CPA's obligation of confidentiality does not extend to improper disclosure in the financial statements, however. If auditors discover that a lawsuit levied against the client is unsettled at the report date and that its possible consequences are material to the financial statements under examination, they should reveal this finding in their report.

The Rule on Confidential Client Information

ET Section 301.01 (Rule 301) states that:

A member shall not disclose any confidential information obtained in the course of a professional engagement except with the consent of the client.

This rule shall not be construed (a) to relieve a member of his obligation under Rules 202 and 203, (b) to affect in any way his compliance with a validly issued subpoena or summons enforceable by order of a court, (c) to prohibit review of a member's professional practices as a part of voluntary quality review under Institute authorization or (d) to preclude a member from responding to any inquiry made by the ethics division or Trial Board of the Institute, by a duly constituted investigative or disciplinary body of a state CPA society, or under state statutes.

Members of the ethics division and Trial Board of the Institute and professional practice reviewers under Institute authorization shall not disclose any confidential client information which comes to their attention from members in disciplinary proceedings or otherwise in carrying out their official responsibilities. However, this prohibition shall not restrict the exchange of information with an aforementioned duly constituted investigative or disciplinary body.

Confidentiality and Violations of the Law

A major problem arises when a CPA learns, during the course of an engagement, that a client apparently has broken a law. For instance, assume that a CPA discovers, while preparing a tax return, that the client has failed to file returns for previous years. What should the CPA do? Is there a "higher" duty to breach the rule on confidentiality and inform the Internal Revenue Service of this omission? According to official guidelines of the AICPA, the answer is "no" unless the client grants permission. (This answer always raises interesting questions among auditing students.) The following comments describe the position.[11]

A CPA shall advise his client promptly upon learning of an error in a previously filed return, or upon learning of a client's failure to file a required return. His advice should include a recommendation of the measures to be taken. Such advice may be given orally. The CPA is neither obligated to inform the Internal Revenue Service nor may he do so without his client's permission.

[11]TX Section 161.04 of *AICPA Professional Standards, Volume 1 (Statement on Responsibilities in Tax Practice No. 6)*, issued by the Division of Federal Taxation of the AICPA.

Also, as a practical matter, the CPA would be well advised to withdraw from an engagement if he advises his client to go voluntarily to the IRS and the client refuses to do so.

In a related matter, CPAs often represent their clients in IRS administrative proceedings where tax matters are in dispute. If CPAs know of an error in the client's tax return, should they reveal this information to the IRS during the course of the proceedings? The answer is the same. TX Section 171.04 *(Statement on Responsibilities in Tax Practice No. 7)* contains these additional comments:

> *When the CPA is representing a client in an administration proceeding in respect of a return in which there is an error known to the CPA that has resulted or may result in a material understatement of tax liability, he should request the client's agreement to disclose the error to the Internal Revenue Service. Lacking such agreement, the CPA may be under a duty to withdraw from the engagement.*

Confidential and Privileged Communications

Generally, CPAs do not enjoy the same protection on confidentiality as that possessed by members of certain other professions (legal, medical, etc.). CPAs can and have been required to disclose confidential client information in court proceedings. On the other hand, professions enjoying "privileged communications" cannot be required to reveal information received from a client, even in a court of law.

The Concept of Contingent Fees

Another aspect of CPA-client relationships is the question of appropriate fees. As members of a profession, CPAs are expected to provide services in a competent manner and to charge fees commensurate with the nature of these services. Unlike other professionals, though, CPAs are not allowed to make charges contingent on their findings or on the result of their services.

As pointed out in Chapter 1, a wide variety of net income figures can be produced in financial statements, depending on which accounting methods are followed. Even though the statements belong to the client, auditors are in a position to influence the financial results. For example, auditors will ask the client to enter adjusting entries on the books if they believe that the amounts or classifications of items in the financial statements are improper and are material. If the audit fee were based on financial statement results (net income), one can easily imagine the temptation for an auditor to "overlook" an adjustment that lowers profit.

Such a rule is unique to the accounting profession because of the concept of independence that auditors are required to maintain. Lawyers are allowed to collect a percentage of the damages gained for their client in civil suits. This practice is permissible because lawyers act as advocates for their clients and are expected to place the clients' interests above the interests of other parties. In contrast, auditors are expected to report on the fairness of financial statements read by third-party groups that have an economic interest in such statements.

Text of Rule on Contingent Fees

ET Section 302.01 (Rule 302) reads:

Professional services shall not be offered or rendered under an arrangement whereby no fee will be charged unless a specified finding or result is attained, or where the fee is otherwise contingent upon the findings or results of such services. However, a member's fees may vary depending, for example, on the complexity of the service rendered.

Fees are not regarded as being contingent if fixed by courts or other public authorities or, in tax matters, if determined based on the results of judicial proceedings or the findings of governmental agencies.

Explanation of the Contingent Fee Rule Exception

The last paragraph in the rule serves the practical purpose of allowing accountants and lawyers to work together on tax cases. In situations of this sort, an audit is not performed; therefore, the strict rule on independence does not apply. If accountants are aiding their clients in a hearing before the Internal Revenue Service or a court, the question of tax liability is settled by a ruling from the IRS or the court. Thus, there is no question of impropriety because accountants are not in a position to decide the final results.

Relationship with Colleagues

Encroachment—The Rule and General Discussion

Although it is generally assumed that CPAs will show each other professional courtesy, the *Code of Professional Ethics* contains guidelines on proper conduct in this area. ET Section 401.01 (Rule 401) reads:

A member shall not endeavor to provide a person or entity with a professional service which is currently provided by another public accountant except:

1. He may respond to a request for a proposal to render services and may furnish service to those who request it. However, if an audit client of another independent public accountant requests a member to provide professional advice on accounting or auditing matters in connection with an expression of opinion on financial statements, the member must first consult with the other accountant to ascertain that the member is aware of all the available relevant facts.

2. Where a member is required to express an opinion on combined or consolidated financial statements which include a subsidiary, branch or other component audited by another independent public accountant, he may insist on auditing any such component which in his judgment is necessary to warrant the expression of his opinion.

A member who receives an engagement for services by referral from another public accountant shall not accept the client's request to extend his service beyond the specific engagement without first notifying the referring accountant, nor shall he seek to obtain any additional engagement from the client.

Though it is perfectly acceptable and understandable for CPAs to develop their practice, they should not do it by directly encroaching upon the practice of colleagues. To engage in what is known popularly as "cut-throat" competition may earn a CPA short-term benefits, but in the long run it degrades the image of the profession in the public's eyes and ultimately impairs his self-interest as well.

Of course, there is nothing unethical about furnishing services to those *who request them,* even though such a request comes from the client of a colleague. To do otherwise would deny the public access to the expertise it desires and CPAs the opportunity to enlarge their business.

Communication With Predecessor Auditors

Professional courtesy requires that the CPA consult with the predecessor accountant in circumstances in which the client wishes to change public accountants. To do so is not only ethical but prudent, because the reason for a change may bear investigation.

For example, communication with the predecessor auditor might reveal that there was a dispute with the client on a material deviation from generally accepted accounting principles in the proposed financial statements. If such is the case, the successor auditor should ascertain whether the client is "shopping around" for someone who will go along with what the predecessor auditor was unwilling to accept. It is important that auditors be circumspect in their selection of clients. Communication with the predecessor auditor can be helpful in ascertaining lack of integrity in the client's management.

The predecessor auditor is bound by the confidential information provisions of the AICPA *Code of Professional Ethics.* Therefore, it is necessary for the successor auditor to obtain permission from his prospective client before making inquiries of the predecessor auditor.

The Audit of a Branch or Subsidiary

Part 2 of the rule on encroachment gives the principal auditors of a consolidated group of companies the right to audit any subsidiary, branch, or component that they consider necessary to furnish sufficient evidence for an opinion. In most cases, the principal auditor serves the parent of the consolidated group.

Other Concepts and Rules

The Rationale for a Rule on Advertising

The accounting profession's position on advertising is one of the most difficult ethical concepts for the layman to understand. The natural inclination is to consider unrestricted advertising a perfectly legitimate method of selling services and to regard its limitation as an unnecessary restriction.

For organizations that sell products, and even for some organizations that render services, unrestricted advertising is a logical (and perhaps necessary) business activity. But for some groups that regard themselves as professionals, this is not the case. One characteristic of professions is adherence to moral and ethical standards that are higher than those required by law. Violation of a state board of public accountancy ethical rule on advertising may result in suspension or revocation of a CPA's license to practice public accountancy in that state.

Accounting is not the only profession that prohibits certain forms of advertising. Both the legal and medical professions have rules on the subject, although the details of such rules are different. These professionals believe that the nature of their personal services dictates this position. Doctors work with their clients' health; lawyers work with their clients' civil or criminal situation; public accountants concern themselves with their clients' financial affairs.

This is not to say that CPAs are prohibited from selling their services. There are prescribed ways in which public accountants can build their practices by truthful advertising and by circulating information about their services. The acts that a CPA should avoid in the area of advertising are those that are false, misleading, or deceptive.

Text and Interpretations of the Rule on Advertising

The text of the rule on solicitation and advertising is very general. ET Section 502.01 (Rule 502) reads:

> *A member shall not seek to obtain clients by advertising or other forms of solicitation in a manner that is false, misleading, or deceptive. A direct, uninvited solicitation of a specific potential client is prohibited.*

But the application of the rule can be very complex. For this reason, interpretations have been issued on such matters as newspaper ads, telephone directory listings, and business cards. The general tone of these interpretations is that announcements, publications, and other forms of publicity are acceptable if they conform to good taste and are confined to the dissemination of basic information about the firm.

For example, CPAs should not send out material that deprecates other CPAs or lists their own accomplishments. Publications should not refer to fields of specialization.

In the area of listings, the firm may supply basic information about names, addresses, services offered, fees, and professional attainments. Prohibited activities are those that (1) create false or unjustified expectations of favorable results, (2) contain self-laudatory statements not based on verifiable facts, (3) make comparisons with other CPAs, or (4) contain testimonials or endorsements.

CPAs may give speeches, conduct seminars, and distribute professional literature as long as such actions do not involve direct uninvited solicitation of a specific potential client.

A self-designation as "expert" or "specialist" is prohibited. Methods for recognizing specialized fields of competence have not been sufficiently developed.

In general, the theme that runs through the rule is that advertising should be restrained and in good taste.

The courts have taken a hard look at advertising prohibitions in the professions. A United States Supreme Court ruling now allows a limited amount of advertisement by lawyers. In the future, the accountant's rule on advertising may be modified again. Until 1978 advertising in general was prohibited by the AICPA.

The Rule on Commissions

Because CPAs are restricted in obtaining clients by solicitations, the main avenue for growth of their practice is the reputation they achieve. Ideally, this reputation spreads in such a way that business is acquired through the referrals of professional acquaintances and others.

Certain ethical limits are placed on such referrals. It is acceptable for CPAs to take referrals from someone in order to obtain a client. Likewise, it is proper for CPAs to make referrals to clients for someone else's services. However, it is considered unethical for a CPA to pay or to receive a commission for referrals.

ET Section 503.01 (Rule 503) reads:

A member shall not pay a commission to obtain a client, nor shall he accept a commission for a referral to a client of products or services of others. This rule shall not prohibit payments for the purchase of an accounting practice or retirement payments to individuals formerly engaged in the practice of public accounting or payments to their heirs or estates.

The purpose of this type of rule is to prevent CPAs from charging the client for something other than services. If commissions were allowed for referrals, the amount of the commission would probably be added to the total fee. Also it is in the public interest for auditors to be selected on the basis of their reputation and ability rather than the payment of a fee or commission.

The exception to the rule on commissions applies to payments that CPAs make to a referring CPA who renders *services* to the client. For example, assume that a CPA performs a small accounting service for a client who, in turn, asks him to perform other services. This first CPA is unable to do so but he refers the work to a second CPA. The second CPA pays the first CPA for the accounting service performed for the client and then adds this amount to the total bill. This procedure is acceptable because the client is asked to pay only for accounting services.

Incompatible Occupations

Persons who practice public accountancy are encouraged to uphold a positive image by avoiding certain associations that might create a conflict of interest. It is believed that damage could be done to this image if CPAs concurrently engage in a business that is considered to be incompatible. ET Section 504.01 (Rule 504) reads:

> *A member who is engaged in the practice of public accounting shall not concurrently engage in any business or occupation which would create a conflict of interest in rendering professional services.*

There is no general consensus within the accounting profession as to whether the practice of law *is* compatible with the practice of public accounting. Possibly this lack of consensus is due to the fact that lawyers and CPAs have worked together for many years. Many individuals acquire both the CPA title and a law degree, primarily for the purpose of engaging in tax practice. But aside from this obvious bias, there appears to be some logic for placing the practice of law in a different category than other businesses. Both professions offer services that overlap in certain aspects, and both groups abide by codes of ethics that have some similarity. Many accountants believe, however, that one firm should not engage in both functions. Certainly, the large CPA firms consider the lawyers employed by them to be engaged in the practice of *accountancy*.[12]

Discussion of the Rule on the Form of Practice and Name

The last provision of the *Code of Professional Ethics* is a rule that specifies the forms and names that CPAs may use in their organizational structure. The revised Code of 1973 combines several subjects previously covered in separate rules. One subject is the question of whether CPA firms should be organized in partnership or corporate form. Another topic is the joint ownership of a public accounting firm by CPAs and non-CPAs. In the early years of the century, it was

---

[12]For additional views, see Stephen E. Loeb and Burt A. Leete, "The Dual Practitioner: CPA, Lawyer, or Both?" *The Journal of Accountancy,* August 1973, pp. 57–63.

common practice for public accountants to organize as corporations, and in some cases a majority of the stock was held by non-accountants. But this type of ownership of public accounting corporations disturbed accountants. A resolution was passed by the American Institute of Accountants (predecessor to the AICPA) in 1919 that prohibited members from conducting their practice in corporate form unless *all* of the stockholders, directors, and officers were practicing public accountants.

The resolution later was changed to a general comment that audit corporations were not desirable for the accounting profession. In an attempt to rid corporations of non-accountants, the profession then went even further and issued a condemnation of the corporate form. Finally, in 1938, a definite rule prohibiting members from practicing in corporate form was adopted.

This rule was largely unchallenged for many years. Then, in the late 1950s and early 1960s, pressure began to build for the modification of this prohibition. Attitudes had begun to change for several reasons.

1. Some members desired to form "service bureau" types of corporations in order to provide computer services to customers.

2. Certain tax advantages were available to public accountants practicing in corporate form.

3. Members sought protection against the liability threat as a result of the increase in lawsuits during the early 1960s.

Other members argued against the rule change on the basis that the public image might be damaged and that the quality of the work might suffer if unlimited liability were taken away.

The forces of change prevailed, and in 1969 the rule was changed to allow practice by CPA firms in corporate form if certain stipulations are met. Some of the more important stipulations are that all shareholders must be CPAs, transfer of shares can be made only to a CPA, and non-CPA directors or officers cannot exercise any professional authority. Many state boards of public accountancy have rules prohibiting practice in corporate form. As a result, the corporate structure is used primarily by CPA firms with practices in one or a few states.

The Rule on Form of Practice and Name

ET Section 505.01 (Rule 505) reads:

A member may practice public accounting, whether as an owner or employee, only in the form of a proprietorship, a partnership or a professional corporation whose characteristics conform to resolutions of Council . . .

A member shall not practice under a firm name which includes any fictitious name, indicates specialization or is misleading as to the type of organization (proprietorship, partnership or corporation). However, names of one or more past partners or shareholders may be included in the firm name of a successor partnership or corporation. Also, a partner surviving the death or withdrawal of all other partners may continue to practice under the partnership name for up to two years after becoming a sole practitioner.

A firm may not designate itself as "Members of the American Institute of Certified Public Accountants" unless all of its partners or shareholders are members of the Institute.

It is important to note that the rule does not prohibit joint ownership of businesses between CPAs and non-CPAs. But if these businesses perform services of a type performed by public accountants (auditing, tax, management advisory, write-up, etc.), they are required to conform to the AICPA *Code of Professional Ethics* practiced by CPA firms if the CPA(s) actively participates in the operations of the organization. In recent years, partnerships and corporations have been formed between CPAs and non-CPAs for the purpose of offering consulting services in the area of computer systems. The AICPA has designated these businesses as organizations that perform services of a type performed by public accountants. Therefore, though the non-CPAs do not come under the jurisdiction of the Code of Ethics, the CPAs are responsible for the acts of the company.

If CPAs only have *investments* in businesses of the type described, the rule is not as strict. ET Section 505.02 (Interpretation 505-1) has been issued, which reads (paraphrased): A member may have a financial interest in a corporation which performs services of a type performed by public accountants and whose characteristics do not conform to the resolutions of the Council of the Institute if: (1) the interest is not material to the corporation's net worth, and (2) the member is only an investor.

The AICPA continues to study the relationships between practicing CPAs and non-CPAs who perform accounting services. Significant changes in this area may be forthcoming in the future.

Competitive Bidding

Until a few years ago, a provision in the AICPA's Code of Ethics prohibited competitive bidding by members. The rule stated:

A member ... shall not make a competitive bid for a professional engagement. Competitive bidding for public accounting services is not in the public interest, is a form of solicitation, and is unprofessional.

The purpose of this stipulation was to prevent the type of price competition that might damage the public image of CPAs. In addition, it was believed that attempts by one CPA to acquire clients by under-pricing another CPA might result in substandard service, because it might be necessary to reduce the quality of the services to meet the lower price.

In the 1960s and early 1970s, it became evident that enforcement of this rule would subject the AICPA to charges of anti-trust law violation by the U.S. Justice Department. Although the rule stayed on the books for several years, no attempt was made to implement it. Finally, this provision was dropped when the Code of Ethics was revised in 1973.

Despite the absence of a rule against competitive bidding, there is no evidence that "price wars" have emerged among CPAs.[13] Perhaps this outcome validates the long-standing assumption that members of a profession follow high ethical standards regardless of the absence of prohibitive rules.

Summary

1. CPAs, as well as members of other professions, operate in a special environment; the personal nature of their services makes them subject to special rules.

2. It is not enough for CPAs merely to act "within the law," for many actions that they consider unethical are legal. Restrictions on unlimited advertising are prime examples.

3. In many cases, the guidelines by which CPAs are expected to conduct themselves are not simple and clear-cut, and judgment must be exercised. To illustrate, consider the decision auditors should make in disclosing certain sensitive information in their report, when the rule on confidentiality must be balanced against the full disclosure requirement.

One impression that should *not* be left (one that is difficult to avoid in discussing material of this sort) is that public accountants are frequently subjected to disciplinary measures. This is not the case. CPAs can and do avoid most ethical problems simply by using good professional judgment in their relations with clients, colleagues, and others.

If public accountants have any doubt as to the propriety of their actions, they can consult the AICPA, their state society, or their state board and usually receive an answer. If their past actions are discovered to have been improper, immediate corrective measures normally will be sufficient.

The purpose of this chapter (and of Chapters 1 and 3) is to provide background for the technical material that follows by discussing the environment in which auditors operate. In succeeding chapters, these concepts are drawn upon in order to place the function of auditing in its proper perspective.

[13]Many state boards of public accountancy still prohibit competitive bidding.

Appendix A to Chapter 2

The Concepts of Professional Ethics and
Rules of Conduct of the
AICPA Code of Professional Ethics*

Introduction

A man should be upright; not be kept upright.

Marcus Aurelius

.01 A distinguishing mark of a professional is his acceptance of responsibility to the public. All true professions have therefore deemed it essential to promulgate codes of ethics and to establish means for ensuring their observance.

.02 The reliance of the public, the government and the business community on sound financial reporting and advice on business affairs, and the importance of these matters to the economic and social aspects of life impose particular obligations on certified public accountants.

.03 Ordinarily those who depend upon a certified public accountant find it difficult to assess the quality of his services; they have a right to expect, however, that he is a person of competence and integrity. A man or woman who enters the profession of accountancy is assumed to accept an obligation to uphold its principles, to work for the increase of knowledge in the art and for the improvement of methods, and to abide by the profession's ethical and technical standards.

.04 The ethical Code of the American Institute emphasizes the profession's responsibility to the public, a responsibility that has grown as the number of investors has grown, as the relationship between corporate managers and stockholders has become more impersonal and as government increasingly relies on accounting information.

.05 The Code also stresses the CPA's responsibility to clients and colleagues, since his behavior in these relationships cannot fail to affect the responsibilities of the profession as a whole to the public.

.06 The Institute's Rules of Conduct set forth minimum levels of acceptable conduct and are mandatory and enforceable. However, it is in the best interests of the profession that CPAs strive for conduct beyond that indicated merely by prohibitions. Ethical conduct, in the true sense, is more than merely abiding by the letter of explicit prohibitions. Rather it requires unswerving commitment to honorable behavior, even at the sacrifice of personal advantage.

.07 The conduct toward which CPAs should strive is embodied in five broad concepts stated as affirmative Ethical Principles:

Independence, integrity and objectivity. A certified public accountant should maintain his integrity and objectivity and, when engaged in the practice of public accounting, be independent of those he serves.

Competence and technical standards. A certified public accountant should observe the profession's technical standards and strive continually to improve his competence and the quality of his services.

Responsibilities to clients. A certified public accountant should be fair and candid with his clients and serve them to the best of his ability, with professional concern for their best interests, consistent with his responsibilities to the public.

Responsibilities to colleagues. A certified public accountant should conduct himself in a manner which will promote cooperation and good relations among members of the profession.

Other responsibilities and practices. A certified public accountant should conduct himself in a manner which will enhance the stature of the profession and its ability to serve the public.

.08 The foregoing Ethical Principles are intended as broad guidelines as distinguished from enforceable Rules of Conduct. Even though they do not provide a basis for disciplinary action, they constitute the philosophical foundation upon which the Rules of Conduct are based.

.09 The following discussion is intended to elaborate on each of the Ethical Principles and provide rationale for their support.

Independence, Integrity and Objectivity

A certified public accountant should maintain his integrity and objectivity and, when engaged in the practice of public accounting, be independent of those he serves.

.01 The public expects a number of character traits in a certified public accountant but primarily integrity and objectivity and, in the practice of public accounting, independence.

Independence has always been a concept fundamental to the accounting profession, the cornerstone of its philosophical structure. For no matter how competent any CPA may be, his opinion on financial statements will be of little value to those who rely on him—whether they be clients or any of his unseen audience of credit grantors, investors, governmental agencies and the like—unless he maintains his independence.

.02 Independence has traditionally been defined by the profession as the ability to act with integrity and objectivity.

.03 Integrity is an element of character which is fundamental to reliance on the CPA. This quality may be difficult to judge, however, since a particular fault of omission or commission may be the result either of honest error or a lack of integrity.

.04 Objectivity refers to a CPA's ability to maintain an impartial attitude on all matters which come under his review. Since this attitude involves an individual's mental processes, the evaluation of objectivity must be based largely on actions and relationships viewed in the context of ascertainable circumstances.

.05 While recognizing that the qualities of integrity and objectivity are not precisely measurable, the profession nevertheless constantly holds them up to members as an imperative. This is done essentially by education and by the Rules of Conduct which the profession adopts and enforces.

.06 CPAs cannot practice their calling and participate in the world's affairs without being exposed to situations that involve the possibility of pressures upon their integrity and objectivity. To define and proscribe all such situations woud be impracticable. To ignore the problem for that reason, however, and to set no limits at all would be irresponsible.

.07 It follows that the concept of independence should not be interpreted so loosely as to permit relationships likely to impair the CPA's integrity or the impartiality of his judgment, nor so strictly as to inhibit the rendering of useful services when the likelihood of such impairment is relatively remote.

.08 While it may be difficult for a CPA always to appear completely independent even in normal relationships with clients, pressures upon his integrity or objectivity are offset by powerful countervailing forces and restraints. These include the possibility of legal liability, professional discipline ranging up to revocation of the right to practice as a CPA, loss of reputation and, by no means least, the inculcated resistance of a disciplined professional to any infringement upon his basic integrity and objectivity. Accordingly, in deciding which types of relationships should be specifically prohibited, both the magnitude of the threat posed by a relationship and the force of countervailing pressures have to be weighed.

.09 In establishing rules relating to independence, the profession uses the criterion of whether reasonable men, having knowledge of all the facts and taking into consideration normal strength of character and normal behavior under the circumstances, would conclude that a specified relationship between a CPA and a client poses an unacceptable threat to the CPA's integrity or objectivity.

.10 When a CPA expresses an opinion on financial statements, not only the fact but also the appearance of integrity and objectivity is of particular importance. For this reason, the profession has adopted rules to prohibit the expression of such an opinion when relationships exist which might pose such a threat to integrity and objectivity as to exceed the strength of countervailing forces and restraints. These relationships fall into two general categories: (1) certain financial relationships with clients and (2) relationships in which a CPA is virtually part of management or an employee under management's control.

.11 Although the appearance of independence is not required in the case of management advisory services and tax practice, a CPA is encouraged to avoid the proscribed relationships with clients regardless of the type of services being rendered. In any event, the CPA, in all types of engagements, should refuse to subordinate his professional judgment to others and should express his conclusions honestly and objectively.

.12 The financial relationships proscribed when an opinion is expressed on financial statements make no reference to fees paid to a CPA by a client. Remuneration to providers of services is necessary for the continued provision of those services. Indeed, a principal reason for the development and persistence in the professions of the client-practitioner relationship and of remuneration by fee (as contrasted with an employer-employee relationship and remuneration by salary) is that these arrangements are seen as a safeguard of independence.

.13 The above reference to an employer-employee relationship is pertinent to a question sometimes raised as to whether a CPA's objectivity in expressing an opinion on financial statements will be impaired by his being involved with his client in the decision-making process.

.14 CPAs continually provide advice to their clients, and they expect that this advice will usually be followed. Decisions based on such advice may have a significant effect on a client's financial condition or operating results. This is the case not only in tax engagements and management advisory services but in the audit function as well.

.15 If a CPA disagrees with a client on a significant matter during the course of an audit, the client has three choices—he can modify the financial statements (which is usually the case), he can accept a qualified report or he can discharge the CPA. While the ultimate decision and the resulting financial statements clearly are those of the client, the CPA has obviously been a significant factor in the decision-making process. Indeed, no responsible user of financial statements would want it otherwise.

.16 It must be noted that when a CPA expresses an opinion on financial statements, the judgments involved pertain to whether the results of operating decisions of the client are fairly presented in the statements and not on the underlying wisdom of such decisions. It is highly unlikely therefore that being a factor in the client's decision-making process would impair the CPA's objectivity in judging the fairness of presentation.

.17 The more important question is whether a CPA would deliberately compromise his integrity by expressing an unqualified opinion on financial statements which were prepared in such a way as to cover up a poor business decision by the client and on which the CPA has rendered advice. The basic character traits of the CPA as well as the risks arising from such a compromise of integrity, including liability to third parties, disciplinary action and loss of right to practice, should preclude such action.

.18 Providing advice or recommendations which may or may not involve skills logically related to a client's information and control system, and which may affect the client's decision-making, does not in itself indicate lack of independence.

However, the CPA must be alert to the possibility that undue identification with the management of the client or involvement with a client's affairs to such a degree as to place him virtually in the position of being an employee, may impair the appearance of independence.

.19 To sum up, CPAs cannot avoid external pressures on their integrity and objectivity in the course of their professional work, but they are expected to resist these pressures. They must, in fact, retain their integrity and objectivity in all phases of their practice and, when expressing opinions on financial statements, avoid involvement in situations that would impair the credibility of their independence in the minds of reasonable men familiar with the facts.

Competence and Technical Standards

A certified public accountant should observe the profession's technical standards and strive continually to improve his competence and the quality of his services.

.01 Since accounting information is of great importance to all segments of the public, all CPAs, whether in public practice, government service, private employment or academic pursuits, should perform their work at a high level of professionalism.

.02 A CPA should maintain and seek always to improve his competence in all areas of accountancy in which he engages. Satisfaction of the requirements for the CPA certificate is evidence of basic competence at the time the certificate is granted, but it does not justify an assumption that this competence is maintained without continuing effort. Further, it does not necessarily justify undertaking complex engagements without additional study and experience.

.03 A CPA should not render professional services without being aware of, and complying with, the applicable technical standards. Moreover, since published technical standards can never cover the whole field of accountancy, he must keep broadly informed.

.04 Observance of the rule on competence calls for a subjective determination by a CPA with respect to each engagement. Some engagements will require a higher level of knowledge, skill and judgment than others. Competence to deal with an unfamiliar problem may be acquired by research, study or consultation with a practitioner who has the necessary competence. If a CPA is unable to gain sufficient competence through these means, he should suggest, in fairness to his client and the public, the engagement of someone competent to perform the needed service, either independently or as an associate.

.05 The standards referred to in the rules are elaborated and refined to meet changing conditions, and it is each CPA's responsibility to keep himself up to date in this respect.

Responsibilities to Clients

A certified public accountant should be fair and candid with his clients and serve them to the best of his ability, with professional concern for their best interests, consistent with his responsibilities to the public.

.01 As a professional person, the CPA should serve his clients with competence and with professional concern for their best interests. He must not permit his regard for a client's interest, however, to override his obligation to the public to maintain his independence, integrity and objectivity. The discharge of this dual responsibility to both clients and the public requires a high degree of ethical perception and conduct.

.02 It is fundamental that the CPA hold in strict confidence all information concerning a client's affairs which he acquires in the course of his engagement. This does not mean, however, that he should acquiesce in a client's unwillingness to make disclosures in financial reports which are necessary to fair presentation.

.03 Exploitation of relations with a client for personal advantage is improper. For example, acceptance of a commission from any vendor for recommending his product or service to a client is prohibited.

.04 A CPA should be frank and straightforward with clients. While tact and diplomacy are desirable, a client should never be left in doubt about the CPA's position on any issue of significance. No truly professional man wil subordinate his own judgment or conceal or modify his honest opinion merely to please. This admonition applies to all services including those related to management and tax problems.

.05 When accepting an engagement, a CPA should bear in mind that he may find it necessary to resign if conflict arises on an important question of principle. In cases of irreconcilable difference, he will have to judge whether the importance of the matter requires such an action. In weighing this question, he can feel assured that the practitioner who is independent, fair and candid is the better respected for these qualities and will not lack opportunities for constructive service.

Responsibilities to Colleagues

A certified public accountant should conduct himself in a manner which will promote cooperation and good relations among members of the profession.

.01 The support of a profession by its members and their cooperation with one another are essential elements of professional character. The public confidence and respect which a CPA enjoys is largely the result of the cumulative accomplishments of all CPAs, past and present. It is, therefore, in the CPA's own interest, as well as that of the general public, to support the collective efforts of colleagues through professional societies and organizations and to deal with fellow practitioners in a manner which will not detract from their reputation and well-being.

.02 Although the reluctance of a professional to give testimony that may be damaging to a colleague is understandable, the obligation of professional courtesy and fraternal consideration can never excuse lack of complete candor if the CPA is testifying as an expert witness in a judicial proceeding or properly constituted inquiry.

.03 A CPA has the obligation to assist his fellows in complying with the Code of Professional Ethics and should also assist appropriate disciplinary authorities in enforcing the Code. To condone serious fault can be as bad as to commit it. It may be even worse, in fact, since some errors may result from ignorance rather than intent and, if let pass without action, will probably be repeated. In situations of this kind, the welfare of the public should be the guide to a member's action.

.04 While the Code proscribes certain specific actions in the area of relationships with colleagues, it should be understood that these proscriptions do not define the limits of desirable intraprofessional conduct. Rather, such conduct encompasses the professional consideration and courtesies which each CPA would like to have fellow practitioners extend to him.

.05 It is natural that a CPA will seek to develop his practice. However, in doing so he should not seek to displace another accountant in a client relationship, or act in any way that reflects negatively on fellow practitioners.

.06 A CPA may, of course, provide service to those who request it, even though they may be served by another practitioner in another area of service, or he may succeed another practitioner at a client's request. In such circumstances it is desirable before accepting an engagement that the CPA who has been approached should advise the accountant already serving the client. Such action is indicated not only by considerations of professional courtesy but by good business judgment.

.07 A client may sometimes request services requiring highly specialized knowledge. If the CPA lacks the expertise necessary to render such services, he should call upon a fellow practitioner for assistance or refer the entire engagement to another. Such assistance or referral brings to bear on the client's needs both the referring practitioner's knowledge of the client's affairs and the technical expertise of the specialist brought into the engagement. The rules encourage referrals by helping to protect the client relationships of the referring practitioner.

Other Responsibilities and Practices

A certified public accountant should conduct himself in a manner which will enhance the stature of the profession and its ability to serve the public.

.01 In light of the importance of their function, CPAs and their firms should have a keen consciousness of the public interest and the needs of society. Thus, they should support efforts to achieve equality of opportunity for all, regardless of race, religious background or sex, and should contribute to this goal by their own service relationships and employment practices.

.02 The CPA is a beneficiary of the organization and character of his profession. Since he is seen as a representative of the profession by those who come in contact with him, he should behave honorably both in his personal and professional life and avoid any conduct that might erode public respect and confidence. . . .

.06 In determining fees, a CPA may assess the degree of responsibility assumed by undertaking an engagement as well as the time, manpower and skills required to perform the service in conformity with the standards of the profession. He may also take into account the value of the service to the client, the customary charges of professional colleagues and other considerations. No single factor is necessarily controlling.

.07 Clients have a right to know in advance what rates will be charged and approximately how much an engagement will cost. However, when professional judgments are involved, it is usually not possible to set a fair charge until an engagement has been completed. For this reason CPAs should state their fees for proposed engagements in the form of estimates which may be subject to change as the work progresses.

.08 Other practices prohibited by the Rules of Conduct include using any firm designation or description which might be misleading, or practicing as a professional corporation or association which fails to comply with provisions established by Council to protect the public interest.

.09 A member, while practicing public accounting, may not engage in a business or occupation which is incompatible therewith. While certain occupations are clearly incompatible with the practice of public accounting, the profession has never attempted to list them, for in most cases the individual circumstances indicate whether there is a problem. . . .

.10 Paying a commission is prohibited in order to eliminate the temptation to compensate anyone for referring a client. Receipt of a commission is proscribed since practitioners should look to the client, and not to others, for compensation for services rendered. The practice of paying a fee to a referring CPA irrespective of any service performed or responsibility assumed by him is proscribed because there is no justification for a CPA to share in a fee for accounting services where his sole contribution was to make a referral.

.11 Over the years the vast majority of CPAs have endeavored to earn and maintain a reputation for competence, integrity and objectivity. The success of these efforts has been largely responsible for the wide public acceptance of accounting as an honorable profession. This acceptance is a valuable asset which should never be taken for granted. Every CPA should constantly strive to see that it continues to be deserved.

Rules of Conduct

Definitions (Section 91)

.01 The following definitions of terminology are applicable wherever such terminology is used in the Rules and Interpretations.

.02 Client. The person(s) or entity which retains a member or his firm, engaged in the practice of public accounting, for the performance of professional services.

.03 Council. The Council of the American Institute of Certified Public Accountants.

.04 Enterprise. Any person(s) or entity, whether organized for profit or not, for which a CPA provides services.

.05 Firm. A proprietorship, partnership, or professional corporation or association engaged in the practice of public accounting, including individual partners or shareholders thereof.

.06 Financial statements. Statements and footnotes related thereto that purport to show financial position which relates to a point in time or changes in financial position which relate to a period of time, and statements which use a cash or other incomplete basis of accounting. Balance sheets, statements of income, statements of retained earnings, statements of changes in financial position, and statements of changes in owners' equity are financial statements.

.07 Incidental financial data included in management advisory services reports to support recommendations to a client, and tax returns and supporting schedules do not, for this purpose, constitute financial statements; and the statement, affidavit, or signature of preparers required on tax returns neither constitutes an opinion on financial statements nor requires a disclaimer of such opinion.

.08 Institute. The American Institute of Certified Public Accountants.

.09 Interpretations of Rules of Conduct. Pronouncements issued by the division of professional ethics to provide guidelines as to the scope and application of the Rules of Conduct.

.10 Member. A member, associate member, or international associate of the American Institute of Certified Public Accountants.

.11 Practice of public accounting. Holding out to be a CPA or public accountant and at the same time performing for a client one or more types of services rendered by public accountants. The term shall not be limited by a more restrictive definition which might be found in the accountancy law under which a member practices.

.12 Professional services. One or more types of services performed in the practice of public accounting.

Applicability of Rules (Section 92)

.01 The Institute's Code of Professional Ethics derives its authority from the bylaws of the Institute which provide that the Trial Board may, after a hearing, admonish, suspend, or expel a member who is found guilty of infringing any of the bylaws or any provisions of the Rules of Conduct.

.02 The Rules of Conduct which follow apply to all services performed in the practice of public accounting including tax and management advisory services except (a) where the wording of the rule indicates otherwise and (b) that a member who is practicing outside the United States will not be subject to discipline for departing from any of the rules stated herein so long as his conduct is in accord with the rules of the organized accounting profession in the country in which he is practicing. However, where a member's name is associated with financial statements in such a manner as to imply that he is acting as an independent public accountant and under circumstances that would entitle the reader to assume that United States practices were followed, he must comply with the requirements of Rules 202 and 203.

.03 A member may be held responsible for compliance with the Rules of Conduct by all persons associated with him in the practice of public accounting who are either under his supervision or are his partners or shareholders in the practice.

.04 A member engaged in the practice of public accounting must observe all the Rules of Conduct. A member not engaged in the practice of public accounting must observe only Rules 102 and 501 since all other Rules of Conduct relate solely to the practice of public accounting.

.05 A member shall not permit others to carry out on his behalf, either with or without compensation, acts which, if carried out by the member, would place him in violation of the Rules of Conduct.

Independence, Integrity and Objectivity

Rule 101—Independence. A member or a firm of which he is a partner or shareholder shall not express an opinion on financial statements of an enterprise unless he and his firm are independent with respect to such enterprise. Independence will be considered to be impaired if, for example:

A. During the period of his professional engagement, or at the time of expressing his opinion, he or his firm

    1. a. Had or was committed to acquire any direct or material indirect financial interest in the enterprise; or

       b. Was a trustee of any trust or executor or administrator of any estate if such trust or estate had or was committed to acquire any direct or material indirect financial interest in the enterprise; or

2. Had any joint closely held business investment with the enterprise or any officer, director or principal stockholder thereof which was material in relation to his or his firm's net worth; or

3. Had any loan to or from the enterprise or any officer, director or principal stockholder thereof. This latter proscription does not apply to the following loans from a financial institution when made under normal lending procedures, terms and requirements:

   (a) Loans obtained by a member or his firm which are not material in relation to the net worth of such borrower.

   (b) Home mortgages.

   (c) Other secured loans, except loans guaranteed by a member's firm which are otherwise unsecured.

B. During the period covered by the financial statements, during the period of the professional engagement or at the time of expressing an opinion, he or his firm

1. Was connected with the enterprise as a promoter, underwriter or voting trustee, a director or officer or in any capacity equivalent to that of a member of management or of an employee; or

2. Was a trustee for any pension or profit-sharing trust of the enterprise.

The above examples are not intended to be all-inclusive.

Rule 102—Integrity and objectivity. A member shall not knowingly misrepresent facts, and when engaged in the practice of public accounting, including the rendering of tax and management advisory services, shall not subordinate his judgment to others. In tax practice, a member may resolve doubt in favor of his client as long as there is reasonable support for his position.

Competence and Technical Standards

Rule 201—General standards. A member shall comply with the following general standards as interpreted by bodies designated by Council, and must justify any departures therefrom.

a. *Professional Competence.* A member shall undertake only those engagements which he or his firm can reasonably expect to complete with professional competence.

b. *Due Professional Care.* A member shall exercise due professional care in the performance of an engagement.

c. *Planning and Supervision.* A member shall adequately plan and supervise an engagement.

d. *Sufficient Relevant Data.* A member shall obtain sufficient relevant data to afford a reasonable basis for conclusions or recommendations in relation to an engagement.

e. *Forecasts.* A member shall not permit his name to be used in conjunction with any forecast of future transactions in a manner which may lead to the belief that the member vouches for the achievability of the forecast.

Rule 202—Auditing standards. A member shall not permit his name to be associated with financial statements in such a manner as to imply that he is acting as an independent public accountant unless he has complied with the applicable generally accepted auditing standards promulgated by the Institute. Statements on Auditing Standards issued by the Institute's Auditing Standards Executive Committee are, for purposes of this rule, considered to be interpretations of the generally accepted auditing standards, and departures from such statements must be justified by those who do not follow them.

Rule 203—Accounting principles. A member shall not express an opinion that financial statements are presented in conformity with generally accepted accounting principles if such statements contain any departure from an accounting principle promulgated by the body designated by Council to establish such principles which has a material effect on the statements taken as a whole, unless the member can demonstrate that due to unusual circumstances the financial statements would otherwise have been misleading. In such cases his report must describe the departure, the approximate effects thereof, if practicable, and the reasons why compliance with the principle would result in a misleading statement.

Rule 204—Other technical standards. A member shall comply with other technical standards promulgated by bodies designated by Council to establish such standards, and departures therefrom must be justified by those who do not follow them.

Responsibilities to Clients

Rule 301—Confidential client information. A member shall not disclose any confidential information obtained in the course of a professional engagement except with the consent of the client.

This rule shall not be construed (a) to relieve a member of his obligation under Rules 202 and 203, (b) to affect in any way his compliance with a validly issued subpoena or summons enforceable by order of a court, (c) to prohibit review of a member's professional practices as a part of voluntary quality review under Institute authorization or (d) to preclude a member from responding to any inquiry made by the ethics division or Trial Board of the Institute, by a duly constituted investigative or disciplinary body of a state CPA society, or under state statutes.

Members of the ethics division and Trial Board of the Institute and professional practice reviewers under Institute authorization shall not disclose any confidential client information which comes to their attention from members in disciplinary proceedings or otherwise in carrying out their official responsibilities. However, this prohibition shall not restrict the exchange of information with an aforementioned duly constituted investigative or disciplinary body.

Rule 302—Contingent fees. Professional services shall not be offered or rendered under an arrangement whereby no fee will be charged unless a specified finding or result is attained, or where the fee is otherwise contingent upon the findings or results of such services. However, a member's fees may vary depending, for example, on the complexity of the service rendered.

Fees are not regarded as being contingent if fixed by courts or other public authorities or, in tax matters, if determined based on the results of judicial proceedings or the findings of governmental agencies.

Responsibilities to Colleagues

Rule 401—Encroachment. A member shall not endeavor to provide a person or entity with a professional service which is currently provided by another public accountant except:

1. He may respond to a request for a proposal to render services and may furnish service to those who request it. However, if an audit client of another independent public accountant requests a member to provide professional advice on accounting or auditing matters in connection with an expression of opinion on financial statements, the member must first consult with the other accountant to ascertain that the member is aware of all the available relevant facts.

2. Where a member is required to express an opinion on combined or consolidated financial statements which include a subsidiary, branch or other component audited by another independent public accountant, he may insist on auditing any such component which in his judgment is necessary to warrant the expression of his opinion.

A member who receives an engagement for services by referral from another public accountant shall not accept the client's request to extend his service beyond the specific engagement without first notifying the referring accountant, nor shall he seek to obtain any additional engagement from the client.

Other Responsibilities and Practices

Rule 501—Acts discreditable. A member shall not commit an act discreditable to the profession.

Rule 502—Advertising and other forms of solicitation. A member shall not seek to obtain clients by advertising or other forms of solicitation in a manner that is false, misleading, or deceptive. A direct uninvited solicitation of a specific potential client is prohibited. . . .

Rule 503—Commission. A member shall not pay a commission to obtain a client, nor shall he accept a commission for a referral to a client of products or services of others. This rule shall not prohibit payments for the purchase of an accounting practice or retirement payments to individuals formerly engaged in the practice of public accounting or payments to their heirs or estates.

Rule 504—Incompatible occupations. A member who is engaged in the practice of public accounting shall not concurrently engage in any business or occupation which would create a conflict of interest in rendering professional services.

Rule 505—Form of practice and name. A member may practice public accounting, whether as an owner or employee, only in the form of a proprietorship, a partnership or a professional corporation whose characteristics conform to resolutions of Council. . . .

A member shall not practice under a firm name which includes any fictitious name, indicates specialization or is misleading as to the type of organization (proprietorship, partnership or corporation). However, names of one or more past partners or shareholders may be included in the firm name of a successor partnership or corporation. Also, a partner surviving the death or withdrawal of all other partners may continue to practice under the partnership name for up to two years after becoming a sole practitioner.

A firm may not designate itself as "Members of the American Institute of Certified Public Accountants" unless all of its partners or shareholders are members of the Institute.

Appendix B to Chapter 2

Professional Corporations or Associations

The following resolution of Council was approved at the spring meeting of Council on May 6, 1969 and amended at the fall meeting of Council on October 12, 1974:

RESOLVED, that members may be officers, directors, stockholders, representatives or agents of a corporation offering services of a type performed by public accountants only when the professional corporation or association has the following characteristics:

1. *Name.* The name under which the professional corporation or association renders professional services shall contain only the names of one or more of the present or former shareholders or of partners who were associated with a predecessor accounting firm. Impersonal or fictitious names, as well as names which indicate a speciality, are prohibited.

2. *Purpose.* The professional corporation or association shall not provide services that are incompatible with the practice of public accounting.

3. *Ownership.* All shareholders of the corporation or association shall be persons engaged in the practice of public accounting as defined by the Code of Professional Ethics. Shareholders shall at all times own their shares in their own right, and shall be the beneficial owners of the equity capital ascribed to them.

4. *Transfer of Shares.* Provision shall be made requiring any shareholder who ceases to be eligible to be a shareholder to dispose of all of his shares within a reasonable period to a person qualified to be a shareholder or to the corporation or association.

5. *Directors and Officers.* The principal executive officer shall be a shareholder and a director, and to the extent possible, all other directors and officers shall be certified public accountants. Lay directors and officers shall not exercise any authority whatsoever over professional matters.

6. *Conduct.* The right to practice as a corporation or association shall not change the obligation of its shareholders, directors, officers and other employees to comply with the standards of professional conduct established by the American Institute of Certified Public Accountants.

7. *Liability.* The stockholders of professional corporations or associations shall be jointly and severally liable for the acts of a corporation or association, or its employees—except where professional liability insurance is carried, or capitalization is maintained, in amounts deemed sufficient to offer adequate protection to the public. Liability shall not be limited by the formation of subsidiary or affiliated corporations or associations each with its own limited and unrelated liability.

In a report approved by Council at the fall 1969 meeting, the Board of Directors recommended that professional liability insurance or capitalization in the amount of $50,000 per shareholder/officer and professional employee to a maximum of $2,000,000 would offer adequate protection to the public. Members contemplating the formation of a corporation under this rule should ascertain that no further modifications in the characteristics have been made.

Appendix C to Chapter 2

Ethics Rulings

The Rules of Conduct of the AICPA *Code of Professional Ethics* are very general and subject to different interpretations. To provide some guidance to CPAs in the conduct of their daily practice, ethics rulings have been issued by the AICPA Professional Ethics division's Executive Committee. These rulings are answers to questions sent to this committee by CPAs and others who are in doubt about the application of a provision of the *Code of Professional Ethics* to particular situations incurred in practice.

Listed hereafter are some rulings that relate to the Rules of Conduct discussed earlier in the text. The listings are categorized according to ethics topics to facilitate use with the material in the main part of the chapter.

Independence (Section 191)

1. Acceptance of a Gift

Question   Would the independence of a member's firm be considered to be impaired if an employee or partner accepts a gift or other unusual consideration from a client?

Answer   If an employee or partner accepts more than a token gift from a client, even with the knowledge of the member's firm, the appearance of independence may be lacking.

2. Member as Stockholder in Country Club

Question   A member belongs to a country club in which membership requirements involve the acquisition of a pro rata share of equity or debt securities. Would the independence of the member's firm be considered to be impaired with respect to the country club?

Answer   Independence of the member's firm would not be considered to be impaired since membership in such a club is essentially a social matter. Accordingly, such equity or debt ownership is not considered to be a direct financial interest within the meaning of Rule 101. However, the member should not take part in the management of the club.

3. Member as City Council Chairman

Question   A member is the chairman of a city council. Would the independence of the member's firm be impaired with respect to state governmental agencies and other governments within the state?

Answer   Independence of the member's firm would not be considered to be impaired with respect to any governmental unit except those under the council's control.

4. Financial Interest by Employee

Question   A professional employee of a member's firm owns stock in an audit client. Would the independence of the member's firm be considered to be impaired with respect to this client?

Answer   The appearance of independence of the member's firm would be considered to be impaired if the professional staff employee of the member's firm was either involved in the engagement or located in the office of the firm participating in a significant portion of the audit. An employee of a member's firm who is not in such circumstances might have an immaterial financial interest in the audit client of his firm without impairing the appearance of independence of his firm.

5. Member's Employee as Treasurer of a Client

Question   Would the independence of a member's firm be considered to be impaired if an employee serves as treasurer of a client which is a charitable organization?

Answer   Independence of the member's firm would be considered to be impaired since management functions are involved.

6. Faculty Member as Auditor of a Student Fund

Question   A tenured member on the faculty of a university is asked to audit the financial statements of the Student Senate Fund. The university has the following connections with this fund:

1. The basic faculty-administration-student relationship.

2. It acts as a collection agent for student fees and remits them to the Student Senate.

3. It requires that a member of the administration approve Student Senate checks by signing them.

Would the independence of the member's firm be considered to be impaired under these circumstances? Would independence be considered to be impaired if the member was in public practice and served as a part-time faculty member?

Answer   Under either situation posed, independence of the member's firm would be considered to be impaired with respect to the Student Senate Fund since the member would be auditing several of the management functions performed by the university, his employer.

7. Past Due Fees

Question   A member's client has been unable to meet his current obligations. As a result, substantially all amounts due the member's firm for the preceding year are unpaid and past due. Would the independence of the member's firm be considered to be impaired with respect to the client for the current year?

Answer   Independence of the member's firm may be impaired if more than one year's fees due from a client for professional services remain unpaid for an extended period of time. Such amounts, when they are long past due according to a firm's normal billing terms, take on some of the characteristics of a loan within the meaning of Rule 101. Under these conditions, it may appear that the practitioner is providing working capital for his client and that the collection of past due amounts may depend on the nature of the auditor's report on the client's financial statements.

At the time a member issues a report on a client's financial statements, the client should not be indebted to the member for more than one year's fees. Accordingly, unless the amounts involved are clearly insignificant to both the client and the member, independence is considered to be impaired if fees for all professional services rendered for prior years are not collected before the issuance of the member's report for the current year.

Technical Standards (Section 291)

1. Opinion by Member Not in Public Practice

Question A member has become an employee of a corporation with extensive outside interests. The employer asks the member to perform examinations of these corporate interests and to express an opinion for internal purposes only. Would there be any violation of the Code if these examinations were made and an opinion were expressed by the member?

Answer The member is not in practice as a public accountant and may perform services required by his employer, including performing an examination of outside corporate interests. He may use his CPA designation in his internal reports if his status as an employee is made clear. The internal reports should be issued on his employer's letterhead and should make no reference to generally accepted auditing standards. If the internal reports are made available to third parties, the member should use only his title as an employee and omit any reference to his CPA designation.

2. Letterhead

Question A member performs accounting services on a gratis basis for a private club of which he is treasurer. His firm does no work for the club. Would it be proper for him to issue financial statements in connection with his accounting services for the club on his firm letterhead with a disclaimer for lack of independence?

Answer It would be preferable for the stationery of the club to be used for presentation of the financial statements with an indication that the auditor is acting as treasurer. However, should he use his firm's letterhead, section 517 of Statement on Auditing Standards 1 (AU Section 517) is applicable.

Responsibilities to Clients (Section 391)

1. Revealing Names of Employer's Clients

Question   A staff member wishes to submit his resumé to another firm. May he include as part of his experience the names of companies for which he performed audits?

Answer   The mere engagement of a member's firm is often a confidential matter between accountant and client. Unless the company is publicly held, a member should not reveal the fact that he had served on an assignment without the client's permission.

2. Fee as Expert Witness

Question   May a member, as an expert witness in a damage suit, receive compensation based on the amount awarded the plaintiff?

Answer   Such an agreement would violate Rule 302, which prohibits contingent fees. Compensation for expert testimony may be at a standard per diem rate for such services or at a fixed sum previously agreed upon.

3. Fee as a Percentage of Tax Savings

Question   May a member base his fee for preparing a tax return on how much in taxes he can save his client?

Answer   Basing a fee for preparing a tax return on the amount saved in taxes would be a violation of Rule 302. A properly prepared return results in a proper tax liability, and there is no basis for computing a saving. To make a fee contingent upon the amount of taxes saved presumes a tax liability has been established which an accountant is attempting to reduce, whereas all persons concerned with the preparation of a tax return should attempt to determine only the correct tax liability. A member who computes his fee in the manner suggested in the question would also violate Rule 501.

Responsibilities to Colleagues (Section 491)

1. Consulting Engagement

Question   A member gave advice on a tax question to a client that he had served for the past ten years. The client, with the approval of the member, submitted the question to another CPA firm which concurred with the member's advice on the tax question and subsequently sent a general client memorandum dealing with taxes to the member's client. Has the second CPA firm violated Rules 401 and 502 of the Code?

Answer Yes, since at the conclusion of an engagement that is clearly a consulting engagement for a specific question, the client relationship ceases particularly if the client is being served on a regular basis by another accountant.

Chapter 2 References

Carmichael, D. R., and R. J. Swieringa, "The Compatibility of Auditing Independence and Managment Services—An Identification of Issues," *The Accounting Review,* October 1968, pp. 697–705.

Hartley, Ronald V., and Timothy L. Ross, "MAS and Audit Independence: An Image Problem," *The Journal of Accountancy,* November 1972, pp. 42–51.

Helstein, Richard S., "Privileged Communications for CPAs," *The Journal of Accountancy,* December 1970, pp. 39–46.

Higgins, Thomas G., and Wallace E. Olson, "Restating the Ethics Code: A Decision for the Times," *The Journal of Accountancy,* March 1972, pp. 33–39.

Loeb, Stephen E., and Burt A. Leete, "The Dual Practitioner: CPA, Lawyer, or Both?" *The Journal of Accountancy,* August 1973, pp. 57–63.

Schlosser, Robert E., and Associates, "An Historical Approach to the Concept of Independence," *The New York Certified Public Accountant,* July 1969, pp. 517–527.

Willingham, John J., Charles H. Smith, and Martin E. Taylor, "Should the CPA's Opinion Be Extended to Include Forecasts?" *Financial Executive.* September 1970, pp. 80–89.

Chapter 2
Questions Taken from the Chapter

2–1.   What is unique about auditors' relationship with their clients?

2–2.   If high ethical and moral behavior is expected of professionals, why is there a formal code of ethics?

2–3. What procedure is followed when a member is charged with violation of the AICPA *Code of Professional Ethics?*

2–4. Name the three parts of the Code of Ethics. What does each part contain?

2–5. Name two aspects of auditors' independence.

2–6. Briefly trace the history of the AICPA's rule on independence. How does this contrast with the history of the SEC's rule on independence?

2–7. Indicate the major differences between the present SEC and AICPA rules on independence.

2–8. Under certain conditions, can a member accept a gift from a client and not impair the appearance of independence? (Answer in appendix c.)

2–9. Should a member's employee serve as treasurer of a client which is a charitable organization? (Answer in appendix c.)

2–10. Give two examples of indirect interest.

2–11. Under what four conditions may a member audit a client for whom he performs bookkeeping services? What is the SEC rule on this matter?

2–12. What is the AICPA's position on CPAs rendering both audit services and management services to the same client?

2–13. The appearance of independence is not required for two types of accounting services. What are these two types?

2–14. Quote or paraphrase two ethical rules on technical standards.

2–15. The Council of the AICPA has designated three sets of pronouncements as generally accepted accounting principles. What are these three sets?

2–16. Give two examples of justifiable departures from the literal application of pronouncements on accounting principles.

2–17.   Give two examples of circumstances that would *not* qualify for justifiable departures from the literal application of pronouncements on accounting principles.

2–18.   Quote or paraphrase the technical rule on forecasts. What is the purpose of this rule?

2–19.   What four items of information should the member disclose when he is preparing or assisting in preparing forecasts of the results of future transactions?

2–20.   Into what role should CPAs be placed when they assist in forecast preparation?

2–21.   What four exceptions are allowed to the rule on confidential client information?

2–22.   What should CPAs do when they learn that clients have failed to file tax returns for previous years? What is the CPA prohibited from doing?

2–23.   Under what conditions can a CPA reveal to the Internal Revenue Service information about an error in a client's tax return? Under what conditions does the CPA have a duty to withdraw from a tax engagement?

2–24.   What might auditors be tempted to do in regard to their examination of financial statements if there were no rule on contingent fees?

2–25.   Why are lawyers permitted to accept contingent fees though CPAs are not?

2–26.   What exceptions are allowed to the rule on contingent fees? Why?

2–27.   What should a member do if he is asked to provide professional advice to a client of another independent CPA and this advice is on accounting or auditing matters in connection with the expression of an opinion on financial statements?

2–28.   What should a member do if he receives an engagement by referral and is asked to extend his services?

2–29.   In what way might communication with a predecessor auditor be helpful to the successor auditor?

2–30.   What part of the rule on encroachment was added as a last section?

2–31.   Why is the giving or receiving of commissions on referrals considered improper? What exception is allowed?

2–32.   If advertising is legal, why is it restricted in the *Code of Professional Ethics?*

2–33.   Quote or paraphrase the rule on incompatible occupations.

2–34.   Why is the practice of law considered by some members of the accounting profession to *be* compatible with the practice of public accounting?

2–35.   What rule did the AICPA pass, almost forty years ago, in regard to practice in corporate form? What arguments were given for changing the rule? What arguments were given for not changing the rule?

2–36.   Under what conditions can a AICPA member own and actively participate in a business with a non-CPA if that business performs services of a type performed by public accountants? What if the CPA is only an investor?

2–37.   What rule has been removed from the Code because of possible antitrust action? Why was the rule originally included?

Chapter 2
Objective Questions Taken from CPA Examinations

2–38.   Printers, Inc., an audit client of James Frank, CPA, is contemplating the installation of an electronic data-processing system. It would be inconsistent with Frank's independence as the auditor of Printers' financial statements for him to
   a.   Recommend accounting controls to be exercised over the computer.
   b.   Recommend particular hardware and software packages to be used in the new computer center.
   c.   Prepare a study of the feasibility of computer installation.
   d.   Supervise operation of Printers' computer center on a part-time basis.

2–39.   The CPA should not undertake an engagement if his fee is to be based upon
   a.   The findings of a tax authority.

    b.   A percentage of audited net income.

    c.   Per diem rates plus expenses.

    d.   Rates set by a city ordinance.

2–40.   The CPA ethically could

    a.   Perform an examination for a financially distressed client at less than his customary fees.

    b.   Advertise only as to his expertise in preparing income tax returns.

    c.   Base his audit fee on a percentage of the proceeds of his client's stock issue.

    d.   Own preferred stock in a corporation which is an audit client.

2–41.   The CPA should not

    a.   Disclose that he is a CPA in a situation-wanted advertisement.

    b.   Describe himself as a tax expert or management consulting specialist.

    c.   Apply for a position with another firm without informing his present employer.

    d.   Advise clients and professional contacts of the opening of a new office.

2–42.   With respect to examination of the financial statements of the Third National Bank, a CPA's appearance of independence ordinarily would not be impaired by his

    a.   Obtaining a large loan for working capital purposes.

    b.   Serving on the committee which approves the bank's loans.

    c.   Utilizing the bank's time-sharing computer service.

    d.   Owning a few inherited shares of Third National common stock.

2–43.   Mercury Company, an audit client of Eric Jones, CPA, is considering acquiring Hermes, Inc. Jones' independence as Mercury's auditor would be impaired if he were to

    a.   Perform on behalf of Mercury a special examination of the financial affairs of Hermes.

    b.   Render an opinion as to each party's compliance with financial covenants of the merger agreement.

    c.   Arrange through mutual acquaintances the initial meeting between representatives of Mercury and Hermes.

    d.   Negotiate the terms of the acquisition on behalf of Mercury.

2–44.   The concept of materiality will be least important to the CPA in determining the

    a.   Scope of his audit of specific accounts.

    b.   Specific transactions which should be reviewed.

    c.   Effects of audit exceptions upon his opinion.

    d.   Effects of his direct financial interest in Pyzi* upon his independence.

*This refers to the company involved.

2–45.   On its business stationery, a CPA firm should not list
   a.   The firm's name, address, and telephone number.
   b.   Names of deceased partners in the firm name.
   c.   Membership in state CPA society.
   d.   That it has tax expertise.

2–46.   A CPA should reject a management advisory services engagement if
   a.   It would require him to make management decisions for an audit client.
   b.   His recommendations are to be subject to review by the client.
   c.   He audits the financial statements of a subsidiary of the prospective client.
   d.   The proposed engagement is not accounting-related.

2–47.   Triolo, CPA, has a small public accounting practice. One of Triolo's clients desires services which Triolo cannot adequately provide. Triolo has recommended a larger CPA firm, Pinto & Co., to his client, and, in return, Pinto has agreed to pay Triolo 10% of the fee for services rendered by Pinto for Triolo's client. Who, if anyone, is in violation of the AICPA's *Code of Professional Ethics?*
   a.   Both Triolo and Pinto.
   b.   Neither Triolo nor Pinto.
   c.   Only Triolo.
   d.   Only Pinto.

2–48.   The CPA who regularly examines Viola Corporation's financial statements has been asked to prepare pro forma income statements for the next five years. If the statements are to be based upon the Corporation's operating assumptions and are for internal use only, the CPA should
   a.   Reject the engagement because the statements are to be based upon assumptions.
   b.   Reject the engagement because the statements are for internal use.
   c.   Accept the engagement provided full disclosure is made of the assumptions used and the extent of the CPA's responsibility.
   d.   Accept the engagement provided Viola certifies in writing that the statements are for internal use only.

2–49.   Jackson, CPA, has a public accounting practice. He wishes to establish a separate partnership to offer data-processing services to the public and other public accountants.
   a.   Jackson cannot be a partner in any separate partnership which offers data-processing services.
   b.   Jackson may form a separate partnership, but it will be subject to the *Code of Professional Ethics.*
   c.   Jackson may form a separate partnership as long as all partners are CPAs.
   d.   Jackson may form a separate partnership, but he must give up his public accounting practice.

2–50.   Clark, CPA, wishes to express an opinion that the financial statements of Smith Co. are presented in conformity with generally accepted accounting principles: however, the financial statements contain a departure from *APB No. 5*.
   a.   Under any circumstances, Clark would be in violation of the *Code of Professional Ethics* if he were to issue such an opinion.
   b.   Clark should disclaim an opinion.
   c.   Clark may issue the opinion he desires if he can demonstrate that due to unusual circumstances the financial statements of Smith Co. would otherwise have been misleading.
   d.   This specific situation is not covered by the rules established by the *Code of Professional Ethics*.

2–51.   Smith and Jones, CPAs, wish to incorporate their public accounting practice.
   a.   An appropriate name for the new corporation would be the Financial Specialist Corporation.
   b.   Smith and Jones need *not* be individually liable for the acts of the corporation if the corporation carries adequate professional liability insurance.
   c.   Smith's ten-year-old son may own stock in the corporation.
   d.   The corporation may provide services that are incompatible with the practice of public accounting as long as a non-CPA employee performs the services.

2–52.   The *Code of Professional Ethics* considers Statements on Auditing Standards (formerly Statements on Auditing Procedure) issued by the Institute's Auditing Standards Executive Committee (formerly the Committee on Auditing Procedure) to
   a.   Supersede generally accepted auditing standards.
   b.   Be separate and independent of generally accepted auditing standards.
   c.   *Not* be part of the *Code* since specific rules pertaining to technical standards are established by the *Code* itself.
   d.   Be interpretations of generally accepted auditing standards.

2–53.   Poust, CPA, has sold his public accounting practice to Lyons, CPA.
   a.   Poust must obtain permission from his clients before making available working papers and other documents to Lyons.
   b.   Poust must obtain permission from his clients for only audit-related working papers and other documents before making them available to Lyons.
   c.   Poust must return the working papers and other documents to his clients, and Lyons must solicit the clients for his use of the materials.
   d.   Poust must obtain permission from his clients for only tax-related working papers and other documents before making them available to Lyons.

2–54.   Gutowski, a practicing CPA, has written an article which is being published in a professional publication. The publication wishes to inform its readers about

Gutowski's background. The information regarding Gutowski should *not*
a.   List the degrees held by Gutowski.
b.   State with which firm Gutowski is associated.
c.   State that Gutowski is a tax expert.
d.   List other publications by Gutowski.

2–55.   Adams is the executive partner of Adams & Co., CPAs. One of its smaller clients is a large non-profit charitable organization. The organization has asked Adams to be on its board of directors, which consists of a large number of the community's leaders. Membership on the board is honorary in nature. Adams & Co. would be considered to be independent
a.   Under *no* circumstances.
b.   As long as Adams' directorship was disclosed in the organization's financial statements.
c.   As long as Adams was *not* directly in charge of the audit.
d.   As long as Adams does *not* perform or give advice on management functions of the organization.

2–56.   Godette, a non-CPA, has a law practice. Godette has recommended one of his clients to Doyle, CPA. Doyle has agreed to pay Godette 10% of the fee for services rendered by Doyle to Godette's client. Who, if anyone, is in violation of the *Code of Professional Ethics?*
a.   Both Godette and Doyle.
b.   Neither Godette nor Doyle.
c.   Only Godette.
d.   Only Doyle.

2–57.   Which of the following criteria is unique to the auditor's attest function?
a.   General competence.
b.   Familiarity with the particular industry of which his client is a part.
c.   Due professional care.
d.   Independence.

2–58.   A client's management has asked you to consider a special study of a proposed modification of the system of internal control. It wishes to modify the existing system in approximately eighteen months in conjunction with an upgrading of its computer system.

The scope of the study and evaluation to be reported on would be substantially more extensive than that required by a normal financial audit. The special report would be used solely for the internal information of management. Which of the following is the most appropriate response concerning the acceptance or rejection of the proposed engagement?

    a. Explain that you are prohibited from accepting such an engagement because the client is not planning to allow distribution of the report to the general public upon request.

    b. Explain that you can accept the engagement if it will not require decisions on your part which would impair your independence in connection with future audits of the client.

    c. Explain that you are prohibited from accepting such an engagement because it would deal with a proposed rather than an existing system of controls.

    d. Explain that you can accept the engagement but that there is a standard reporting format which you will be required to use when communicating your findings.

2–59.   Pickens and Perkins, CPAs, decide to incorporate their practice of accountancy. According to the AICPA *Code of Professional Ethics,* shares in the corporation can be issued

    a. Only to persons qualified to practice as CPAs.

    b. Only to employees and officers of the firm.

    c. Only to persons qualified to practice as CPAs and members of their immediate families.

    d. To the general public.

2–60.   Fenn & Co., CPAs, has time available on a computer which it uses primarily for its own internal record-keeping. Aware that the computer facilities of Delta Equipment Co., one of Fenn's audit clients, are inadequate for the company needs, Fenn offers to maintain on its computer certain routine accounting records for Delta. If Delta were to accept the offer and Fenn were to continue to function as independent auditor for Delta, then Fenn would be in violation of

    a. SEC, but not AICPA, provisions pertaining to auditors' independence.

    b. Both SEC and AICPA provisions pertaining to auditors' independence.

    c. AICPA, but not SEC, provisions pertaining to auditors' independence.

    d. Neither AICPA nor SEC provisions pertaining to auditors' independence.

2–61.   The certified public accounting firm of Lincoln, Johnson & Grant is the auditor for the Union Corporation. Mr. Lee, President of Union Corp., has asked the firm to perform management advisory services in the area of inventory management. Mr. Lee believes the procedures in this area are inefficient. Considering the dual engagement of the regular audit and the management services assignment, which of the following functions could impair the CPA firm's independence?

    a. Identify the inventory-management problem as caused by the procedures presently operative in the purchasing, receiving, storage, and issuance operations.

  b. Study and evaluate the inventory-management problem and suggest several alternative solutions.

  c. Develop a time schedule for implementation of the solution adopted by Mr. Lee, to be carried out and supervised by Union Corp. personnel.

  d. Supervise management of purchasing, receiving, storage, and issuance operation.

2–62. Marquis, CPA, occasionally undertakes management-advisory-services engagements, although his practice deals primarily with auditing services. The Keller Corporation has recently asked Marquis to make a study of the company's executive-compensation package. Marquis has no prior experience in this area. What would be the most appropriate course of action for Marquis to follow?

  a. Decline the engagement because he cannot expect to complete it without undertaking research in the area.

  b. Accept the engagement and issue a report which may lead to the belief that he vouches for the achievability of the results indicated by the recommended actions.

  c. Accept the engagement and research, study, or consult with knowledgeable experts in order to increase his competence.

  d. Accept the engagement and perform it in accordance with generally accepted auditing standards.

2–63. What is the meaning of the generally accepted auditing standard which requires that the auditor be independent?

  a. The auditor must be without bias with respect to the client under audit.

  b. The auditor must adopt a critical attitude during the audit.

  c. The auditor's sole obligation is to third parties.

  d. The auditor may have a direct ownership interest in his client's business if it is not material.

2–64. Which of the following is prohibited by the AICPA *Code of Professional Ethics?*

  a. Use of a firm name which indicates specialization.

  b. Practice of public accounting in the form of a professional corporation.

  c. Use of the partnership name for a limited period by one of the partners in a public accounting firm after the death or withdrawal of all other partners.

  d. Holding as an investment ten of 1,000 outstanding shares in a commercial corporation which performs bookkeeping services.

2–65. A CPA's report accompanying a cash forecast or other type of projection should

  a. Not be issued in any form because it would be in violation of the AICPA *Code of Professional Ethics.*

  b. Disclaim any opinion as to the forecast's achievability.

  c. Be prepared only if the client is a not-for-profit organization.

    d.   Be a qualified short-form audit report if the business concern is operated for a profit.

2–66.   In which one of the following situations would a CPA be in violation of the AICPA *Code of Professional Ethics* in determining his fee?

    a.   A fee based on whether the CPA's report on the client's financial statements results in the approval of a bank loan.

    b.   A fee based on the outcome of a bankruptcy proceeding.

    c.   A fee based on the nature of the service rendered and the CPA's particular expertise instead of the actual time spent on the engagement.

    d.   A fee based on the fee charged by the prior auditor.

2–67.   The AICPA *Code of Professional Ethics* states that a CPA shall not disclose any confidential information obtained in the course of a professional engagement except with the consent of his client. In which one of the situations given below would disclosure by a CPA be in violation of the Code?

    a.   Disclosing confidential information in order to properly discharge the CPA's responsibilities in accordance with his profession's standards.

    b.   Disclosing confidential information in compliance with a subpoena issued by a court.

    c.   Disclosing confidential information to another accountant interested in purchasing the CPA's practice.

    d.   Disclosing confidential information in a review of the CPA's professional practice by the AICPA Quality Review Committee.

2–68.   During the course of an audit, the client's controller asks your advice on how to revise the purchase journal so as to reduce the amount of time his staff takes in posting. How should you respond?

    a.   Explain that under the AICPA *Code of Professional Ethics* you cannot give advice on management advisory service areas at the same time you are doing an audit.

    b.   Explain that under the AICPA Statement on Management Advisory Services informal advice of this type is prohibited.

    c.   Respond with definite recommendations based on your audit of these records but state that you will not assume any responsibility for any changes unless your specific recommendations are followed.

    d.   Respond as practicable at the moment and express the basis for your response so it will be accepted for what it is.

2–69.   Under the AICPA *Code of Professional Ethics*, a CPA may issue an unqualified opinion on financial statements which contain a departure from generally accepted accounting principles if he can demonstrate that due to unusual circumstances the financial statements would be misleading if the departure were not made. Which of the following is an example of unusual circumstances which could justify such a departure?

a.   New legislation.

b.   An unusual degree of materiality.

c.   Conflicting industry practices.

d.   A theoretical disagreement with a standard promulgated by the Financial Accounting Standards Board.

2–70.   Green, a CPA not in public practice, is an employee in the internal audit department of Bigg Conglomerate Company. The management has asked Green to perform examinations of potential acquisitions and to express an opinion thereon. Bigg will use the reports for internal purposes and to show to its bankers in accordance with certain loan agreements. Your response to Items 70 and 71 should be based on the AICPA *Code of Professional Ethics.*

How should Green sign the report?

a.   Green, CPA.

b.   Green, Internal Auditor.

c.   Green, CPA (Internal Auditor).

d.   Green, Internal Auditor (CPA).

2–71.   If Green performed the same examination as would have been done by the outside auditors, what difference should there be in the opinion Green would render?

a.   None.

b.   Green should not refer to generally accepted auditing standards.

c.   Green should qualify his opinion based on lack of independence.

d.   Green should disclaim an opinion based on lack of independence.

2–72.   *Statements on Auditing Standards* issued by the AICPA's Auditing Standards Executive Committee are

a.   Part of the generally accepted auditing standards under the AICPA *Code of Professional Ethics.*

b.   Interpretations of generally accepted auditing standards under the AICPA *Code of Professional Ethics,* and departures from such statements must be justified.

c.   Interpretations of generally accepted auditing standards under the AICPA *Code of Professional Ethics* and such statements must be followed in every engagement.

d.   Generally accepted auditing procedures that are not covered by the AICPA *Code of Professional Ethics.*

2–73.   Reed, a partner in a local CPA firm, performs free accounting services for a private club of which Reed is treasurer. Which of the following would be the

most preferable manner for Reed to issue the financial statements of the club?
a.   On the firm's letterhead with a disclaimer for lack of independence.
b.   On the firm's letterhead with a disclaimer for unaudited financial statements.
c.   On plain paper with no reference to Reed so that Reed will not be associated with the statements.
d.   On the club's letterhead with Reed signing as treasurer.

2–74.   Upon discovering irregularities in a client's tax return that the client would *not* correct, a CPA withdraws from the engagement. How should the CPA respond if asked by the successor CPA why the relationship was terminated?
a.   "It was a misunderstanding."
b.   "I suggest you get the client's permission for us to discuss all matters freely."
c.   "I suggest you ask the client."
d.   "I found irregularities in the tax return which the client would not correct."

2–75.   Which of the following publications does *not* qualify as a statement of generally accepted accounting principles under the AICPA *Code of Professional Ethics?*
a.   AICPA Accounting Research Bulletins and APB Opinions.
b.   Accounting interpretations issued by the AICPA.
c.   Statements of Financial Standards issued by the FASB.
d.   Accounting interpretations issued by the FASB.

2–76.   Below are the names of four CPA firms and pertinent facts relating thereto. Unless otherwise indicated, the individuals named are CPAs and partners, and there are *no* other partners. Which firm name and related facts indicate a violation of the AICPA *Code of Professional Ethics?*
a.   Green, Lawrence, and Craig, CPAs. (Craig died about five years ago; Green and Lawrence are continuing the firm.)
b.   Clay and Sharp, CPAs. (The name of Andy Randolph, CPA, a third active partner, is omitted from the firm name.)
c.   Fulton and Jackson, CPAs. (Jackson died about three years ago; Fulton is continuing the firm as a sole proprietorship.)
d.   Schneider & Co., CPAs, Inc. (The firm has ten other stockholders who are all CPAs.)

2–77.   A CPA who has given correct tax advice which is later affected by changes in the tax law is *required* to
a.   Notify the client upon learning of any change.
b.   Notify the client only when the CPA is actively assisting with implementing the advice or is obliged to so notify by specific agreement.
c.   Notify the Internal Revenue Service.

    d.   Take no action if the client has already followed the advice unless the client asks the question again.

2–78.   The AICPA *Code of Professional Ethics* states, in part, that a CPA should maintain integrity and objectivity. Objectivity in the Code refers to a CPA's ability

    a.   To maintain an impartial attitude on all matters which come under the CPA's review.

    b.   To independently distinguish between accounting practices that are acceptable and those that are *not*.

    c.   To be unyielding in all matters dealing with auditing procedures.

    d.   To independently choose between alternate accounting principles and auditing standards.

Chapter 2
Discussion/Case Questions

2–79.   The following cases relate to the CPA's management of his accounting practice.

Case 1

Tom Jencks, CPA, conducts a public accounting practice. In 1970, Mr. Jencks and Harold Swann, a non-CPA, organized Electro-Data Corporation to specialize in computerized bookkeeping services. Mr. Jencks supplied 20% of the capital stock. Mr. Swann is the salaried general manager of Electro-Data. Mr. Jencks is affiliated with the Corporation only as a stockholder; he receives no salary and does not participate in day-to-day management. However, he has transferred all of his bookkeeping accounts to the Corporation and recommends its services whenever possible.

Required:

Organizing your presentation around Mr. Jencks' involvement with Electro-Data Corporation, discuss the following:

    a.   In your opinion, is Mr. Jencks in direct violation of any provision of the Code of Ethics?

    b.   What if Mr. Jencks participated in management?

    c.   Without regard to the Code of Ethics, discuss the propriety of the arrangement above.

Case 2

Judd Hanlon, CPA, was engaged to prepare the federal income tax return for the Guild Corporation for the year ended December 31, 1971. This is Mr. Hanlon's first engagement of any kind for the Guild Corporation. In preparing the 1971 return, Mr. Hanlon finds an error on the 1970 return. The 1970 depreciation deduction was overstated significantly. Accumulated depreciation brought forward from 1969 to 1970 was understated; thus, the 1970 base for declining balance depreciation was overstated. Mr. Hanlon reported the error to Guild's controller, the officer responsible for tax returns. The controller stated: "Let the revenue agent find the error." He further instructed Mr. Hanlon to carry forward the material overstatement of the depreciable base to the 1971 depreciation computation. The controller noted that this error also had been made in the financial records for 1970 and 1971 and offered to furnish Mr. Hanlon with a letter assuming full responsibility for this treatment.

Required:
a. Did Mr. Hanlon initially handle the situation correctly?
b. Discuss the additional action that Mr. Hanlon should now undertake. Should he take this information directly to the IRS? (Answer the question only in relation to tax.)

(AICPA adapted)

2–80.   During the course of an audit, Fred Curious, the senior assigned to the job, noticed that several employees were paid at a rate less than that required by the federal minimum wage law. He checked into this matter and found that the company was engaged in interstate commerce, which means that all employees are subject to the federal minimum wage law.

He decided to tell the company controller, feeling that the violation was probably an oversight and could be corrected easily. However, the controller was irritated when told of this situation by the senior. He informed Mr. Curious that this matter had nothing to do with the audit and that it should be dropped.

Mr. Curious took his information to the CPA firm's partner in charge of the audit. The partner showed concern for the senior's feelings, but was reluctant to do anything about it. He suggested that Mr. Curious make a notation of the underpayment in his working papers and assured Mr. Curious that, as partner, he would take full responsibility.

Required:
a. If you were Mr. Curious, what position would you take about the partner's suggestions?
b. What position should the partner have taken when he learned of this situation? What action should he have taken?
c. Answer this question without regard for the *Code of Professional Ethics*. Do you think that Mr. Curious and/or the audit partner should take this matter directly to the applicable federal authorities?

2–81.   For a number of years, Mr. Alservice, a CPA, had performed bookkeeping services for a certain client, including the recording of the journal entries and the posting of the accounts. Mr. Alservice also prepared the tax returns. In partial payment for these services, the client often gave the CPA shares of stock in his company in lieu of cash.

In March 1975, Mr. Alservice sold his accounting practice to Mr. Partserv, another CPA. Mr. Alservice agreed to stay on as a part-time employee of Mr. Partserv and also agreed to continue handling certain accounts, including the client for whom he kept books.

In April 1975, a local bank notified this client that an audit was required in order to continue loans that had been made in the last few years. When Mr. Partserv learned of this audit requirement, he asked Mr. Alservice to perform the audit. Mr. Partserv indicated, however, that there would be two problems with this arrangement. First of all, there was the question of Mr. Alservice's stock holdings in the client company. Second, there was the matter of the bookkeeping that Mr. Alservice did for the client.

Mr. Alservice suggested a solution to both problems. He felt that because Mr. Partserv would sign the audit report as partner, there was no reason to give up his stock. Mr. Alservice also suggested that the client could hire a part-time bookkeeper to record and post the entries. His functions would be confined to such things as making adjusting journal entries, writing up the quarterly payroll reports, and reconciling the bank accounts.

Required:
a. What do you think of Mr. Alservice's first suggestion? His second suggestion?
b. If you find any objections to either of Mr. Alservice's suggestions, propose alternatives that might overcome your objections.
c. If the client planned to "go public" by offering its securities to the public, how would you take Mr. Alservice's suggestions?

2–82.   Gilbert and Bradley formed a corporation called Financial Services, Inc., with each man taking 50% of the authorized common stock. Gilbert is a CPA and a member of the American Institute of CPAs. Bradley is a CPCU (Chartered Property Casualty Underwriter). The Corporation performs auditing and tax services under Gilbert's direction and insurance services under Bradley's supervision.

One of the Corporation's first audit clients was the Grandtime Company. Grandtime had total assets of $600,000 and total liabilities of $270,000. In the course of his examiniation, Gilbert found that Grandtime's building with a book value of $240,000 was pledged as security for a ten-year note in the amount of $200,000. The client's statements did not mention that the building was pledged as security for the ten-year note. Inasmuch as the failure to disclose the lien did not affect either the value of the assets or the amount of the liabilities, and his examination was satisfactory in all other respects, Gilbert rendered an unqualified opinion on Grandtime's financial statements. About two months after the date of his opinion, Gilbert learned that an insurance company was planning to lend Grandtime $150,000 in the form of a first-mortgage note on the building. Realizing that the insurance company was unaware of the existing lien on the building, Gilbert had Bradley notify the insurance company of the fact that Grandtime's building was pledged as security for the term note.

Shortly after the events described above, Gilbert was charged with a violation of professional ethics.

Required:
Identify and discuss the ethical implications of those acts by Gilbert that were in violation of the AICPA *Code of Professional Ethics.*

(AICPA adapted)

2–83. During the course of his audit examination, Mr. Atest discovered that the clerical work in the Accounting Department was very slow and inefficient. It occurred to him that a limited computer system might increase the efficiency of operations. Therefore, he orally recommended such a system.

Mr. Atest's knowledge of computers was very limited, so he referred the client to Mr. Consult, a fellow CPA, who had a good reputation in this area. Arrangements were made and the consulting engagement was started.

The client seemed to be quite pleased with Mr. Consult's work, and when he found out that Mr. Consult also conducted audits, he suggested that Mr. Consult take over both the company audit and any future consulting engagement. Mr. Consult hesitated to do this, but after considerable urging from the client, he accepted the offer.

During the course of the computer study, Mr. Consult virtually supervised the installation of new equipment and wrote up all of the system documentation. Whenever he spoke to the client controller about his recommendations, the controller would reply that the matter was strictly up to Mr. Consult because no one else in the firm knew anything about computers.

Required:
a. Did Mr. Consult violate the Code of Ethics by accepting the audit engagement? Did he violate it in any other manner?
b. Without regard to the Code of Ethics, do you think Mr. Consult should have accepted the audit engagement? Why or why not?

2–84. The president of Shambra Products has consulted you in regard to an audit. Shambra, which has never been audited, has been developing a process to convert swamp gas into natural gas. Although developments to date have been encouraging, numerous problems remain to be resolved.

The Company has expended substantially all of the funds it has been able to raise to date, and an audit is needed in connection with further attempts to raise money from a stock or debt offering. The Company does not have sufficient funds to pay for an audit now, but it will be able to pay if additional financing is acquired. Although the Company is small at the present time, it has the potential for becoming one of the largest companies in the nation if it is successful in developing its process.

Should you accept the audit of Shambra Products? If so, why? If not, why not?

2–85.   John Smith, CPA, recently received a copy of the revised Code of Ethics from the AICPA. After reading it, Smith wonders whether some of his past actions may have been inadvertent violations of the independence requirement. He specifically recalled the following events:

    a.   The president of a client company with whom he is close personal friends hosted a trip to Las Vegas for Smith and his wife. All expenses, including gambling losses, were paid by the president of the client company.

    b.   Smith was interested in serving the community in some fashion. He asked the executive vice-president of another client who was also active in community affairs to help him get an appointment to the board of directors of the United Fund. Smith received the appointment and has devoted many hours to this activity.

    c.   Smith attended a formal dinner party given by a client officer at the local country club.

    d.   The treasurer of another client paid for two of Smith's lunches. Only the client's business was discussed at one lunch; no business was discussed at the other.

For each of the aforementioned events, indicate whether you believe Smith was in violation of the letter or spirit of the Code of Ethics and, if so, why. If not, why not?

2–86.   During the audit of Park Co., a bottler and distributer of "Marshwater" soft drink, your discussion with the financial vice-president reveals that the Company is paying rebates to certain of its larger customers for carrying Marshwater. The financial vice-president is reluctant to discuss the rebates and tells you that information regarding them is confidential and cannot be discussed further. He points out that your Code of Ethics requires you to respect confidentiality.

Discuss what, if any, further action you would take on this matter.

2–87.   A CPA is explaining the concept of auditor's independence to some associates and is encountering some difficult questions. Here are some examples.

    a.   The accounting profession seems to be inconsistent in its concept of independence. Why does it restrict the auditor from owning *any* stock of clients, but allow auditors to keep books for clients?

b. Isn't the biggest impairment to auditors's independence the fees they receive from clients? How can auditors maintain that they are independent from companies that supply their major source of revenue?

c. How can a CPA firm maintain independence when it audits systems that were designed by them during the course of a management services engagement?

Reply to each of these questions.

2–88. The AICPA *Code of Professional Ethics* prohibits its practicing members from jointly engaging in occupations that are incompatible with the practice of public accounting. Indicate whether you consider each of the following to be an incompatible occupation. Give your reasons why.

a. Lawyer
b. Doctor
c. Real estate agent
d. Insurance agent
e. Night club operator
f. Travel agent
g. Management services consultant
h. Stock broker

2–89. Mr. Bland, a practicing CPA, has been asked to accept an audit engagement that has been conducted for several years by Mr. Blue. During the preliminary conversations between Mr. Bland and his potential client, X Company, the question of the predecessor auditor arose. Mr. Bland indicated that he needed to talk to Mr. Blue and would like to review the prior-year working papers. The president of X Company stated that he could provide any needed information about prior-year audits and preferred that no contact with Mr. Blue be made. The president then described the differences that had caused him to replace Mr. Blue as auditor. The president ended the conversation by stating that he would not absolutely veto a review of prior-year working papers, but would prefer that such a review not be conducted.

Required:
What should Mr. Bland do?

Chapter 3

The Auditor's Responsibility— Legal Environment

Learning Objectives *After reading and studying the material in this chapter, the student should*

Know the general format of, and the reasons for, the different audit report styles used during the last sixty years.

Understand the evolution of auditors' responsibilities during the twentieth century.

Understand how and why the SEC was created and its past and present impact on the auditing function.

Know the summary facts and the auditing implications of the major legal cases in auditing during the last fifty years.

As pointed out at the end of Chapter 1, one of the objectives of this text is to explain the risk accountants encounter when they undertake an audit. In this chapter, the evolution of auditors' responsibilities is traced from the early part of the century to the present time. Because some of the changes have been reflected in the audit report, the discussions include some of the various report styles that have been used during the last sixty years.

In addition, several landmark legal cases involving auditors are summarized. These cases are discussed in other chapters where they relate to audit objectives and techniques. The purpose of introducing them at this point is to show how certain external forces and events have increased auditors' risk and changed their view of their responsibilities.

An Example of Today's Report Style

The auditor's report is the end-product of the audit process. It is his message to the users regarding his opinion of management's financial statement representations. Therefore, each word and phrase is extremely important, especially because the style popularly used today has evolved over the last sixty years. The report consists of a scope paragraph (what the auditor did) and an opinion paragraph (the auditor's belief about the result of his endeavors). Here is an example of a current report.

> We have examined the balance sheet of ABC Company as of [at] December 31, 19XX, and the related statements of income and retained earnings and changes in financial position for the year then ended. Our examination was made in accordance with generally accepted auditing standards and, accordingly, included such tests of the accounting records and such other auditing procedures as we considered necessary in the circumstances.
>
> In our opinion, the financial statements referred to above present fairly the financial position of X Company as of [at] December 31, 19XX, and the results of its operations and the changes in its financial position for the year then ended, in conformity with generally accepted accounting principles applied on a basis consistent with that of the preceding year.[1]

The Auditing Function in the Early Years

To provide a comprehensive view of the change that has taken place in the auditors' concept of their responsibility, it is necessary to compare the foregoing report style with the style used in 1915.[2]

> We have audited the books and accounts of the ABC Company for the year ended December 31, 1915, and we certify that, in our opinion, the above balance sheet correctly sets forth its position as at the termination of that year, and that the accompanying profit and loss account is correct.

Quite a difference! One might wonder whether the writers of a 1915 report conducted the same type of audit. In many respects, they did not. Examinations conducted in 1915 were different in the following ways.

---

[1]*AICPA Professional Standards, Volume 1,* Commerce Clearing House, Inc., AU Section 509.07 *(SAS No. 2).*

[2]George Cochrane, "The Auditor's Report: Its Evolution in the U.S.A.," *The Accountant,* Nov. 4, 1950, pp. 448–460.

1. There were no laws requiring an audit, because the SEC had not been established. In many cases, reports were not issued to stockholders. If an audit was needed, the management or board of directors hired the auditor and the report was addressed to them.[3]

2. The audit was performed for management. It was viewed as a type of guarantee or certification as to the correctness of the accounts; hence, the term "certificate."

3. The examination in some cases was a complete one: all transactions were checked. In fact, fraud detection was the avowed purpose of some audits.

4. Because large-scale investment by absentee owners had not reached today's volume, the examination was centered on the balance sheet. For most companies this was the only published financial statement. The net income was one figure added to retained earnings in the balance sheet.

5. The concentration was on "certifying" the account balances that were on the books. Almost complete reliance was placed on use of internal evidence, i.e., evidence that originates and/or is held by the client. It was not always standard practice to obtain independent evidence external to the client's accounting system.

6. Accounting and reporting principles were not defined as clearly as they are today. There was a widespread impression (as there still is) that the figures on financial statements represented exactness. Because the auditor was acting on behalf of management and because management wanted a certificate, it was natural that the auditor would attest to the "correctness" of the statements.[4]

Changes During the 1920s

As firms began to sell more capital stock to the public, ownership became more distinct from management and evolved into a separate class of proprietors called third-party investors. Because the purpose of audits was to satisfy groups that had a monetary interest in the enterprise, it was only natural that auditors' obligations shifted toward those third parties. The increase in size of many corporations made it impractical to check every transaction. As a result of these changes, auditors were forced to rely on tests, and determination of fairness of financial statements began to replace fraud detection as the major purpose of the audit.

[3]Today, the stockholders often theoretically elect the auditors, but as a practical matter they only approve the recommendation of the president or board of directors.

[4]Cochrane, *op. cit.*

In addition, accounting principles began to be developed. In 1929, the AICPA revised a bulletin called *Approved Methods for the Preparation of Balance Sheet Statements,* which was first published in 1917. It stressed that because the income statement was of primary importance to stockholders, investors, and creditors, it should be prepared in detail, including comparative figures from prior years. Also, it pointed out that auditors should study the firm's financial system and use that study as a basis for testing rather than conducting a detailed check of every transaction. In general, this decade can be characterized as an era in which the auditor's responsibility to third parties began to increase.

The *Ultramares* Case

A vivid illustration of this increase in responsibility occurred during the 1920s, when a third-party creditor sued a CPA firm for negligence and fraud. The effect of the court's decision was so far-reaching that even today this case is considered a landmark with regard to third-party action against an accountant under common law. Following is a summary of the facts.

Touche, Niven & Co. (now Touche, Ross & Co.) were the auditors for Fred Stern & Co.[5] An audit was performed for the year ended December 31, 1923. The accountants knew that creditors of some type would use the balance sheet, but they were not aware of the specific creditor. The audit was completed and the following report was issued.

> We have examined the accounts of Fred Stern & Co., Inc. for the year ending December 31, 1923, and hereby certify that the annexed balance sheet is in accordance therewith and with the information and explanations given us. We further certify that, subject to provision for federal taxes on income, the said statement, in our opinion, presents a true and correct view of the financial situation of Fred Stern & Co., Inc., as at Dec. 31, 1923.[6]

According to the audited balance sheet, the asset total was approximately $2,500,000 and the net worth was approximately $1,000,000. On the basis of these figures, the creditor lent the firm about $165,000.

In reality, the net worth of the company was virtually zero. Most of the overstatement was due to a large block of fictitious accounts receivable. Following the normal procedures of the day, the auditors confined their investigation to evidence created and/or held by the client, such as sales invoices. No letters were sent to customers asking them to verify directly to the auditors the existence of the account balances.

---

[5]All large CPA firms at one time or another have been involved in litigation, and the naming of any firm in this book is no reflection on that firm.

[6]*Ultramares Corporation v. Touche* (255 N.Y. 170, 174 N.E. 441, 1931).

When the overstatement was discovered, the third-party creditor sued the auditors for both negligence and fraud. (Ordinary negligence represents carelessness; fraud represents a willful attempt to deceive.)

In the original trial the judge disallowed the charge of negligence because there was a lack of privity or contractual relationship between the auditor and the third-party creditor. In doing so, he followed the common-law concept that had developed to that date, i.e., only the party that contracted with the auditor and was known to him could take action against him for ordinary carelessness. The auditor's lack of knowledge about the name of the creditor and the use that he would make of the statements apparently prevented privity. The charge of fraud was disallowed also.

In the appeals trial the disallowance of the negligence charge was upheld, and the following words were used to justify this decision.

> *If liability for negligence exists, a thoughtless slip or blunder, the failure to detect a theft or forgery beneath the cover of deceptive entries, may expose accountants to a liability in an indeterminate amount for an indeterminate time to an indeterminate class.*[7]

However, the Court of Appeals ordered a new trial on the charge of fraud (willful attempt to deceive). In the opinion of the judge, the facts of the case had opened up a possibility that the negligence might be so serious as to constitute a form of fraud. The judge brought up the question of whether the accountants had any genuine belief in the statements. The exact words of the court were:

> *Our holding does not emancipate accountants from the consequences of fraud. It does not relieve them if their audit has been so negligent as to justify a finding that they had no genuine belief in its adequacy, for this again is a fraud.*[8]

The case ultimately was settled out of court, but important auditing implications were established. Although the common-law precedent that disallows third-party action against an auditor for *ordinary* negligence was upheld, a new precedent was established that apparently allowed third-party action against an auditor for very serious (gross) negligence.

Although the court made no official comment, it is possible that the wording of the original audit report damaged the auditor's case, because it said that the auditor *certified* (guaranteed) the statements to be true and correct (exact). The report might have been taken as a statement of fact, and not opinion.

---

[7] *Ultramares, op. cit.*
[8] *Ultramares, op. cit.*

Shortly after the decision was handed down, the words "certify," "true," and "correct" were taken out of the general report format.[9] A major result of *Ultramares* was the creation of a new theory of negligence under which third-party liability could be sustained.

The Securities Act

The *Ultramares* court decision took away some common-law protection that the auditor previously had enjoyed. In 1933, the U.S. Congress passed the Securities Act, which increased still further the auditor's potential liability to third parties.

The purpose of the Securities Act was to regulate security offerings to the public through the mails or interstate commerce.[10] A registration statement had to be filed with each *new* public offering of stock. Each registration statement was to include a balance sheet for the previous year and income statements for the past three years. The financial statements had to be certified by independent public accountants.[11]

As far as accountants' potential liability is concerned, the significant part of the act is the provision giving third-party investors the right to claim money damages against the auditor for ordinary negligence, regardless of the fact that they are not clients. For legal actions that come under this law, which is limited to financial statements included in registration statements for new public offerings of securities, the Securities Act takes precedence over the common law that was reaffirmed in the *Ultramares* case, i.e., privity of contract is necessary for a third party to successfully sue the auditor in an action based on ordinary negligence.

The money damages are the differences between (1) the amount the investor paid for the security, and (2) the market price at the time of the suit or sales price.[12]

Additional provisions of the act have special applicability to auditors.

1. The third party is not required to prove negligence or fraud or to demonstrate that he relied on the financial statements or information omitted from the statements. His *prima facie* case is an alleged false statement or misleading omission.

2. The burden of proof is on the auditor to show that the third-party losses were caused by other factors.

---

[9]It should be pointed out that a number of current auditing pronouncements resulted from legal cases.

[10]Denzil Y. Causey, Jr., *Duties and Liabilities of the CPA*, Bureau of Business Research, University of Texas at Austin, 1973, p. 88.

[11]Saul Levy, *Accountants' Legal Liability*, American Institute of Accountants, 1954.

[12]Causey, *op. cit.*

3. In addition, the potential liability extends to the effective date of the registration statement, which might extend beyond the date of the audit report. Normally, the latter date is based on the end of the auditor's field work.

4. However, the accountant does have an ordinary negligence defense known as "due diligence" if he can show that after "reasonable investigation" he had reason to believe and did believe that the statements were true as of the effective date of the registration statement ("true" is the term used in the act).[13]

A hypothetical example illustrates these four points. Assume that a CPA firm is engaged to audit the financial statements of a company that plans to sell its stock to the public for the first time. These statements cover a period of three years, from January 1, 1976 to December 31, 1978.

The auditors finish the examination on February 15, 1979, and the audit report bears this date. Because of the time involved in preparation, however, the registration statement becomes effective April 15, 1979. Suppose that sometime in March 1979 the company is sued for a sum that is quite large in relation to the balance sheet total and the net income. No mention is made of this litigation in the financial statements that are included in the registration statement.

If investors buy the stock and the price drops after knowledge of this litigation becomes public, these third parties may sue the auditors to recover their lost investment. Their case is the misleading omission in the financial statements. The auditors cannot use the audit report date as a defense because their responsibility extends to April 15, 1979. But if the auditors can show that they conducted a "reasonable investigation" of events up to April 15 and had no reason to suspect litigation, they may have a "due diligence" defense.

The Securities Exchange Act

When the Securities Act was passed, the Interstate Commerce Commission administered its provisions. In 1934, Congress passed the Securities Exchange Act to regulate public security trading, and created the Securities and Exchange Commission (SEC) to replace the Interstate Commerce Commission as the regulatory agency for those acts.

Under the Securities Exchange Act, certain companies with publicly held securities are required to file with the SEC, on a periodic basis, various types of financial information including financial statements certified by independent public accountants.

In two ways, the Securities Exchange Act is less severe than the Securities Act.[14]

---

[13]Causey, *op. cit.*

[14]Causey, *op. cit.*, pp. 88–89.

1. The 1933 act relates to the registration of *new* public offerings of securities. The 1934 act is a separate piece of legislation and relates to continuous trading of securities with no effective registration date. Thus, there is no problem of determining the scope of audit procedures after the end of the field work.

2. Under the 1933 act, the burden of proof is on the accountant to show that the third party's loss was not due to false or misleading statements. Under the 1934 act, some of the burden is shifted to the third-party plaintiff. He must prove that he relied on the financial statements and that the damages were caused by doing so.

The auditor's defense probably is easier under the 1934 act. If he can prove that he performed in good faith and had no knowledge that the statements were false or misleading, he can avoid liability. The 1934 act does not amend the 1933 act; the acts apply to different situations.

One of the controversial aspects of the 1934 act is the power granted to the Securities and Exchange Commission. In the wake of the criticism leveled against the profession during the early 1930s, Congress gave the SEC the authority to determine accounting principles, for the purpose of financial reporting on securities offered to the public. Although the SEC has influenced many reporting changes in the last forty years, they have chosen to let the profession regulate itself so far.

The *McKesson & Robbins* Case[15]

Although the securities acts of 1933 and 1934 increased the auditor's responsibility to third parties, a famous fraud case in the late 1930s served to remind the profession that their evidence-gathering techniques were somewhat obsolete.

McKesson & Robbins was a wholesale drug company whose securities were listed on the New York Stock Exchange and were registered under the Securities Exchange Act. One of the alleged activities of the firm was a foreign business carried on through a Canadian subsidiary. This separate division showed on its books sizable purchases, sales, and inventories, but, in fact, the transactions of this division were fictitious.

The effort that was expended to perpetrate this fraud stretches the imagination. Documents apparently were drawn up for vendors, customers, bank accounts, and so on. Non-existent warehouses in Canada were supposed to contain the inventory. Obviously, such an operation could not have been carried on for several years without massive collusion on the part of the president of the company and several key officers.

---

[15]Excerpts and paraphrases from the *Report on Investigation of the Securities and Exchange Commission in the Matter of McKesson & Robbins, Inc.,* December, 1940.

The fraud became known in 1939. In 1940, the Securities and Exchange Commission conducted an investigation. During the series of hearings, two questions were raised that had direct relevance both to Price Waterhouse & Co., McKesson & Robbins' auditors, and to the profession in general.

1. To what extent were contemporary acceptable auditing standards and procedures followed?

2. To what extent did the contemporary auditing standards and procedures provide necessary safeguards and insure reliability of the statements?

Concerning the first question, the SEC believed that the auditors were negligent and that gross discrepancies of this type should have been caught. In general, however, it was conceded that the auditors had followed procedures that were acceptable at that time.

On the second question, the SEC was very critical of the scope of audit procedures followed by the entire profession. Although it declined to exercise the full authority granted by Congress, the SEC made several suggestions.

1. In the two important areas of accounts receivable and inventory, the SEC recommended that evidence of validity should be obtained from external and independent sources.

2. The stockholders should hire the auditor.

3. The audit report should be addressed to the stockholders.

4. The auditors should attend stockholder meetings and answer questions from this group.

In their reply to the SEC, the McKesson & Robbins auditors maintained that their examination was not designed to detect fraud, particularly if collusion of this type existed. (This same line of reasoning has been "codified" into the profession's official pronouncements.) But the SEC's first recommendation was followed explicitly in 1940.

The AICPA issued a pronouncement requiring auditors to gather independent and external evidence on accounts receivable and inventory, if the amounts are material. (An amount is considered material if it would cause an informed reader to view the financial statements differently.) The evidence-gathering techniques that the accounting profession used to satisfy its auditing responsibility had undergone a permanent change. Today, it is routine procedure to seek independent verifications of such accounts.

Changes During the 1940s and 1950s

The SEC's other three suggestions have been followed partially. Sometimes stockholders ratify the auditors, but this action is often a mere formality because the recommendation for the auditors comes from the board of directors. There is no SEC requirement concerning the addressee of the audit report, so the addressee can be the stockholders or the board of directors. Although the auditor's attendance at stockholder meetings is still optional, the SEC does require that the auditor be available to attend such meetings or disclose to stockholders that the auditor will not be available.

As pointed out in Chapter 1, significant pronouncements were made on auditing standards during the period between the early 1940s and the late 1950s. For one thing, the members of the profession made it clear that *they* did not regard fraud detection as the major purpose of an audit of the financial statements. The current standard form of the audit report was adopted during this period, and its emphasis is on ascertaining the fairness of the financial statements in accordance with generally accepted accounting principles. By the end of the 1950s, the AICPA's Committee on Accounting Procedure had issued fifty-one pronouncements on accounting principles.

Also, during this period the ten auditing standards (referred to in Chapter 1) were adopted. It was not necessary for the auditors to refer to their procedures in the report. They could merely state that generally accepted auditing standards had been followed in the conduct of their examination.

Except in the case of special audits, testing continued to be an accepted method of gathering evidence on which to base an opinion. The quality of the company's financial control system served as a basis for determining the extent of this testing.

The *Barchris Construction Corporation* Case[16]

The 1960s were characterized by rising prices for many corporate stocks, and many of the prices proved to be inflated. This decade also ushered in the "age of litigation," which is still in progress. As a result of several lawsuits, coupled with an increasingly active role by the SEC, the auditor's expanded responsibility to third parties was confirmed. A major example is the *Barchris* case, tried under the Securities Act of 1933. The importance of this particular lawsuit is that a favorable judgment was rendered for the third-party plaintiffs, who were bringing action under that portion of the law covering false statements and misleading omissions in the registration statement. Because some of the information in the registration statement came directly from the most recently audited financial statements, Peat, Marwick, Mitchell & Co., the auditors, were involved.

---

[16]Excerpts and paraphrases from *Escott* v. *Barchris Construction Corp.*, 283 F. Supp. 643. 701 (S.D.N.Y. 1968).

Barchris Construction Corporation was a builder of bowling facilities. Until 1961 the corporation had financed all of its needs from sources other than publicly traded securities. In the early part of 1961, they filed a registration statement for the public sale of convertible bonds, which placed them under the provisions of the federal securities acts. Figures from the audited financial statements as of December 31, 1960, and the previous three years were included in the registration statement.

Because the effective date of the registration statement was later than the balance sheet date, the auditors conducted a review of events between the balance sheet date and the effective date of the registration statement, in addition to their regular procedures. Everything appeared to be satisfactory.

Late in 1962, Barchris went bankrupt. The third-party purchasers of the bonds filed suit against the auditors, using the 1933 act as their basis for the suit. They claimed that the registration statement contained false statements and material omissions. The auditors used the "due diligence" defense provided for in the act.

During the trial, the focus was on an apparent overstatement of earnings and an overstatement of the current ratio. Some of the alleged earnings overstatement was caused by recording the gain on a sale-leaseback as current revenue, rather than amortizing it over the life of the lease. The current-ratio overstatement was caused by a variety of factors, such as inclusion of an account receivable from a consolidated subsidiary and the omission of a liability.

The court ruled that the registration statement contained false statements of material facts. Most of the "due diligence" defense was disallowed. The case has several important implications.

1. The court made the statement that "accountants should not be held to a higher standard than that recognized in their profession." Yet, in the case of the sale and leaseback, the gain on the sale was treated differently than the judge thought it should be. At the time of the audit, the only authoritative guide was *Accounting Research Bulletin No. 43,* which stated that there should be disclosure of the principal details of any important transactions concerning sale and leaseback. *APB No. 5,* issued by the Accounting Principles Board at a later time, specifies that material gains or losses on sale-leaseback operations should be amortized over the life of the lease.[17] But at the time of the trial, *APB No. 5* had not been issued, and both methods were being followed in practice. Therefore, the judge chose one method over the other.

2. The court took exception to the review of events between the balance sheet date and the registration date. It was believed that more time should have been spent and a deeper and more critical inquiry made of certain important matters.

---

[17] *AICPA Professional Standards, Volume 3,* Commerce Clearing House, Inc., AC Section 5351.21 *(APB Opinion No. 5).*

3. The court also declared the overstatement of net income (14%) to be immaterial because, in its opinion, the earnings from one year to the next would have increased significantly anyway. In contrast, an overstatement of the current ratio (15%) was considered to be material. Accountants may be puzzled by this distinction. But it should be remembered that there were no authoritative standards on materiality at the time of this lawsuit, and thus, the judiciary was simply filling the vacuum.[18]

In retrospect, it can be seen that the *Barchris* decision changed considerably the auditor's working climate. For one thing, it showed that the section of the 1933 Securities Act pertaining to accountants' liability can be used as a basis upon which to sue auditing firms. For another, it demonstrated that the courts are willing to choose between alternative accounting principles if they believe the application of a certain pronouncement does not result in fair statements (the court ruling on the sale and leaseback). This trend toward a separate definition of fairness by non-accountants is continuing.

The *Yale Express* Case[19]

In the 1960s another case reminded the accounting profession of the growing influence of third-party groups. In early 1964, Yale Express Systems, Inc., distributed to the public audited financial statements for the year ending December 31, 1963. The statements also were filed with the SEC in accordance with the 1934 act.

During the same period (early 1964), Peat, Marwick, Mitchell & Co., the auditing firm, changed activities and engaged in a series of management services studies for Yale Express. During the course of these studies, the auditors discovered facts that made it apparent that the assets of the 1963 statements were materially overstated, and thus caused the statements to be misleading.

Unfortunately, the situation was similar to that in the *Barchris* case in that precedent for appropriate action had not been established by the AICPA. The auditors did not report their finding to the SEC until the management services study was released in the middle of 1965. By then it was too late, according to the creditors and stockholders of Yale Express. A suit was filed against the auditors.

The case had at least two complicating factors.

1. First, when the discovery of the overstatement was made, the CPA firm was no longer acting in the capacity of an independent auditor subject to the 1934 act. CPAs performing management services engagements are not acting in an "advocate" role, similar to that of lawyers; they are acting as

---

[18]For a contemporary opinion on the need for audit guidelines in this area, see Leopold A. Bernstein, "Materiality—The Need for Guidelines," *Lybrand Journal,* April 1968, pp. 11–20.

[19]Excerpts and paraphrases from *Fischer* v. *Kletz,* 266 F. Supp 180 (S.D.N.Y. 1967).

consultants. One question that arises, then, is whether the public accountant should reverse roles and assume the status of an independent auditor in situations of this type.

2. Second, according to the CPAs, the discovery was not made until after the report had been issued. At that time, the AICPA had made no pronouncement to the effect that subsequent discovery of facts in existence at the report date imposes an obligation on the auditor to disclose the contents of material discoveries made after the issuance of the report. Since then, the AICPA has issued a pronouncement imposing such an obligation.[20]

The *Continental Vending* Case[21]

Until the 1960s, members of the accounting profession generally assumed that auditors would be charged with criminal fraud only if there were reason to believe that they had overtly conspired with their client to produce false or misleading financial statements. The *Continental Vending* case shattered this belief by demonstrating that auditors could be accused of conspiracy for an important omission that was "conscious" and "willful."

Continental Vending Machine Corporation had an affiliate named Valley Commercial Corporation (Valley). Both companies were dominated by the president, who owned about 25% of the stock of Continental. He had a habit of investing in stock by borrowing sums of money from Valley, which, in turn, borrowed the same amount from Continental. At the balance sheet date, approximately $3,500,000 appeared on the books of Continental as a receivable from Valley (in reality, a receivable from the president).

Adding to this financial complexity was the fact that Continental discounted notes to a bank through Valley. These notes appeared on the books of Continental as a note payable to Valley (actually a note to the bank). At the balance sheet date, the amount was about $1,000,000.

The notes to the financial statements contained an explanation that the $3,500,000 receivable, less the $1,000,000 payable, was secured on the date of the auditor's report by the market value of securities assigned as collateral for the receivable. Coopers & Lybrand, the CPA firm, attested to the fairness of the financial statements.

On the surface, no problem was apparent, despite the unusual nature of the transactions. But most of the president's pledged collateral was stock in Continental. Moreover, the collateral was worth only $2,900,000 on the report date, and because the offset of the $1,000,000 payable against the $3,500,000 receivable was improper, a shortage of approximately $600,000 resulted (the

---

[20]*AICPA Professional Standards, Volume 1,* Commerce Clearing House, Inc., AU Section 561 (*SAS No. 1*).

[21]Excerpts and paraphrases from *U.S.* v. *Simon,* 425 F.2d 796 (1969).

$3,500,000 gross receivable less the $2,900,000 of collateral).[22] Also, the value of the entire pledged collateral was questionable.

If only a civil suit had been filed against the auditors, the accounting profession might not have given this case much attention. But the auditors were charged with conspiring and adopting a scheme to violate federal *criminal* statues by certifying to false and misleading financial statements. The following language was used by the court in stating the case that the government had to prove:

> Not to show that the defendants were wicked men, with designs on anyone's purse, which they obviously were not, but rather that they had certified a statement knowing it to be false.

Moreover, the court declared that proof of conformity with generally accepted accounting principles would constitute only *partial* evidence of statement fairness.

Three auditors of the CPA firm were found guilty and fined. Some members of the accounting profession believed that the verdict of criminal fraud was too harsh for what was essentially a mistake in judgment. Many other members thought that more than a mistake was involved. Nevertheless, the case increased auditors' legal risk by widening the areas in which they might be charged with criminal conduct.[23] No longer is it necessary for the CPA to collaborate actively with the client in order to be found guilty of fraud. Fraud now may occur if the client is allowed to "get away" with something that the auditors, in their expert capacity, actually know and should have disclosed. In addition, the concept of fairness began to broaden beyond the application of generally accepted accounting principles. Such a broadening trend is reflected in AU Section 411 *(SAS No. 5)* and is discussed fully in Chapter 15.

The *1136 Tenants' Corporation* Case[24]

Over the years, CPAs have performed a special type of service which is something between bookkeeping and an audit. Sometimes a client requests the preparation of "unaudited" financial statements in which no opinion as to fairness is rendered. The client may wish to send these statements to a third-party creditor, or he may simply use them internally. Because no audit opinion is given, the CPA usually considered this type of engagement to be a relatively risk-

---

[22]There seems to be an indication that the third general standard was not followed. This standard states that due professional care is to be exercised in the performance of the examination and the preparation of the report.

[23]For further insights into the implications of this case, see David B. Isbell, "The Continental Vending Case: Lessons for the Profession," *The Journal of Accountancy,* August 1970, pp. 33–40.

[24]Excerpts and paraphrases from *1136 Tenants' Corp.* v. *Max Rothenberg & Co.,* 36 A.D. 2d 804, 319 N.Y.S.2d 1007 (1971).

free service. But preparation and issuance of unaudited financial statements became an acute problem several years ago when an unfavorable verdict was rendered by a New York court against a CPA firm.

A cooperative apartment corporation hired the firm to "perform all necessary accounting and auditing services." The purpose of this engagement was to check on the custodianship of the corporation's managing agent.

Apparently, there was a misunderstanding between the corporation and the accountants as to the nature and extent of the services. No letter explaining the terms of the engagement had been written and sent to the owners of the apartment corporation. The owners thought that an audit would be performed, whereas the accountants understood that the services would be confined to "write-up" work consisting of maintenance of the books, preparation of financial statements, and preparation of related tax returns.

The engagement was completed, and the CPA firm issued a transmittal letter to the owners indicating that no independent verifications had been made.

The situation might have remained stagnant except that defalcations were committed by the managing agent. When the discovery was made, the owners sued the accountants for negligence in failing to uncover the fraud. The accountants, in turn, maintained that they had not been engaged to perform an audit.

The court decided that the weight of the argument was in favor of the owners; money damages were awarded.

There were several complicating factors in the case.

1. No written engagement letter was issued by the CPA outlining the scope of the services.

2. No clearly worded report disclaiming an opinion on fairness accompanied the unaudited statements.

3. Because *some* audit procedures were followed by the CPA, the judge apparently construed the services to be an audit and, therefore, ruled that the defalcation should have been caught.

4. The auditor noted several missing invoices, but failed to follow up on them.

What precedent, if any, has been set by this case? What lessons can be learned? After the verdict was rendered, several articles appeared suggesting the use of safeguards to avoid this type of situation in the future. Chief among these safeguards is a clearly worded engagement letter from the CPA to his prospective client stating specifically what procedures he will perform.

What about the question of audit procedures on "write-up" work? CPAs recognize that even in this type of engagement there are some services that could be taken as auditing procedures, e.g., checking the computation of depreciation.

At what point might a court decide that the CPA has "committed" himself to an audit?

Although no definite answers are available, a clearly written engagement letter should help prevent this problem or issue from arising.

Pending any final disposition of these matters, the best that a CPA can do is to follow the pronouncements and guidelines suggested by the AICPA. A CPA no longer can avoid litigation simply because he performs a service that does not require an opinion on fairness.

The case is referred to in Chapter 17 with regard to special reports.

The *Equity Funding* Case

Traditionally, auditors have assumed that their client's top-level management is truthful. In fact, it is difficult to conduct an audit if management is not honest, because much of the evidence for the opinion comes from their representations.

But in recent years a disturbing tendency has arisen (or surfaced) among some top corporation officials, i.e., the deliberate attempt to deceive auditors. A recent example is the *Equity Funding Corporation* case. The fraud in this case, like that in the *McKesson & Robbins* case, consisted of massive collusion on the part of higher management with the intent to falsify assets and earnings.

Equity Funding Corporation of America was formed in 1959. Its major function was to invest in mutual funds the money received on capital stock sales and to sell life insurance back to the stockholders. The insurance business was ultimately shifted to a subsidiary called Equity Funding Life Insurance Company.

During the 1960s, Equity Funding's earnings grew. However, in the late 1960s and early 1970s, mutual funds lost some of their glamour, and interest in this type of investment waned. The management of Equity Funding apparently was unwilling to reflect the consequences of this situation on the company's financial statements. When business failed to materialize, dummy customers were made up, and false information was stored in the computer files.

To make the transactions appear real, a set of computer programs was used which cleverly concealed the deception. It is now apparent that large-scale collusion occurred between officers and computer personnel.

Equity Funding convinced other insurance companies to make advances to them on a co-insurance arrangement. Future premiums were to be used to repay these companies in later years. Because a sizable percentage of the policies never existed and future premiums never materialized, these insurance companies lost much of their advances.

Several CPA firms worked as auditors for Equity Funding at one time or another (Seidman & Seidman were the last auditors, having merged with a local firm that previously performed the audit), yet none was able to detect this fraud. When the facts finally became known in 1973, the case proved to be a source of embarrassment for the accounting profession. It seems clear that the auditor's assumption of top-level honesty may have to be re-examined. AU Section 160.19

(SAS No. 4) contains suggestions on policies and procedures for acceptance and continuance of clients. There is more about this case in Chapters 7 and 14.

The *Hochfelder* Case: A Possible Turning Point

From the early 1960s to the mid-1970s, the frequency of legal liability suits filed by third-party groups against auditors under the Securities Exchange Act of 1934 was increasing. Some major court decisions went *against* the CPA firms. A decision by the U.S. Supreme Court in 1976, however, may have signaled a reversal of the trend of the previous decade.

Suit was brought against Ernst & Ernst, a large CPA firm, by a group of investors that had placed sums of money in what they believed to be "escrow" accounts of First Securities Company of Chicago, a brokerage house. It appears that the funds were used improperly by the president of First Securities Company and that no accountability existed for them. The fraud was perpetrated by a so-called "mail rule" prohibiting anyone but the president from opening incoming mail addressed to him. No record of the receipts was placed on the company's books. The CPAs, Ernst & Ernst, failed to detect the defalcation.

When the fraud was uncovered, the investors sued the CPA firm for negligence under the Securities Exchange Act of 1934. Ernst & Ernst was not accused of participating willingly in the fraud but of negligent conduct because of their failure to discover the material weakness in internal control resulting from the mail rule. After the lower court and appeals court trials, the suit was heard by the U.S. Supreme Court. The court ruled for the CPA firm, and the suit was disallowed.

It may be more difficult in the future for third-party groups to secure damages from CPAs under the Securities Exchange Act of 1934. Apparently, willful attempt to defraud the investors must be proved, rather than simple negligence. Also, because the investors did not use the financial statements and auditors' opinion, they were unable to prove that they had relied on the statements and that such reliance resulted in their financial loss (see the earlier summary of the provision of the Securities Exchange Act of 1934.)[25]

Whether the ruling in this case proves to be a real landmark decision or not remains to be seen. In any event, it appears likely that it will be more difficult in the future for third-party groups to sue successfully on the basis of Section 10(b) of the Securities Exchange Act of 1934, unless more than simple negligence can be proved.

The Present-Day Situation

After almost a century of audit experience, where does the accounting profession in this country now stand?

[25]Jon Ekdahl, "Hochfelder—A Significant Decision for the Accounting Profession," *The Arthur Andersen Chronicle,* April 1976, pp. 34–39.

Today, auditors are faced with the distinct possibility of legal action from clients, third parties, or regulatory agencies. It is no longer feasible for auditors to *examine* the books, *check* all transactions, and certify the truthfulness of the balance sheet to management. Instead, it is necessary for them to *examine* the financial statements, follow generally accepted auditing *standards,* make appropriate *tests,* and, as a result, give their *opinion* on the *fairness of the statements,* in accordance with *generally accepted accounting principles,* consistently applied from one accounting period to the next.

The public accountant is also facing responsibilities and risks that were unknown thirty or forty years ago. Despite the official pronouncements of the profession disclaiming automatic responsibility for fraud detection, the public appears unwilling to accept this arrangement. Lawsuits are common, some as the result of a defalcation that remained hidden during the course of an audit. If public pressure forces CPAs to make fraud detection a primary objective, there will have been, in some ways, a full-circle return to the objective of examinations for fraud during the early part of the century. However, modern-day audits must, of necessity, be based on *testing,* which increases the risk significantly.

Courts are taking a more "hard-line" position on criminal cases levied against auditors. If the trend established in the *Continental Vending* case continues, auditors can expect to be charged with fraud for overlooking their clients' deception (omission rather than commission).[26] In addition, the accounting profession may have to devise new procedures to detect or discourage deliberate client misrepresentation on financial statements. In fact, AU Section 327 *(SAS No. 16),* an AICPA pronouncement on detection of errors and irregularities, suggests that the auditor should design his audit programs appropriately to provide himself with a reasonable basis for believing that the financial statements taken as a whole are not misstated materially.

The definition of "fairness" appears to be broadening. Auditors in the future may not be able to assume that the proper application of generally accepted accounting principles automatically assures fair statements.[27]

The U.S. Senate and House of Representatives have raised serious questions about the independence of auditors and the gap that might exist between the quality of audit services and the expectations of financial statement users.

One should not be left with the impression that the evolution of the audit process has been nothing more than a steady succession of court cases. It is true that legal cases and SEC actions have changed significantly auditors' concept of their responsibilities. These responsibilities have shifted, relatively, from clients to third-party investors. But the accounting profession has responded to the challenge and has increased the quality of its services. For every "landmark" case that has proved the auditors' fallibility, there are thousands of "clean" audits that

---

[26]However, recent SEC disciplinary actions against CPA firms (e.g., prohibitions on accepting public companies as new clients for a certain period of time, required quality-control reviews, etc.) may signal a change in direction. The SEC may begin using these disciplinary actions as a substitute for referring criminal charges to the Justice Department.

[27]In fact, there is sentiment within the accounting profession to delete "fair" from the auditor's report and to substitute more explicit phrases in its place.

have given the financial community much needed reliance on financial statements.

One of the messages that should come from a discussion of the cases is that auditing is not like "other" jobs. This function involves unique risks and requires unique personal characteristics; both factors are covered in greater depth in other chapters. For each example of poor judgment (which frequently results in national publicity), there are thousands of instances where good judgment appears to have been used. For every case in which the client was able to compromise the auditor's independence, there are many more in which the auditors successfully asserted their independence.

Litigation Guidelines

In view of the recent litigation and the changing public attitudes toward auditors, what can be done to lower the chances of major lawsuits in the future? Several options are open to auditors.

1. Care should be taken in selection and retention of clients. Companies that have a history of financial difficulties are more likely than those without such difficulties to take measures to falsify earnings and hide events that should be disclosed in the financial statements. In general, CPA firms are giving attention to this matter and are "screening" prospective clients more carefully by investigating the backgrounds both of the company and of its officers.

2. Auditors should make certain that generally accepted auditing standards are followed. From the study of the cases in the chapter, it is obvious that one or more of these standards were violated in several instances. Auditors should always remember that the quality of their work is subject to review and will be scrutinized in the event of litigation.

3. Auditors should have a clear understanding with the client as to the scope of the engagement, the services that will be rendered, and the auditor's responsibility. Engagement letters (discussed in Chapter 4) help auditor and client to avoid misunderstandings.

Chapter 3 References

Bernstein, Leopold A., "Materiality—The Need for Guidelines," *Lybrand Journal,* April 1968, pp. 11–20.

Brown, R. Gene, "Changing Audit Objectives and Techniques," *The Accounting Review,* October 1962, pp. 696–703.

Causey, Denzil Y., Jr., *Duties and Liabilities of the CPA,* Bureau of Business Research, University of Texas at Austin, 1973.

Cochrane, George, "The Auditor's Report: Its Evolution in the U.S.A.," *The Accountant,* Nov. 4, 1950.

Isbell, David B., "The Continental Vending Case: Lessons for the Profession," *The Journal of Accountancy,* August 1970, pp. 33–40.

Levy, Saul, *Accountants' Legal Liability,* American Institute of Accountants, 1954.

Reiling, Henry B., and Russell A. Taussig, "Recent Liability Cases—Implications for Accountants," *The Journal of Accountancy,* September 1970, pp. 39–53.

Appendix to Chapter 3

A Summary of Auditor's Responsibilities Under Common Law and Securities Acts of 1933 and 1934

	Action that can be taken under		
Improper Act by Auditor	*Common Law*	*The Securities Act*	*The Securities Exchange Act*
1. Ordinary Negligence			
A. Client takes action	Action against auditor is allowed.	Action against auditor is allowed.	Action against auditor is allowed.
B. Third party takes action	Action against auditor is not allowed.	Action against auditor is allowed for alleged false statement or misleading omission. Auditor must show that losses were caused by other factors.	Action against auditor is allowed if third party can prove reliance on financial statements and that damage from false statement or misleading omission is caused by such reliance.
2. Fraud (Gross Negligence)			
A. Client or third party takes action	Action against auditor is allowed.	Action against auditor is allowed.	Action against auditor is allowed.

Note: The details of the auditor's responsibilities under common law and the Securities Acts are contained in the body of the chapter.

Chapter 3
Questions Taken from the Chapter

3–1.   What is the difference between the scope paragraph and the opinion paragraph of the standard short-form report?

3–2.   Write a standard short-form report.

3–3.   Indicate five major differences between the report style used today and the report style used in the early 1900s.

3–4.   What changes were made during the 1920s that caused auditors to modify the scope of their examination?

3–5.   Briefly list the facts of the *Ultramares* case. Paraphrase the report issued by the auditors.

3–6.   Distinguish between negligence and fraud.

3–7.   What precedent was set in the *Ultramares* case?

3–8.   Why was ordinary negligence disallowed in the *Ultramares* case? Would the auditor's knowledge of the specific creditor have made any difference?

3–9.   What common-law relationship was changed with the Securities Act of 1933?

3–10.   When is a registration statement filed? What does it include?

3–11.   What is the difference between the Securities Act of 1933 and the Securities Exchange Act of 1934?

3–12.   What is the "due diligence" defense that can be used by auditors under the Securities Act of 1933? What is the "good faith" and "no knowledge" defense that can be used under the Securities Exchange Act of 1934?

3–13.   Briefly summarize the facts of the *McKesson & Robbins* case.

3–14.   What recommendations were made by the SEC as a result of their hearings conducted on the *McKesson & Robbins* case?

3–15.   What pronouncement was issued by the AICPA as a result of the *McKesson & Robbins* case?

3–16.   Briefly summarize the facts of the *Barchris Construction Corporation* case.

3–17.   In the *Barchris* case, how did the judiciary inject itself into the matter of accounting principles?

3–18.   In the *Barchris* case, why did the judiciary believe that the overstatement of earnings was immaterial?

3–19.   How did the *Barchris* case change the auditor's working climate?

3–20.   Briefly summarize the facts of the *Yale Express* case. What two roles were assumed by the CPAs?

3–21.   Briefly summarize the facts of the *Continental Vending* case. Why were criminal charges brought against the auditors?

3–22.   In what way did the *Continental Vending* case change the auditor's legal risk?

3–23.   Briefly summarize the facts of the *1136 Tenants' Corporation* case.

3–24.   What is the difference between "audited" and "unaudited" financial statements?

3–25.   In the *1136 Tenants'* case, what was the difference between the client's and the CPA's view of the purpose of the accounting services?

3–26.   What questions were left unanswered by the decision in the *1136 Tenants'* case?

3–27.   Why is it very difficult to conduct an audit without truthful representations from the client's top-level management?

3–28.   What did the management of Equity Funding Corporation do to falsify its records and cover up the fraud?

3–29.   Name four risks and responsibilities that have been created or enlarged for the auditor within the last forty years.

3–30.   This question is designed to test the ability to distinguish the differences among facts and ramifications of the various cases discussed in the chapter. Beside each description, list the letter of the appropriate case(s) taken from the following list (each case can be listed more than once).

a. *Ultramares*
b. *McKesson & Robbins*
c. *Barchris*
d. *Yale Express*
e. *Continental Vending*
f. *1136 Tenants'*
g. *Equity Funding*

    (1)   Applied the appropriate provisions of the Securities Act of 1933 concerning the effective date of the registration statement.

    (2)   Set a precedent on the application of common law for instances of gross negligence on the part of the auditors.

    (3)   The question of "write-up" work for clients.

    (4)   Fraud was committed with the aid of computers.

    (5)   Earnings were falsified through non-existent accounts receivable and inventories.

    (6)   The judiciary decided on the materiality of overstated earnings.

    (7)   Criminal charges were filed against the auditors.

    (8)   The common-law precedent regarding third-party action against auditors for ordinary negligence was upheld.

    (9)   Brought up the question of auditors' responsibilities after the issuance of the report.

    (10)   Brought out the problem of the honesty of top-level management.

    (11)   Brought out the importance of a letter of engagement.

    (12)   Brought up the question of whether statements could adhere to generally accepted accounting principles and still not be fair.

    (13)   Resulted in AICPA pronouncement concerning evidence-gathering on accounts receivable and inventories.

    (14)   Relates to the question of proper disclosure.

Chapter 3
Objective Questions Taken from CPA Examinations

3–31.   When compared to the auditor of fifty years ago, today's auditor places *less* relative emphasis upon

a. Confirmation.
b. Examination of documentary support.
c. Overall tests of ratios and trends.
d. Physical observation.

3–32. The most significant aspect of the *Continental Vending* case was that it
   a. Created a more general awareness of the auditor's exposure to criminal prosecution.
   b. Extended the auditor's responsibility for financial statements of subsidiaries.
   c. Extended the auditor's responsibility for events after the end of the audit period.
   d. Defined the auditor's common-law responsibilities to third parties.

3–33. Which of the following best describes a trend in litigations involving CPAs?
   a. A CPA cannot render an opinion on a company unless the CPA has audited all affiliates of that company.
   b. A CPA may not successfully assert as a defense that the CPA had no motive to be part of a fraud.
   c. A CPA may be exposed to criminal as well as civil liability.
   d. A CPA is primarily responsible for a client's footnotes in an annual report filed with the SEC.

3–34. The independent auditor of 1900 differs from the auditor of today in that the 1900 auditor was more concerned with the
   a. Validity of the income statement.
   b. Determination of fair presentation of financial statements.
   c. Improvement of accounting systems.
   d. Detection of irregularities.

3–35. The *1136 Tenants'* case was chiefly important because of its emphasis upon the legal liability of the CPA when associated with
   a. A review of interim statements.
   b. Unaudited financial statements.
   c. An audit resulting in a disclaimer of opinion.
   d. Letters for underwriters.

Chapter 3
Discussion/Case Questions

3–36. Mr. Clyde Neglent, CPA, was engaged to perform an audit of Hidden Records, Inc., a retail department store whose securities were traded actively on the New York Stock Exchange. Mr. Neglent had made a review of the business and its owners prior to accepting the engagement and had found the following.

Hidden Records, Inc. securities were popular, particularly during the last two years when total assets had risen by 20% and earnings by 15%. The Company appeared to have the characteristics of a solid commercial establishment. But one thing bothered Mr. Neglent. The president and other key officers had unstable records of employment. In addition, Mr. Neglent was puzzled by the sudden rise in profits during a period when retail sales in that region of the country had dropped. Nevertheless, the audit was taken and work soon commenced.

The Company kept most of its records on computer files. The accounts receivable were maintained on magnetic tape. Mr. Neglent's knowledge of computer systems was minimal (another reason for his hesitation in accepting the engagement). Therefore, he decided to consult with the company's data-processing manager to determine the best way to audit the records stored in computer files.

Mr. Clev, the data-processing manager, made several suggestions on the various techniques that could be used to extract audit evidence. Most of them involved printouts of information stored in the computer. These suggestions all appeared to be reasonable and were followed by Mr. Neglent.

Mr. Neglent wished to send confirmation letters to a sample of the customers whose accounts made up the accounts receivable figure, because this amount represented approximately 20% of the total assets. Mr. Clev stated that he had a sampling plan that he used when it was necessary to select customer accounts randomly for various reasons. He offered to make this sample selection for Mr. Neglent and to print out confirmation requests. These requests would be given to Mr. Neglent, who, in turn, would mail them independently to the customers. Mr. Neglent agreed, and this procedure was carried out. Replies received from all the customers to whom letters were sent indicated that the account balances were correct.

All other phases of the examination went smoothly, and the audit report was issued routinely. The report contained an unqualified opinion. Several months later, Hidden Records, Inc., declared bankruptcy, and its securities were taken off the market. It appears that the Company had inflated its earnings and assets by creating false sales and accounts receivable.

Required:
a. What legal action can be taken against the auditor by the investors in the securities of Hidden Records, Inc.? In your opinion, would this legal action be successful?
b. On what basis could criminal charges be filed against the auditor? In your opinion, would these charges be sustained by the court?
c. Would the investors' legal recourse be different if Hidden Records, Inc., did not sell securities on an open exchange?
d. Name any auditing standard(s) that you believe may have been violated by the auditor. Why?
e. What defense(s) can the auditor use against a civil suit by investors? Against criminal charges? In your opinion, are these defenses valid?

3–37. XYZ Oil Co. had been a problem for its auditor, Joe Jones, CPA, for several years. The Company's stock was traded in the over-the-counter market, and management was concerned with keeping net income as high as possible to support the stock price. Each year, the controller would refuse to record any of Jones' proposed entries if he could establish that they were not material. Accordingly, each year's net income was generally overstated by 10% to 15%, which Jones accepted on the basis that it was not material.

Of particular concern to Jones was a $1 million investment in an oil venture in South America; however, he had been unsuccessful to date in proving to the controller's satisfaction that a loss should be recorded on this investment.

During the current year's audit, Jones discovered a report indicating that the South American oil venture was worthless and had been for two years. When he discussed this report with the controller, the controller suggested that the investment be written off in equal amounts over the next ten years to prevent a significant effect in any year and to prevent embarrassment or liability to Jones from disclosing that past financial statements were misstated. The controller stated that recognition of the loss in the current year would mean bankruptcy for XYZ Oil Co. and certain liability for Jones.

Discuss the action you would take if you were Jones. What liability does Jones face if he accepts the controller's solution? What liability does he face if he doesn't accept the controller's solution?

3–38. A colleague has read about the legal cases with which auditors have been involved in recent years. During a discussion, you consider it beneficial to summarize some highlights of auditors' legal liability during the past fifty years. Select five cases you think have had a major impact on audit practice and auditors' responsibility. In two or three sentences each, explain the impact of these cases.

3–39. Refer to the first two pages of the chapter. Compare the current type of auditor's report with the type used in 1915. List what you consider to be the important differences between the two report styles and indicate the reasons why you think these differences exist.

3–40. It seems that every time a legal challenge is made to an audit, the CPAs either lose the case or are embarrassed about the facts uncovered during the audit.

Reply to the statement above.

3–41. Is it possible to audit a company's statements if top management decides they will be dishonest with the auditors?

3-42. Can financial statements be in accord with generally accepted accounting principles and not be fair?

3-43. In view of the changes during the last forty years, what has happened to the precedents set in the *Ultramares* case? How has the *Hochfelder* case reaffirmed or destroyed the precedents?

3-44. Alan Jason, CPA, had been auditing the financial statements of Rosfeld Stores, Inc., a large department store, for several years. During the last two years, a nationwide economic recession had affected Rosfeld's sales and net income severely. The company, looking for ways to lower operating expenses, constantly asked Jason how the audit fee could be reduced.

One day Rosfeld's president called Jason and suggested a method for saving money on the forthcoming audit. The president suggested that his personnel select the customer accounts on which confirmation letters would be sent. The same personnel would write and mail the confirmation letters. All of these procedures would be performed under the supervision of one of Jason's auditors.

Jason agreed to this procedure. As the auditor watched, two of Rosfeld's employees selected every fifth account, wrote a confirmation letter, and mailed the letters with the CPA firm's return address on the envelope. All of the confirmation letters were returned by the customers directly to the auditor with indications that the amount of accounts receivable on Rosfeld's books was correct. Jason issued a report with an unqualified opinion.

The next year it was discovered that Rosfeld Stores, Inc. had been "inflating" its sales by making up "dummy" customers, complete with assumed records and files. Rosfeld declared bankruptcy, and its creditors filed suit against Jason for negligence and fraud.

a. Which auditing standards, if any, were violated by the auditors in the conduct of the examination?

b. Will Rosfeld's creditors win their case for negligence? Fraud? Assume that Rosfeld Stores, Inc. sells its securities on the New York Stock Exchange.

c. Is there any way that Jason could have satisfied Rosfeld's need to save money on the confirmation process, and yet have avoided the described problem?

Chapter 4

The Framework of an Audit

Learning Objectives *After reading and studying the information in this chapter, the student should:*

Know the process of arranging audit work with the client.

Understand the economics of public accounting.

Know the form and function of an audit program.

Understand how audit work can be scheduled at various times during the year.

Understand general evidence-gathering concepts.

Understand the business approach to an audit.

This chapter summarizes the framework within which audit theory is applied and describes how the many audit procedures set forth elsewhere in this book are performed. The material in the chapter provides a background for consideration of internal control (starting in Chapter 5) and evidence-gathering (starting in Chapter 9). Also, it shows how procedures discussed in those chapters are interrelated to form the total audit function.

Arrangement of the Audit

Sometimes one wonders how accounting firms obtain clients, particularly in view of the partial limitation on advertising discussed in Chapter 2. Although clients can be obtained by acquiring the practice of or merging with another accounting firm, the most important source of new clients is the reputation of the accounting firm and the individual members thereof for the performance of quality work. Satisfied clients who recognize the expertise of their auditors are the principal contributors to the development of a good reputation. This reputation, although not generally known by the public, is common knowledge to bankers, attorneys, investment advisors, and others in a position to recommend or influence the selection of auditors. Needless to say, the poor reputation of an accounting firm also becomes known among the business community, and such a firm's long-range prospects are limited.

Acceptance of Clients

Although most auditors are eager to obtain new clients, both prudence and generally accepted auditing standards dictate that some investigation be made of prospective clients before they are accepted.

As a matter of prudence, many auditors discuss the business reputation of a prospective client with their acquaintances in the business community such as bankers and attorneys. Any indication of improper conduct on the part of a company or its officers should cause the auditor to consider rejecting that company as a prospective client.

The matter of acceptance of new clients is mentioned in AU Section 160.19 *(SAS No. 4)* and AU Section 315 *(SAS No. 7)*. The former section outlines the quality-control considerations of auditors and includes the establishment of policies and procedures for deciding whether to accept or continue a professional relationship with a client to minimize the likelihood of associating with a company whose management lacks integrity. The latter section states the requirements for communications between predecessor and successor auditors when a change of auditors occurs. It places the initiative for the communications with the successor auditor, who is required to make specific inquiry (after obtaining permission from the prospective client) of the predecessor auditor as to such matters as the integrity of management, disagreements with management about accounting principles, auditing standards, and other significant matters, and the reason for the change in auditors. The predecessor auditor should respond promptly and fully; any limitation on the response must be noted. A limited response or one that reflects adversely on the management of a prospective client must be given serious consideration in the decision to accept or reject the client.

Auditors have become more selective in recent years in their acceptance of clients because lack of management integrity was a major factor in certain important lawsuits against auditors (discussed in Chapter 3). A painful lesson learned from these criminal and civil suits was that an early evaluation of management integrity is as important to the assurance of fair financial statements as the time spent in performing audit procedures. The *Equity Funding* and *Continental Vending* cases discussed in Chapter 3 are prime examples.

Fee Arrangements

Most accounting firms bill their clients on a per diem basis, i.e., at hourly or daily rates. These rates depend on the experience and expertise of the individuals working on the engagement (partners have the highest billing rates and new staff members have the lowest), and the type of work being performed (work on a registration statement requiring knowledge of SEC rules and regulations may be billed at a higher rate than regular audit work).

As noted in Chapter 2, because of antitrust considerations the AICPA rule against competitive bidding was eliminated, although some state boards of public accountancy have continued the prohibition.[1] Nevertheless, most accounting firms refrain from competitive bidding (except where required by law, as for certain government contracts). There is, however, a difference between a competitive bid and an estimate. An estimate is subject to adjustment upward or downward, depending on the actual time spent on the engagement by the auditor and the actual per diem costs incurred, whereas a bid is a fixed price. It is generally believed that the public is entitled to some estimate of the cost of auditing services; however, the rules of state boards of accountancy vary widely on this matter, and auditors should have a clear understanding of the rules in each of the states in which they practice before giving a bid or an estimate to a prospective client.

Engagement Letters

In Chapter 3, the discussion of the *1136 Tenants' Corporation* case illustrates the danger of a misunderstanding between the accountant and the client when unaudited financial statements are prepared. To confirm the CPA's responsibility, an engagement letter should be sent to the client. Engagement letters also should be prepared for audit engagements. These letters, normally addressed to the chairman of the board of directors, the chairman of the audit committee, or the chief executive officer, include (1) a confirmation of the audit engagement for the current year, (2) a disclaimer of responsibility to detect fraud, (3) fee and billing arrangements, and (4) other matters, if applicable, such as reviews of financial statements included in SEC filings, income tax returns, etc. One form that an audit engagement letter might take is shown in Figure 4.1.

Some auditors ask that the letter be returned to them after the recipient has signed an acknowledgement of the arrangements in a space at the end of the letter; others merely retain a copy of the engagement letter in their files.

---

[1]The U. S. Department of Justice has brought antitrust litigation against at least one state board of accountancy (Texas).

PARTNER & CO.
999 VERIFY STREET
HYPOTHETICAL, ARKANSAS 70000

May 1, 19X8

Mr. Robert K. Luckey
Chairman of the Audit Committee
X Co.
122 West Avenue
Hypothetical, Arkansas   70000

Dear Mr. Luckey,

This will confirm our arrangements with you to examine the financial statements of X Co. for the year 19X8. Our examination will be performed in accordance with generally accepted auditing standards, and accordingly will include such tests of the accounting records and such other auditing procedures as we consider necessary in the circumstances.

Generally accepted auditing standards will not necessarily reveal all errors or irregularities because of inherent limitations in the auditing process, and our examination will not include a detailed audit of transactions for any portion of the year such as would normally be required to disclose irregularities such as fraud, defalcations, etc. A strong system of internal control and adequate fidelity bond insurance usually constitutes the most effective and economical safeguard against irregularities. Although we cannot undertake to discover irregularities, any coming to our attention will be promptly reported to you.

The charges for our services will be made at our regular per diem rates plus out-of-pocket expenses.

We appreciate this opportunity to be of service to you.

Sincerely,

Figure 4.1   An Audit Engagement Letter

Scheduling the Work

Efficient scheduling of audit work is the key to maximizing the effectiveness and monetary return of an accounting firm. This fact becomes clear when one considers the economics of a public accounting practice.

The Economics of Public Accounting

Because a significant majority of businesses prepare financial statements on a calendar-year basis, there is a greatly disproportional demand on auditors' time during the early months of the year. Figure 4.2 illustrates how an auditor's workload might vary during a year.

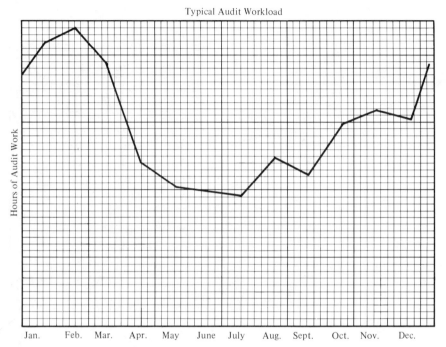

Typical Audit Workload

Figure 4.2   Typical Audit Workload

From Figure 4.2 it is easy to see why an auditor can anticipate some long workdays during the first three months of a year. An accounting firm must employ enough personnel to meet the peak demands of its clients, which occur in February in the illustration. Consequently, in other months, such as May and June, the accounting firm will be paying a number of personnel who are not producing revenue. Other functions can be performed during these months, such as professional development and training, planning the next year's audits, and vacations; however, any audit time that can be moved out of the peak period and into another time of the year, as illustrated in Figure 4.3, will increase the efficiency and monetary return of the accounting firm. Not all audit procedures must be performed after the end of the period being audited, and most accounting firms strive to perform as much work at a preliminary date as possible. This practice allows the accounting firm to reduce its staff requirements and the clients to issue financial statements and annual reports at an earlier date. Examples of audit procedures that can be carried out at different times of the year are given in a separate section of this chapter.

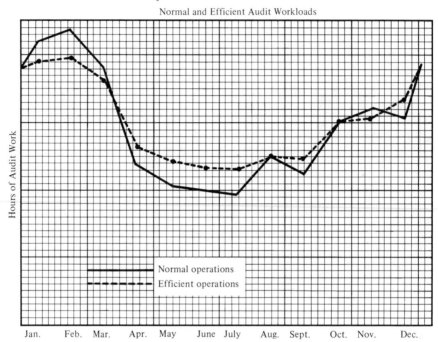

Normal and Efficient Audit Workloads

Figure 4.3 Normal and Efficient Audit Workloads

The Audit Team

The number and experience of the individual auditors assigned to an engagement vary with the size and complexity of the audit.[2] In most cases, however, the audit team will consist of the following personnel:

1. One or more staff accountants, usually having three years' experience or less, who perform audit procedures as directed by a senior accountant.

2. A senior or in-charge accountant with from three to six years' experience who performs difficult audit procedures requiring more subjective judgment, and who supervises and reviews the daily work of the staff

---

[2]The experience levels given in the following descriptions are generalizations; they vary widely in actual practice.

accountants. The senior accountant may have special training or experience in the client's industry.

3. A manager or supervisor who may have from 4 to 10 years' experience. The manager is responsible for arranging and requesting staff for each audit, reviewing the audit work performed by the staff and senior auditors, and billing and collecting the audit fee.

4. A partner with perhaps more than 10 years' experience who has overall and final responsibility for the audit. The partner reviews the audit work of the staff, senior, and manager, resolves audit problems with the client, and approves the form of and signs the audit report.

There is normally some continuity of the audit team for a particular client from year to year, although terminations and promotions may result in some changes. To obtain a fresh look at each client periodically, many accounting firms require a complete change in the audit team for a particular client every five years. Some critics of the accounting profession have maintained that this change is not sufficient and that accounting firms should be rotated periodically, but this idea has not received wide acceptance.

Audit Staff Assignments

A personnel or office manager is usually responsible for assigning individuals on the audit staff to particular engagements. This person must make the assignments in a manner that will meet client requirements (including time deadlines and industry and technical specialization), maximize the number of staff assigned to audit engagements at any point in time, minimize overtime requirements, and provide beneficial experience to the staff. The staff assignment process can become very complex as the number of staff members and audit engagements grows; because the process is basically an optimization problem, some firms have found mathematical programming models helpful.

Planning and Programming

Planning and programming are essential to the efficient conduct of an audit, regardless of its size. In addition, the first standard of field work requires that the work be planned adequately.

In planning an examination, an auditor must consider the following matters.

1. General business and industry conditions and peculiarities and the entity's accounting policies and procedures.

2. Anticipated reliance on internal control and an estimate of materiality levels.

3. Potential problem areas in the financial statements and conditions that may require an extension or modification of anticipated audit procedures.

4. The type of audit report anticipated.

Information regarding these matters may be obtained from a review of prior-year reports and audit working paper files, discussions with client personnel, review of interim financial statements, consideration of recent professional pronouncements, etc.

As pointed out in the section on the economics of public accounting, it is advantageous to both the auditor and the client to perform as much of the audit as possible outside the peak audit demand period. Several audit procedures can be performed at certain times other than after the end of the client's year.

The audit work can be divided in various ways, one of which is to split the work into the following phases.

Phase I   —Scheduling, planning, and programming
Phase II —Preliminary audit work
Phase III—Year-end work
Phase IV—Final audit work
Phase V   —Post-audit work

In Phase I, the auditor should consider more efficient audit methods and procedures, opportunities for shifting more work out of the busy audit period, and the elimination of non-essential work, with primary emphasis on preparing or updating the audit program and reviewing prior-year audit working papers to find areas where the effectiveness of the audit work could be improved. The timing of this work is flexible, but it should be completed before the beginning of the preliminary audit work. This work should be performed in the client's office, if possible; in practice, it is often done in the auditor's office.

More efficient audit methods and procedures might include the use of a generalized computer audit program (discussed in Chapter 14), more clerical assistance from clients, statistical sampling techniques (discussed in Chapters 6 and 13), and better organization of audit working papers (discussed in Chapter 9).

The shifting of additional work to a preliminary period requires understanding and very careful evaluation of internal control. If internal control is very strong, and the auditor's experience with the reliability of the accounting records and management's integrity has been good, some audit procedures normally performed after the end of the year can be moved to a preliminary date.

Non-essential work tends to become embedded in an audit program over a period of time. Client procedures and internal controls change, and certain audit steps become obsolete. However, unless the audit program is reviewed critically and revised, the unnecessary audit steps may continue to be performed for many years.

If the aforementioned factors are considered during the planning phase, the result should be a higher-quality audit with a more efficient use of personnel.

Phase II, performed perhaps one to four months before the end of the client's year, consists of the execution of audit procedures that need not be performed at year-end or during the final audit work. Remember that it is the auditor's objective to move as much work as possible into Phase II. There are no specific rules or guidelines as to exactly what work can be performed at this time. Each engagement must be studied and the timing of the audit steps worked out after careful consideration of the accounting procedures, internal controls, and other factors. Nevertheless, there are audit procedures that *usually* are done in Phase II, some that *may* be done, and others that *seldom* are done.

The review and tests of accounting procedures and internal control usually are performed in Phase II, because the extent and nature of subsequent audit procedures depend on the results of these reviews and tests. One may wonder whether the auditor is required to perform the same reviews and tests during Phase IV to cover the period from the preliminary audit work to year-end, inasmuch as the entire year is the subject of the audit. In practice, auditors do make inquiries and limited reviews of procedures and controls during Phase IV, but not in as great depth as in Phase II. To be sure that transactions from the preliminary date to year-end are subjected to the same tests as those performed in Phase II, some auditors choose as the population for their tests all transactions from one preliminary date to the next; thus, they include the last few months of the prior year in the tests. In addition to covering all transactions over a period of time, this technique should give the auditor additional assurance of consistency of accounting procedures.

Other procedures usually followed in Phase II include a review of minutes of meetings of stockholders and board of directors (and any subcommittees thereof) and audit tests of additions to and reductions of accounts with balances that tend to carry forward rather than turn over rapidly. Examples are property and equipment, deferred charges, long-term debt, and stockholders' equity. Transactions in these accounts can be tested through a preliminary date and then tested from that date to year-end in Phase IV. This approach is much more efficient than auditing transactions for the entire year in Phase IV.

The audit procedures that *may* be performed in Phase II depend on the auditor's evaluation of the client's accounting procedures and internal controls. If controls are strong, such audit procedures as confirmation of accounts receivable, observation of physical inventories, and confirmation of accounts payable may be performed in Phase II. If such procedures are performed in Phase II, the auditor must review transactions in the accounts from the confirmation date to year-end to be satisfied of the validity of the account balances at year-end.

Some audit procedures are seldom performed in Phase II, because they would duplicate work done in Phase IV. For example, tests of accrued liabilities, review of loan agreements for compliance with restrictions, and examination of the stockbook normally are done at year-end, and work done at a preliminary date would be an unnecessary duplication.

Phase III involves work that should be performed at the end of a client's year. Examples of such procedures include observation of physical inventories, count of cash funds, and inspection of marketable or investment securities. This work often must be done on the specific year-end date, because otherwise it may not be possible to determine that the asset was actually on hand at that date. Certain techniques, such as the use of seals on cash boxes and the limitation of access to bank safety deposit boxes, may allow some of these procedures to be moved from year-end to other dates.

Phase IV can begin as soon as the client has prepared and posted the final accounting entries and totaled, balanced, and closed the accounting records. It consists of execution of all audit procedures not previously performed, and preparation and issuance of the audit report. Although sometimes pressured by a client to issue an audit report before all procedures have been completed, the auditor must insist on completing all aspects of the examination, no matter how minor, before issuing the report.

Very little work remains for Phase V. Some auditors use the period following the end of the peak demand period to prepare staff evaluation reports, send billings to clients, etc., although others argue that these matters should be handled at the end of Phase IV. This time period also may be used to perform some of the procedures included in Phase I, because audit programs and ideas for increasing audit efficiency are fresh in the minds of those who performed the work.

The Audit Program

The most important control mechanism in an audit is the audit program. It should outline all of the audit procedures that are considered necessary for an auditor to express an opinion on the client's financial statements. Some auditors use preprinted programs for their audits, whereas others believe that an audit program must be specifically tailored to each audit engagement.

An audit program that is properly prepared and used serves the following purposes:

1. Provides evidence of proper planning of the work and allows a review of the proposed scope of the audit. The program gives the partner, manager, and other members of the audit team an opportunity to review the proposed scope of the audit *before* the work is performed, when there is still time to modify the proposed audit procedures.

2. Provides guidance to less experienced staff members. The specific audit steps to be performed by each staff person are indicated in the program.

3. Provides evidence of work performed. As each audit step is performed, the staff person signs or initials in a space beside that step on the program to indicate that it has been completed.

4. Provides a means of controlling the time spent on an engagement. The audit program usually includes the estimated time required to perform each audit step and a space in which the actual time required can be shown. Thus, a staff member knows approximately how much time a certain audit step should require and can ask the senior for assistance if appreciably more or less time appears to be needed.

5. Provides evidence of the consideration of internal control in relation to the proposed audit procedures. Many programs include a brief summary of the important internal control features in each section of the audit and some overall evaluation of the strengths and weaknesses of these controls. Thus, audit procedures can be restricted where controls are strong and expanded where they are weak.

The staff must keep in mind that the audit program is only a *tentative* program based on assumptions about the client's accounting procedures and internal controls. When the work begins, if the conditions are not as anticipated, the auditor may need to revise the audit program on the basis of conditions actually found. For example, if an auditor designs the audit program to confirm only a few accounts receivable because internal control is believed to be strong (on the basis of past experience or a preliminary review), the program must be changed to confirm more accounts receivable if the anticipated strengths are not found.

An illustration of the property and equipment and accumulated depreciation section of an audit program is shown in the appendix to Chapter 11. The program has four major sections: (1) account description, which states briefly the nature of the items included in the account, (2) evaluation of internal control, which summarizes important strong and weak points of internal control and includes an overall evaluation of the controls, (3) audit procedures, which outline the specific audit steps to be performed, who should perform them, and an estimate of the required time, and (4) conclusion, representing the opinion of the staff as to whether the audit objectives have been accomplished.

The illustration indicates that internal control over property additions is strong, but that controls over property retirements are relatively weak, because no periodic review is made for unreported retirements, and individual pieces of equipment are not tagged or otherwise specifically identified as belonging to the company. The strong control over additions does not mean that they need not be audited, but it does allow the auditor to restrict the extent of the procedures (by exclusion of property additions of less than $2,000, in this case). In contrast, an extended search for unrecorded liabilities is performed because of weaknesses of controls in this area.

Finally, the audit program in the appendix to Chapter 11 illustrates the division of the audit work between preliminary and final. Some procedures are performed at a preliminary date and are updated through the end of the year during the final audit work (tests of property additions and maintenance expense by examination of vendor invoices, etc.). Other procedures are performed at only one time (preparation of lead schedule) to avoid unnecessary duplication.

Evidence-Gathering

The collective purpose of all audit procedures is to gather sufficient competent evidence to form an opinion on the fairness of the financial statements taken as a whole. In a narrower sense, however, the individual audit procedures depend on the nature of the account(s) under examination, the objectives of auditing that account, and the documents or records used to compile the account balance.

For example, certain accounts, such as sales and expenses, represent totals for a certain period, usually a year. Other accounts, such as property and equipment, represent accumulated balances from several years. Still other accounts represent balances with a rapid turnover, such as cash and accounts receivable. Other aspects of an account that influence audit procedures are volume of transactions (usually small for capital stock and long-term debt and large for cash and accounts payable), and verifiability (such as physical observation of inventories and calculation of deferred charges).

A discussion of audit procedures, then, must include consideration of the following factors:

1. The means of gathering audit evidence.

2. The characteristics of such evidence.

3. The objectives of auditing a particular account.

4. The audit procedures required to satisfy the objectives.

The first two factors which are generally applicable to the audit of any account are discussed below, whereas the last two which are unique to each account are discussed in subsequent chapters.

The Means of Gathering Audit Evidence

Generally, evidence-gathering techniques can be put into the following categories:

1. Observation.

2. Confirmation.

3. Calculation.

4. Analysis.

5. Inquiry.

6. Inspection.

7. Comparison.

For the audit of certain accounts, it is possible that all seven techniques will be used. The inventory account serves as an example.

1. Normally, inventory quantities are counted by the client at year-end. The evidence as to the validity of these quantities can be obtained by *observation*.

2. Some of the year-end inventory may be on consignment or in public warehouses. Evidence of the existence of this merchandise can be acquired by *confirmation*.

3. Evidence that inventory listings have been properly extended and totaled is acquired by *calculation*.

4. For inventory to be stated fairly at the balance sheet date, a proper year-end cutoff (a determination that only goods on hand at the end of the year are included in inventory and that all liabilities for goods on hand are recorded properly) should be made by the client. Purchases should be included in inventory only if title to the goods has passed to the client at the balance sheet date, and sales should be excluded from inventory only if title has passed from the seller at the same date. To determine that a proper inventory cutoff has been made, the auditors use *analysis*.

5. During the process of observing the inventory count, auditors sometimes have questions about the manner in which various quantities are determined and the possible obsolescence of certain items. The auditors may attempt to obtain answers to these questions by *inquiry*.

6. The testing of inventory transactions represented by vendor invoices, material issue tickets, shipping records, etc., is a necessary part of determining the reliability of the records that support the inventory account. These documents can be tested by *inspection*.

7. Inventory should be priced at the lower of cost or market. Auditors can gather evidence that this procedure is used by making *comparisons*.

The auditor uses the seven techniques to design the specific audit procedures to be applied in each area of the audit.

The Characteristics of Audit Evidence

It is important that the auditor appreciate the different characteristics of audit evidence and the reliance that can be placed on each type. Audit evidence can be characterized as:

1. Generated and held by the client.

2. Received from outside parties and held by the client.

3. Received directly by the auditor by independent means or from independent or quasi-independent parties.

The first type consists of the client's books and records (the general ledger, etc.) as well as corroborating documents such as sales invoices, disbursement checks, and work orders. Minutes of meetings of the board of directors and company committees also are examples of this type of evidence.

The second category of evidence includes purchase orders from customers, purchase invoices from vendors, bank statements, titles to properties, insurance policies, etc. The fact that independent parties prepare the documents and send them to the client adds reliability to this type of evidence. Client custodianship of these documents takes away some of their reliability.

Generally, the third classification of evidence is considered the most reliable, because the auditors procure this evidence independently and control its use. The most common type of independent evidence consists of letters secured from customers, creditors, insurance companies, lawyers, and others concerning the status of certain financial information. Other types of independent evidence are answers to oral inquiries and inspection of facilities or documents.

An auditor normally prefers evidence of the third type if it is available, but in many situations it is impossible or impracticable to obtain (as in the audit of deferred income taxes, payroll expense, unamortized debt discount, etc.). In using evidence of the first type, the auditor must be particularly alert and consider, among other things, the source from which the evidence originated within the client company. For example, more reliance can be placed on a sales invoice supported by a shipping ticket from the shipping department than on a journal entry prepared within the accounting department. Similarly, an auditor can place more reliance on an explanation for a variation in annual sales received from a plant manager than on a similar explanation received from the chief accountant. The reason for this bias is that the chief accountant generally will give an explanation to support the amounts shown in the accounting records, whereas the plant manager, who may be unaware of what is shown in the accounting records, is more likely to base his explanation on actual operating factors. The auditor must attempt to obtain the most independent evidence that is available.

In the case of an examination of inventory, the auditors could use the third type of evidence to satisfy the objective of ascertaining the physical existence of the items that constitute this account. However, proof of ownership and proper valuation would require some use of the first and second evidence categories.

A Business Approach to Auditing

It is important that an auditor maintain an overall perspective of the financial statements he is examining. It is easy to become involved in the details and forget to ask, "Does this answer or presentation make good sense in light of present industry and economic conditions?" To be able to answer that question adequately, the auditor must understand the client's business operations and have some knowledge of industry and economic conditions. This means that the auditor must be more than a number checker; he must develop business judgment and knowledge equal to or greater than those of the clients he serves.

This knowledge will help the auditor to identify audit areas needing special consideration and to determine the appropriateness of accounting principles and the adequacy of disclosure. Also, he will be in a better position to evaluate the reasonableness of estimates and representations made by management.

The auditor acquires an understanding of the client's business operations by discussions with operating personnel, as well as with financial personnel. During the audit, he should not isolate himself in the accounting department, but should seek explanations for variations in the client's operations from engineers, operation managers, marketing managers, and others. After all, it is the client's business operations that are reflected in the financial statements. Who could be more qualified to discuss and explain them than the people directly involved? Such discussions are useful not only to provide audit evidence, but also to broaden the auditor's understanding of the operation of business in general and the client's business in particular. This understanding allows him to use such techniques as operating and financial ratio analysis as more effective audit tools and to detect financial statement items that appear unusual. Audit work then can be concentrated on these items.

An auditor has many sources from which to gain knowledge of a particular industry and general economic conditions. Periodicals and other publications are available for almost every industry, and government agencies publish a wide range of business statistics. Knowledge of general economic conditions can be secured from such publications as *The Wall Street Journal* and *Business Week.*

The application of operating, industry, and general economic knowledge to the audit of financial statements sometimes is referred to as "the business approach" to an audit. Some examples of how it could be useful follow.

1. You are reviewing the audit trial balance of a sugar refiner at the beginning of an audit, and you are surprised by a large increase in sales during the current year. Because of your knowledge of the industry, you know the approximate average price per pound of sugar during the year; dividing this price into total sales gives you approximately 10 million pounds sold during the year. You recall that the production capacity of

your client's plant is only 6 million pounds per year. Though there may be a reason for this difference (reduction of inventory, purchase of refined sugar from others for resale, etc.), you make a note to consider expanding your audit work in the sales area.

2. In your audit of a manufacturing company, you note that the gross profit ratio has increased from 20% to 30%, but you know that because of overcapacity in the industry, several sales price reductions were made during the year which alone would result in a decline in the gross profit ratio. When the client is unable to explain the increase satisfactorily, you revise your audit program to increase tests of inventory (for possible overstatement), accounts payable (for possible unrecorded purchases), and sales (for possible price allowances).

3. While reviewing the audit trial balance, you note that interest income on short-term investments has increased from that of the prior year. You mentally calculate that the average rate of interest received during the year, based on an average of the beginning and ending amounts invested, was 9%; however, from your knowledge of general economic conditions, you know that short-term interest rates did not exceed 7% during the entire year. You recompute interest income for the year because you believe it is overstated and find that the client did make an error by accruing interest receivable for the last quarter of the year twice. Your new assistant thinks that you have some strange, mystical, extra-sensory power.

Ratio Analysis

One specific application of a business approach to auditing is the calculation and evaluation of certain financial statement ratios. These data should not be computed and appraised in a "mechanical" fashion without regard for the business and industry environment in which the client operates. But if such computations are used properly and are integrated with pertinent information obtained from other sources, they can be valuable tools in allocating more efficiently the auditor's time and efforts.

As an illustration, assume the calculation of the following ratios for the client's current and prior years' financial statements (the current year's statements are preliminary, and audit adjustments have not been made).

Ratio	19X8	19X9
Current ratio	3.0 to 1	2.2 to 1
Accounts receivable turnover	5.5	4.2
Inventory turnover	6.3	7.2
Earnings per share	$5.10	$5.30

What audit significance can be attached to these ratios? Taken in the abstract, they probably mean little. But coupled with business information obtained from other sources, they could show the auditor where to concentrate his efforts. Here are some examples.

1. The client has a note payable with a restriction requiring the company to maintain at least a 2 to 1 current ratio. The current ratio on the 19X9 balance sheet is close to the minimum and showed a decrease from 19X8. Given these facts, it might be prudent for the auditor to scrutinize the assets included in the current category and the liabilities placed in the long-term category. Also, the auditor should consider more testing for unrecorded current liabilities.

2. Accounts receivable turnover decreased from 19X8 to 19X9. In 19X9, credit sales increased, credit terms remained the same, and a new system of more rapid billing was introduced. Under these conditions, the auditor concludes that accounts receivable turnover *should* have increased in 19X9 unless slow-paying customers caused collections to falter. More attention than usual should be paid to checking the client's formulation of allowance for doubtful accounts.

3. The inventory turnover increased in 19X9. But considering the introduction of a new line of high-turnover merchandise, the increase appears modest. During the inventory observation, it was noted that there were unusually large stocks of certain lines of merchandise on hand. The auditor should devote special attention to procedures designed to detect obsolete inventory.

4. The earnings per share increased at a time when the industry in which the client's company operates experienced a slump. The earnings per share of competitors decreased from 19X8 to 19X9. Although it might be difficult to specify specific audit areas that need attention, the auditor could take a close look at the client's capitalization policies on expenditures. The auditor should also be alert especially for signs of asset overstatement and liability understatement.

The foregoing examples are only a few of the many that could be presented. Effective use of ratio analysis can help to point out critical areas and, thus, result in a better-quality audit.

Regression Analysis

Another specific application of the business approach to auditing is the use of regression analysis to predict the dollar amount of an account. If this predicted amount is materially different from the actual figure recorded on the client's records, special audit effort is devoted to that account. We might recall that regression analysis measures the rate by which a dependent variable changes in relation to an independent variable. For simple linear regression, the analysis is illustrated in the following chart.

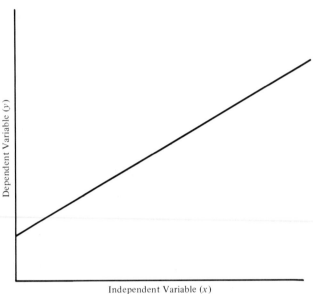

Independent Variable (*x*)

The independent variable can be expressed as number of units, number of hours, number of dollars, etc. The dependent variable can also be expressed as one of several measures. Some examples of independent and dependent variables are (1) machine hours and repair expenses, (2) direct labor costs and indirect labor costs, and (3) sales dollars and delivery expenses. We will use the third example as the model for the illustration.

A first-order regression function[3] may be expressed as:

$$E(Y) = B_0 + B_1X$$

where,

E(Y) is the expected value of Y

B_0 is often thought of as the fixed element
of the dependent variable but it may have no

---

[3]For information regarding statistical regression techniques, see John Neter and William Wasserman, *Applied Linear Statistical Models,* (Homewood, Ill.: Richard D. Irwin, Inc., 1974).

particular meaning as a separate term in the
regression function.

B₁ is the rate of change of Y for each unit change
in X.

The auditor may assume that there should be a close relationship between the client's dollar sales for the year and the delivery expenses and decide to predict the amount of delivery expense by using simple linear regression analysis.

To predict the delivery expense rate of variability in relation to dollar sales, the auditor collects the actual dollar amounts of both accounts for the previous ten years. It is assumed that the dollar relationships that exist between these accounts for the 10-year period are typical of the relationship that should exist in the current year.

The sales and delivery expenses for the 10-year period are listed hereafter.

Year	Sales (in thousands of dollars)	Delivery Expense (in thousands of dollars)
19W9	$10,019	$ 78
19X0	11,904	94
19X1	13,638	110
19X2	15,622	127
19X3	19,431	151
19X4	22,774	175
19X5	26,058	183
19X6	31,354	210
19X7	35,900	255
19X8	41,895	317

By the use of the least-squares statistical technique, the auditor calculates the estimated fixed portion of delivery expense at $10,000 and the estimated variable portion at $.007 per sales dollar. The auditor could plot the data on a chart and draw the calculated variable expense line. Such a chart follows.[4]

An examination of the chart shows that the correlation between sales and delivery expense appears to be very high (the coefficient of determination is .98). Unless the auditor has other evidence to contradict the appearance of high correlation, he would assume that $10,000 plus $.007 per sales dollar is a good predictor of delivery expense.

---

[4]Because of the aggregate relationships, the auditor might choose to draw a chart and "freehand" a straight line rather than use the least-squares technique.

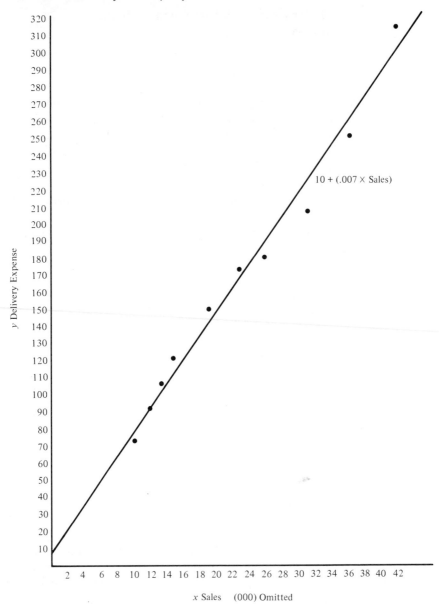

10 + (.007 × Sales)

y Delivery Expense

x Sales   (000) Omitted

The next step is to calculate the predicted amount of delivery expense and compare it with the actual figure on the client's 19X9 income statement. If, in the auditor's judgment, the difference is too large, more than the usual amount of audit effort should be directed at this account.

Assume that 19X9 sales are $45,000,000 and the 19X9 delivery expense is $300,000. The predicted delivery expense is $325,000 [$10,000 + (.007 ×

$45,000,000)]. The auditor must make a judgment on the materiality of this difference.[5] If he considers the $25,000 to be material, he might examine a significant portion of the evidence supporting the delivery expense and related asset accounts. Such an examination could reveal misclassifications of expenditures between prepaid expenses and delivery expense.

It should be pointed out that a decision to concentrate audit effort in this direction might have been made by a cursory examination of delivery expense and sales for the prior 10 years. Regression analysis techniques refine the process, however. The methodology is easy to apply and with the advent of sophisticated minicomputers, the calculations can be made quickly.

Relationships Among Accounts

An auditor must be aware of the natural flow of funds through a set of financial statements while performing his audit work. This flow of funds and the resulting relationships are such that evidence gathered to support the propriety of one account often contributes to the accomplishment of the audit objectives for another account. For example, audit procedures performed and evidence gathered in the audit of inventory, cash, and accounts payable contribute to the audit of cost of goods sold. Likewise, the auditor who has satisfactorily tested the client's procedures for recording sales and cash transactions has a certain amount of evidence in support of accounts receivable before he performs any audit procedures listed in that section. Table 4.1 illustrates pairs of accounts on which audit work often is performed simultaneously.

ACCOUNTS ON WHICH AUDIT WORK OFTEN IS PERFORMED SIMULTANEOUSLY

Balance Sheet Accounts	*Income Statement Accounts*
Accounts receivable	Sales and bad debt expense
Inventories	Cost of goods sold
Investment securities	Other income (dividends and interest)
Property and equipment	Maintenance and depreciation expense
Prepaid expenses	Selling and administrative expenses
Notes payable	Interest expense
Accrued income taxes	Income tax expense

Table 4.1 Accounts on Which Audit Work Often is Performed Simultaneously

Many of the balance sheet accounts listed in Table 4.1 can be confirmed by outside parties (accounts receivable and notes payable) or observed by the auditor (inventories and investment securities). Many of the income statement accounts, because they represent a large number of individual items flowing

[5]See the materiality discussion in Chapter 9.

through the accounts during the year (sales and cost of goods sold), require great reliance on tests of the procedures for recording income and expense and the system of internal control. These two approaches, however, are complementary; confirmation and observation contribute to the audit of the income statement accounts, and the tests of procedures add assurance to the audit of the balance sheet accounts. In fact, the double-entry accounting system allows accounts to be audited from two directions.

Consider sales, for example. This account can be, and to some extent is, audited by examining evidence to support the amounts recorded in the account, such as sales invoices and shipping documents. This is only a portion of the evidence gathered for the audit of sales, however. The following formula:

$$\begin{array}{c}\text{Accounts} \\ \text{receivable} \\ \text{at beginning} \\ \text{of a period}\end{array} + \begin{array}{c}\text{Credit} \\ \text{sales} \\ \text{for a} \\ \text{period}\end{array} - \begin{array}{c}\text{Cash} \\ \text{collections} \\ \text{of accounts} \\ \text{receivable}\end{array} = \begin{array}{c}\text{Accounts} \\ \text{receivable} \\ \text{at end of a} \\ \text{period}\end{array}$$

can be rewritten as:

$$\begin{array}{c}\text{Credit} \\ \text{sales} \\ \text{for a} \\ \text{period}\end{array} = \begin{array}{c}\text{Accounts} \\ \text{receivable} \\ \text{at end of} \\ \text{a period}\end{array} + \begin{array}{c}\text{Cash} \\ \text{collections} \\ \text{of accounts} \\ \text{receivable}\end{array} - \begin{array}{c}\text{Accounts} \\ \text{receivable} \\ \text{at beginning} \\ \text{of a period.}\end{array}$$

Because it is a common practice in many companies to flow all sales through accounts receivable, credit sales are often equivalent to total sales. In these cases, by auditing the accounts receivable balance at the end of a period (assuming that the balance at the beginning of the period was audited in the prior year), and by testing the procedures for recording cash receipts, the auditor adds evidence to the audit of sales without gathering any evidence in direct support of the amounts recorded in the sales account.

Other Audit Procedures

Examples of audit procedures are given in Chapters 10, 11, and 12. The auditor must remain keenly alert while performing these procedures, no matter how routine a step appears to be. One brief period of carelessness can have a devastating effect on the quality of the audit. Although the senior, manager, and partner will review the audit working papers prepared by the staff, none of these people will see the invoice, canceled check, or other document that actually was examined. Therefore, a basic responsibility for the audit rests with the staff.

The staff is assisted in meeting this responsibility by senior supervision. The senior should explain the purpose and means of accomplishing the audit steps assigned to the staff. Though the staff may be encouraged to develop their own solutions to problems encountered on the engagement, the senior should review the soundness of these solutions with the staff and assist them when significant difficulties are encountered.

Each firm should establish a formal set of procedures to be followed if differences of opinion concerning accounting or auditing issues arise among firm personnel involved in an examination. These procedures should enable the staff to document the reason for disagreement and, if necessary, to be disassociated from the resolution of the issue.

The review process provides the means for controlling the performance of the audit procedures. The audit working papers should be reviewed at each level of responsibility from the senior through the partner. In other words, the senior reviews the working papers prepared by the staff, the manager reviews the working papers prepared by the staff and the senior, and the partner reviews all of the working papers, including any prepared by the manager. As the reviews are conducted at each level, the reviewer makes notes or points about the audit work to be clarified, answered, or cleared either personally or by someone at a lower level. Satisfactory disposition must be made of all such points before the audit can be considered complete.

Issuing the Audit Report

Although the auditor prefers that the financial statements of the client be completed in final form before the audit commences, this is seldom the case in practice. Even if the auditor has no adjustments to make in the financial records, certain financial statement classifications and disclosures may evolve from and during the audit. Thus, at the end of the audit, the auditor must review the financial statements to determine whether they are in accordance with generally accepted accounting principles and whether they contain all information necessary for a clear understanding of the company's operations and its financial position. If the company is subject to SEC regulations, the auditor also must determine that the financial statements comply with the SEC accounting rules. It is not unusual for an auditor to recommend additional footnote disclosure or clarification of footnote wording as a result of the review of the financial statements. Only after the financial statements are in final form can the auditor decide on the type of audit report to be issued (discussed in Chapters 15 and 16).

Once the financial statements and audit report are in final form, many accounting firms subject them to some type of quality-control review. Quality-control reviews take many forms. They often include (1) proofreading and clerical check, (2) tracing amounts and disclosures in the financial statements to the audit working papers to be sure that they are the same in both documents and have been subjected to audit, and (3) reading the audit report and financial statements by another audit partner who was not associated with the engagement (some accounting firms carry this step further by requiring a complete review of the audit working papers by the second audit partner).

The first procedure listed involves a comparison of the final typed or printed copy of the audit report and financial statements with the draft copy to detect the omission of words or sentences and misspellings, and a check of the clerical accuracy of all totals. The second procedure is performed to determine that there is support in the audit working papers for all amounts and disclosures in the financial statements. This step is usually done by a senior or manager who has had no prior association with the engagement. The final procedure is sometimes thought of as a test to determine whether an individual who is knowledgeable about accounting, but unfamiliar with the particular financial statements being reviewed, believes that the financial statements are presented in a clear and understandable manner.

The auditor's report is signed and released to the client only after all of the foregoing procedures have been performed and all questions or comments raised as a result thereof have been cleared.

Systems Evaluation Approach

The Appendix I to the text contains descriptions of a new audit approach to the planning phase used by the accounting firm of Peat, Marwick, Mitchell & Co.

Chapter 4
Questions Taken from the Chapter

4–1.   How do accounting firms obtain new clients?

4–2.   Who are the principal contributors to the reputation of an accounting firm?

4–3.   Why should auditors make some investigation of prospective clients before accepting them?

4–4.   Does the predecessor or successor auditor have the responsibility to initiate communications when a change of auditors occurs?

4–5.   What specific inquiries must the successor auditor make of the predecessor auditor?

4–6.   Why have auditors become more selective in recent years in their acceptance of clients?

4–7.   What basis do most accounting firms use to bill clients for their services?

4–8.   Explain the difference between a competitive bid and an estimate of an accounting fee.

4–9.   List the matters normally included in an audit engagement letter.

4–10.   What is the key to maximizing the effectiveness and monetary return of an accounting firm?

4–11.   Why is there a greatly disproportionate demand on an auditor's time during the early months of the year?

4–12.   Why do auditors attempt to perform as much audit work as possible at a preliminary date?

4–13.   Describe the composition of the audit team and the responsibilities of its members.

4–14.   Explain how accounting firms obtain a fresh look at each client periodically. What other approach is suggested by some critics of the accounting profession?

4–15.   What does the personnel manager attempt to achieve when assigning the audit staff to particular engagements?

4–16.   What should the auditor attempt to achieve during the planning phase?

4–17.   List four possibilities for improving the efficiency of audit methods and procedures.

4–18.   What factors must be considered in determining whether additional audit work can be shifted to a preliminary period?

4–19.   Explain the tendency for non-essential work to become embedded in an audit program over a period of time.

4–20.   What purposes are served by an audit program that is properly prepared and used?

4–21.   Why is an audit program considered tentative? What could cause it to be changed?

4–22. What are the major sections of an audit program?

4–23. What are the five phases into which audit work can be divided?

4–24. When is Phase I performed and what work is involved?

4–25. When is Phase II performed? Give examples of audit procedures that usually, may be, and seldom are, done in Phase II.

4–26. How can auditors subject transactions from the preliminary date to year-end to the same tests as those performed in Phase II?

4–27. Give three examples of procedures performed in Phase III and state when they are performed.

4–28. Describe two techniques for moving certain audit procedures from Phase III to other dates.

4–29. What is the collective purpose of all audit procedures?

4–30. What are some of the determinants of individual audit procedures?

4–31. List seven techniques for gathering audit evidence. State how each could be applied to the audit of inventory.

4–32. List the three characteristics of audit evidence and give an example of each.

4–33. Which characteristic of audit evidence is considered the most reliable? Why?

4–34. Why doesn't the auditor always obtain evidence with the characteristic that is most reliable?

4–35. Why might an auditor place more reliance on an explanation received from a plant manager than on one from the chief accountant?

4–36. What is the "business approach" to an audit?

4–37. How does an auditor gain knowledge of his client's business operations?

4–38.　Where does an auditor learn of industry or general economic conditions?

4–39.　How is ratio analysis used in an audit?

4–40.　How does the auditor use regression analysis?

4–41.　State and describe each term in a first-order regression function.

4–42.　Why should an auditor be aware of the natural flow of funds through a set of financial statements?

4–43.　List six pairs of accounts on which audit work often is performed simultaneously.

4–44.　Explain how audit evidence relating to sales can be obtained without examining documents in support of amounts recorded in the sales account.

4–45.　When can Phase IV begin and what audit work is performed?

4–46.　Discuss the work that may be performed in Phase V.

4–47.　Explain the review process for audit working papers.

4–48.　Explain three procedures often included in an accounting firm's quality-control review.

Chapter 4
Objective Questions Taken from CPA Examinations

4–49.　Fox, CPA, is succeeding Tyrone, CPA, on the audit engagement of Genesis Corporation. Fox plans to consult Tyrone and to review Tyrone's prior-year working papers. Fox may do so if
a.　Tyrone and Genesis consent.
b.　Tyrone consents.
c.　Genesis consents.
d.　Tyrone and Fox consent.

4–50.   In order to achieve effective quality control, a firm of independent auditors should establish policies and procedures for
  a.   Determining the minimum procedures necessary for unaudited financial statements.
  b.   Setting the scope of audit work.
  c.   Deciding whether to accept or continue a client.
  d.   Setting the scope of internal control study and evaluation.

4–51.   As generally conceived, the "audit committee" of a publicly held company should be made up of
  a.   Representatives of the major equity interests (bonds, preferred stock, common stock).
  b.   The audit partner, the chief financial officer, the legal counsel, and at least one outsider.
  c.   Representatives from the client's management, investors, suppliers, and customers.
  d.   Members of the board of directors who are not officers or employees.

4–52.   The auditor interviews the plant manager. The auditor is most likely to rely upon this interview as primary support for an audit conclusion on
  a.   Capitalization vs. expensing policy.
  b.   Allocation of fixed and variable costs.
  c.   The necessity to record a provision for deferred maintenance cost.
  d.   The adequacy of depreciation expense.

4–53.   Which of the following types of documentary evidence should the auditor consider to be the most reliable?
  a.   A sales invoice issued by the client and supported by a delivery receipt from an outside trucker.
  b.   Confirmation of an account-payable balance mailed by and returned directly to the auditor.
  c.   A check issued by the company and bearing the payee's endorsement which is included with the bank statement mailed directly to the auditor.
  d.   A working paper prepared by the client's controller and reviewed by the client's treasurer.

4–54.   Rosenberg, CPA, is auditing Bulmash Hospital, a private not-for-profit organization. What is Rosenberg's best source of evidence to substantiate the classification of restricted funds?
  a.   Minutes of the board of directors.
  b.   Correspondence from executors or donors.
  c.   State laws and regulations.
  d.   Confirmation with trust company.

4–55.   Auditing standards differ from auditing procedures in that procedures relate to
    a.   Measures of performance.
    b.   Audit principles.
    c.   Acts to be performed.
    d.   Audit judgments.

4–56.   The ordinary examination of financial statements is *not* primarily designed to disclose defalcations and other irregularities, although their discovery may result. Normal audit procedures are more likely to detect a fraud arising from
    a.   Collusion on the part of several employees.
    b.   Failure to record cash receipts for services rendered.
    c.   Forgeries on company checks.
    d.   Theft of inventories.

4–57.   Audit programs generally include procedures necessary to test actual transactions and resulting balances. These procedures are *primarily* designed to
    a.   Detect irregularities that result in misstated financial statements.
    b.   Test the adequacy of internal control.
    c.   Gather corroborative evidence.
    d.   Obtain information for informative disclosures.

4–58.   Although the validity of evidential matter is dependent on the circumstances under which it is obtained, there are three general presumptions which have some usefulness. The situations given below indicate the relative reliability a CPA has placed on two types of evidence obtained in different situations. Which of these is an *exception* to one of the general presumptions?
    a.   The CPA places more reliance on the balance in the scrap sales account at Plant A, where the CPA has made limited tests of transactions because of good internal control, than at Plant B, where the CPA has made extensive tests of transactions because of poor internal control.
    b.   The CPA places more reliance on the CPA's computation of interest payable on outstanding bonds than on the amount confirmed by the trustee.
    c.   The CPA places more reliance on the report of an expert on an inventory of precious gems than on the CPA's physical observation of the gems.
    d.   The CPA places more reliance on a schedule of insurance coverage obtained from the company's insurance agent than on one prepared by the internal audit staff.

4–59.   Omaha Sales Company asked a CPA's assistance in planning the use of multiple regression analysis to predict district sales. An equation has been estimated based upon historical data, and a standard error has been computed. When regression analysis based upon past periods is used to predict for a future period, the standard error associated with the predicted value, in relation to the standard error for the base equation, will be

    a.   Smaller.

    b.   Larger.

    c.   The same.

    d.   Larger or smaller, depending upon the circumstances.

4-60.   A CPA establishes quality-control policies and procedures for deciding whether to accept or continue a client. The primary purpose for establishing such policies and procedures is

    a.   To enable the auditor to attest to the integrity or reliability of a client.

    b.   To comply with the quality-control standards established by regulatory bodies.

    c.   To lessen the exposure to litigation resulting from failure to detect irregularities in client financial statements.

    d.   To minimize the likelihood of association with clients whose managements lack integrity.

4-61.   An auditor uses analytical review during the course of an audit. The most important phase of this review is the

    a.   Computation of key ratios such as inventory turnover and gross profit percentages.

    b.   Investigation of significant variations and unusual relationships.

    c.   Comparison of client-computed statistics with industry data on a quarterly and full-year basis.

    d.   Examination of the client data that generated the statistics that are analyzed.

4-62.   At interim dates an auditor evaluates a client's internal accounting control procedures and finds them to be effective. The auditor then performs a substantial part of the audit engagement on a continuous basis throughout the year. At a minimum, the auditor's year-end audit procedures must include

    a.   Determination that the client's internal accounting control procedures are still effective at year-end.

    b.   Confirmation of those year-end accounts that were examined at interim dates.

    c.   Tests of compliance with internal control in the same manner as those tests made at the interim dates.

    d.   Comparison of the responses to the auditor's internal control questionnaire with a detailed flow chart at year-end.

4-63.   The first standard of field work requires, in part, that audit work be properly planned. Proper planning as intended by the first standard of field work would occur when the auditor

    a.   Eliminates the possibility of counting inventory items more than once by arranging to make extensive test counts.

b. Uses negative accounts receivable confirmations instead of positive confirmations because the latter require mailing of second requests and review of subsequent cash collections.

c. Compares all cash as of a particular date to avoid performing time-consuming cash cutoff procedures.

d. Physically observes the movement of securities already counted to guard against the substitution of such securities for others which are *not* actually on hand.

Chapter 4
Discussion/Case Questions

4–64. Joe Melton, CPA, has been contacted by the president of Mudalum Co. (a company developing a process to turn mud into aluminum) and asked to perform the company's audit for the current year. During a meeting with the president, Joe learned that the Securities and Exchange Commission had a suit pending against the Company charging it with an illegal distribution of securities. The president explained that it was an honest mistake by Mudalum because neither its former attorneys nor its former auditors informed him of the SEC registration requirements. For this reason, the president decided to change auditors and has threatened the former auditors with a lawsuit. Because of the threatened litigation, the former auditors refuse to discuss their audit of Mudalum Co. with Joe.

If you were Joe, would you accept Mudalum Co. as a client? Discuss your reasoning.

If you wished to make a further investigation of the situation, what would you investigate and what evidence would you gather?

May the former auditors properly refuse to discuss their audit of Mudalum Co. with Joe?

4–65. Bill Grote, CPA, has conducted audits for more than forty years and, because of his extensive experience, he does not believe he needs an audit program in order to know what audit procedures to perform. Tom Prince, another CPA who recently became Bill's partner, believes that he should try to convince Bill to use an audit program.

Is Tom right? If not, why not? If so, what reasons should he use?

4–66. In late spring of 197X, you are advised of a new assignment as in-charge accountant of your CPA firm's recurring annual audit of a major client, the

Lancer Company. You are given the engagement letter for the audit covering the calendar year December 31, 197X, and a list of personnel assigned to this engagement. It is your responsibility to plan and supervise the field work for the engagement.

Required:

Discuss the necessary preparation and planning for the Lancer Company annual audit *prior* to beginning field work at the client's office. In your discussion include the sources you should consult, the type of information you should seek, the preliminary plans and preparation you should make for the field work, and any actions you should take relative to the staff assigned to the engagement. *Do not write an audit program.*

(AICPA adapted)

4–67. In a properly planned examination of financial statements, the auditor coordinates his reviews of specific balance sheet and income statement accounts.

Required:

Why should the auditor coordinate his examinations of balance sheet accounts and income statement accounts? Discuss and illustrate by examples.

(AICPA adapted)

4–68. In your audit of June Co., you are investigating an increase in sales during the current year and have made inquiry of the sales manager, who tells you the increase is due to a sales price increase during the year. When you ask to see a current price list to verify the increase, the sales manager becomes angry and asks if you don't trust him. How would you reply?

4–69. State how each of the seven evidence-gathering techniques could be used in the audit of cash.

4–70. Your knowledge of general economic conditions during the year ended December 31, 19X9, included the facts that inflation had been recorded at a high 12% rate which caused a serious slowdown in business (real GNP declined 6%) and a steep drop in the stock market (approximately 25%). Unemployment was high, even though unions were demanding large wage increases. Because of high interest rates and the stock market decline, money supplies were tight, and long-term financing practically unavailable.

Discuss the effects of these conditions on your audits of the following companies for the year ended December 31, 19X9:

a. A construction contractor specializing in small office buildings on fixed-price contracts.

b. An investment company with 50% of its portfolio in common stocks and 50% in long-term bonds.

c. A manufacturing company that has always been only marginally profitable and whose 4% bonds mature January 1, 19X0.

d. A small finance company specializing in consumer loans.

e. A manufacturer whose warranty costs have historically averaged 5% of sales.

4–71. A CPA has been asked to audit the financial statements of a publicly held company for the first time. The preliminary verbal discussions and inquiries have been completed between the CPA, the company, the predecessor auditor, and all other necessary parties. The CPA is now preparing an engagement letter.

Required:
List the *items* that should be included in the typical engagement letter in these circumstances and describe the *benefits* to be derived from preparing an engagement letter.

(AICPA adapted)

4–72. Classify the following items of audit evidence from the most reliable (1) to least reliable (19). Discuss the reasons for your classifications.
a. Canceled payroll check.
b. Client sales invoice.
c. Request for a travel advance.
d. Client-prepared receiving report for merchandise.
e. Client-prepared depreciation worksheet.
f. Journal entry to correct an account classification.
g. Board of director minutes signed by the corporate secretary.
h. Unsigned board of director minutes.
i. Vendor invoice for merchandise purchased.
j. Client-prepared purchase order for merchandise.
k. Copy of client articles of incorporation.
l. Bank statement.
m. Cutoff bank statement received directly by the auditor.
n. Positive accounts receivable confirmation.
o. Negative accounts receivable confirmation.
p. Client-prepared inventory count sheet.
q. Oral client representations.
r. Written client representations.
s. Representation letter from client's attorney.

4–73. Trent and Jones, CPAs, were planning their initial audit of Kargo Corporation, whose stock was traded actively on the New York Stock Exchange. Although the partners had consulted with the predecessor auditor and had reviewed prior-year working papers, they still had a number of questions about the areas that should be given special attention in the engagement. To provide them with more insight, Mr. Trent and Mr. Jones decided to take the published financial statements of Kargo Corporation for the last two years and the statements for the first quarter of this year and develop some ratios. They calculated the following amounts.

	19X6	*19X7*	*1st Quarter 19X8*
Current ratio	2.1 to 1	2.0 to 1	1.8 to 1
Accounts receivable turnover	8.3	8.4	8.6*
Inventory turnover	7.4	9.2	10.8*
Times interest earned	1.7	1.6	1.4
Earnings per share	10.50	11.62	3.45
Debt/Equity ratio	.95	1.20	1.24
Dividends per share	2.50	3.00	.90

*Annualized

In addition to these ratios, the following information is available.

1. Kargo's credit terms are 30 days net.
2. The Kargo Corporation has a loan restriction that requires them to maintain at least a 2.0 to 1 current ratio.
3. The "normal" inventory turnover for the industry in which Kargo Corporation operates is 10.

Required:
a. Indicate the areas where the auditors should concentrate special effort. Consider only the areas revealed by the ratio analysis. Do not name specific audit procedures.
b. Discuss actions that the management of Kargo might be inclined to take to cover up any possible adverse effects of these listed ratios. What could the auditors do to provide reasonable assurance that these actions were found by them?

4–74. During the planning phase of an audit engagement, it was decided that special attention should be paid to the time spent testing some of the income statement accounts. The auditors believed that some accounts required considerable attention and others required very little work. To gain more insight into this matter, the auditors decided to test an account that had taken a considerable number of hours to analyze in the prior-year audit. The account selected was maintenance expense. In previous years, support for all charges to this account in excess of $50 had been examined to determine if the account contained a material amount that belonged in fixed-asset additions. The balance in the account last year was $5,000, and the net income for the same year was $25,000.

The auditors decided to use regression analysis to ascertain whether they had put too much effort into the analysis of this account last year. To do so, they decided to use regression analysis and predict last year's maintenance expense on the basis of the number of machine-hours spent the same year. This predicted figure will be compared with the $5,000 amount on last year's income statement to determine whether there was a difference significant enough to justify the audit effort they devoted to this account.

The auditors collected the number of machine-hours and maintenance expense amounts by month for the 12-month period two years prior to the current fiscal year. A regression analysis revealed that maintenance expense is fully variable and is expected to change $.05 for each machine-hour spent. In the prior year, 80,000 machine-hours were spent.

Required:
a. Were the auditors justified in auditing maintenance expense in the described manner? Why or why not?
b. What is your answer if we assume that 60,000 machine-hours were spent? 20,000 hours?
c. What additional factors should be considered in deciding on the scope of the maintenance expense examination?
d. Why might regression analysis provide misleading data in making this decision?

Chapter 5

The General Nature of Internal Control

Learning Objectives   *After reading and studying the material in this chapter, the student should*

Understand the concept of internal control.

Understand why internal control is necessary for almost all types of organizations.

Know the objectives of internal control.

Be able to read flow-chart documentation for internal control systems.

Understand the AICPA model for studying and evaluating internal control.

Know the difference between reviewing and testing a system of internal control.

Be able to apply tests of compliance and tests of transactions to internal control systems.

A quick glance at today's standard short-form report shows that no mention is made of internal control in the scope paragraph. In light of the importance of the subject matter, this seems to be a significant omission. In earlier years, reports did include comments on internal control, but today the reference to generally accepted auditing standards is assumed to be adequate.

In the early part of the century, when business organizations were smaller, a more comprehensive audit could be performed. In later years, it became apparent that only a relatively small percentage of the transactions and balances could be scrutinized. As responsibilities to third-party groups began to accelerate, auditors began to look for guidelines to help them reduce the risks involved in testing a company's financial statement figures. (There is a risk that errors or irregularities will occur, and a risk that they will not be detected by the auditor.) The obvious approach was to examine the quality of internal control. If the internal control system were satisfactory, the tests of balances could be restricted. If the system were weak, the extent of the tests of balances would be larger.

We will explore the essentials of internal control as they relate to the auditors' testing responsibilities and the control of the audit risks. Concrete examples are included, because knowledge of the system is needed for an understanding of the material in Chapters 6, 7, and 8.

The Features of Internal Control

Most organizations that engage in financial transactions use some type of control. For example, religious and charitable institutions that accept currency and coins from the public have more than one person counting the funds, and often they use a locked box in the collection process.

In the business world, many companies that use payroll checks require two signatures before these checks can be cashed by the bank. Another payroll control, when individuals receive wages on the basis of time spent on the job, is to require that someone in a supervisory capacity approve the work records before money is disbursed.

Without controls of this type, it would be very difficult for any organization to protect its assets, rely on its records, or, in general, operate in an efficient manner. The extent to which firms need control systems depends on the complexity of the record-keeping process, the cost and effort involved in setting up controls, and the possible consequences of omitting these controls. It is management's responsibility, *not the auditor's,* to ensure that a proper structure is provided to minimize the chances of undetected errors or irregularities.

AU Section 320.10 of *AICPA Professional Standards,* Volume 1 *(SAS No. 1)* defines internal control as the organizational plan and other measures designed to accomplish the following objectives:

1. Safeguard assets,

2. Check the accuracy and reliability of accounting data,

3. Promote operational efficiency, and

4. Encourage adherence to managerial policies.

This definition is broad, but several examples illustrate the differences among the four objectives. Assume that a company assigns to one person the responsibility for (1) sorting the cash received through the mail, (2) recording the credits to the customer ledger accounts, and (3) preparing the bank deposit slip. Clearly, an arrangement of this type is undesirable (although some organizations use it of necessity). Cash could be taken and the postings to the customer accounts manipulated to cover the shortage. Assets can be safeguarded by separating the responsibilities (Objective No. 1). Doing so would force the individuals involved to engage in a collusive effort in order to perpetrate an irregularity. The probability of such an occurrence is lower if a multiple effort is required to accomplish it.

In another area, balancing the accounts receivable detail with the control account should help to ensure the accuracy of accounting data. Even if no irregularity occurs, errors might be detected if record-keeping accountability were checked by another person (Objective No. 2).

Even if assets are safeguarded and the accounting data are accurate and reliable, administrative controls may still be weak. These types of controls can be strengthened in several ways. One is implementation of a sound standard cost system that helps to promote managerial efficiency by pinpointing wasteful spending (Objective No. 3). Another is a clear delineation of job responsibilities that encourages adherence to managerial policies (Objective No. 4). Accounting controls accomplish Objectives 1 and 2. Administrative controls accomplish Objectives 3 and 4.

Though acknowledging the importance of administrative controls, most members of the profession generally take the position that accounting controls are more important for the stated purpose of ensuring fairness. It should be explained that in many audit situations it is very difficult to distinguish clearly among the four control objectives. One way to differentiate the objectives (although it is an oversimplification) is to say that a firm could be grossly inefficient and sustain a heavy loss, and yet the loss could be judged to be fair. In some audit situations, both types of controls might be examined, although accounting controls normally warrant more attention.

Specific Concepts of Internal Control

In a more specific sense, the accounting control system should have a number of individual characteristics to give the company reasonable assurance that accounting controls are functioning properly. AU Section 320.35–.48 *(SAS No. 1)* groups such characteristics into the following classifications.

1. Personnel should have competence and integrity.

2. There should be no incompatible functions such that any person is in a position both to perpetrate and conceal irregularities in the normal course of his duties. To accomplish a proper segregation of duties, the system, insofar as possible, should

provide for different individuals to perform the functions of (a) authorizing a transaction, (b) recording a transaction, (c) maintaining custody of the assets that result from a transaction, and (d) comparing assets with the related amounts recorded in the accounting records.

3. Authorization for transactions should be issued by persons acting within the scope of their authority and the transactions should conform to the terms of the authorizations.

4. Transactions should be recorded at the amounts and in the accounting periods in which they were executed. The transactions should be recorded in proper accounts.

5. Access to assets should be limited to authorized personnel.

6. There should be independent comparisons of assets with the recorded accountability of these assets.

In the remainder of the text we will study the methods that the independent auditor uses to ascertain the extent to which these characteristics of a good accounting control system are in effect.

Study of Internal Control

Why Internal Control Should Be Studied

The study of internal control is the logical starting point for the ordinary examination of financial statements. For a variety of reasons, its importance has increased significantly within the last thirty to forty years. First, testing has long since replaced a "complete audit" as standard operating procedure. The study of internal control provides the auditor with the basis for reliance on the records and for determining the nature, extent, and timing of audit tests.

In addition, there is the ever-present possibility of fraud. The profession has incorporated into its official guidelines a statement that the responsibility for fraud detection may be fulfilled if generally accepted auditing standards are used.

Despite this pronouncement, there is no real evidence that the public has ever accepted this viewpoint. If the current trend continues, courts and regulatory agencies may force CPA firms to accept some limited responsibility for fraud detection. In the meantime, the CPA is well advised to take extra care in this phase of the audit, because the issue of responsibility for fraud detection has not been settled.

When Internal Control Should Be Studied

There are no rigid rules about the manner and the time frame in which the study of internal control should be performed. Most CPA firms prefer to make the study before the end of the client's fiscal year. This is obviously the ideal time, because one of the major purposes of the work is to determine the nature, timing, and extent of other audit tests, and it also serves the practical purpose of spreading the workload.

Steps in Reviewing and Evaluating Internal Control

Although the specific techniques for reviewing and evaluating internal control differ among auditors, there is a general model that is followed by many of them.

1. Documentation and review of the system—This step consists of the auditor obtaining or reaffirming (in the case of a repeat engagement) an understanding of how the system functions. The methods for acquiring this knowledge are (a) narrative descriptions of the system, (b) flow-chart descriptions of the system, (c) internal control questionnaires filled out by the auditor, (d) oral inquiries of the client, and (e) observations by the auditor. In some cases the auditor might wish to add assurances to his understanding of the system by selecting a few transactions of each type and tracing their flow through the system.

2. Preliminary evaluation of the system—The auditor then assesses the strengths and weaknesses of the system and the major areas of concern. An opinion is formed on whether reliance can be placed on the controls. On the basis of the preliminary evaluation, a tentative decision will be made on any necessary modification to the audit program for substantive tests.

3. Tests of compliance—The purpose of this step is to ascertain whether the system is functioning as prescribed. This step is the testing of the system to see that it works as described in the narratives, flow charts, etc.

4. Re-evaluation of the system—After the tests of compliance are completed, the auditor will reassess the system to ascertain whether the tests have changed his evaluation. A final decision is made on the reliability of the controls for the purpose of modifying the audit program for substantive tests.

Within this model, there are a number of ways to study and evaluate internal control. We will present the AICPA approach, the general outline of which is described above. Another approach is the interim phase of the Systems Evaluation Approach (SEA) described in Appendix I of the text.

Review of the System

The Different Ways to Review a System

Suppose one wishes to find out as much as possible about an organizational structure. What methods would be appropriate? To begin with, it would be helpful to read some written material about the system. The information should be arranged logically so that readers can make judgments about the system and compare its attributes with some standard.

It might also be beneficial to examine a pictorial graph or flow chart of the structure. Such graphic materials might aid the readers' comprehension of the overall information flow in a way that a narrative description cannot.

Finally, one would want to ask questions about the system. Some of the answers might come from company employees, whereas other answers would come from observations made during the course of the review. Other questions could be answered by looking at the system documentation.

Depending on one's preferences, any one or all three of the methods described could be used as long as one satisfied the objective of becoming sufficiently informed about the characteristics of the system.

From an auditor's standpoint, the review of the system of internal control is the process of obtaining information about the organization and the prescribed procedures. To put it simply, it is the technique of finding out how the system is *supposed* to work. The documentation created or obtained by the auditor is some combination of:

1. A narrative description of the system,

2. A systems flow chart, and

3. A questionnaire.

These three types of documentation can be illustrated by describing a hypothetical system which is not difficult to visualize—the route that checks might take when they are received by a firm. This example is deliberately oversimplified and taken out of context, but it enables one to focus on certain characteristics of an accounting control system.

A Narrative Description of Cash Receipts on Account

Assume that checks from customers are received in the mail room each morning (receipt of currency is very uncommon). The checks are accompanied by a remittance advice, which is simply the top half of the invoice originally sent to the customer. It contains the customer's name, address, account number, and amount of payment.

The mail clerk endorses the checks restrictively and then makes two copies of a listing of the receipts. The purpose of the listing is to establish an early and comprehensive record of the money received.

The cash receipts are turned over to the cashier. The first copy of the cash receipts listing is sent with the remittance advices to the accounts receivable department. The second copy is sent to the general ledger department.

By requiring two copies and routing them independently to the accounts receivable and general ledger departments, the company provides a double check on the accuracy of each department's clerical processing.

In the accounts receivable department, the cash receipts listing and the remittance advices are reconciled. This step provides an additional clerical check of the accuracy and completeness of both records. The cash credits then are posted to the customer ledger accounts. The listing and remittance advices are filed.

In the general ledger department the total of the cash receipts listing is verified clerically and then posted to the control account. The listing is filed. Once a month the control total is balanced with the detail of the customer ledger accounts.

The cashier in the cash receipts department takes the checks received in the mail, combines them with the cash taken from the register, and makes out a deposit slip. The advantage of routing the checks directly from the mail clerk to the cashier is that they are not processed in either the accounts receivable or the general ledger department, where the records of cash receipts are kept. The cashier then takes the money to the local bank and makes the deposit. The deposit slip is returned by the bank to the cashier.

A Flow-Chart Description of Cash Receipts on Account

If the auditor *designs* a flow chart of the system, it will be necessary for him to visualize the information flow and the documents being processed. Using standard symbols, the flow chart should then be constructed in such a way that meaningful information about the system can be picked up by a reviewer who is familiar with the symbols. If the auditor uses a client-constructed flow chart, he should possess the ability to read it, interpret the symbols, and draw useful conclusions about the system depicted in the flow chart.

For a number of years, standard flow chart symbols have been used. These symbols are in the chart shown hereafter. Following this illustration is a systems flow chart for the receipts on account system previously described in narrative form.

STANDARD FLOW CHART SYMBOLS

Basic Symbols

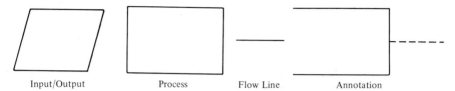

| Input/Output | Process | Flow Line | Annotation |

Specialized Input/Output Symbols

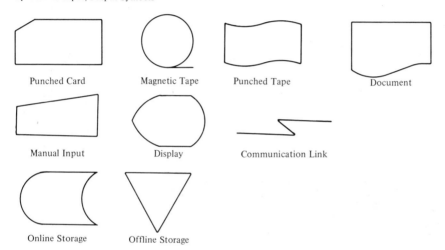

Punched Card   Magnetic Tape   Punched Tape   Document

Manual Input   Display   Communication Link

Online Storage   Offline Storage

Specialized Processing Symbols

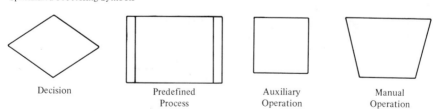

| Decision | Predefined Process | Auxiliary Operation | Manual Operation |

Additional Symbols

Connector   Terminal

SYSTEMS FLOW CHART FOR RECEIPTS ON ACCOUNT

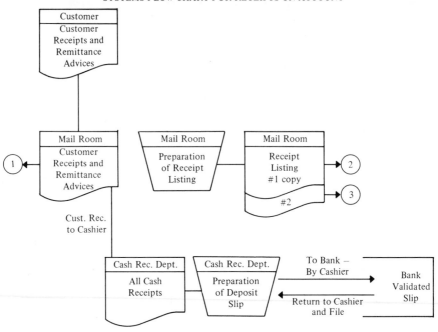

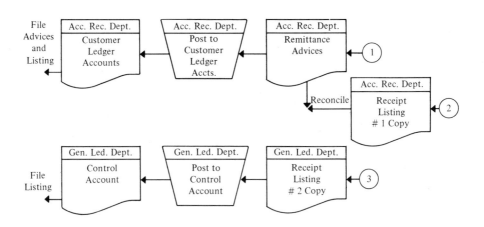

Discussion and Comparison of Narrative and Flow Chart

In some ways a narrative description and a flow chart complement each other. A systems flow chart (in the authors' opinion) should give a pictorial image of the general flow of information from one document to another, from one location to another, and, if applicable, from one machine to another. By viewing this documentation, the auditor should be able to arrive at some tentative conclusions about the order (or lack of order) in the system.

The narrative description should fill in the detail. For example, the narrative explains that currency seldom is received and that the checks are endorsed restrictively. In a complicated system, there may be some omissions in the flow chart that can be described in the narrative.

The Internal Control Questionnaire

The auditor's appraisal is not complete until he asks himself a series of critical questions and then proceeds to answer them. (In some cases the completion of the questionnaire is the initial step.) The information for the questionnaire can come from several sources.

1. The narrative description or the flow chart.

2. Conversations with the client's personnel. Actually, these two sources are interrelated, because some of the information for the narrative description and the flow chart comes from such conversations.

3. Tests of the system. Tests of compliance are discussed in detail in another part of this chapter.

Traditionally, internal control questionnaires have been an important element in the review of internal control systems. Although newer methods such as flow-charting have gained wide acceptance in recent years, the questionnaire approach is still popular among many CPA firms. Generally, the form calls for the auditor to record a *yes* or *no* answer to the questions on the form. This is a practical approach. But a low-quality review of the system may result, if auditors thoughtlessly base their answers on last year's examination. The answers should be determined only after proper investigation of flow charts, narratives, results of conversations, etc.

A review of internal control is highly subjective. Decisions about the quality of the system should be made only after all available evidence is gathered and thought is given to the matter. *Yes* and *no* answers to an internal control questionnaire provide good but not total evidence of the quality of the system. Auditors might wish to investigate certain parts of the system on the basis of the answers they develop. Also, as explained later, the system must be tested.

Listed hereafter is a completed questionnaire that pertains directly to the system of cash receipts on account. It is not a complete set of questions; many others could be asked.

An Internal Control Questionnaire for Cash Receipts on Account

Question	Answer	Alternate Procedure or Additional Information
1. Does the person who handles the cash have access to the records of cash receipts?	No	
2. Is immediate control established over mail receipts?	Yes	
3. Are deposits made intact daily?	Yes	
4. Do different individuals handle the detail customer ledger accounts and control accounts?	Yes	
5. Is the clerical accuracy of the cash receipts and remittance advices checked?	Yes	
6. Is the bank-validated deposit slip returned to someone other than the cashier?	No	The bank reconciliation is prepared in the general ledger department.
7. Is the total of the deposit slip reconciled with the remittance advice and receipt listing total before the deposit is made?	No	No compensating strength found.

Preliminary Evaluation

After the review of internal control is completed, it is advisable for the auditor to make a preliminary evaluation of the system and to form a tentative conclusion as to whether the control procedures used by the client are satisfactory. The conclusion is tentative, because the system has not been tested to ascertain whether the client's prescribed control procedures as documented in the flow charts, narratives, questionnaires, etc. are actually complied with in the periodic routines of the business.

According to AU Section 320.68 *(SAS No. 1)*, the conclusion should state whether the controls are satisfactory for the auditor's purpose. The following criteria of satisfaction are appropriate.

> The procedures and compliance should be considered satisfactory if the auditor's review and tests disclose no condition he believes to be a material weakness for his purpose. In this context a material weakness means a condition in which the auditor believes the prescribed procedures or the degree of compliance with them does not provide reasonable assurance that errors or irregularities in amounts that would be material in the financial statements being audited would be prevented or detected within a timely period by employees in the normal course of performing their assigned functions.

If a material weakness is found, the auditor might choose not to rely on internal control for that class of transactions (sales, payroll, etc.). This choice would necessitate almost total dependence on tests of details of transactions and balances in that area and on analytical reviews to obtain the evidence needed to support an opinion of the fairness of the financial statements.[1]

A satisfactory internal control system in the area under review allows the auditor to place *some* reliance on the system. The amount of reliance probably depends on the immaterial weaknesses found during the review. AU Section 320 *(SAS No. 1)* provides no specific guidance on this matter other than to suggest that the extent of substantive tests can vary inversely with the auditor's reliance on internal control. If the auditor considers the internal control to be of sufficient quality, the substantive tests can be less extensive than they would be otherwise.

An Example of a Preliminary Evaluation

With the discussion of the preceding section as a background, consider an example of a preliminary evaluation of the system of cash receipts on account. The steps in making this evaluation are to:

1. Consider the errors or irregularities that could occur in the system.

2. Consider control procedures (if any) that should prevent or detect these errors or irregularities.

3. Determine whether these needed internal control procedures are part of the system.

Such an evaluation could be made in the form shown on page 174.

The system weaknesses can be selected easily from the third column of the evaluation form illustrated.

1. One clerk opens the mail, endorses the checks, and prepares the receipt listings.

2. The bank-validated deposit slip is returned directly to the cashier.

3. The totals of the deposit slip and receipt listing are not reconciled before deposit of money in the bank.

Would the auditor conclude that these weaknesses are material, as defined in *SAS No. 1,* and thus render the system of internal control in the area of cash receipts unsatisfactory? This decision is a matter of professional opinion.

---

[1] These tests, which are called substantive tests, are discussed in Chapters 9–14. They include such procedures as examination of documents supporting account balances, letters from independent parties verifying account balances, etc.

Errors or Irregularities That Could Occur in the System	Internal Control Procedures (if any) That Should Prevent or Detect These Errors or Irregularities	Internal Control Procedures That are Documented as Part of the System
1. The cashier could take money and conceal the shortage by altering the cash receipt records.	1. Separate the recording of the cash receipts and the physical custody of the money.	1. According to the system documentation, these two functions are separated.
2. Some of the cash receipts might be lost during processing.	2. Establish immediate control over cash receipts.	2. The mail clerk prepares a separate listing of cash receipts when they are received each day.
3. Checks could be endorsed illegally by the mail clerk and the amounts omitted from the receipt listing.	3. Separate these duties between two people, one to endorse checks and the other to prepare the receipt listing. Detection might occur if customers complained about lack of credit for payments.	3. Only one mail clerk is employed. Checks are endorsed and receipt listings are made by this individual. However, customers use a separate accounts receivable department for questions about their accounts.
4. The cashier might take money and substitute a "fake" deposit slip to make it appear that an actual deposit has been made.	4. Require bank-validated deposit slip to be returned to and inspected by someone other than the cashier. Detection might occur if a party other than the cashier received and reconciled the monthly bank statement.	4. The bank-validated deposit slip is returned directly to the cashier. However, an independent party receives and reconciles the bank statement.
5. Errors in compilation of the customer account balances could result in inaccurate accounts receivable amounts.	5. Require the accounts receivable department and ledger control to derive their amounts independently of each other.	5. The accounts receivable subsidiary ledger and general ledger control are maintained separately. Each department obtains separate copies of the receipt listings. Periodically, their totals are reconciled.
6. Errors could be made in the deposit slip and/or the daily records of cash receipts.	6. Reconcile the deposit slip and receipt listing totals before the deposit of money in the bank.	6. This reconciliation is not performed. (The bank statement is reconciled independently.)

However, the auditor probably would consider "compensating strengths" before reaching a decision. For example, customer complaints about failure to obtain proper credit for payments are made to a separate accounts receivable department. This procedure might mitigate the potential problems of employing only one mail clerk. Also, the fact that a party other than the cashier reconciles the bank statement offsets the potential irregularity that might occur because the bank-validated deposit slip is returned directly to the cashier.

If the system is deemed unsatisfactory because the weaknesses are material, the auditor might rely only on substantive tests to gather evidence supporting the financial statement account balances affected by the system of cash receipts. A good example is accounts receivable, which certainly is affected by cash receipt transactions. With some exceptions, the auditor sends letters directly to a sample of the client's customers. These letters request the customers to reply directly to the auditor their agreement or disagreement with the accounts receivable amount on the client's books as of the balance sheet date. In the absence of any reliance on internal control over cash receipts transactions, the auditor might wish to enlarge the sample of such letters (i.e., place more reliance on substantive tests).

If the system is considered satisfactory because the weaknesses are immaterial, the auditor would rely, to some degree, on internal control over cash receipts. This reliance would reduce the need to use substantive tests and might allow the auditor to retain or lower the usual sample size of these types of tests.

The relationships between the review and evaluation of internal control and other auditing procedures are shown in the chart on page 176.

Some Additional Thoughts

There is a *cost* of installing a good system of internal control. At some point, the cost may not be justified by the *benefits* to be obtained from installing a particular feature. For example, the addition of an extra mail clerk probably would help the control of incoming receipts. However, auditors might conclude that management is justified in not hiring another person because built-in safeguards make this unnecessary. In other areas, auditors may strongly recommend the additional cost, i.e., hiring a cashier rather than allowing the accounts receivable clerk to handle the cash receipts.

Internal control involves a *risk*. There is no system that could not break down under certain conditions. The general ledger clerk might ignore the receipts listing and obtain the control total directly from the accounts receivable department. The cashier might find out, in advance, the total amount recorded for cash receipts and adjust the deposit record accordingly. Auditors try to *reduce* the probability of undetected errors and irregularities by determining that good procedures have been established and that they are being followed as prescribed. They generally take the position that collusive efforts on the part of employees or unusual breakdowns are beyond the scope of the auditor's function.

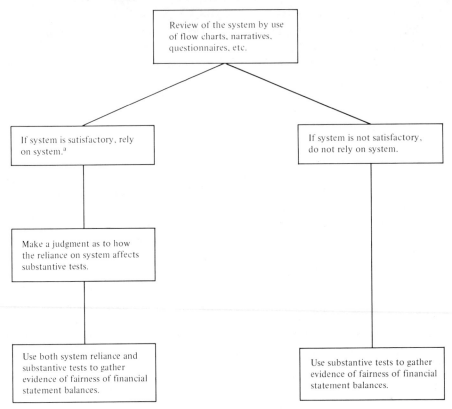

Review of the system by use of flow charts, narratives, questionnaires, etc.

If system is satisfactory, rely on system.[a]

If system is not satisfactory, do not rely on system.

Make a judgment as to how the reliance on system affects substantive tests.

Use both system reliance and substantive tests to gather evidence of fairness of financial statement balances.

Use substantive tests to gather evidence of fairness of financial statement balances.

[a] Presume that the tests of compliance will affirm the opinions developed during the review.

Tests of Compliance

As previously pointed out, the auditor's review and preliminary evaluation of internal control may lead to the conclusion that the system for a particular class of transactions (cash receipts, payroll, etc.) is satisfactory. If so, some degree of reliance is placed on the system, and substantive auditing procedures can be adjusted. However, relative assurance must be strengthened by *testing* the system to ascertain that the client's prescribed procedures noted during the review processes *actually* are applied as indicated in the flow charts, narratives, conversations, etc. In fact, tests of compliance are not only desirable, but also necessary if the auditor is to rely on the system in determining the nature, extent, and timing of substantive audit tests. In other words, the auditor must be reasonably certain that the system functions in the alleged manner.

What if the auditor's review leads to the conclusion that the system is unsatisfactory and that no reliance is to be placed on it? In such a case, the auditor need not conduct tests of compliance, because almost total reliance will be placed on substantive tests in gathering evidence on the related financial statement balances. Thus, the audit effort can be directed more efficiently toward these substantive tests (letters of confirmation to accounts receivable customers, for instance).

In some cases, the auditor may consider the effort of conducting tests of compliance to be too costly and time consuming in relation to the potential benefit of reducing substantive tests for the same class of transactions. For example, certain audit clients may have large dollar amounts but a very small number of accounts receivable balances. Letters of confirmation will be sent to practically all customers anyway, and other substantive tests are easy to apply. Therefore, very little reduction in audit effort will result from conducting tests of compliance in the areas of cash receipts and sales.

In audits of small firms, much of the review and compliance testing may be bypassed if the substantive tests include examination of a very large percentage of transactions and balances. Careful judgment must be used by the auditor, however, before the review and evaluation of internal control is curtailed significantly or bypassed.

What if the tests of compliance do *not* support the opinion of internal control formed in the review process? In other words, what if the prescribed system is not being complied with in a manner satisfactory to the auditor? In such an event, the auditor might be forced to draw the same conclusions about the system that would have been formed if the review had revealed material weaknesses. In both cases the system is not satisfactory for the auditor's purposes.

The tests of compliance might fail to support the conclusions of the review process by showing that (1) there is no compliance, or (2) the degree of compliance is unacceptable. Both of these conditions and the various types of tests of compliance are discussed in the two following sections.

A Walk-Through

One example of a test of compliance is a "walk-through" of a transaction or a set of transactions from beginning to end. The transaction in this system is:

Cash_____

Accounts Receivable_____

It starts in the mail room and ends in the bank, in the general ledger department, and in the accounts receivable department. A walk-through might take the following form.

1. Observe (perhaps on a surprise basis) the handling of the cash and remittance advices. Make sure that a restrictive endorsement stamp is placed on each check. See that the cash is sent to the cashier and that the remittance advices and receipt listing go to the accounts receivable department.

2. Observe the preparation of the deposit slip. Make certain that there is limited access to the area in which this preparation takes place. Ascertain that *all* of the day's receipts are being deposited intact with no substitutions. It is standard practice to reconcile the deposit slip with the receipt listing; see that this is done.

3. Accompany the cashier to the bank. In some cases the auditor may want to leave a confirmation letter. This letter will be returned directly to the auditor and will contain a certification that all of the deposit was credited to the client's account.

4. Follow through and make sure that the detail and total of the deposit slip matches the receipt listings processed in the general ledger department and the accounts receivable department.

In some cases it might be more feasible to walk-through a transaction that already has been recorded. The point is that the auditor is testing the *existence* of compliance rather than the *degree* of compliance. If the walk-through is satisfactory, it is assumed that this part of the system functions adequately.

Some auditors use a walk-through as a technique for developing their understanding of the system and verifying the accuracy of the flow charts, narratives, and other documentation. In such cases, a walk-through might be classified more appropriately as a part of the review process. Additional inquiries and observations might be made during compliance testing.

Testing Both Compliance and Dollar Validity

The other example of a test of compliance is broader in scope. The auditor does not merely ascertain the *existence* of compliance, but ascertains that the degree of compliance is at a tolerable level. The following type of tests might be applicable.

1. Select a sample of receipt listings and trace the names, dates, and dollar amounts to the validated deposit slips. The purpose of this test is to ascertain that money initially received and recorded in the mail room was deposited to the client's accounts (no loss of assets). It should be noted that in this type of test the auditor is not only checking compliance with the system, but also ascertaining the validity of dollar representations.

2. Select a sample of deposit slips and trace the names, dates, and dollar amounts to the posted entries in the individual customer accounts. The purpose of this test is to provide the auditor with reasonable assurance that the cash receipts were credited properly to the customers. (*Note:* One may wish to consider other tests of compliance at this point.)

The Desired Degree of Certainty and the Tolerance Level

If the auditor accepts the view that the degree of compliance is significant, he is faced with two critical decisions that will have direct effect on the size of his test.

1. What risk is the auditor willing to take that his estimate of the degree of compliance is wrong because the sample characteristics do not provide a good indication of the population characteristics? Phrased in another way, how certain does the auditor need to be? In a population of 250 receipt listings, how many should be traced to deposit slips? There are, of course, no definite answers to these questions. The auditor uses judgment and any information that he has gained during the course of the audit. Unless statistical sampling is used (as discussed in the next chapter), this judgment rarely is expressed as a percentage. One thing is definite, however: as the auditor's need for certainty increases, so does the sample size necessary to satisfy this need. Reference to the system of cash receipts reveals a feature that could help the auditor make a more informed judgment. The cash and the receipt listings are handled in two separate departments. For "lapping" to occur, either collusion or some unusual breakdown would be necessary.[2] This safeguard of separation might enable the auditor to change his sample size, particularly if successful compliance with this system had been observed in past audit engagements. (Some auditors maintain that the sample sizes for tests of compliance should be increased because of prescribed strong internal control features. The auditor should be relatively sure that the features are strong).

2. What error tolerance is the auditor willing to accept and still conclude that compliance is satisfactory? For example, if he were tracing receipt listings to validated deposit slips, what percentage of errors would he be willing to accept? Again, the auditor's judgment must be based, in part, on the importance and materiality of the test. As his error tolerance decreases, the necessary sample size will increase.

An Example of a Test of Compliance

A Description of the Test

Even though it is accepted practice for the auditor to conduct tests of compliance, there are no rigid rules regarding the type, the timing, or the size of

---

[2]Lapping is a procedure in which the individual handling the money withholds some portion of the day's receipts and also fails to record proper credit to certain customers' accounts. When subsequent receipts come in, some portion of these receipts is used to credit the previous customers' accounts. This process continues in the same manner on a day-to-day basis. It is difficult for such an irregularity to be carried out successfully unless the custody and recording of cash are handled by the same person or there is collusion between the people performing these two functions.

Description of Receipts on Account

Customer Number	Name	Receipt Number	Date Received in Mail Room	Amount
0053	Leven Bros. TV & Furn. Co.	02359	1-2-X9	$ 590.00
0007	Alpha Wholesale Co.	02360	1-3-X9	3,525.00
0056	Wolsey St. Furn. Display	02380	2-14-X9	480.00
0081	Eastgate TV and Furn. Store	02381	2-15-X9	347.00
0036	Keller-Goodman Fine Furn.	02382	2-18-X9	480.00
0012	Gamma Furn. Co., Inc.	02417	4-5-X9	5,210.00
0043	Bart Bros. Hardware and Furn.	02418	4-8-X9	288.00
0008	Griffin Bros. Wholesale	02463	6-10-X9	750.00
0057	Cardwell Wholesale Furn.	02464	6-11-X9	1,020.00
0063	Mann's	02465	6-12-X9	2,820.00
0073	North Blvd. Furn. Co.	02514	8-20-X9	2,800.00
0094	Ardmore Wholesale Furn. Co.	02515	8-21-X9	306.00
0092	Church St. Bargain Furn.	02524	9-3-X9	119.00
0077	Gray Furn. Supply	02525	9-4-X9	822.00
0099	Peters Furn. Display	02526	9-5-X9	1,020.00
0089	Colidge St. Discount Furn.	02551	10-10-X9	980.00
0082	Westend Furn.	02552	10-11-X9	5,220.00
0074	Quirk's Furn.	02581	11-21-X9	357.00
0069	Central Wholesale Furn.	02582	11-22-X9	1,720.00
0062	Rolling Hills Furn. Store	02593	12-9-X9	1,000.00
0017	Bargin Discount House	02594	12-10-X9	1,550.00
0077	Gray Furn. Supply	02608	12-31-X9	1,000.00

the tests. If it is practical, the auditor might take a sample of the year-to-date transactions. Another possibility, although a less desirable one, is to select a sample of a month's entries.

In the past, some auditors have used a method known as the "test-month" approach in which one or more months are selected as the sample period, and all transactions are tested for that given time frame. Under this procedure, known as a "cluster sample," full attention is given only to certain elements of a population. When it is used, care is taken to choose different months each time the audit is conducted. The theory behind this approach is that one month's events should be typical of those of the entire year. Even though use of the test month is very practical, it may be excessively risky, because no transactions are examined in other months. (Many firms have abandoned the test-month approach and are selecting a sample from all or most of the year's transactions.)

In the illustration above, the sample comes from an entire year's transactions. The auditor traces each receipt in the sample to the validated deposit slip. According to prescribed procedures, deposits are made intact daily. Therefore, the name, date, and amount should be the same in both documents.

The assumption is made that there are 250 working days in the year, and that a deposit is made each day. For brevity's sake, only a few receipts and

Description of Bank Deposits

Date of Deposit	Name	Amount
1-2-X9	Leven Brothers TV and Furniture Company	$ 590.00
1-3-X9	Alpha Wholesale Company	3,525.00
2-15-X9	Wolsey Street Furniture Display	480.00*
2-15-X9	Eastgate TV and Furniture Store	347.00
2-18-X9	Keller-Goodman Fine Furniture	480.00
4-5-X9	Gamma Furniture Company, Inc.	5,210.00
4-8-X9	Bart Brothers Hardware and Furniture	288.00
6-10-X9	Griffin Brothers Wholesale	750.00
6-12-X9	Cardwell Wholesale Furniture	1,020.00*
6-12-X9	Mann's	2,820.00
8-20-X9	North Boulevard Furniture Company	2,800.00
8-21-X9	Ardmore Wholesale Furniture Company	306.00
9-3-X9	Church Street Bargain Furn.	119.00
9-5-X9	Gray Furniture Supply	822.00*
9-5-X9	Peters Furniture Display	1,020.00
10-10-X9	Colidge Street Discount Furniture	980.00
10-11-X9	Westend Furniture	5,220.00
11-22-X9	Quirk's Furniture	357.00*
11-22-X9	Central Wholesale Furniture	1,720.00
12-10-X9	Rolling Hills Furniture Store	1,000.00*
12-10-X9	Bargin Discount House	1,550.00
12-31-X9	Gray Furn. Supply	1,000.00

*Deposited one day later than recorded on the books.

deposits are shown. For the days of the year that are omitted, assume that no errors exist.

Evaluation of the Results

Of 250 receipts processed during the year, deposits were late on 5. This is a 2% error rate in the population, but the auditor, of course, would not know it. It is interesting to examine some different types of sampling plans that the auditor might have used and see what results would have been obtained.

If the test-month approach had been used, a lag would have been discovered *if* some combination of the months of February, June, September, November, and December had been chosen. Had any other months been selected, a test of all the periods' receipts would have given the auditors a false conclusion that there was compliance with prescribed procedures. Thus, we can see why the test-month approach is considered by many to be a weak sampling method. Chapter 6 contains discussions of statistical sampling and the use of a simple random selection of test items.[3]

[3]When simple random sampling is used, each element of the population has an equal chance of selection.

What if the auditor had selected every fifth receipt and the examination was to find an error percentage in the sample that was representative of the error percentage in the population? Starting with the first receipt, the selected numbers would be:

02359	02419	02469	02524	02574
02364	02424	02474	02529	02579
02369	02429	02479	02534	02584
02374	02434	02484	02539	02589
02379	02439	02489	02544	02594
02384	02444	02494	02549	02599
02389	02449	02499	02554	02604
02394	02454	02504	02559	
02399	02459	02509	02564	
02404	02464	02514	02569	
02409		02519		
02414				

A deposit lag would be found in Receipt No. 02464. The amount was put on the books on 6-11 and was deposited 6-12. The error rate is 2% (1 error in 50 receipts), which is the same as the population error rate.

Despite this accuracy, there are some lingering questions. On what basis was a sample size of 50 determined? If auditors find one deposit lag in 50 receipts, what significance can be drawn from this finding, considering the fact that they would not know the true error rate? These questions are answered in the next chapter where statistical sampling is discussed.

Auditors also would consider other factors, such as the period of the time lags, the size of the amounts involved, and their prior experience with such non-compliances.

Evaluation of Internal Control

After the auditors have reviewed and tested the system of internal control, they should make a final evaluation of their findings. The major purpose of this step is to put the finishing touch to the first phase of the audit. Because much of the remaining examination will consist of substantive tests of income statement and balance sheet accounts, this evaluation will be of some significance in determining the extent of future tests.[4]

Sometimes the auditor will suggest to the client certain system changes that should be made to lower the probability of inaccuracies or inefficiencies. Such suggestions normally are made in a formal letter addressed to the board of directors. More is said about this procedure in Chapter 8.

---

[4]For a more comprehensive treatment of this subject, see R. K. Mautz and Donald L. Mini, "Internal Control Evaluation and Audit Program Modification," *The Accounting Review,* April 1966, pp. 283–291.

The system of cash receipts illustrated in this chapter is basically sound, but, as was pointed out, it has certain weaknesses. Whether or not the future level of substantive tests or tests of balances should be altered is largely a matter of opinion. In Chapter 9, more attention is given to this matter.

Reliance on Internal Auditors

Chapter 1 contains a discussion of the role of internal auditors within the client's organization. Although internal auditors technically constitute part of the structure of internal control, they serve a special function in that they check the work of other members of the client's organization. Internal auditors also enjoy a measure of independence from the company's operating personnel, because they often report to the financial vice-president or, in some cases, to the audit committee of the board of directors.

Partly for the foregoing reasons, the AICPA has given its sanction to the independent auditor's use of the internal audit function in making a study and evaluation of internal control. AU Section 322.04 *(SAS No. 9)* contains the following comments on the subject.

> The work performed by internal auditors may be a factor in determining the nature, timing, and extent of the independent auditor's procedures. If the independent auditor decides that the work performed by internal auditors may have a bearing on his own procedures, he should consider the competence and objectivity of internal auditors and evaluate their work.

How would the independent auditor determine the competence and objectivity of the internal auditors? One approach is to check the company's hiring criteria and the organizational level to which the internal auditor reports. It is important that the internal auditor not be placed under the supervision of the individuals whose operations he audits and reports upon.

To evaluate the performance of internal auditors, the independent auditor must examine the work produced by this group and compare this work with some type of standard. One possible standard is similar work performed by the independent auditors themselves.

Obviously, care must be taken to avoid placing too much reliance on the work of the internal auditors, regardless of their semi-independent status within the client's organization. It is the *independent* auditors who are responsible for conducting the examination and signing the audit report.

Systems Evaluation Approach

Appendix I to the text contains descriptions of a new audit approach to the interim phase used by the accounting firm of Peat, Marwick, Mitchell & Co.

Summary

In modern-day audits of medium and large institutions, the study and evaluation of internal control are virtual necessities. Without these procedures, there would be little guidance for controlling the audit risk or for determining the basis for future testing. The "image" of the company's accounting system carries through the rest of the audit and aids in many decisions.

Within broad guidelines, CPAs can conduct the details of the study by using a combination of several methods. But they must be ready to defend their findings in the event of challenge from other parties. Therefore, auditors must document their study with completed internal control questionnaires, memos, and programs that show the amount of testing and the results obtained.

A thorough study of internal control sets the framework for quality performance in the other phases of the audit. The use of statistical sampling to design sample plans for, and to evaluate the results of, tests of compliance is explored in Chapter 6.

Chapter 5 References

Bower, James B., and Robert E. Schlosser, "Internal Control—Its True Nature," *The Accounting Review,* April 1965, pp. 338–344.

Freedman, Martin S., Jr., "A Primer on Fraud and Embezzlement," *Management Accounting,* October 1973, pp. 36–40.

Konrath, Larry F., "The CPA's Risk in Evaluating Internal Control," *The Journal of Accountancy,* October 1971, pp. 53–56.

Mautz, R. K., and Donald L. Mini, "Internal Control Evaluation and Audit Program Modification," *The Accounting Review,* April 1966, pp. 283–291.

Appendix to Chapter 5

Study and Evaluation of Other Systems[1]

Introduction

To provide exposure to other internal control systems and the organizational framework in which they function, three organization charts are reproduced. These charts are accompanied by procedural flow charts drawn up in summary fashion to provide a general overview of segregation of functions within the organizational framework.

[1]The case study that is available with this text contains descriptions of credit sales, cash receipts, sales returns, and cash sales.

Each flow chart is accompanied by (1) a general narrative description of the system, (2) a partial internal control questionnaire stressing segregation of functions, (3) comments about how the system might prevent errors or irregularities, (4) some examples of tests of compliance and transactions applicable to that system, and (5) descriptions of how the system characteristics might affect substantive tests.

Narrative of Purchases

1. When the Stores Department needs merchandise, two copies of a requisition are prepared. One copy is sent to the Accounts Payable Department and the other copy to the Purchasing Agent.

2. Upon receipt of the requisition, the Purchasing Agent, after obtaining a competitive bid, initiates a five-copy purchase order.

3. One copy of the purchase order is sent back to the Stores Department, where the price and other descriptions are checked.

4. Copies of the purchase order are sent to the Accounts Payable and Receiving Departments to serve as file copies and as notification that merchandise has been ordered.

5. Two copies of the purchase order are sent to the vendor, who acknowledges receipt by sending one copy back to the Purchasing Agent.

6. The vendor ships the merchandise to the Receiving Department, where an independent count of the merchandise is made and three copies of a receiving report are filled out.

7. The merchandise, after being checked, is sent to the Stores Department, along with two copies of the receiving report. The Stores Department checks the merchandise with the listings on the receiving report and sends one copy of the receiving report to the Accounts Payable Department.

8. The third copy of the receiving report is sent from the Receiving Department to the Purchasing Agent, where terms and other details are checked.

9. Two copies of an invoice are sent from the vendor to the Purchasing Agent. The Purchasing Agent retains one copy and sends the other copy to the Accounts Payable Department.

10. The Accounts Payable Department compares the details of the requisition, the purchase order, the invoice, and the receiving report. All four documents are filed together as documentation of the transaction.

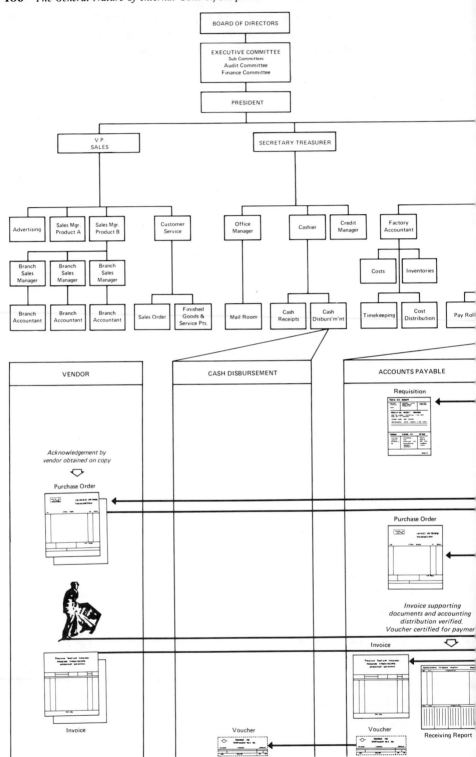

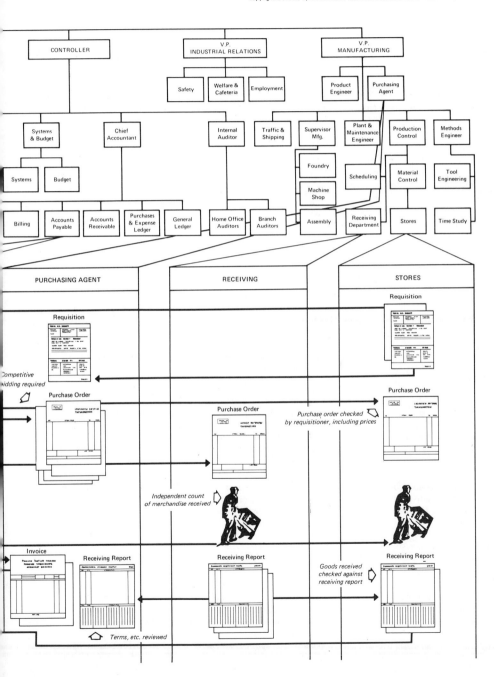

Internal Control Questionnaire for Purchases

Question	Answer	Comments
1. Is there separation between authorizing, recording, and custody of merchandise purchases?	Yes	The Stores Department requests the merchandise, the Purchasing Agent orders it, the Receiving Department receives it, and the Accounts Payable Department records the transaction.
2. Are steps taken to ensure the best price for merchandise?	Yes	Competitive bids are required.
3. Is immediate control established over merchandise received from vendors?	Yes	A receiving report is made out when merchandise is received.
4. Are receiving reports made out after an independent count?	Yes	However, a copy of the purchase order is held in the Receiving Department.
5. Are procedures used to ensure that merchandise ordered is received?	Yes	The receiving report is compared with the purchase order.
6. Are procedures used to ensure that merchandise invoiced by the vendor has been received?	Yes	The receiving report is compared with the invoice.

Evaluation of Purchases System

Possible Errors or Irregularities	Procedures that Might Prevent or Detect Errors or Irregularities	Procedures Documented in the System
1. The merchandise requested by the Stores Department could be different from that ordered and/or received.	1. The Stores Department could check both the merchandise and the order and receipt documents against their request.	1. This procedure is not followed. However, the Accounts Payable Department compares the documents.
2. Merchandise could be taken by individuals in the Receiving Department.	2. Other departments could check the billing from the vendor against the merchandise received.	2. The Accounts Payable Department matches the invoice with one of the Stores Department's copies of the receiving report. The Stores Department matches the receiving report with the merchandise.
3. The vendor could be overpaid for the merchandise.	3. Different departments could count the merchandise and process the billing from the vendor.	3. The Receiving Department independently counts the merchandise but has no access to the invoice.

Some Tests of Compliance and Transactions for the Purchases System

Test	*Purpose of the Test*
1. Compare the Stores Department's purchase requisitions with their receiving reports.	1. To determine that the items requested by the Stores Department were received from the vendor.
2. Compare invoice price with bids submitted by vendors.	2. To determine that competitive bidding procedures are followed.
3. Ascertain that the receiving reports are initialed or signed properly.	3. To determine that someone is taking responsibility for documenting independent counts.
4. Compare the vendor's invoice with a copy of the receiving report.	4. To ascertain that merchandise billed by the vendor was received by the company.
5. Compare the requisition, purchase order, invoice, and receiving report held in the Accounts Payable Department.	5. To ascertain that the entire transaction is accounted for properly.

The Effect of Purchases Evaluation on Substantive Tests

System Characteristics	*Substantive Tests Possibly Affected by This Characteristic*
1. Competitive bids are used on purchases.	1. Tests of inventory pricing.
2. Independent receiving reports are filled out.	2. Tests of inventory cutoffs.
3. Invoices are checked against the receiving reports.	3. Tests for unrecorded liabilities.
4. Prices on the purchase order are checked in the Stores Department.	4. Tests of inventory pricing.
5. The Accounts Payable Department compares the detail of the requisition, purchase order, invoice, and receiving report.	5. Confirmation of accounts payable.
6. Both copies of the vendor's invoice are sent directly to the purchasing agent.	6. Confirmation of accounts payable.
7. A copy of the purchase order is sent to the Receiving Department.	7. Comparison of physical and perpetual inventory.

Narrative of Cash Disbursements

1. When a cash disbursement is due, two copies of a voucher are made out in the Accounts Payable Department. After the supporting documents (invoice, purchase order, requisition, and receiving report) and the accounting distribution are verified, the voucher is certified for payment.

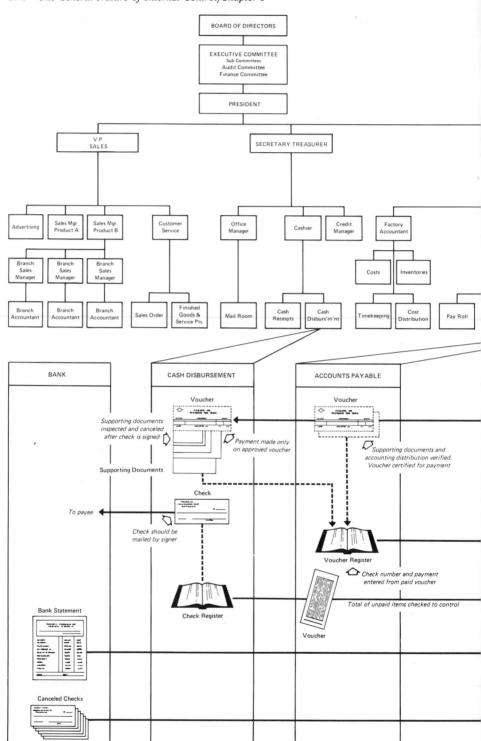

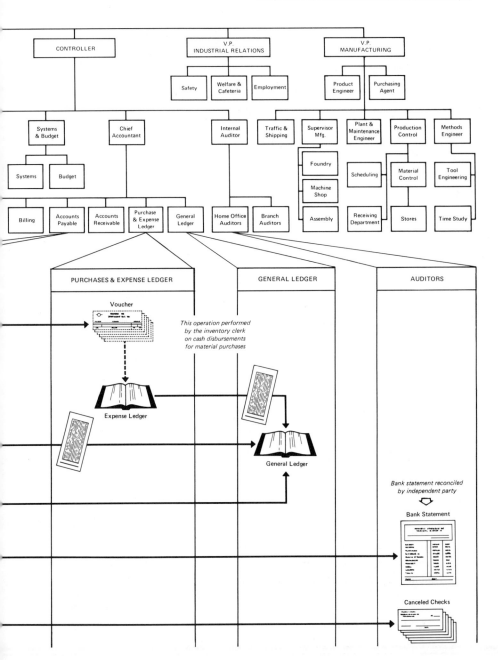

CASH DISBURSEMENTS

PROCEDURAL FLOW CHART SHOWN IN RELATION
TO ORGANIZATION CHART TO PORTRAY
THE CONTROL OBTAINED THROUGH SEGREGATION
OF FUNCTIONAL RESPONSIBILITY

Internal Control Questionnaire for Cash Disbursements

Question	Answer	Comments
1. Is there separation between authorizing and recording a disbursement?	Yes	Payment is authorized in the Cash Disbursements Department on the basis of an approved voucher sent from the Accounts Payable Department. The check is made out and mailed by a separate individual.
2. Are steps taken to ensure that merchandise and/or services have been received before payment is authorized and made?	Yes	Supporting documents are forwarded with the voucher from the Accounts Payable to the Cash Disbursements Department. There, the supporting documents must be inspected prior to payment authorization.
3. Are steps taken to prevent overpayments?	Yes	Supporting documents are canceled after the check is signed.
4. Are steps taken to ensure that the check is not diverted improperly?	Yes	The check signer mails the check to the payee.
5. Is the bank statement returned to the department from which the check was made out and recorded?	No	The bank statement is sent directly to the Internal Audit Department.
6. Is control maintained over vouchers payable?	Yes	Vouchers are recorded in a voucher register and the unpaid items are checked with the general ledger control account.
7. Are general ledger control totals maintained?	Yes	The General Ledger Department keeps control totals of expense charges and vouchers payable. However, the debits and credits to vouchers payable come directly from the voucher register, rather than the documents themselves.

2. The amount of the voucher is entered in the voucher register, which is kept in the Accounts Payable Department. One copy of the voucher is sent to the Purchase and Expense Ledger Department, where the amount of the voucher is posted to the expense ledger. One copy of the voucher along with the supporting documents is sent to the Cash Disbursements Department.

3. In the Cash Disbursements Department, the supporting documents are examined and the voucher is approved for payment.

Evaluation of Cash Disbursements System

Possible Errors or Irregularities	Procedures that Might Prevent or Detect Errors or Irregularities	Procedures Documented in the System
1. Checks could be written to improper payees and not recorded in the check register.	1. Checks should be signed and mailed by an individual other than the approver of vouchers. The bank statement should be reconciled independently.	1. These procedures are followed.
2. Duplicate payments could be made to vendors.	2. Some type of payment notation should be placed on documents.	2. Supporting documents are canceled after the check is signed.
3. Payment could be made without proper authorization.	3. Payment could be made only after inspection of proper documents and approval.	3. Payment authorization is made only after supporting documents and the voucher from the Accounts Payable Department are inspected.
4. There could be an inaccurate record of unpaid vouchers.	4. The total of unpaid vouchers could be checked with the general ledger control account.	4. This procedure is followed.

4. A check is made out and signed after inspection of the approved voucher. The check is mailed to the payee by the signer. The amount of the check is entered in the check register. The supporting documents are canceled after the check is signed.

5. The check number and the amount of the paid voucher are sent to the Accounts Payable Department and entered in the voucher register.

6. The General Ledger Department receives control totals from the expense ledger in the Purchases and Expense Ledger Department, the voucher register in the Accounts Payable Department, and the check register in the Accounts Payable Department. Periodically, the total of unpaid items in the voucher register is checked with the voucher control account kept in the General Ledger Department.

7. The bank sends the bank statement and canceled checks directly to the Internal Audit Department, where an independent bank reconciliation is performed.

Some Tests of Compliance and
Transactions for the Cash Disbursements System

Test	*Purpose of the Test*
1. Trace entries in the check register, voucher register, and expense ledger to appropriate documents (checks and vouchers).	1. To ascertain the validity of the entries in these books by examining the supporting documents.
2. Compare vouchers with supporting documents (invoices, purchase orders, receiving reports, and requisitions).	2. To determine that voucher preparation is based on adequate evidence of merchandise and/or services received.
3. Examine vouchers for indication of payment approval.	3. To determine that there is evidence of the examination of vouchers and approval prior to issuing checks to creditors.
4. Examine supporting documents for indication of cancellation.	4. To ascertain that the probability of overpayment to creditors has been minimized.
5. Trace entries in the check register, voucher register, and expense register to the general ledger.	5. To check the accuracy of postings from books of original entry to the general ledger.

The Effect of Cash Disbursements Evaluation on Substantive Tests

System Characteristics	*Substantive Tests Possibly Affected by This Characteristic*
1. Supporting documents and accounting distribution are verified before voucher is approved for payment.	1. Tests of the accounting treatment of merchandise purchases, fixed-asset additions, etc.
2. The bank statement is sent directly to and is reconciled by the internal auditors.	2. Bank reconciliations.
3. Supporting documents are canceled after the check is signed.	3. Confirmation of accounts payable.
4. A voucher register is used.	4. Tests for unrecorded liabilities.
5. The unpaid items in the voucher register are reconciled with the general ledger control account.	5. Confirmation of accounts payable.

Narrative of Payroll

1. The Timekeeping Department keeps two sets of time records. Employee clock cards, which originate in the Timekeeping Department, show the number of hours worked by each employee. Job time tickets, which come from the Shops Department, show the amount of time for each employee.

Internal Control Questionnaire for Payroll

Question	*Answer*	*Comments*
1. Is there separation among authorizing of the payroll, recording the payroll, and disbursing the payroll checks?	Yes	The voucher is prepared in the Accounts Payable Department. The payroll is recorded in the Payroll Department. The checks are disbursed separately, often by internal auditors.
2. Is an imprest payroll account used?	Yes	One check is written to the payroll account. Individual employees' checks are written against this account.
3. Are steps taken to guard against paying checks to improper or fictitious employees?	Yes	The internal auditors periodically make "payoffs" by disbursing the checks and requiring employee identification.
4. Are steps taken to ensure that the total dollar amount of the payroll is accounted for and distributed properly to the labor accounts?	Yes	A payroll clearing account is kept in the General Ledger Department. Both the labor distribution journal and the voucher are posted to this account.
5. Are steps taken to ensure that the payroll amounts are accurate?	Yes	The job time tickets are checked with the employee clock cards. The employee clock cards and the employment, rate, and deduction records are used to prepare the payroll.
6. Are employment records kept in a department separate from payroll preparation?	Yes	Employment and rate authorization records are kept in the Employment Department.

2. The Timekeeping Department checks the job time tickets with the employee clock cards. The job time tickets then are sent to the Cost Distribution Department. The employee clock cards are sent to the Payroll Department.

3. The Cost Distribution Department records the information from the job time tickets in a labor distribution journal. This journal is sent to the General Ledger Department, where an entry is made to a payroll clearing account.

4. The Payroll Department secures employment and rate authorization records and deduction slips from the Employment Department. These records and the employee clock cards are used to prepare the payroll record. The payroll record is forwarded to the Accounts Payable Department.

5. The Accounts Payable Department uses the payroll record and any other applicable supporting documents to prepare a voucher that authorizes the writing of the payroll check. The voucher information is forward-

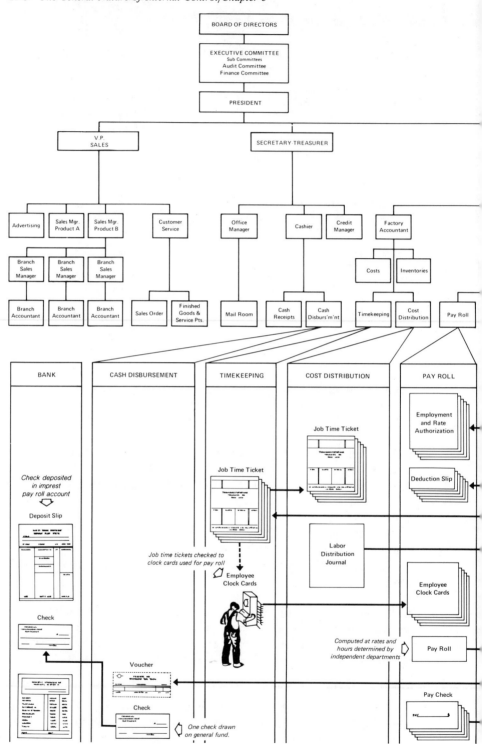

5–17.   What is a test month?

5–18.   What is simple random sampling?

5–19.   Under what conditions is it acceptable for independent auditors to obtain direct assistance on the engagement from internal auditors?

Chapter 5
Objective Questions Taken from CPA Examinations

5–20.   Normally, the auditor does not rely upon his study and testing of the client's system of internal control to
a.   Evaluate the reliability of the system.
b.   Uncover embezzlements of the client's assets.
c.   Help determine the scope of other auditing procedures to be followed.
d.   Gain support for his opinion as to the accuracy and fairness of the financial statements.

5–21.   The CPA reviews Pyzi's payroll procedures. An example of an internal control weakness is to assign to a department supervisor the responsibility for
a.   Distributing payroll checks to subordinate employees.
b.   Reviewing and approving time reports for subordinates.
c.   Interviewing applicants for subordinate positions prior to hiring by the personnel department.
d.   Initiating requests for salary adjustments for subordinate employees.

·5–22.   The CPA tests sales transactions. One step is tracing a sample of sales invoices to debits in the accounts receivable subsidiary ledger. Upon the basis of this step, he will form an opinion as to whether
a.   Each sales invoice represents a bona fide sale.
b.   All sales have been recorded.
c.   All debit entries in the accounts receivable subsidiary ledger are properly supported by sales invoices.
d.   Recorded sales invoices have been properly posted to customer accounts.

5–23.   Of the following, the best statement of the CPA's primary objective in reviewing internal control is that the review is intended to provide
a.   Reasonable protection against client fraud and defalcations by client employees.

b. A basis for reliance on the system and determining the scope of other auditing procedures.

c. A basis for constructive suggestions to the client for improving his accounting system.

d. A method for safeguarding assets, checking the accuracy and reliability of accounting data, promoting operational efficiency, and encouraging adherence to prescribed managerial policies.

5–24. A company holds bearer bonds as a short-term investment. Custody of these bonds and submission of coupons for interest payments normally is the responsibility of the

a. Treasury function.

b. Legal counsel.

c. General-accounting function.

d. Internal-audit function.

5–25. Matching the supplier's invoice, the purchase order, and the receiving report normally should be the responsibility of the

a. Warehouse-receiving function.

b. Purchasing function.

c. General-accounting function.

d. Treasury function.

5–26. A CPA learns that his client has paid a vendor twice for the same shipment, once based upon the original invoice and once based upon the monthly statement. A control procedure that should have prevented this duplicate payment is

a. Attachment of the receiving report to the disbursement support.

b. Prenumbering of disbursement vouchers.

c. Use of a limit or reasonableness test.

d. Prenumbering of receiving reports.

5–27. For good internal control, the person who should sign checks is the

a. Person preparing the checks.

b. Purchasing agent.

c. Accounts-payable clerk.

d. Treasurer.

5–28. The authorization for write-off of accounts receivable should be the responsibility of the

a. Credit manager.

b. Controller.

c. Accounts receivable clerk.

d. Treasurer.

5–29. The actual operation of an internal control system may be most objectively evaluated by
   a. Completing a questionnaire and flow chart related to the accounting system in the year under audit.
   b. Review of the previous year's audit work papers to update the report of internal control evaluation.
   c. Selection of items processed by the system and determination of the presence or absence of errors and compliance deviations.
   d. Substantive tests of account balances based on the auditor's assessment of internal control strength.

Items 5–30 through 5–33 are based on the following information:

The following sales procedures were encountered during the regular annual audit of Marvel Wholesale Distributing Company.

Customer orders are received by the sales-order department. A clerk computes the dollar amount of the order and sends it to the credit department for approval. Credit approval is stamped on the order and returned to the sales-ord department. An invoice is prepared in two copies, and the order is fi' tne "customer order" file.

The "customer copy" of the invoice is sent to the billing department and held in the "pending" file awaiting notification that the order was shipped.

The "shipping copy" of the invoice is routed through the warehouse and the shipping department as authority for the respective departments to release and ship the merchandise. Shipping department personnel pack the order and prepare a three-copy bill of lading: the original copy is mailed to the customer, the second copy is sent with the shipment, and the other is filed in sequence in the "bill of lading" file. The invoice "shipping copy" is sent to the billing department.

The billing clerk matches the received "shipping copy" with the customer copy from the "pending" file. Both copies of the invoice are priced, extended, and footed. The customer copy is then mailed directly to the customer, and the "shipping copy" is sent to the accounts receivable clerk.

The accounts receivable clerk enters the invoice data in a sales-accounts-receivable journal, posts the customer's account in the "subsidiary customers' accounts ledger," and files the "shipping copy" in the "sales invoice" file. The invoices are numbered and filed in sequence.

5–30. In order to gather audit evidence concerning the proper credit approval of sales, the auditor would select a sample of transaction documents from the population represented by the
   a. "Customer order" file.
   b. "Bill of lading" file.

    c. "Subsidiary customers' accounts ledger."
    d. "Sales invoice" file.

5–31. In order to determine whether the system of internal control operated effectively to minimize errors of failure to post invoices to customers' accounts ledger, the auditor would select a sample of transactions from the population represented by the
    a. "Customer order" file.
    b. "Bill of lading" file.
    c. "Subsidiary customers' accounts ledger."
    d. "Sales invoice" file.

5–32. In order to determine whether the system of internal control operated effectively to minimize errors of failure to invoice a shipment, the auditor would select a sample of transactions from the population represented by the
    a. "Customer order" file.
    b. "Bill of lading" file.
    c. "Subsidiary customers' accounts ledger."
    d. "Sales invoice" file.

5–33. In order to gather audit evidence that uncollected items in customers' accounts represented valid trade receivables, the auditor would select a sample of items from the population represented by the
    a. "Customer order" file.
    b. "Bill of lading" file.
    c. "Subsidiary customers' accounts ledger."
    d. "Sales invoice" file.

5–34. The auditor recognizes that a "system" of internal control extends beyond those matters which relate directly to the functions of the accounting and financial departments. Which one of the following would the auditor generally consider least a part of a manufacturing company's "system" of internal control?
    a. Time and motion studies which are of an engineering nature.
    b. Quarterly audits by an insurance company to determine the premium for workmen's compensation insurance.
    c. A budgetary system installed by a consulting firm other than a CPA firm.
    d. A training program designed to aid personnel in meeting their job responsibilities.

5–35. As an in-charge auditor, you are reviewing a write-up of internal control weaknesses in cash receipt and disbursement procedures. Which one of the following weaknesses, standing alone, should cause you the least concern?
    a. Checks are signed by only one person.

b. Signed checks are distributed by the controller to approved payees.

c. Treasurer fails to establish *bona fides* of names and addresses of check payees.

d. Cash disbursements are made directly out of cash receipts.

5–36.   From the standpoint of good procedural control, distributing payroll checks to employees is best handled by the

a. Acccounting department.

b. Personnel department.

c. Treasurer's department.

d. Employee's departmental supervisor.

5–37.   In a company whose materials and supplies include a great number of items, a fundamental deficiency in control requirements would be indicated if

a. Perpetual inventory records were not maintained for items of small value.

b. The storekeeping function were to be combined with production and record-keeping.

c. The cycle basis for physical inventory taking were to be used.

d. Minor supply items were to be expensed when purchased.

5–38.   In testing the payroll of a large company, the auditor wants to establish that the individuals included in a sample actually were employees of the company during the period under review. What will be the *best* source to determine this?

a. Telephone contacts with the employees.

b. Tracing from the payroll register to the employees' earnings records.

c. Confirmation with the union or other independent organization.

d. Examination of personnel department records.

5–39.   An auditor is testing sales transactions. One step is to trace a sample of debit entries from the accounts receivable subsidiary ledger back to the supporting sales invoices. What would the auditor intend to establish by this step?

a. All sales have been recorded.

b. Debit entries in the accounts receivable subsidiary ledger are properly supported by sales invoices.

c. All sales invoices have been properly posted to customer accounts.

d. Sales invoices represent bona fide sales.

5–40.   The primary purpose of internal control relating to heavy equipment is

a. To ascertain that the equipment is properly maintained.

b. To prevent theft of the equipment.

c. To determine when to replace the equipment.

d. To promote operational efficiency of the dollars invested in the equipment.

5-41.   Before relying on the system of internal control, the auditor obtains a reasonable degree of assurance that the internal control procedures are in use and operating as planned. The auditor obtains this assurance by performing
a.   Substantive tests.
b.   Transaction tests.
c.   Compliance tests.
d.   Tests of trends and ratios.

5-42.   Which of the following activities would be *least* likely to strengthen a company's internal control?
a.   Separating accounting from other financial operations.
b.   Maintaining insurance for fire and theft.
c.   Fixing responsibility for the performance of employee duties.
d.   Carefully selecting and training employees.

5-43.   Which of the following is an invalid concept of internal control?
a.   In cases where a person is responsible for all phases of a transaction, there should be a clear designation of that person's responsibility.
b.   The recorded accountability for assets should be compared with the existing assets at reasonable intervals, and appropriate action should be taken if there are differences.
c.   Accounting control procedures may appropriately be applied on a test basis in some circumstances.
d.   Procedures designed to detect errors and irregularities should be performed by persons other than those who are in a position to perpetrate them.

5-44.   A well-designed system of internal control that is functioning effectively is most likely to detect an irregularity arising from
a.   The fraudulent action of several employees.
b.   The fraudulent action of an individual employee.
c.   Informal deviations from the official organization chart.
d.   Management fraud.

5-45.   The financial management of a company should take steps to see that company investment securities are protected. Which of the following is *not* a step that is designed to protect investment securities?
a.   Custody of securities should be assigned to persons who have the accounting responsibiltiy for securities.
b.   Securities should be properly controlled physically in order to prevent unauthorized usage.
c.   Access to securities should be vested in more than one person.
d.   Securities should be registered in the name of the owner.

5–46.   In comparison with the external auditor, an internal auditor is more likely to be concerned with
    a.   Internal administrative control.
    b.   Cost accounting procedures.
    c.   Operational auditing.
    d.   Internal accounting control.

5–47.   An auditor is planning the study and evaluation of internal control for purchasing and disbursement procedures. In planning this study and evaluation, the auditor will be *least* influenced by
    a.   The availability of a company manual describing purchasing and disbursement procedures.
    b.   The scope and results of audit work by the company's internal auditor.
    c.   The existence within the purchasing and disbursement area of internal control strengths that offset weaknesses.
    d.   The strength or weakness of internal control in other areas, e.g., sales and accounts receivable.

5–48.   Which of the following statements relating to compliance tests is most accurate?
    a.   Auditing procedures cannot concurrently provide both evidence of compliance with accounting control procedures and evidence required for substantive tests.
    b.   Compliance tests include physical observations of the proper segregation of duties which ordinarily may be limited to the normal audit period.
    c.   Compliance tests should be based upon proper application of an appropriate statistical sampling plan.
    d.   Compliance tests ordinarily should be performed as of the balance sheet date or during the period subsequent to that date.

5–49.   Transaction authorization within an organization may be either specific or general. An example of specific transaction authorization is the
    a.   Establishment of requirements to be met in determining a customer's credit limits.
    b.   Setting of automatic re-order points for material or merchandise.
    c.   Approval of a detailed construction budget for a warehouse.
    d.   Establishment of sales prices for products to be sold to any customer.

5–50.   An example of an internal control weakness is to assign to a department supervisor the responsibility for
    a.   Reviewing and approving time reports for subordinate employees.
    b.   Initiating requests for salary adjustments for subordinate employees.
    c.   Authorizing payroll checks for terminated employees.
    d.   Distributing payroll checks to subordinate employees.

5–51. It would be appropriate for the payroll accounting department to be responsible for which of the following functions?
   a. Approval of employee time records.
   b. Maintenance of records of employment, discharges, and pay increases.
   c. Preparation of periodic governmental reports as to employees' earnings and withholding taxes.
   d. Distribution of pay checks to employees.

5–52. Which of the following best describes the principal advantage of the use of flow charts in reviewing internal control?
   a. Standard flow charts are available and can be effectively used for describing most company internal operations.
   b. Flow charts aid in the understanding of the sequence and relationships of activities and documents.
   c. Working papers are not complete unless they include flow charts as well as memoranda on internal control.
   d. Flow-charting is the most efficient means available for summarizing internal control.

5–53. In connection with the study of internal control, an auditor encounters the following flow-charting symbols:

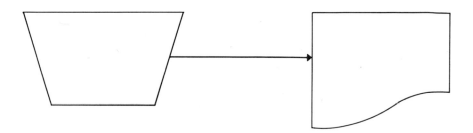

The auditor would conclude that:
   a. A document has been generated by a manual operation.
   b. A master file has been created by a computer operation.
   c. A document has been generated by a computer operation.
   d. A master file has been created by a manual operation.

5–54. In connection with the study and evaluation of internal control during an examination of financial statements, the independent auditor
   a. Gives equal weight to internal accounting and administrative control.
   b. Emphasizes internal administrative control.
   c. Emphasizes the separation of duties of client personnel.
   d. Emphasizes internal accounting control.

5–55. A CPA auditing an electric utility wishes to determine whether all customers are being billed. The CPA's best direction of test is from the
a. Meter department records to the billing (sales) register.
b. Billing (sales) register to the meter department records.
c. Accounts receivable ledger to the billing (sales) register.
d. Billing (sales) register to the accounts receivable ledger.

5–56. An important purpose of the auditor's review of the client's procurement system should be to determine the effectiveness of the procedures to protect against
a. Improper materials handling.
b. Unauthorized persons issuing purchase orders.
c. Mispostings of purchase returns.
d. Excessive shrinkage or spoilage.

5–57. The basic concept of internal accounting control which recognizes that the cost of internal control should *not* exceed the benefits expected to be derived is known as
a. Reasonable assurance.
b. Management responsibility.
c. Limited liability.
d. Management by exception.

5–58. Internal control over cash receipts is weakened when an employee who receives customer mail receipts also
a. Prepares initial cash receipts records.
b. Records credits to individual accounts receivable.
c. Prepares bank deposit slips for all mail receipts.
d. Maintains a petty cash fund.

5–59. Contact with banks for the purpose of opening company bank accounts should normally be the responsibilty of the corporate
a. Board of directors.
b. Treasurer.
c. Controller.
d. Executive committee.

5–60. Effective internal control in a small company that has an insufficient number of employees to permit proper division of responsibilities can *best* be enhanced by
a. Employment of temporary personnel to aid in the separation of duties.
b. Direct participation by the owner of the business in the record-keeping activities of the business.
c. Engaging a CPA to perform monthly "write-up" work.
d. Delegation of full, clear-cut responsibilty to each employee for the functions assigned to each.

5–61.   Which of the following procedures would normally be performed by the auditor when making tests of payroll transactions?

    a.   Interview employees selected in a statistical sample of payroll transactions.

    b.   Trace number of hours worked as shown on payroll to timecards and time reports signed by the foreman.

    c.   Confirm amounts withheld from employees salaries with proper governmental authorities.

    d.   Examine signatures on paid salary checks.

5–62.   Effective internal control over the purchasing of raw materials should usually include all of the following procedures *except*

    a.   Obtaining third-party written quality and quantity reports prior to payment for the raw materials.

    b.   Determining the need for the raw materials prior to preparing the purchase order.

    c.   Systematic reporting of product changes which will affect raw materials.

    d.   Obtaining financial approval prior to making a commitment.

5–63.   Effective internal control over purchases generally can be achieved in a well-planned organizational structure with a separate purchasing department that has

    a.   The ability to prepare payment vouchers based on the information on a vendor's invoice.

    b.   The responsibility of reviewing purchase orders issued by user departments.

    c.   The authority to make purchases of requisitioned materials and services.

    d.   A direct reporting responsibility to the controller of the organization.

5–64.   Which of the following audit procedures is most effective in testing credit sales for understatement?

    a.   Age accounts receivable.

    b.   Confirm accounts receivable.

    c.   Trace sample of initial sales slips through summaries to recorded general ledger sales.

    d.   Trace sample of recorded sales, from general ledger to initial sales slip.

5–65.   Which of the following is the *least* likely reason for the auditor's study and evaluation of internal control?

    a.   To determine the extent of audit testing.

    b.   To serve as a basis for reliance on the controls.

    c.   To determine the nature of transactions.

    d.   To serve as a basis for constructive service suggestions.

5–66. A company has additional temporary funds to invest. The board of directors decided to purchase marketable securities and assigned the future purchase and sale decisions to a responsible financial executive. The best person(s) to make periodic reviews of the investment activity should be
    a. An investment committee of the board of directors.
    b. The chief operating officer.
    c. The corporate controller.
    d. The treasurer.

5–67. Which of the following *best* describes proper internal control over payroll?
    a. The preparation of the payroll must be under the control of the personnel department.
    b. The confidentiality of employee payroll data should be carefully protected to prevent fraud.
    c. The duties of hiring, payroll computation, and payment to employees should be segregated.
    d. The payment of cash to employees should be replaced with payment by checks.

Chapter 5
Discussion/Case Questions

5–68. During your review of cash receipts and cash disbursements of the Joppa Corporation, the following situations arise:
    a. Remittance advices are missing for one of the days selected for testing of cash receipts. Two receipts from unrelated customers in the amounts of $45 and $80, respectively, are recorded in the cash receipts journal, but these amounts do not appear on the validated duplicate deposit slip. The only amount on the deposit slip unaccounted for is a check for $125. What further investigation would you make?
    b. The client could find no support in the paid invoice file for the following disbursements. What type of support should be located and examined in order for you to be satisfied as to the propriety of such disbursements?
        (1) Initial payment of rent for parking spaces.
        (2) Deposit on purchase of real estate.
        (3) Payment of real estate taxes.
        (4) Payment of estimated federal income taxes.
        (5) Initial payment on a conditional sales contract.

(Used with permission of Ernst & Ernst)

5–69. Assume that an independent auditor is examining the financial statements of a small not-for-profit institution. He finds that the accounting system is handled by one individual who writes checks, records and deposits the cash receipts, and records and posts all transaction entries. The independent auditor suggests that at least *one* additional person is needed to provide minimum control for the accounting system.

    a. The directors of the institution agreed to hire an additional person. Suggest how the bookkeeping duties can be divided between the two individuals so as to maximize the probability that assets will be safeguarded and the accounting records will be reliable.

    b. The directors reject the suggestion of hiring another individual. What alternative courses of action can be followed by the independent auditor?

5–70. The following is a list of "tasks" that the independent auditor might ask the internal auditor to perform during the course of an audit. Indicate whether you think each task is a proper or improper function for the internal auditor. Give reasons for your answers.

    a. Footing and crossfooting the cash disbursements journal (the same procedure to be performed on a test basis later by the independent auditor).

    b. Designing a sample plan for examining payroll checks (the examination to be made by the independent auditor).

    c. Testing individual perpetual inventory items based on sample selection made by the independent auditor (this same testing will not be performed by the independent auditor).

    d. Constructing a systems flow chart that will be used later by the independent auditor in his internal control review process.

    e. Providing answers to the independent auditor's internal control questionnaire (the independent auditor will test the records later).

5–71. An audit client proposes that the internal auditors develop all the systems flow charts and give copies to the independent auditors for review purposes. One member of the audit team favors this proposal because of the time saved. Another member of the audit team opposes the idea because he believes that the review of internal control will become a mechanical process.

Required:
Give as many reasons as you can in support of and against the client's proposal. Do you think the proposal should be followed?

5–72. Listed hereafter are eight responsibilities that might be performed by individuals in an accounting system. Assume that four persons are employed for the purpose of handling these eight responsibilities. No more than two jobs will be assigned to one person.

    a. Responsible for the general ledger.

    b. Responsible for the accounts receivable subsidiary ledger.

c. Deposit the cash receipts in the bank.

d. Responsible for the purchases journal.

e. Write checks to creditors.

f. Responsible for the cash disbursements journal.

g. Responsible for the cash receipts journal.

h. Responsible for the payroll register.

Required:

(1) Assuming that no employee will perform more than two jobs, list three combinations of two jobs each for the three employees. List in order of preferability, considering the combinations that will safeguard assets and ensure reliability of the records. Support your listings with reasons.

(2) List the *worst* combination of jobs that you can think of, and support your reasoning.

5–73.   Listed hereafter are weaknesses in internal control that might be found either through the review process or by conducting tests of compliance. For each weakness, indicate its *potential* effect on the nature, timing, and extent of substantive tests in that same area. For example, significant violations of the credit policy might cause the auditor to place more audit effort on the client's method for determining allowance for doubtful accounts.

a. Many large dollar extension errors are found in the sales invoices.

b. The payroll supervisor distributes payroll checks to the employees.

c. The same person writes checks, records entries in the cash disbursement book, and reconciles the bank statement.

d. An unacceptable number of instances were found in which the purchase invoice was not canceled when payment was made to the vendor.

e. Independent receiving reports are not prepared when merchandise from vendors is received.

5–74.   As part of the tests of transactions, the in-charge accountant asks you to perform the payroll test. Factory workers work a 40-hour week, are paid time and a half for over 40 hours per week, and punch timecards which must be approved by the crew foreman at the end of the week (overtime must also be approved by the factory superintendent). Office workers work 35 hours per week, are paid time and a half for over 35 hours per week (no work on Saturdays or Sundays), and prepare weekly time sheets which must be approved by the office manager. The company withholds amounts from pay for savings bonds, charitable contributions, and repayments of loans granted to employees. These are withheld upon authority of signed savings bond deduction forms, contribution forms, and notes stipulating repayment terms, respectively.

Upon hiring an employee, the plant manager (factory) or controller (office) authorizes the initial per-hour pay rate and subsequent increments in the spaces provided on the "employee application form" filed in the individual's personnel file.

Payroll payments are made by check drawn on the imprest payroll bank account.

The schedule on page 215 was prepared by the in-charge accountant, and he has asked you to perform the procedures necessary to ascertain the propriety of all of the payroll information, including distribution of wages to the various factory and office departmental expense accounts. Reproduce the payroll test schedule on the opposite page, list the procedures you would perform, and use tick marks to refer them to the information in the schedule.

Required:

If you found the following situations to exist during your payroll test, indicate your reasons for discussing them with the controller, and include the possible undesirable circumstances which are indicated. List other courses of action that should be taken.

a. Employees' timecards show missed time clock recordings and are written in by employee—no indication of approval by foreman.

b. Employees' timecards show in excess of 8 hours per day, entitling them to overtime pay—not approved by foreman or factory superintendent.

c. During your examination of canceled payroll checks, you notice several double endorsements—the payee on the check and the payroll clerk.

d. In checking out the various deductions from gross pay, you notice several deductions for loan repayments but cannot locate the notes signed by the employees.

(Used with permission of Ernst & Ernst)

5–75. The company's sales department receives all sales orders (mail or phone), prepares the sales order form, and forwards a copy to the credit manager for credit approval. After being approved, the sales order copy is returned to the sales department, where the shipping order copy is sent to the shipping department for assembling the order. After the order is prepared for shipping, a prenumbered bill of lading is prepared for the shipment.

Upon shipment of the order, the shipping order copy is returned with a copy of the bill of lading to the sales department, where it is matched with the copy in the open sales order file. All the documents are forwarded to the accounting department.

From the information supplied by the sales department, the prenumbered sales invoices are prepared in the accounting department. Two copies are mailed to the customer, and the invoice information is entered in detail in the sales journal. Invoice amounts are posted to the customers' accounts receivable cards at the same time they are entered in the sales journal. Totals from the sales journal are posted to the general ledger at month-end.

Required:

a. List the procedures you would perform to test the recorded sales.

PAYROLL TEST
JOPPA CORPORATION

Employee	Dept. Charged	No. of Exempt.	Hours Worked	Hourly Rate	Gross Wages	Fed. W/H	FICA W/H	State W/H	Other Deductions			Net Pay	Check No.
									Bonds	Charity	Loan		
R. Jackson	Paint	8	46	$2.45	$120.08	$ .30	$ 5.76	$-0-	$ 5.00	$1.00	$ -0-	$108.02	110
A. Palmer	Varnish	1	73	2.60	232.70	34.00	11.17	3.58	10.00	1.00	20.00	152.95	136
B. Elliott	Receiving	3	52	2.70	156.60	15.50	7.52	3.12	-0-	2.00	5.00	123.46	148
D. Howard	Shipping	3	40	2.75	110.00	8.70	5.28	1.27	5.00	-0-	-0-	89.75	166
D. Stacey	Office	5	50	2.87	165.03	13.10	7.92	2.77	5.00	1.50	25.00	109.74	201

b. Since this company maintains a file of unfilled sales orders in the sales department, list the recommendations you would give to the client for review of the file and pursuit of old unfilled sales orders.

(Used with permission of Ernst & Ernst)

5–76.   You were engaged to examine the financial statements of Blank City Newspapers, Inc., for the year ended December 31. The Company publishes a newspaper with a daily circulation of approximately 65,000.

During your examination of accounts receivable, certain unusual transactions were noted and discussed with Company executives. When questioned about the transactions, the cashier admitted to a defalcation and gave Company executives 298 remittance advices consisting of either the top half of the newspaper's monthly statement, the voucher portion of the customers' checks, or the cashier's memo of payment. The cashier said these remittance advices (amounting to $74,437.38) were the record of payments made by customers for which the checks had been deposited in the bank, but no credit had been given in the customers' accounts. She had diverted some previous checks received by mail to an unauthorized bank account that she established in the company's name using a forged corporate resolution. Funds could be withdrawn upon her signature.

The Company was shocked because the cashier was an old and faithful employee who was never sick and so interested in her work that she never took a vacation.

To conceal the defalcation, the cashier was lapping accounts receivable. As the total misappropriated grew to a sizable amount, the process became so involved that several customers were receiving credit up to two weeks late.

A review of the Company's accounting procedures and internal controls for cash receipts disclosed the following:

The cashier reported directly to the controller.

Mail was opened by another employee of the Company. This employee did not prepare a record of the money and checks received before he distributed them.

All mail and over-the-counter receipts (cash sales and payments on accounts) were given to the cashier. Receipts included approximately $100 to $200 daily in currency.

The cashier prepared a daily report showing customer and amount of payment. She gave this report, together with remittance advices, to the accounts receivable department for posting to the customers' accounts receivable ledger cards.

The cashier prepared the bank deposit and took the deposit to the bank. She obtained an authenticated deposit slip from the bank.

The cashier gave the authenticated deposit slip to the controller, who compared the total shown thereon with the total shown on the daily cash receipts report. These totals were always in agreement.

Your audit procedures included the confirmation of accounts receivable as of December 31. Most of the replies to the requests for confirmation were returned reporting no difference. However, the following replies were received from various customers:

"Less our ch #54818—$6,358.04, dated 12/23"

"Corrected statement received 1/8"

"$5,902.77 paid on our check #235043 dated 12/18"

"Payment on December 19th—$4,626.15 covering November"

Required:
a. List the weaknesses noted in the Company's accounting procedures and internal controls for cash receipts.
b. List the audit procedures that might have disclosed the defalcation by the cashier.
c. What should you do when you discover unusual items that might indicate a defalcation by one of the Company's employees?
d. What alternatives would you recommend management consider to prevent a recurrence of this situation?

(Used with permission of Ernst & Ernst)

5–77. Certain audit tests consist of tracing dollar amounts from one document or record to another. The purpose of the audit test depends on the direction of the tracing. For each of the following sets of audit tests, indicate the difference in the purpose of each test.
a. Tracing amounts from the sales invoice to the accounts receivable subsidiary ledger, and vice versa.
b. Tracing amounts from the purchase invoice to the purchases journal, and vice versa.
c. Tracing amounts from the receipts listing to the duplicate deposit slip, and vice versa.
d. Tracing amounts from the payroll checks to the payroll register, and vice versa.
e. Tracing amounts from receiving reports to purchase invoices, and vice versa.

5–78. A feature of internal control is considered weak if errors or irregularities could occur as a result of the feature's existence. For each of the following situations, indicate the error or irregularity that could occur.

    a. Furnishing copies of fully completed purchase orders to the clerks that count incoming merchandise and fill out receiving reports.

    b. Allowing the payroll supervisor to disburse payroll checks.

    c. No reconciliation between the accounts receivable subsidiary ledger and the general ledger control account.

    d. No credit check made before sales are made to customers.

    e. Cancellation stamps are not placed on purchase invoices.

    f. Payments are made by vendor monthly statements rather than vendor invoices.

    g. Employees are paid in cash.

    h. The person that handles cash also approves accounts receivable write-offs.

    i. Extensions on sales invoices are not checked by another person.

    j. The cashier reconciles the bank statement.

5–79. In a good system of internal control, the following functions are separated: (1) authorizing a transaction, (2) recording a transaction, (3) maintaining custody of assets that result from a transaction, and (4) comparing assets with the related amounts recorded in the accounting records. In each of the following situations, indicate which functions are combined.

    a. The payroll supervisor disburses the payroll checks.

    b. The general ledger clerk keeps the accounts receivable subsidiary ledger.

    c. The cashier keeps the cash receipts journal.

    d. The accounts payable supervisor writes and mails the checks.

    e. The storeroom clerks count the physical inventory.

    f. The cashier reconciles the bank statements.

5–80. Examine the flow chart of raw material purchases on page 219 and do the following.

    a. Explain the function that is performed for each of the lettered symbols A through E.

    b. List the weaknesses that you see in the system as depicted in the flow chart.

    c. List the features of good internal control that should exist in a purchases system of this type but are not shown in the flow chart (for example, the purchase requisition should be approved).

(AICPA adapted)

5–81. You have been engaged by the management of Alden, Inc., to review its internal control over the purchase, receipt, storage, and issue of raw materials. You have prepared the following comments to describe Alden's procedures.

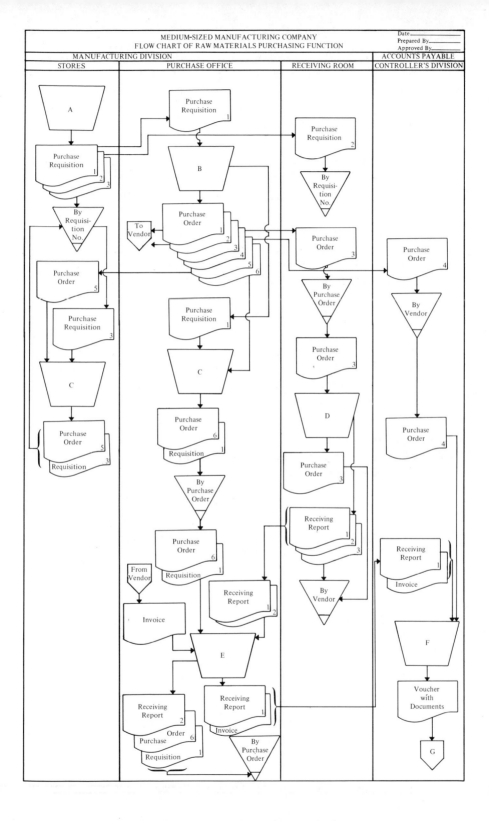

MEDIUM-SIZED MANUFACTURING COMPANY
FLOW CHART OF RAW MATERIALS PURCHASING FUNCTION

Date___
Prepared By___
Approved By___

MANUFACTURING DIVISION

ACCOUNTS PAYABLE

STORES | PURCHASE OFFICE | RECEIVING ROOM | CONTROLLER'S DIVISION

Raw materials, which consist mainly of high-cost electronic components, are kept in a locked storeroom. Storeroom personnel include a supervisor and four clerks. All are well trained, competent, and adequately bonded. Raw materials are removed from the storeroom only upon written or oral authorization of one of the production foremen.

There are no perpetual inventory records; hence, the storeroom clerks do not keep records of goods received or issued. To compensate for the lack of perpetual records, a physical inventory count is taken monthly by the storeroom clerks, who are well supervised. Appropriate procedures are followed in making the inventory count.

After the physical count, the storeroom supervisor matches quantities counted against a predetermined reorder level. If the count for a given part is below the reorder level, the supervisor enters the part number on a materials-requisition list and sends this list to the accounts payable clerk. The accounts payable clerk prepares a purchase order for a predetermined reorder quantity for each part and mails the purchase order to the vendor from whom the part was last purchased.

When ordered materials arrive at Alden, they are received by the storeroom clerks. The clerks count the merchandise and agree the counts to the shipper's bill of lading. All vendors' bills of lading are initialed, dated, and filed in the storeroom to serve as receiving reports.

Required:
Describe the weaknesses in internal control and recommend improvements of Alden's procedures for the purchase, receipt, storage, and issue of raw materials. Organize your answer sheet as follows:

Weaknesses	Recommended Improvements

(AICPA adapted)

Chapter 6

The Application of Statistical Sampling to the Study and Evaluation of Internal Control[1]

Learning Objectives *After reading and studying the material in this chapter, the student should*

Know the definition of the statistical terms associated with estimation sampling for attributes.

Know the difference between testing compliance by use of judgment sampling and by use of statistical sampling.

Know the steps involved in using estimation sampling for attributes.

Be able to read the AICPA attribute and discovery sampling tables, determine appropriate sample sizes, and evaluate sample results.

Have acquired an appreciation of the relationship between statistical sampling and the study and evaluation of internal control.

Know the steps involved in using discovery sampling.

---

[1]Users may wish to read the appendix to Chapter 13 before reading this chapter. The appendix is a primer on statistical sampling.

Basically, this chapter is a continuation of the material in Chapter 5, the concepts underlying the study and evaluation of internal control. Included here is a set of statistical techniques designed for tests of compliance (one phase of the study and evaluation process).

Recall that auditors generally ascertain how a system is *supposed* to function by origination and/or examination of various types of documentation. AU Section 320A.22 of *AICPA Professional Standards,* Volume 1 *(SAS No. 1)* contains the following comments about the relationship of statistical sampling to this phase of the audit.

> The auditor's knowledge of the procedures prescribed by the client ordinarily is obtained by inquiry or reference to written instructions, and his understanding of their function and limitations is based on his training, experience and judgment. On this basis, the auditor makes a preliminary evaluation of the effectiveness of the prescribed procedures, assuming that compliance with them is satisfactory. Statistical sampling is not applicable to this phase of the evaluation.

Auditors then determine whether the system is functioning as prescribed by conducting tests of compliance. AU Section 320A.22–.23 *(SAS No. 1)* also contains comments on the role of statistical sampling in this phase.

> As to the second phase, statistical sampling may be applied to test compliance with internal control procedures that leave an audit trail in the form of documentary evidence of compliance. . . . On the other hand, statistical sampling generally is not applicable to tests of compliance with internal control procedures that depend primarily on appropriate segregation of duties and leave no audit trail of documentary evidence in this respect.

It should be established from the outset, then, that the use of statistical sampling in the study and evaluation of internal control is limited, but potentially effective. Therefore, in the context of the application described above, we will explain the techniques for and the advantages of using certain types of statistical sampling to conduct tests of compliance on procedures that leave a documentary trail of evidence. To provide continuity with Chapter 5, the same system of cash receipts is used in this chapter and the same tests are illustrated.

The remainder of the chapter contains discussions of (1) the general concepts of statistical sampling as it applies to the study and evaluation of internal control, (2) methodology of estimation sampling for attributes, and (3) an illustration of estimation sampling for attributes. The appendix contains an explanation of discovery sampling.

Auditor's Inherent Sampling Problems

If it were practicable for auditors to determine compliance with a system by checking all documents, postings, and so on, there would be no need to consider the use of statistical sampling, because no testing would be necessary. But a complete examination is not possible for most audits, and tests must be conducted by the best sampling techniques applicable in the circumstances.

Auditors (or anyone else taking samples) must cope with two uncertainties when making tests.

1. Except in rare situations, the characteristics of a sample will differ, in some respect, from the characteristics of a population, and in the case of audit testing, there is a need to know how much difference is probable. For example, assume that 1000 sales invoices are processed during the year, and auditors select a sample of 200 to check the pricing accuracy. If there are 20 pricing errors in the population of 1000 (a 2% error rate), the auditors ideally would find 4 errors in their sample of 200 (a 2% error rate). Because this is unlikely, auditors can only hope that the difference between the sample and population error rates is predictable. Otherwise, they will be unable to draw accurate inferences about the population from their sample. Generally, auditors are satisfied if they have a reasonably good idea about the differences between the sample and population characteristics, and if this difference is tolerable. Statistically, the range within which the sample result is expected to be accurate is called *precision*.

2. The other uncertainty is the question of how sure auditors can be that the sample results are within a given precision range of the population characteristics. Again, auditors are satisfied if they can predict and control this uncertainty. Statistically, the percentage of assurance that sample results are within a given precision range of the population characteristics is called *reliability* or *confidence level*.

Advantages of Statistical Sampling

If auditors are concerned about the *degree* of compliance, and if there is a trail of documentary evidence, statistical sampling is appropriate. By using this method, auditors are not giving up their judgment; they are merely refining it. No longer do they state that they are reasonably certain that the actual percentage of pricing errors in the population is within a *tolerable* range of their estimate. Instead, the use of statistical sampling might result in such a statement as: "The auditors have *90%* reliability that the actual percentage of pricing errors in the population is between *3%* and *7%*." If the best estimate of the population error rate is 5%, the precision is ±2%, the precision range is 4% (3% to 7%), and the upper limit of the precision range is 7%. It is this upper range that we will concentrate on in the chapter.

The major advantage of statistical sampling, then, is that its use enables auditors to place scientifically determinable limits on the uncertainties and risks that are inherent in the testing process.

In line with this philosophy, the AICPA has given its sanction to statistical sampling and indicates that its use is permitted (and optional) under generally accepted auditing standards. AU Section 320A.04 *(SAS No. 1)* states:

> In determining the extent of a particular audit test and the method of selecting items to be examined, the auditor might consider using statistical sampling techniques which have been found to be advantageous in certain instances. The use of statistical sampling does not reduce the use of judgment by the auditor but provides certain statistical measurements as to the results of audit tests, which measurements may not otherwise be available.

In reality, statistical sampling aids in only two areas of compliance testing. These are (1) the determination of an appropriate sample size, given the auditor's judgment on precision, reliability, and other factors, and (2) a *quantitative* evaluation of the sample results to which the auditor applies additional judgments on the audit significance of such results. Therefore, the auditor applies judgment before *and* after the use of statistical sampling.

The Application to Tests of Compliance

Despite the scientific advantages of statistical sampling, auditors tend to limit the particular tests of compliance to which these methods are applied. For instance, auditors would not consider it appropriate to use statistical sampling if their prime interest were to determine that a certain feature of internal control merely *exists* in practice. The observance of a proper segregation of duties is an example. Also, in certain situations, auditors may be satisfied if they see that proper initials or signatures are placed on one or a few documents.

In some of the many tests of compliance that normally are conducted, however, auditors are likely to be concerned about the *percentage* of errors. In a given situation, an auditor may have reason to believe that the failure of the company to place a cancellation stamp on paid creditors' invoices is serious if it occurs in a certain percentage of the cases (6%, for example). Any percentage of errors above this maximum is a signal that compliance with this control feature is not adequate and that double payment of invoices is probable.

In this case, the auditor will want to know with a certain percentage of assurance (reliability) if the true error rate (non-cancellation of invoices) exceeds 6% (the upper limit of the precision). By using certain statistical sampling techniques, the auditor can obtain a sample size that will furnish this information.

Also, certain tests have a dual characteristic in that they are designed both to ascertain compliance with the system of internal control and to provide evidence as to the validity of dollar figures in the firm's financial statements. Some examples are the testing of footings and extensions in books of original entry, the testing of postings from one set of records to another, and certain types of accounts receivable confirmation procedures. It is likely that auditors will wish to satisfy themselves as to the *degree* of compliance (and, in some cases, a dollar estimation will be desired, as discussed in Chapter 13).[2]

Reliability and Precision Levels

The AICPA Position

If it is true that the levels of reliability and precision are matters of auditing judgment, then it follows that there should be some basis upon which auditors form these judgments. The AICPA has chosen to provide only general guidelines and to leave the choice of specific percentages to individual auditors. In accordance with this viewpoint, the following comments are contained in AU Section 320B.18 *(SAS No. 1)*:

> In considering the precision desired for compliance tests, it is important to recognize the relationship of procedural deviations to (a) the accounting records being audited, (b) any related accounting control procedures, and (c) the purpose of the auditor's evaluation.

It is also pointed out that certain types of compliance errors may have little or no effect on the dollar amount of the accounting records, whereas others may have considerable effect. Examples of the former type of error are missing cancellation stamps on invoices and missing signature approvals on time records. Examples of the latter are an improper amount on a check and an incorrect posting. The implication is that the tolerable precision limit can be higher in tests for errors that may not affect the dollar amount of the records.

However, the AICPA, as well as some CPA firms, seems to favor high reliability levels for compliance testing. This opinion also is expressed in AU Section 320B.23 *(SAS No. 1)*:

> In contrast, the reliability level is related to the probability that the auditor's conclusion based on this precision will be correct. Thus the choice of reliability level establishes the level of confidence the auditor desires; it is the complement of the level of sampling risk that he is willing to assume that his conclusion will be incorrect.

On the basis of an example of a 95% reliability level and a desired upper precision limit of 6%, AU Section 320B.24 concludes that the following meaning could be construed in a test of compliance.

---

[2]For two distinct viewpoints on tests of compliance, see: Howard F. Stettler, "Some Observations on Statistical Sampling in Auditing," *The Journal of Accountancy,* April 1966, pp. 55–60; D. R. Carmichael, "Tests of Transactions—Statistical and Otherwise," *The Journal of Accountancy,* February 1968, pp. 36–40.

If the actual but unknown rate of procedural deviations in the population exceeds 6 percent, at least 95 percent of all possible samples of the same size that could be selected from the same population would result in upper precision limits that would exceed 6 percent. Therefore, at least 95 percent of such samples would protect the auditor against the risk of overevaluating the degree of compliance with the procedures.

As indicated hereafter, this protection guards against an acceptance error (the error of accepting compliance when one should not) and is considered vitally important to control. For this reason, auditors prefer to keep the reliability level fairly high on tests of compliance.

Other Ideas

There is a school of thought among some auditors that the reliabilty percentage needed in compliance tests is, in part, a function of the preliminary evaluation of internal control obtained through examination of the documentation. According to this viewpoint, a very favorable preliminary evaluation calls for a *large* reliability percentage in the subsequent tests of compliance (the opposite of what one might conclude intuitively). The reasoning behind this viewpoint is that a strong system of internal control makes it possible to do less subsequent testing on financial statement balances. But to be reasonably sure that the modification of future tests is a valid step, auditors need a high reliability on tests of compliance because their purpose is to prove the internal control system is *really* as strong as prescribed.

By the same line of reasoning, an average system of internal control calls for a lower reliability level for compliance tests because there is less to be verified. An extremely weak system does not call for any compliance testing, because there it almost nothing to verify; future tests must be large, anyway.

Sampling in an Age of Litigation

Although no landmark legal case has been tried with statistical sampling as a central issue, there is potential for litigation. When auditors choose to use statistical sampling as a tool for testing the validity of any important element such as internal control, they tend to be cautious about judgments that are expressed in definite fashion (reliability of 95%, precision of 6%). This tendency may account for conservative reliability levels that guard against the error of accepting internal control compliance as satisfactory when it is not.

As pointed out in Chapter 1, auditors document their evidence-gathering techniques in working papers; statistical sampling judgments are no exception. Later, the auditors may be required to defend these judgments in court. Thus, more care is needed in designing sample plans than was necessary when such methods were not used. Perhaps this necessary caution is another advantage of statistical sampling as an audit tool.

Estimation Sampling for Attributes

With the foregoing discussion as a conceptual background, we can now turn to a description of estimation sampling for attributes and its relationship to an audit situation. The following steps are used.

1. Determine the sampling plan. This consists of

   A. Determining the audit procedure to use in the sample.

   B. Determining the population and defining an error.

   C. Determining the sampling methodology (usually random or systematic sampling).

2. Determine the sample size.

3. Take the sample and quantitatively determine the sample results.

4. Evaluate the quantitative sample results.

The Sampling Plan

The Audit Procedure

Determination of the audit procedure logically is the first step because all of the other variables in the attribute-sampling application depend on this selection. The population characteristics, the assessment of the significance of the test, and the resulting sample size of a verification of invoice approvals will differ from a test of invoice postings to a journal.

There are a number of tests of compliance that are applicable to every part of an internal control system. The auditor should apply attribute sampling only to those tests that are suitable for this type of statistical sampling application, e.g., those that leave a trail of documentary evidence. In general, a test of compliance will have the necessary characteristics if the mechanics of conducting the test require (1) inspection of signatures, initials, or cancellation stamps, (2) checking footings or extensions, or (3) tracing dates, names, amounts, etc. from one document to another.

Sometimes two or more attributes can be checked with the same audit tests. For example, amounts on checks can be traced to a journal, and the endorsements can be examined on the same set of checks. However, if two or more attributes are examined on the same records, the auditor should regard each attribute as a separate test with separate criteria and different sample sizes. The auditor should choose the largest sample of multiple attribute documents if the sample sizes for each attribute differ.

The Determination of the Population and the Definition of an Error

Before auditors proceed with the sampling, they must decide what constitutes the population and their definition of an error. This appears to be a simple task, but an illustration shows why it is not. Assume that auditors plan to test the extensions on 400 invoices and that each invoice contains an average of 5 lines. The population can be defined in one of two ways, either as 400 invoices or as 2000 lines. If the former criterion is used, an error consists of one or more extension mistakes made on 1 invoice. If the latter criterion is employed, an error consists of one extension mistake made on 1 line; several errors could appear on 1 invoice.

Auditors may use either definition, provided that it is consistent with their sampling procedures. If it is probable that a clerk would incorrectly extend only one line of an invoice, the difference between the two techniques might be insignificant.

Sampling Methodology—Random Numbers

Attribute-sampling methods are techniques for estimating the number of errors in a population. They are designed under the assumption that the selection of items will be made in a random fashion. As pointed out in Chapter 5, "unrestricted random selection" simply means that every element in the population has an equal chance of being taken. In the population of 200 lines, each one has a $\frac{1}{200}$ chance of appearing in the sample. If the method is to be scientifically defensible, a tested table of random numbers should be used. Such a table is illustrated hereafter. The array of numbers is "random" without any discernible pattern. The user should discover this by reading several sets of numbers.

The table can be read vertically, horizontally, or diagonally, and the user can pick the left-hand, middle, or right-hand portion of the number. These methods are illustrated below in a partial reproduction of the random number table.

		Can Be Read							
		Vertically			*Horizontally*		*Diagonally*		
10480	15011	104	048	480	104	150	104	048	480
22368	46573	223	236	368	048	501	465	657	573
24130	48360	241	413	130	480	011			

However, care must be taken to discard non-usable numbers read from the table. For example, in a population of 200 lines, three-digit sets should be used, and any three-digit number over 200 should be discarded, because it exceeds the population total. Assume that the first number selected from the table is 192; this

Col. Line	(1)	(2)	(3)	(4)	(5)	(6)	(7)	(8)	(9)	(10)	(11)	(12)	(13)	(14)
1	10480	15011	01536	02011	81647	91646	69179	14194	62590	36207	20969	99570	91291	90700
2	22368	46573	25595	85393	30995	89198	27982	53402	93965	34095	52666	19174	39615	99505
3	24130	48360	22527	97265	76393	64809	15179	24830	49340	32081	30680	19655	63348	58629
4	42167	93093	06243	61680	07856	16376	39440	53537	71341	57004	00849	74917	97758	16379
5	37570	39975	81837	16656	06121	91782	60468	81305	49684	60672	14110	06927	01263	54613
6	77921	06907	11008	42751	27756	53498	18602	70659	90655	15053	21916	81825	44394	42880
7	99562	72905	56420	69994	98872	31016	71194	18738	44013	48840	63213	21069	10634	12952
8	96301	91977	05463	07972	18876	20922	94595	56869	69014	60045	18425	84903	42508	32307
9	89579	14342	63661	10281	17453	18103	57740	84378	25331	12566	58678	44947	05585	56941
10	85475	36857	53342	53988	53060	59533	38867	62300	08158	17983	16439	11458	18593	64952
11	28918	69578	88231	33276	70997	79936	56865	05859	90106	31595	01547	85590	91610	78188
12	63553	40961	48235	03427	49626	69445	18663	72695	52180	20847	12234	90511	33703	90322
13	09429	93969	52636	92737	88974	33488	36320	17617	30015	08272	84115	27156	30613	74952
14	10365	61129	87529	85689	48237	52267	67689	93394	01511	26358	85104	20285	29975	89868
15	07119	97336	71048	08178	77233	13916	47564	81056	97735	85977	29372	74461	28551	90707
16	51085	12765	51821	51259	77452	16308	60756	92144	49442	53900	70960	63990	75601	40719
17	02368	21382	52404	60268	89368	19885	55322	44819	01188	63255	64835	44919	05944	55157
18	01011	54092	33362	94904	31273	04146	18594	29852	71585	85030	51132	01915	92747	64951
19	52162	53916	46369	58586	23216	14513	83149	98736	23495	64350	94738	17752	35156	35749
20	07056	97628	33787	09998	42698	06691	76988	13602	51851	46104	88916	19509	25625	58104
21	48663	91245	85828	14346	09172	30168	90229	04734	59193	22178	30421	61666	99904	32812
22	54164	58492	22421	74103	47070	25306	76468	26384	58151	06646	21524	15227	96909	44592
23	32639	32363	05597	24200	13363	38005	94342	28728	35806	06912	17012	64161	18296	22851
24	29334	27001	87637	87308	58731	00256	45834	15398	46557	41135	10367	07684	36188	18510
25	02488	33062	28834	07351	19731	92420	60952	61280	50001	67658	32586	86679	50720	94953
26	81525	72295	04839	96423	24878	82651	66566	14778	76797	14780	13300	87074	79666	95725
27	29676	20591	68086	26432	46901	20849	89768	81536	86645	12659	92259	57102	80428	25280
28	00742	57392	39064	66432	84673	40027	32832	61362	98947	96067	64760	64584	96096	98253
29	05366	04213	25669	26422	44407	44048	37937	63904	45766	66134	75470	66520	34693	90449
30	91921	26418	64117	94305	26766	25940	39972	22209	71500	64568	91402	42416	07844	69618
31	00582	04711	87917	77341	42206	35126	74087	99547	81817	42607	43808	76655	62028	76630
32	00725	69884	62797	56170	86324	88072	76222	36086	84637	93161	76038	65855	77919	88006
33	69011	65795	95876	55293	18988	27354	26575	08625	40801	59920	29841	80150	12777	48501
34	25976	57948	29888	80604	67917	48708	18912	82271	65424	69774	33611	54262	85963	03547
35	09763	83473	73577	12908	30883	18317	28290	35797	05998	41688	34952	37888	38917	80050
36	91567	42595	29758	30134	04024	86385	29880	99730	55536	84855	29080	09250	79656	73211
37	17955	56349	90999	49127	20044	59931	06115	20542	18059	02008	73708	83517	36103	42791
38	46503	18584	18845	49618	02304	51038	20655	58727	28168	15475	56942	53389	20562	87338
39	92157	89634	94824	78171	84610	82834	09922	25417	44137	48413	25555	21246	35509	20468
40	14577	62765	35605	81263	39667	47358	56873	56307	61607	49518	89656	20103	77490	18062
41	98427	07523	33362	64270	01638	92477	66969	98420	04880	45585	46565	04102	46880	45709
42	34914	63976	88720	82765	34476	17032	87589	40836	32427	70002	70663	88863	77775	69348
43	70060	28277	39475	46473	23219	53416	94970	25832	69975	94884	19661	72828	00102	66794
44	53976	54914	06990	67245	68350	82948	11398	42878	80287	88267	47363	46634	06541	97809
45	76072	29515	40980	07391	58745	25774	22987	80059	39911	96189	41151	14222	60697	59583
46	90725	52210	83974	29992	65831	38857	50490	83765	55657	14361	31720	57375	56228	41546
47	64364	67412	33339	31926	14883	24413	59744	92351	97473	89286	38931	04110	23726	51900
48	08962	00358	31662	25388	61642	34072	81249	35648	56891	69352	48373	45578	78547	81788
49	95012	68379	93526	70765	10592	04542	76463	54328	02349	17247	28865	14777	62730	92277
50	15664	10493	20492	38391	91132	21999	59516	81652	27195	48223	46751	22923	32261	85653

Source: Interstate Commerce Commission, *Table of 105,000 Random Decimal Digits* (Washington, D.C.: Bureau of Transport, Economics and Statistics, 1949).

Figure 6.1   Table of Random Numbers

means that the 192nd line of the array of population items is selected. If the next number is 419, it should be omitted. This process continues until the sample is chosen. An alternate procedure is to subtract the population total, or a multiple of this total, from three-digit numbers that are too high. For example, if the three-digit number 220 is chosen, it becomes 20 (220 minus 200). If 419 is chosen, it becomes 19 (419 minus 400). By this method, no three-digit number is discarded.

What if a number is selected again? If the table for this type of test calls for "sampling with replacement" (replacing the item before continuing with the sample), the number should be reused. If the table calls for "sampling without replacement" (not replacing the item before continuing with the sample), the number should be ignored.[3]

Sampling Methodology—Systematic Sampling

Simple random sampling is not the only method available for selecting items to be examined. As indicated in Chapter 5, systematic sampling can be used and every *n*th element selected. If such a technique is used, however, the auditor should ascertain that the population from which the sample is taken does not contain unusual items that appear at fixed intervals.

For instance, assume that during certain days of each week, invoices are processed by a part-time employee who makes an unusually large number of mistakes. As a result, approximately every 10th invoice contains a clerical error. If the auditors selected every 10th invoice as their sampling interval, the sampling results would not provide a good estimate of the error condition in the population.

Systematic sampling is easy to use and might provide results comparable to those gained from simple random sampling. But care should be taken by the auditor to avoid using the method with populations that might contain non-random errors.

A certain type of systematic sampling can be used, however, that will avoid the problem of fixed internal errors and will retain some practical advantages as well. This technique is called systematic random sampling. Rather than select every 10th item for examination, the auditor can make simple random selections in such a way that the sampling intervals will *average* every 10th item.

As an example, assume that the auditors wish to select a sample of 100 from a population of 1000. The sampling process should produce an average interval of 10. By drawing random numbers between 1 and 20 and letting the selected number represent the sampling interval, the auditor's objective can be accomplished. The average interval is 10 (approximately), and some randomness is introduced into the selection process.

Refer to the table of random numbers for an illustration.

---

[3]The analogy of drawing marbles from a jar is often used to illustrate these terms. Assume that a jar contains ten marbles, and one is drawn out. If sampling with replacement is used, the marble is replaced in the jar before another sample is drawn, so ten marbles are always left in the jar. If sampling without replacement is used, the marble is not replaced, and the number of marbles in the jar is lowered when a subsequent sample is drawn.

Choose two-digit numbers between 1 and 20. Reading the two left-hand digits in Column 1, we have a partial selection of (read top to bottom and left to right):

10	01	17	06
09	07	14	14
10	02	08	12
07	05	15	20
02	09	15	04[4]

Thus, the first item chosen is the 10th item in the population. The second item is the 19th item in the population, etc.

Systematic random sampling is advantageous if the selection of a simple random sample would otherwise be performed manually. But with the advent of computer programs that select random numbers and sort them in desired order, much of the advantage of systematic random sampling has diminished. Also, a simple random sample furnishes a better basis for making a valid statistical evaluation.

Determination of the Sample Size

In estimation sampling for attributes, three variables must be estimated or known before an initial sample size is selected.

1. First, the auditors must make a judgment on the desired reliability level. This decision should be based on the maximum risk they are willing to take that in this area compliance will be accepted as satisfactory when the error rate is too high. A 5% risk calls for 95% reliability level on a given test of compliance with the preliminary evaluation of internal control. A stated percentage of risk does not mean that auditors are *guessing* at their conclusions. They merely are quantifying their degrees of assurance and risk, both of which are inherent in the auditing process.

2. Next, the auditors need to make a judgment on the precision to be allowed. For example, they may estimate that 4% of the documents in the population are canceled improperly, and this might be the same error percentage that is found in the sample. But it is recognized that the true error percentage is probably different from this estimate. If the auditors

---

[4]The average interval for a large selection should be about 10. For a small selection, the average is slightly different.

allow a precision of ±2%, they are stating that compliance will be considered satisfactory even though the true error rate interval is from 2% to 6% (a total range of 4%).[5] Because it is assumed that auditors are primarily interested in the upper precision limit (or how high the error rate may go), the tables are constructed on a "one-tailed" basis without regard to the lower precision limit.

3. Finally, the auditors need to estimate the actual error rate in the population. Several methods can be used.

(a) Auditors may know from specific experience what the rate, or at least a range of rates, is likely to be.

(b) Auditors may estimate an error rate on the basis of their general audit knowledge and the nature of the test. With a sound system of internal control, it is logical to assume that the "critical" parts of the system contain a lower percentage of errors. For example, the error rate for footings and extensions should be lower than the error rate for failure to cancel paid vendor invoices.

(c) The error rate found in a preliminary sample of approximately 50 yields a usable estimate of the error rate of the population. A sample of 50 may appear to be small, but in statistical terms it is a large sample.

The Use of Attribute-Sampling Tables to Determine Sample Size

Many sets of tables are available from which one can derive a sample size based on the three items of information discussed in the previous section. The tables illustrated here (Table 6.1 set first, and Table 6.2 set next) are merely examples of the types available. They are constructed on the assumption of an infinite population, but either sampling with replacement or sampling without replacement can be used. If the latter method is employed, the attained precision percentages are slightly different from those shown in the tables.

The Table 6.1 set is a graphic display of the relationships among the anticipated error rate, the upper precision limit, and the sample size. There is a different Table 6.1 graph for each reliability level. The dark, heavy lines drawn downward and to the right are the routes that users follow. Assume, for example, that the 90% graph is selected and a 2% error rate in the population is estimated. Locate the 2% column at the top of the graph and follow the heavy line down to the upper precision limit percentage that is tolerable for the particular type of test. Move left across that row to find the sample size. If a 2% anticipated error rate and a 6% upper precision limit are selected, the necessary sample size is slightly more than 75.

---

[5]Often the range is asymmetrical (4% + 2% − 3%) because of the use of the binomial distribution.

RELIABILITY LEVEL – 90 PERCENT

Anticipated Rate of Occurrence

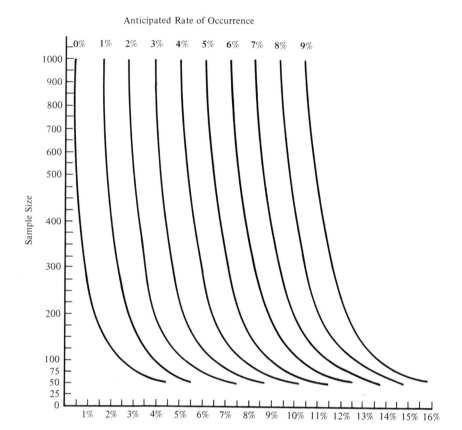

Table 6.1A Attribute Sampling—Determination of Sample Size—
Graphical Form

The 95% and 99% graphs in the Table 6.1 set are used in the same manner. A 2% occurrence rate and a 6% upper precision limit in the 95% graph require a sample size of slightly more than 100. The same occurrence rate and upper precision limit in the 99% graph correspond with a sample size of slightly more than 175. (If there is doubt about the intersect of the heavy black line with the sample size row, choose the larger and more conservative size.)

One distinct advantage of the Table 6.1 set over other types of attribute-sampling tables is the depiction of certain relationships in easy-to-read graphic form. The rightward curve of the heavy black lines indicates that with a given sample size the upper precision limit is always higher than the anticipated rate of

RELIABILITY LEVEL – 95 PERCENT

Anticipated Rate of Occurrence

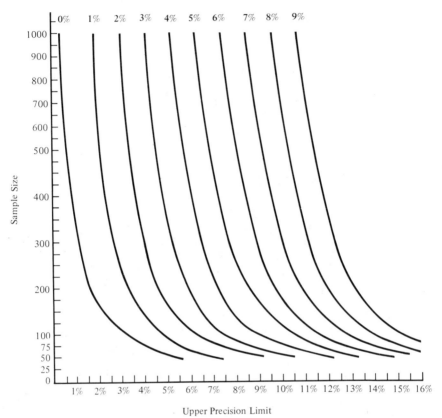

Table 6.1B Attribute Sampling—Determination of Sample Size—Graphical Form

occurrence. (One can easily test this feature; the sample size row of 100 for the 90% graph intersects the heavy black line at a 4% rate of occurrence and 8% precision, at a 7% rate of occurrence and 11½% precision, etc.)

This type of table is simply a graphic display of the concept of precision. If a 4% error rate is anticipated, and a sample size of 100 also results in a 4% error rate, one cannot say that there is a 90% probability that the true error rate in the population is 4%. The limit of the statement is that there is a 90% probability that the true error rate does not exceed 8%.

The Table 6.2 set serves the same purpose as the Table 6.1 set, but its form is discrete and numerical rather than continuous and graphic. The sample-size

RELIABILITY LEVEL – 99 PERCENT

Anticipated Rate of Occurrence

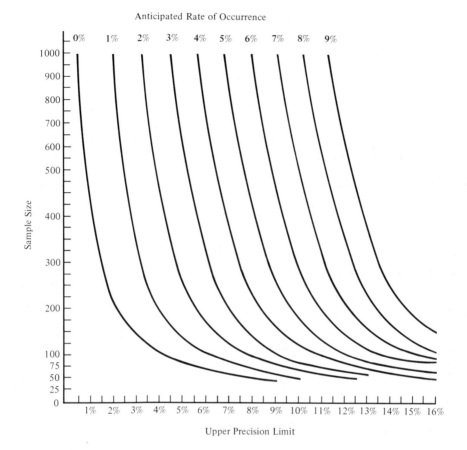

Table 6.1C   Attribute Sampling—Determination of Sample Size—
Graphical Form

column has numerical gaps, that is, it shows sizes of 50, 100, 150, and so on.
Faced with a choice of numbers, the user should select the higher or more
conservative one.

For example, assume that auditors wish to use the 90% table with an
estimated occurrence rate of 2% and an upper precision limit of 6%. The table is
read by first finding the occurrence rate column and reading the numbers down
that column. These numbers in each occurrence rate column represent the upper
precision limit percentages. After finding 6% or the next lowest percentage under
6%, the user reads left across the row and finds the sample size. The next lowest
precision percentage below 6% in the 2% occurrence rate column is 5.2%, which

RELIABILITY LEVEL – 90.0 PERCENT

Sample Size	Occurrence Rate 0.0	.5	1.0	2.0	3.0	4.0	5.0	6.0	7.0	8.0	9.0	10.0	12.0	14.0	16.0	18.0	20.0	25.0	30.0	40.0	50.0
50	4.5			7.6		10.3		12.9		15.4		17.8	20.1	22.7	24.7	27.2	29.1		39.8	50.0	59.9
100	2.3		3.3	5.2	6.6	7.8	9.1	10.3	11.7	12.7	14.0	15.0	17.3	19.6	21.7	24.0	26.1	31.4	36.6	46.9	56.8
150	1.5			4.4		6.9		9.3		11.6		13.9	16.1	18.4	20.5	22.7	24.8		35.2	45.5	55.4
200	1.1		2.6	4.0	5.2	6.4	7.6	8.8	10.0	11.0	12.2	13.3	15.5	17.7	19.8	22.0	24.0	29.3	34.5	44.4	54.4
250	.9			3.7		6.1		8.4		10.7		12.9	15.1	17.2	19.3	21.5	23.6		33.7	43.7	53.7
300	.8		2.2	3.5	4.7	5.9	7.0	8.2	9.3	10.4	11.5	12.6	14.7	16.9	19.0	21.1	23.2	28.2	33.2	43.2	53.2
350	.7			3.3		5.7		8.0		10.2		12.3	14.5	16.7	18.8	20.9	22.8		32.8	42.8	52.8
400	.6		2.0	3.2	4.4	5.6	6.7	7.8	8.9	10.0	11.1	12.2	14.3	16.5	18.5	20.5	22.5	27.5	32.5	42.5	52.5
450	.5			3.1		5.5		7.7		9.9		12.0	14.2	16.3	18.3	20.3	22.3		32.3	42.3	52.2
500	.5		1.8	3.1	4.2	5.4	6.5	7.6	8.7	9.8	10.9	11.9	14.1	16.1	18.1	20.1	22.1	27.1	32.1	42.1	52.0
550	.4			3.0		5.3		7.5		9.7		11.8	13.9	15.9	17.9	19.9	21.9		31.9	41.9	51.9
600	.4		1.7	2.9	4.1	5.2	6.3	7.4	8.5	9.6	10.7	11.7	13.7	15.7	17.7	19.7	21.7	26.7	31.7	41.7	51.7
650	.4			2.9		5.2		7.4		9.5		11.6	13.6	15.6	17.6	19.6	21.6		31.6	41.6	51.6
700	.3		1.7	2.9	4.0	5.1	6.2	7.3	8.4	9.5	10.5	11.5	13.5	15.5	17.5	19.5	21.5	26.5	31.5	41.5	51.5
750	.3			2.8		5.1		7.3		9.4		11.4	13.4	15.4	17.4	19.4	21.4		31.4	41.4	51.4
800	.3		1.6	2.8	3.9	5.0	6.1	7.2	8.3	9.3	10.3	11.3	13.3	15.3	17.3	19.3	21.3	26.3	31.3	41.3	51.3
850	.3			2.8		5.0		7.2		9.2		11.2	13.2	15.3	17.3	19.3	21.3		31.3	41.3	51.3
900	.3		1.6	2.7	3.9	5.0	6.0	7.1	8.2	9.2	10.2	11.2	13.2	15.2	17.2	19.2	21.2	26.2	31.2	41.2	51.2
950	.2			2.7		4.9		7.1		9.1		11.1	13.1	15.1	17.1	19.1	21.1		31.1	41.1	51.1
1000	.2		1.5	2.7	3.8	4.9	6.0	7.1	8.1	9.1	10.1	11.1	13.1	15.1	17.1	19.1	21.1	26.1	31.1	41.1	51.1
1500	.2		1.4	2.5	3.6	4.7	5.7	6.7	7.7	8.7	9.7	10.7	12.7	14.7	16.7	18.7	20.7	25.7	30.7	40.7	50.7
2000	.1		1.3	2.5	3.5	4.5	5.5	6.5	7.5	8.5	9.5	10.5	12.5	14.5	16.5	18.5	20.5	25.5	30.5	40.6	50.6
2500	.1		1.3	2.4	3.4	4.4	5.4	6.4	7.4	8.4	9.4	10.4	12.4	14.4	16.4	18.4	20.4	25.4	30.4	40.4	50.4
3000	.1		1.3	2.4	3.4	4.4	5.4	6.4	7.4	8.4	9.4	10.4	12.4	14.4	16.4	18.4	20.4	25.4	30.4	40.4	50.4
4000	.1		1.2	2.3	3.3	4.3	5.3	6.3	7.3	8.3	9.3	10.3	12.3	14.3	16.3	18.3	20.3	25.3	30.3	40.3	50.3
5000	.0		1.2	2.3	3.2	4.2	5.2	6.2	7.2	8.2	9.2	10.2	12.2	14.2	16.2	18.2	20.2	25.2	30.2	40.2	50.2

Table 6.2A Determination of Sample Size—Tabular Form—One-Sided Upper Precision Limits

RELIABILITY LEVEL – 95.0 PERCENT

Sample Size	Occurrence Rate																				
	0.0	.5	1.0	2.0	3.0	4.0	5.0	6.0	7.0	8.0	9.0	10.0	12.0	14.0	16.0	18.0	20.0	25.0	30.0	40.0	50.0
50	5.8			9.1		12.1		14.8		17.4		19.9	22.3	25.1	27.0	29.6	31.6		42.4	52.6	62.4
100	3.0		4.7	6.2	7.6	8.9	10.2	11.5	13.0	14.0	15.4	16.4	18.7	21.2	23.3	25.6	27.7	33.1	38.4	48.7	56.6
150	2.0			5.1		7.7		10.2		12.6		15.0	17.3	19.6	21.7	24.0	26.1		36.7	47.0	56.8
200	1.5	2.4	3.1	4.5	5.8	7.1	8.3	9.5	10.8	11.9	13.1	14.2	16.4	18.7	20.9	23.1	25.2	30.5	35.7	45.7	55.6
250	1.2			4.2		6.7		9.1		11.4		13.7	15.9	18.1	20.3	22.4	24.6		34.8	44.8	54.7
300	1.0		2.6	3.9	5.2	6.4	7.6	8.8	10.0	11.1	12.2	13.3	15.5	17.7	19.8	22.0	24.1	29.1	34.1	44.1	54.1
350	.9			3.7		6.2		8.5		10.8		13.0	15.2	17.4	19.5	21.7	23.6		33.6	43.6	53.6
400	.7	1.6	2.3	3.6	4.8	6.0	7.2	8.3	9.5	10.6	11.7	12.8	15.0	17.2	19.2	21.2	23.2	28.2	33.2	43.2	53.2
450	.7			3.5		5.9		8.2		10.4		12.6	14.8	16.8	18.9	20.9	22.9		32.9	42.9	52.9
500	.6		2.1	3.4	4.6	5.8	6.9	8.0	9.2	10.3	11.4	12.5	14.6	16.7	18.6	20.7	22.6	27.6	32.6	42.6	52.6
550	.5			3.3		5.7		7.9		10.1		12.3	14.4	16.4	18.4	20.4	22.4		32.4	42.4	52.4
600	.5	1.3	2.0	3.2	4.4	5.6	6.7	7.8	9.0	10.0	11.2	12.2	14.2	16.2	18.2	20.2	22.2	27.2	32.2	42.2	52.2
650	.5			3.2		5.5		7.7		10.0		12.1	14.1	16.1	18.1	20.1	22.1		32.1	42.1	52.1
700	.4		1.9	3.1	4.3	5.4	6.6	7.7	8.8	9.9	10.8	11.9	13.9	15.9	17.9	19.9	21.9	26.9	31.9	41.9	51.9
750	.4			3.1		5.4		7.6		9.8		11.8	13.8	15.8	17.8	19.8	21.8		31.8	41.8	51.8
800	.4	1.1	1.8	3.0	4.2	5.3	6.4	7.5	8.7	9.7	10.7	11.7	13.7	15.7	17.7	19.7	21.7	26.7	31.7	41.7	51.7
850	.4			3.0		5.3		7.5		9.6		11.6	13.6	15.6	17.6	19.6	21.6		31.6	41.6	51.6
900	.3		1.7	3.0	4.1	5.2	6.3	7.5	8.5	9.5	10.5	11.5	13.5	15.5	17.5	19.5	21.5	26.5	31.5	41.5	51.5
950	.3			2.9		5.2		7.4		9.4		11.4	13.4	15.5	17.4	19.5	21.4		31.5	41.5	51.5
1000	.3	1.0	1.7	2.9	4.0	5.2	6.3	7.4	8.4	9.4	10.4	11.4	13.4	15.4	17.4	19.4	21.4	26.4	31.4	41.4	51.4
1500	.2		1.5	2.7	3.8	4.9	5.9	6.9	7.9	8.9	9.9	10.9	12.9	14.9	16.9	18.9	20.9	25.9	30.9	40.9	50.9
2000	.1	.8	1.4	2.6	3.7	4.7	5.7	6.7	7.7	8.7	9.7	10.7	12.7	14.7	16.7	18.7	20.7	25.7	30.7	40.7	50.7
2500	.1		1.4	2.6	3.6	4.6	5.6	6.6	7.6	8.6	9.6	10.6	12.6	14.6	16.6	18.6	20.6	25.6	30.6	40.6	50.6
3000	.1	.8	1.4	2.5	3.6	4.6	5.6	6.6	7.6	8.6	9.6	10.5	12.5	14.5	16.5	18.5	20.5	25.5	30.5	40.5	50.5
4000	.1	.7	1.3	2.4	3.4	4.4	5.4	6.4	7.4	8.4	9.4	10.4	12.4	14.4	16.4	18.4	20.4	25.4	30.4	40.4	50.4
5000	.1	.7	1.3	2.3	3.3	4.3	5.3	6.3	7.3	8.3	9.3	10.3	12.3	14.3	16.3	18.3	20.3	25.3	30.3	40.3	50.3

Table 6.2B Determination of Sample Size—Tabular Form—One-Sided Upper Precision Limits

corresponds to a sample size of 100 (one can read the graph in Table 6.1 as between 75 and 100).

Reproduced below is a section of the 90% table showing the route illustrated in the preceding paragraph (dotted lines and arrows supplied by authors).

OCCURRENCE RATE

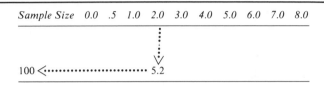

Sample Size	0.0	.5	1.0	2.0	3.0	4.0	5.0	6.0	7.0	8.0
100				5.2						

Both sets of tables illustrate certain numerical relationships among reliability, precision, and sample size. As the user moves down the 2% occurrence rate column, the upper precision limit percentage becomes smaller, and the sample size associated with the precision becomes larger. These shifts are merely a tabular expression of the fact that the user must take larger samples in order to achieve more narrow precision.

Also note that in the 90% table a 2% occurrence rate and a sample size of 100 give the user 5.2% precision. But on the 95% table a 2% occurrence rate and a sample size of 100 furnish the user 6.2% precision, which is not as good. With the same sample size, the user must accept wider precision if increased reliability is desired.

The Use of Attribute-Sampling Tables
to Evaluate Sample Results

The following Table 6.3 sets are used for determining the precision that is actually *attained* when the sampling results are tabulated. It should be noted that the upper precision limit on the percentage of errors is the *only* information that can be derived from these tables; nothing is stated about the nature or dollar amount of the errors. If auditors are concerned about any other characteristics, they must use more subjective methods of evaluation (discussed in another part of this chapter).

Following the example of the previous section, assume that a sample of 100 is selected for the purpose of ascertaining, at a reliability of 90%, that the true error rate in the population is no greater than 5.2%. Reference to Table 6.3 shows the sample size block of 100. The first column represents the number of occurrences or errors found in the sample of 100. The second, third, and fourth columns are the precision levels that are attained for 90%, 95%, and 99%

	Number of Occur-rences	UPL* RELIABILITY LEVEL 90%	95%	99%
Sample Size 25	0	8.80	11.29	16.82
	1	14.69	17.61	23.75
	2	19.91	23.10	29.59
	3	24.80	28.17	34.88
	4	29.47	32.96	39.79
Sample Size 50	0	4.50	5.82	8.80
	1	7.56	9.14	12.55
	2	10.30	12.06	15.77
	3	12.88	14.78	18.72
	4	15.35	17.38	21.50
	5	17.76	19.88	24.15
	6	20.11	22.32	26.71
	8	24.69	27.02	31.61
Sample Size 75	0	3.02	3.92	5.96
	1	5.09	6.17	8.53
	2	6.94	8.16	10.74
	3	8.69	10.01	12.78
	4	10.38	11.79	14.70
	5	12.02	13.51	16.55
	6	13.62	15.18	18.34
	7	15.20	16.82	20.08
	8	16.75	18.42	21.77
	12	22.78	24.63	28.25
Sample Size 100	0	2.28	2.95	4.50
	1	3.83	4.66	6.45
	2	5.23	6.16	8.14
	3	6.56	7.57	9.70
	4	7.83	8.92	11.17
	5	9.08	10.23	12.58
	6	10.29	11.50	13.95
	7	11.49	12.75	15.29
	8	12.67	13.97	16.59
	9	13.83	15.18	17.87
	10	14.99	16.37	19.13
	11	16.13	17.55	20.37
	15	20.61	22.15	25.18
Sample Size 125	0	1.83	2.37	3.62
	1	3.08	3.74	5.19
	2	4.20	4.95	6.55
	3	5.27	6.09	7.81
	4	6.29	7.17	9.00
	5	7.29	8.23	10.15
	6	8.27	9.25	11.26
	7	9.24	10.26	12.34
	8	10.19	11.25	13.40
	9	11.13	12.23	14.44
	10	12.06	13.19	15.47
	11	12.98	14.15	16.48
	12	13.89	15.09	17.47
	13	14.80	16.03	18.45
	19	20.14	21.50	24.16

	Number of Occur-rences	UPL RELIABILITY LEVEL 90%	95%	99%
Sample Size 150	0	1.52	1.98	3.02
	1	2.57	3.12	4.34
	2	3.51	4.14	5.49
	3	4.40	5.09	6.54
	4	5.26	6.00	7.54
	5	6.10	6.88	8.50
	6	6.92	7.74	9.44
	7	7.72	8.59	10.35
	8	8.52	9.42	11.24
	9	9.31	10.24	12.12
	10	10.09	11.05	12.98
	11	10.86	11.85	13.83
	12	11.62	12.64	14.67
	13	12.39	13.43	15.50
	14	13.14	14.21	16.32
	15	13.89	14.98	17.13
	16	14.64	15.75	17.94
	23	19.79	21.02	23.42
Sample Size 175	0	1.31	1.70	2.60
	1	2.20	2.68	3.73
	2	3.01	3.55	4.72
	3	3.78	4.37	5.63
	4	4.52	5.15	6.49
	5	5.24	5.91	7.32
	6	5.94	6.65	8.12
	7	6.63	7.38	8.91
	8	7.32	8.10	9.68
	9	8.00	8.80	10.43
	10	8.67	9.50	11.18
	11	9.33	10.19	11.91
	12	9.99	10.87	12.64
	13	10.65	11.55	13.36
	14	11.30	12.22	14.07
	15	11.95	12.89	14.77
	16	12.59	13.55	15.47
	17	13.23	14.21	16.16
	18	13.87	14.87	16.85
	27	19.51	20.64	22.84
Sample Size 200	0	1.14	1.49	2.28
	1	1.93	2.35	3.27
	2	2.64	3.11	4.14
	3	3.31	3.83	4.93
	4	3.96	4.52	5.69
	5	4.59	5.18	6.42
	6	5.21	5.83	7.13
	7	5.82	6.47	7.82
	8	6.42	7.10	8.50
	9	7.01	7.72	9.16
	10	7.60	8.33	9.82
	11	8.18	8.94	10.46
	12	8.76	9.54	11.10
	13	9.34	10.14	11.73
	14	9.91	10.73	12.36
	15	10.48	11.31	12.98
	16	11.04	11.90	13.59
	17	11.61	12.48	14.20
	18	12.17	13.05	14.81
	19	12.73	13.63	15.41
	20	13.28	14.20	16.01
	21	13.84	14.77	16.60
	30	18.75	19.79	21.82

*Upper Precision Limit

Table 6.3   Evaluation of Sample Results

reliability, respectively. If two errors were found in a sample of 100, the attained precision, at a 90% reliability, is 5.23%. Reproduced below is a partial section of Table 6.3 showing this result (dotted lines and arrows supplied by authors).

RELIABILITY LEVEL

	No. of Occurrences	90%	95%	99%
Sample	0	2.28	2.95	4.50
Size	1	3.83	4.66	6.45
100	2 ·················>	5.23	6.16	8.14
	3	6.56	7.57	9.70

For the user to attain a precision limit of 5.23% or below, at a reliability of 90%, two or fewer errors must be found in the sample of 100. If three errors are discovered, the attained precision is 6.56%, which is higher or wider than desired.

A moment's reflection leads to an interesting conclusion. The sample size of 100 was derived in Table 6.1 or 6.2 by assuming an actual error rate of 2%. In order for the attained precision to be 5.23% or lower, a maximum of two errors must be discovered from the sample of 100 (a 2% rate). Thus, if the error rate in the sample is greater than the estimated error rate in the population, the attained precision may be higher than desired.

What should the user do if three errors are found in a sample of 100? The desired precision might be attained by expanding the sample to 150. If no additional errors are found, the attained precision might be 5.23% or less.

What is the theoretical justification for increasing the sample size from 100 to 150? The reasoning is that when three errors are found, one of two possible situations is indicated.

1. The sample result may not be typical of the population characteristics. The user has incorrectly rejected the population as having an error rate higher than is tolerable when the true error rate in the population is tolerable (the rejection error).

2. The true error rate in the population may be larger than is tolerable.

The extra sampling of 50 furnishes the user with better evidence as to which of the two situations is correct. If only three errors are found in a sample of 150 (a 2% error rate), Situation 1 may be correct.

This expansion of the sample is not acceptable to some auditors. The arguments against enlarging the sample are (1) that it is a useless step because the error percentage in the sample probably will not change, and (2) the user is ignoring the results of a valid sample. However, if the risk of incorrectly rejecting compliance is large, auditors certainly are tempted to expand the sample.

It should be noted that Table 6.3 also shows certain relationships among precision, reliability, and sample size. Taking any sample size row and reading the precision percentages from left to right, one sees that as the desired reliability goes up, the attained precision becomes higher or wider. This is why AU Section 320A *(SAS No. 1)* states that precision and reliability are statistically inseparable.

Critique of Quantitative Sampling Results

The output of Table 6.3 is confined to the upper precision limit that is attained from counting the number of errors. It does not indicate the characteristics of these errors, matters that may be important to auditors. For example, auditors might be more concerned with two errors of several thousand dollars each than with five errors of a few dollars each. This is one area in which attribute sampling fails to provide needed information, because each error is evaluated the same, regardless of its dollar amount.

Auditors might overcome some of this weakness by defining an error as any discrepancy over a certain dollar amount (say, $100). Thus, the upper precision limit is defined as a percentage of mistakes over this minimum dollar figure. Another possibility is to use a different type of statistical test for any phase of the audit examination in which the dollar amount of errors is particularly important.

Of course, tests of compliance that are not related directly to dollar figures are well suited for the application of these quantitative results. A typical example is the examination of signatures and initials on documents. But even here certain questions could arise if the errors appeared to have something other than random characteristics, for example, if all the mistakes were made by one clerk. In this case, auditors may wish to look beyond the table results and make a separate investigation of the source of the errors.

An Application of Statistical Sampling for Attributes

The Sampling Plan

In Chapter 5 is an example of a test of compliance for part of a system of cash receipts. The illustrated procedure is the tracing of a sample of receipt listings to validated deposit tickets, one purpose being to prove that cash received in the mail room is deposited intact daily.

It is assumed that 5 days' receipts in a yearly total of 250 were deposited late (a 2% error rate in the population). Two methods are explained. One is a block or cluster sample of one or more months' receipts, and the other is a systematic sample in which every fifth receipt is selected for testing, starting with the first receipt. The results of the systematic sample are repeated hereafter.

Receipt No.	Late Deposit of Receipts	Systematic Sample Selection	Receipt No.	Late Deposit of Receipts	Systematic Sample Selection
02359		✔	02400		
60			01		
61			02		
62			03		
63			04		✔
64		✔	05		
65			06		
66			07		
67			08		
68			09		✔
69		✔	10		
70			11		
71			12		
72			13		
73			14		✔
74		✔	15		
75			16		
76			17		
77			18		
78			19		✔
79		✔	20		
80	✔		21		
81			22		
82			23		
83			24		✔
84		✔	25		
85			26		
86			27		
87			28		
88			29		✔
89		✔	30		
90			31		
91			32		
92			33		
93			34		✔
94		✔	35		
95			36		
96			37		
97			38		
98			39		✔
99		✔	40		

As shown in the foregoing example, this *particular* systematic sampling plan would have detected one deposit lag in 50 examined deposit slips, or an error rate in the sample equal to the error rate in the population. If the purpose of tracing receipt listings to deposit slips were to obtain a sample with characteristics similar to those of the population, this plan certainly would have been successful. But two observations should be made.

Receipt No.	Late Deposit of Receipts	Systematic Sample Selection	Receipt No.	Late Deposit of Receipts	Systematic Sample Selection
41			83		
42			84		✔
43			85		
44		✔	86		
45			87		
46			88		
47			89		✔
48			90		
49		✔	91		
50			92		
51			93		
52			94		✔
53			95		
54		✔	96		
55			97		
56			98		
57			99		✔
58			02500		
59		✔	01		
60			02		
61			03		
62			04		✔
63			05		
64	✔	✔	06		
65			07		
66			08		
67			09		✔
68			10		
69		✔	11		
70			12		
71			13		
72			14		✔
73			15		
74		✔	16		
75			17		
76			18		
77			19		✔
78			20		
79		✔	21		
80			22		
81			23		
82			24		✔

1. The five late deposits are scattered throughout the population in an *apparently* random fashion. The receipt numbers are 02359–02608, and the deposit lags are on 02380, 02464, 02525, 02581, and 02593. In this situation, systematic sampling can produce usable results. Note that the errors are on receipt numbers ending with 0, 4, 5, 1, and 3. Therefore, any

Receipt No.	Late Deposit of Receipts	Systematic Sample Selection	Receipt No.	Late Deposit of Receipts	Systematic Sample Selection
25	✓		67		
26			68		
27			69		✓
28			70		
29		✓	71		
30			72		
31			73		
32			74		✓
33			75		
34		✓	76		
35			77		
36			78		
37			79		✓
38			80		
39		✓	81	✓	
40			82		
41			83		
42			84		✓
43			85		
44		✓	86		
45			87		
46			88		
47			89		✓
48			90		
49		✓	91		
50			92		
51			93	✓	
52			94		✓
53			95		
54		✓	96		
55			97		
56			98		
57			99		✓
58			02600		
59		✓	01		
60			02		
61			03		
62			04		✓
63			05		
64		✓	06		
65			07		
66			08		

systematic sampling plan in which every fifth receipt is selected will produce one deposit lag if the starting point in the population is a receipt number ending in 0, 4, 5, 1, or 3. But what if the lags were on numbers 02359, 02389, 02424, 02536, and 02594? Then the systematic sample illustrated above would produce 4 errors from a sample of 50 (reference to the illustration shows that each receipt with a number ending in 4 or 9 is selected in the sample). The sample error rate of 8% (4 ÷ 50) might lead auditors to a gross overestimate of the true error rate in the population, which would result either in an unnecessary amount of extra sampling at this point or in excessive testing on dollar balances later.

2. Auditors have no way of measuring the success of their sampling results unless they use statistical methods that make it possible to state, with a certain reliability percentage, that the true error rate is no higher than a given upper precision limit figure. These techniques are based on the use of simple random sampling, which we will use.

Attribute Sampling to Determine Sample Size

One now can apply the statistical tables to the test of compliance described above. The 95% graph of Table 6.1 is selected to determine sample size, and Table 6.3 is used to evaluate sample results. The upper precision limit is 8%, and the estimated error rate is 2%.

It is not necessary to repeat in detail the AICPA's suggested criteria for selecting the reliability and precision percentages. One method of setting the desired reliability on compliance tests is to relate it to the prescribed internal control system. According to the documentation shown in Chapter 5, cash receipts are deposited intact daily; this procedure seems to call for a high reliability on the test of compliance to provide relative certainty that the procedure actually is being carried out.

With regard to the precision, one should remember that client deviations from control procedures are more critical if the deviation definitely results in dollar errors in the financial statements (incorrect checks, failure to bill customers, etc.). But because deposit lags may or may not result in dollar errors, auditors might tend to raise the upper precision limit in this case.

Reference to the 95% graph of Table 6.1 shows that with an estimated error rate of 2% and an upper precision limit of 8%, the required sample is slightly fewer than 75. The applicable part of the 95% graph of Table 6.1 is shown below.

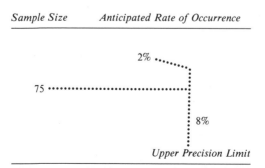

Sample Size   Anticipated Rate of Occurrence

2%

75

8%

Upper Precision Limit

If the sampling results are satisfactory, the auditors would be able to state that at a 95% reliability level, the percentage of deposit lags in the population of receipt listings is no greater than 8%. This statement means that if the actual percentage of deposit lags in the population is above 8%, the sampling results will show, in 95% of the cases, an upper precision limit in excess of 8%. There is a 5% (or less) risk that compliance will be accepted as adequate when it is not.

A look at the 75-sample bloc of Table 6.3 shows that if auditors find no lag or one lag in the sample, the precision will be 8% or below. In contrast, discovery of two or more deposit delays will produce a precision of more than 8%, and the sampling results will be unacceptable. This situation is shown below by a reproduced section of the 75-sample bloc of Table 6.3.

	No. of Occur- rences	Reliability Level		
		90%	95%	99%
Sample	0		3.92	
Size	1		6.17	
75	2		8.16	
	3		10.01	

Note: Although it is not technically correct to do so, auditors might accept a precision of 8.16% based on 2 occurrences in 75.

But what is the probability of finding two or more deposit lags, and thus rejecting the compliance as being unacceptable, when the actual percentage of deposit lags in the receipts is no more than 8%? One may be surprised to learn that the probability of this occurrence is fairly high. The following chart of binomial probabilities shows the probability of finding different numbers of errors in a sample of 75, given that the actual error rate in the population is 2% (which is known).

$n = 75$

r/P	.01	.02	.03	.04	.05	.06	.07	.08	.09	.10
0		.2198								
1		.3364								
2		.2540								
3		.1261								
4		.0463								
5		.0134								
6		.0032		n = sample size						
7		.0006		p = actual error rate in the population						
8		.0001		r = number of errors found in sample						

By adding the probability of finding zero errors (.2198) to the probability of finding one error (.3364), one derives a cumulative probability of finding fewer than two errors of .5562. Thus, the probability of discovering two or more errors is .4438, very large. Auditors can reduce this risk by selecting larger samples, which may or may not be worth the cost and effort.

Illustration of Sample Selection and Quantitative Results

The sample of 75 is selected from the random number table by reading down the leftmost three-digit numbers starting in Column 1, continuing in Column 2, and so on; only numbers between 1 and 250 are chosen. The 75 numbers then are restructured into ascending sequential order to expedite the sampling process. The samples are pulled *without* replacement to avoid the possible problem of having the same error discovered more than once.

Reference to Table 6.3 shows that with one error in a sample of 75, a precision of 6.17% is achieved at a reliability of 95%. On the basis of these results, auditors would accept compliance with this deposit procedure as adequate. The sample numbers are shown on pages 248–249.

Critique of the Quantitative Results

Knowing that the true percentage of deposit lags in the population of receipt listings is 2%, one would expect the foregoing results. But auditors do not have this advance knowledge and therefore must depend on the sample results to give them some measure of assurance.

What if the sample results showed two compliance deviations in a sample of 75? According to Table 6.3, 8.16% precision is attained at a reliability of 95%; 8% is required for auditors to conclude that compliance is satisfactory. Is this level acceptable? Statistically, it is not, but auditors might conclude that because the attained precision shown in the table is so close, the sampling results could be accepted.

What if the sample results showed three compliance deviations? Table 6.3 shows that attained precision is 10.01%, far above the maximum of 8% considered tolerable. In this case, auditors might take one of several courses of action.

1. An investigation could be made by the auditors, or the client under the auditors' supervision, to determine any special reasons for such a large error. One possibility is that a temporary employee handled the deposits for certain periods of time.

2. A conclusion could be reached that compliance is *not* satisfactory, and that a change in the testing of financial statement balances will be needed at a later time. (These types of tests and their relationship to internal control evaluation are discussed in Chapter 9.)

Numbers Pulled from Table	Numbers Rearranged into Ascending Sequential Order	Corresponding Receipt Number	Deposit Lag
104	002	02360	
223	003	61	
241	005	63	
103	007	65	
071	010	68	
023	015	73	
010	020	78	
070	023	81	
024	024	82	
007	034	92	
053	041	99	
005	042	02400	
097	045	03	
179	047	05	
145	048	06	
089	053	11	
156	054	12	
150	055	13	
069	058	16	
143	061	19	
127	062	20	
213	066	24	
205	069	27	
042	070	28	
047	071	29	
185	073	31	
075	075	33	
003	079	37	
015	081	39	
225	089	47	
062	097	55	
110	099	57	
054	102	60	
224	103	61	
055	104	62	
048	110	68	
188	113	02471	
204	127	85	
020	129	87	
166	136	94	
079	139	97	
102	141	99	
034	143	02501	
081	145	03	
099	150	08	
242	151	09	
073	156	14	
129	163	21	
163	166	24	
209	170	28	
181	176	34	

139	179	37
198	181	39
041	183	41
066	185	43
002	186	44
208	187	45
183	188	46
170	189	47
244	198	56
045	204	62
219	205	63
151	206	64
186	208	66
189	209	67
061	213	71
206	219	77
113	223	81
229	224	82
141	225	83
248	229	87
187	241	99
58	242	02600
176	244	02
136	248	06

3. The sample size could be enlarged in an attempt to achieve the desired precision. In practice, this step would necessitate the use of conditional probability tables, the discussion of which is beyond the scope of this chapter.

It should be emphasized, however, that the sample size is expanded because the anticipated benefits outweigh the estimated cost. Auditors might consider this step desirable if the alternative is an excessive amount of testing later.

However, as mentioned earlier, the concept of sample expansion when the initial sampling does not produce the desired upper precision limit is not acceptable to some auditors. Volume 6 of the *Auditor's Approach to Statistical Sampling,* published by the AICPA, contains the following opinion:

> This approach is not advisable because if we found 8 errors in the first 150 items examined, we would also expect to find approximately 8 errors in the *next* 150 items examined. . . . Since internal control in this area cannot be relied on to the extent expected, other work must be expanded.

Summary

With certain types of compliance tests, estimation sampling for attributes provides auditors with a limited, but useful, tool. It is not necessarily true that a major advantage is the reduction of sample size. In some situations, the use of attribute sampling might reduce the sample size, particularly if the population consists of several thousand elements. But with a small population of 250 there is not likely to be much difference between the sample sizes used in judgment and statistical sampling.

For a population of 250 receipt listings, intuition may lead auditors to believe that the tracing of 50 deposits, selected 5 at a time, is sufficient. A look at the 95% graph of Table 6.1 shows that a sample of 50 would achieve only about 9% precision with an estimated error rate of 2%. Thus, use of statistical methods might call for a *larger* sample size than is used in judgment sampling.

It should be noted also that auditors use a statistically determined test result for the same purpose as a test result obtained from judgment sampling, to determine whether compliance with prescribed procedures is adequate. The advantage of the methods illustrated in this chapter is that they provide auditors with more useful information for making this determination.

In Chapter 13 are examples of other types of statistical sampling that are useful for estimating the dollar amount of financial statement balances.

Appendix to Chapter 6

Discovery Sampling for Attributes

Introduction

Chapter 6 concerns attribute sampling for the purpose of estimating the rate of compliance deviations in the population and setting precision intervals around this estimate. There is another form of attribute sampling in which the purpose is to "discover" at least one compliance deviation if the percentage of such deviations in the population is at or above a certain level (generally low). "Discovery sampling," as it is popularly known, is designed to be used in the testing of critical compliance deviations, such as improper payroll endorsements or missing documents.

Because discovery sampling for attributes has not gained much appeal in public practice, it is not used as commonly as estimation sampling. Nevertheless, auditing students should become familiar with these techniques in order to increase their academic knowledge of statistical sampling.

An Example of the Methodology

The methodology for using the discovery sampling tables is similar to that employed for the estimation sampling tables. The major difference is that separate tables are used for different population sizes.

The following steps are based on the use of Table 6.4A. However the same procedures can be followed with Table 6.4B, although different numerical answers are derived.

Required Sample Size	If the Population Occurrence Rate is:							
	.1%	.2%	.3%	.4%	.5%	.75%	1%	2%

The Probability of Discovering at Least One
Occurrence in the Sample is:

Required Sample Size	.1%	.2%	.3%	.4%	.5%	.75%	1%	2%
50	5%	10%	14%	18%	22%	31%	40%	64%
60	6	11	17	21	26	36	45	70
70	7	13	19	25	30	41	51	76
80	8	15	21	28	33	45	55	80
90	9	17	24	30	36	49	60	84
100	10	18	26	33	40	53	64	87
120	11	21	30	38	45	60	70	91
140	13	25	35	43	51	65	76	94
160	15	28	38	48	55	70	80	96
200	18	33	45	56	64	78	87	98
240	22	39	52	62	70	84	91	99
300	26	46	60	70	78	90	95	99+
340	29	50	65	75	82	93	97	99+
400	34	56	71	81	87	95	98	99+
460	38	61	76	85	91	97	99	99+
500	40	64	79	87	92	98	99	99+
600	46	71	84	92	96	99	99+	99+
700	52	77	89	95	97	99+	99+	99+
800	57	81	92	96	98	99+	99+	99+
900	61	85	94	98	99	99+	99+	99+
1,000	65	88	96	99	99	99+	99+	99+
1,500	80	96	99	99+	99+	99+	99+	99+
2,000	89	99	99+	99+	99+	99+	99+	99+

Note: 99+ indicates a probability of 99.5% or greater. Probabilities
in these tables are rounded to the nearest 1%.

Copyright © (1968) by the American Institute of Certified Public
Accountants, Inc.

Table 6.4A   Discovery Sampling for Attributes for Populations Between
5,000 and 10,000

Required Sample Size	If the Population Occurrence Rate is:							
	.3%	.4%	.5%	.6%	.8%	1%	1.5%	2%

The Probability of Discovering at Least One
Occurrence in the Sample is:

Required Sample Size	.3%	.4%	.5%	.6%	.8%	1%	1.5%	2%
50	14%	18%	22%	26%	33%	40%	53%	64%
60	17	21	26	30	38	45	60	70
70	19	25	30	35	43	51	66	76
80	22	28	33	38	48	56	70	80
90	24	31	37	42	52	60	75	84
100	26	33	40	46	56	64	78	87
120	31	39	46	52	62	70	84	91
140	35	43	51	57	68	76	88	94
160	39	48	56	62	73	80	91	96
200	46	56	64	71	81	87	95	98
240	52	63	71	77	86	92	98	99
300	61	71	79	84	92	96	99	99+
340	65	76	83	88	94	97	99+	99+
400	71	81	88	92	96	98	99+	99+
460	77	86	91	95	98	99	99+	99+
500	79	88	93	96	99	99	99+	99+
600	85	92	96	98	99	99+	99+	99+
700	90	95	98	99	99+	99+	99+	99+
800	93	97	99	99	99+	99+	99+	99+
900	95	98	99	99+	99+	99+	99+	99+
1,000	97	99	99+	99+	99+	99+	99+	99+

Note: 99+ indicates a probability of 99.5% or greater. Probabilities
in these tables are rounded to the nearest 1%.

Table 6.4B Discovery Sampling for Attributes for Populations Between
2,000 and 5,000

1. Decide what reliability is needed (assume 95%).

2. Decide what maximum error rate in the population is acceptable at a
reliability of 95% (assume .5%). Normally, the rate will be lower than the

one used in estimation sampling for attributes because the errors are more serious.

3. Auditors will use this method only if they believe that the actual error rate in the population is low. Therefore, no preliminary sample is necessary.

4. Refer to Table 6.4A. Look down the .5% column until the next highest percentage over 95 is found (96). Look across that row and find the necessary sample size (answer—600).

5. Take the sample of 600. The sampling method is *without* replacement. If no errors are found, the following statements can be made (reading from the .5% column to the right).

(a)   There is a 96% probability that the error rate in the population is less than .5%.

(b)   There is a 99% chance that the error rate in the population is less than .75%.

(c)   There is a 99% chance that the error rate in the population is less than 1%.

(d)   There is a 99+% chance that the error rate in the population is less than 2%.

6. If an error is found, some other auditing technique should be applied. Therefore, no table of evaluation of sample results is used with this method.

The following example is part of Table 6.4A showing the route that is taken to locate the sample size for a 96% reliability and a .5% occurrence rate.

Required Sample Size	*Occurrence Rate:*		
	.3%	*.4%*	*.5%*
50			
60			
600			96

Although the discovery sampling tables have limited use, a study of the numbers is helpful academically because it shows the relationships among sample size, reliability, and expected occurrence rates.

Chapter 6 References

Arkin, Herbert, *Handbook of Sampling for Auditing and Accounting,* 2nd edition, McGraw-Hill Book Company, New York, 1974.

Burns, David C., and James K. Loebbecke, "Internal Control Evaluation: How the Computer Can Help," *The Journal of Accountancy,* August 1975, pp. 60–70.

Carmichael, D. R., "Tests of Transactions—Statistical and Otherwise," *The Journal of Accountancy,* February 1968, pp. 36–40.

Roberts, Donald M., "A Statistical Interpretation of SAP No. 54," *The Journal of Accountancy,* March 1974, pp. 47–53.

Stettler, Howard F., "Some Observations on Statistical Sampling in Auditing," *The Journal of Accountancy,* April 1966, pp. 55–60.

Taylor, Robert G., "Error Analysis in Audit Tests," *The Journal of Accountancy,* May 1974, pp. 78–82.

Chapter 6
Questions Taken from the Chapter

6–1.   Which types of tests of compliance are suitable for statistical sampling? Which types are not?

6–2.   Name two uncertainties with which auditors must cope when conducting tests.

6–3.   Define precision and reliability.

6–4.   What limitations are auditors working under when they use judgment sampling?

6–5.   What is the AICPA's position on the use of statistical sampling?

6–6.   Name three tests that ascertain compliance with internal control and provide evidence on the validity of dollar figures.

6–7.   What guidelines are contained in *SAS No. 1* on the determination of precision for compliance tests? On the determination of reliability for compliance tests?

6–8.   According to Appendix B of *SAS No. 1*, a 95% reliability level furnishes what type of protection?

6–9.   Why do some auditors believe that a strong system of internal control calls for a high reliability level on tests of compliance with the system? Why would a weak system of internal control call for very few or no tests of compliance?

6–10.   Why does the use of statistical sampling call for more cautious judgment on the part of auditors?

6–11.   What is the purpose of estimation sampling for attributes?

6–12.   What three factors must auditors estimate or know in order to determine the necessary sample size in estimation sampling for attributes?

6–13.   Indicate two ways in which a population of invoices can be defined for purposes of statistical sampling in tests of compliance.

6–14.   What is unrestricted random sampling?

6–15.   What is meant by the term "sampling with replacement"? "Without replacement"?

6–16.   How can a table of random numbers be used with unnumbered documents?

6–17.   Briefly describe the methods for using Table sets 6.1, 6.2, and 6.3.

6–18.   What are the two reasons why auditors might achieve unsatisfactory results in their sampling?

6–19.   What alternatives are available to auditors if a higher upper precision limit is attained in the sampling results than is considered tolerable?

6–20.   What criticisms have been made of the concept of expanding the sample when auditors do not obtain satisfactory results with the initial sample?

6–21.   In what way does the objective of discovery sampling for attributes differ from the objective of estimation sampling for attributes?

6–22.   Name one major criticism of discovery sampling for attributes.

Chapter 6
Objective Questions Taken from CPA Examinations

6–23.   A CPA specifies that a sample shall have a confidence level of 90%. The specified confidence level assures him of
  a.   A true estimate of the population characteristic being measured.
  b.   An estimate that is at least 90% correct.
  c.   A measured precision for his estimate.
  d.   How likely he can estimate the population characteristic being measured.

6–24.   A CPA specifies that a sample shall have a precision of 5%. The specified precision assures him of
  a.   A precise measure of the population characteristic being measured.
  b.   An estimate that is at least 5% correct.
  c.   A measured reliability for his estimate.
  d.   The range within which the population characteristic being measured is likely to be.

6–25.   If all other factors specified in a sampling plan remain constant, changing the specified reliability from 90% to 95% would cause the required sample size to
  a.   Increase.
  b.   Remain the same.
  c.   Decrease.
  d.   Become indeterminate.

6–26.   If all other factors specified in a sampling plan remain constant, changing the specified precision from 8% to 12% would cause the required sample size to
  a.   Increase.

    b.   Remain the same.

    c.   Decrease.

    d.   Become indeterminate.

6–27.   If all other factors specified in a sampling plan remain constant, changing the estimated occurrence rate from 2% to 4% would cause the required sample size to

    a.   Increase.

    b.   Remain the same.

    c.   Decrease.

    d.   Become indeterminate.

6–28.   In the evaluation of the results of a sample of a specified reliability and precision, the fact that the occurrence rate in the sample was the same as the estimated occurrence rate would cause the reliability of the sample estimate to

    a.   Increase.

    b.   Remain the same.

    c.   Decrease.

    d.   Become indeterminate.

6–29.   A CPA's test of accuracy of inventory counts involves two storehouses. Storehouse A contains 10,000 inventory items, and Storehouse B contains 5,000 items. The CPA plans to use sampling without replacement to test for an estimated 5% error rate. If the CPA's sampling plan calls for a specified reliability of 95% and a maximum tolerable error occurrence rate of 7.5% for both storehouses, the ratio of the size of the CPA's sample from Storehouse A to the size of the sample from Storehouse B should be

    a.   More than 1:1 but less than 2:1.

    b.   2:1.

    c.   1:1.

    d.   More than .5:1 but less than 1:1.

6–30.   Approximately 5% of the 10,000 homogeneous items included in Barletta's finished-goods inventory are believed to be defective. The CPA examining Barletta's financial statements decides to test this estimated 5% defective rate. He learns that by sampling without replacement a sample of 284 items from the inventory will permit specified reliability (confidence level) of 95% and specified precision of $\pm.025$. If specified precision is changed to $\pm.05$, and specified reliability remains 95%, the required sample size is

    a.   72.

    b.   335.

    c.   436.

    d.   1,543.

6–31.   As the specified reliability is increased in a discovery sampling plan for any given population and maximum occurrence rate, the required sample size
   a.   Increases.
   b.   Decreases.
   c.   Remains the same.
   d.   Cannot be determined.

6–32.   To satisfy the auditing standard to make a proper study and evaluation of internal control, Harvey Jones, CPA, uses statistical sampling to test compliance with internal control procedures. Why does Jones use this statistical sampling technique?
   a.   It provides a means of measuring mathematically the degree of reliability that results from examining only a part of the data.
   b.   It reduces the use of judgment required of Jones because the AICPA has established numerical criteria for this type of testing.
   c.   It increases Jones' knowledge of the client's prescribed procedures and their limitations.
   d.   It is specified by generally accepted auditing standards.

6–33.   For a large population of cash disbursement transactions, Smith, CPA, is testing compliance with internal control by using attribute-sampling techniques. Anticipating an occurrence rate of 3%, Smith found from a table that the required sample size is 400 with a desired upper precision limit of 5% and reliability of 95%. If Smith anticipated an occurrence rate of only 2%, but wanted to maintain the same desired upper precision limit and reliability, the sample size would be closest to
   a.   200.
   b.   400.
   c.   533.
   d.   800.

6–34.   Which of the following is an advantage of systematic sampling over random number sampling?
   a.   It provides a stronger basis for statistical conclusions.
   b.   It enables the auditor to use the more efficient "sampling with replacement" tables.
   c.   There may be correlation between the location of items in the population, the feature of sampling interest, and sampling interval.
   d.   It does not require establishment of correspondence between random numbers and items in the population.

6–35.   An example of sampling for attributes would be estimating the
   a.   Quantity of specific inventory items.
   b.   Probability of losing a patent infringement case.

   c.   Percentage of overdue accounts receivable.

   d.   Dollar value of accounts receivable.

6–36.   The purpose of tests for compliance is to provide reasonable assurance that the accounting control procedures are being applied as prescribed. The sampling method that is *most* useful when testing for compliance is

   a.   Judgment sampling.

   b.   Attribute sampling.

   c.   Unrestricted random sampling with replacement.

   d.   Stratified random sampling.

Chapter 6
Discussion/Case Questions and Problems

6–37.   The use of statistical sampling techniques in an examination of financial statements does not eliminate judgmental decisions.

Required:

   a.   Identify and explain four areas where judgment may be exercised by a CPA in planning a statistical sampling test.

   b.   Assume that a CPA's sample shows an unacceptable error rate. Describe the various actions that he may take based upon this finding.

   c.   A non-stratified sample of 80 accounts payable vouchers is to be selected from a population of 3,200. The vouchers are numbered consecutively from 1 to 3,200 and are listed, 40 to a page, in the voucher register. Describe four different techniques for selecting a random sample of vouchers for review.

(AICPA adapted)

6–38.   In a given test of compliance, auditors seek 90% reliability and 6% precision. Reproduced below are two sections of Table 6.3 (evaluation of sample results) and two tables of binomial probabilities, one for a sample size of 75 and the other for a sample size of 100.

TABLE 6.3 EXTRACTS

	No. of Occurrences	Reliability Level		
		90%	95%	99%
Sample	0	3.02	3.92	5.96
Size	1	5.09	6.17	8.53
75	2	6.94	8.16	10.74
	3	8.69	10.01	12.78
Sample	0	2.28	2.95	4.50
Size	1	3.83	4.66	6.45
100	2	5.23	6.16	8.14
	3	6.56	7.57	9.70

Binomial Probabilities

n = 75				n = 100				
r/p	.02	.04	.07	r/p	.02	.04	.07	
0	.2198	.0468	.0043	0	.1326	.0169	.0007	(These are
1	.3364	.1463	.0244	1	.2707	.0703	.0053	partial tables)
2	.2540	.2255	.0680	2	.2734	.1450	.0198	
3	.1261	.2287	.1246	3	.1823	.1973	.0486	

n = sample size p = actual error rate in population
r = number of errors found in sample

Required:
a. Assume that the true error rate is 2%. What is the risk of a reject error if:
    (1) the sample size is 75?
    (2) the sample size is 100?
b. Assume that the true error rate is 7%. What is the risk of an accept error if:
    (1) the sample size is 75?
    (2) the sample size is 100?

Hint—Determine how many errors it would take in each sample size bloc for auditors to accept and reject compliance as being satisfactory. Then reference the list of binomial probabilities. For example, if the sample size is 75, and the true error rate in the population is .02, the probability of finding 1 or more errors is .7802 (100.00 − .2198).

6–39.   Refer back to Ques. 6–38. What can be said about the relationship between sample size and the control of the risk of a reject error?

6–40. Refer back to Ques. 6–38. Even though it is not shown on the table of binomial probabilities, do you think the risk of an accept error is more or less when the actual error rate is .08, as compared with .07? Why?

6–41. Refer back to Ques. 6–38. On the basis of your reading of both tables, why do you think the statement is made that a 90% reliability on tests of compliance means that the maximum risk of an accept error is 10%?

6–42. Refer back to Ques. 6–38. Why is it very unlikely that auditors would know the exact probability of a reject and an accept error in the use of estimation sampling for attributes?

6–43. Refer back to Ques. 6–38. In tests of compliance using these tables, what could auditors do to bring the risk of both a reject and an accept error to approximately the same percentage?

6–44. Using the attribute sample size and sample evaluation tables in the body of the chapter, fill in the blank in each of the following situations.

	A	B	C	D
Desired reliability	95%	90%	99%	90%
Desired upper limit of precision range	6%	8%	6%	10%
Error rate	2%	3%	2%	5%
Sample size	___	___	___	___
Number of errors found in sample	2	1	3	2
Actual or achieved upper limit of precision range	___	___	___	___

6–45. Using the discovery sampling table in the appendix of the chapter, fill in the blank in each of the following situations.

	A	B	C	D
Desired reliability	91%	95%	85%	99%
Population size	2,000–5,000	2,000–5,000	5,000–10,000	5,000–10,000
Estimated population occurrence rate	.5%	.3%	.4%	.2%
Sample size	___	___	___	___

6–46.   An auditor is applying estimation sampling for attributes to the testing of extensions on sales invoices. There are 250 invoices with an average of four sales on each invoice. The auditor uses 1,000 as the population total and classifies each extension mistake as an error. The auditor decides to use 90% reliability, an upper limit to the precision range of 5.2%, and an estimated occurrence rate of 2%.

Assume that the following error condition exists in the population (the invoices are numbered 1–250, number the lines 1–1,000).

Invoice No.	Line No.	Amount of Error	Error Results in Overstatement (O) or Understatement (U) of Sales
10	39	$ 100	U
21	81	350	O
51	202	900	O
53	220	700	O
61	240	950	O
70	291	300	U
102	410	410	U
103	413	850	O
150	600	1,000	O
170	674	150	O
192	798	500	O
203	840	350	O
210	855	520	U
215	890	925	O
224	906	820	O
225	908	1,000	O
231	930	800	O
280	971	900	O

Required:
   a.   Calculate the sample size.
   b.   Take the sample using random selection. Identify the numbers in the random number set with the lines 1–1,000. For example, random number 102 is line 102, etc. If you select a line number listed in the error chart shown above, assume that an error is located.
   c.   Quantitatively evaluate your sample results. If you incorrectly rejected compliance, why did this happen?
   d.   Would the dollar errors you found change the evaluation of your results? Why or why not?

6–47.   Levelland, Inc., a client of your firm for several years, uses a voucher system for processing all cash disbursements, which number about 500 each month. After carefully reviewing the company's internal controls, your firm decided to statistically sample the vouchers for eleven specific characteristics to test operating compliance of the voucher system against the client's representations as to the system's operation. Nine of these characteristics are non-critical; two are critical. The characteristics to be evaluated are listed on the voucher test worksheet.

Pertinent client representations about the system follow:

1. Purchase orders are issued for all goods and services except for recurring services such as utilities, taxes, etc. The controller issues a check request for the latter authorizing payment. Receiving reports are prepared for all goods received. Department heads prepare a services-rendered report for services covered by purchase orders. (Services-rendered reports are subsequently considered receiving reports.)

2. Copies of purchase orders, receiving reports, check requests, and original invoices are forwarded to accounting. Invoices are assigned a consecutive voucher number immediately upon receipt by accounting. Each voucher is rubber-stamped to provide spaces for accounting personnel to initial when (a) agreeing invoice with purchase order or check request, (b) agreeing invoice with receiving report and (c) verifying mathematical accuracy of the invoice.

3. In processing each voucher for payment, accounting personnel match each invoice with the related purchase order and receiving report or check request. Invoice extensions and footings are verified. Debit distribution is recorded on the face of each invoice.

4. Each voucher is recorded in the voucher register in numerical sequence, after which a check is prepared. The voucher packets and checks are forwarded to the treasurer for signing and mailing the checks and canceling each voucher packet.

5. Canceled packets are returned to accounting. Payment is recorded in the voucher register, and the voucher packets are filed numerically.

Following are characteristics of the voucher population already determined by preliminary statistical testing. Assume that each characteristic is randomly distributed throughout the voucher population.

1. Eighty percent of vouchers are for purchase orders; 20% are for check requests.
2. The average number of lines per invoice is four.
3. The average number of accounts debited per invoice is two.

Appropriate statistical sampling tables and the worksheet follow. For values not provided in the tables, use the next value in the table that will yield the most conservative result.

Reliability (Confidence Level): 95%

Sample Size	Upper Precision Limit Percentage					
	1	2	3	4	5	6
90				0	0	
120			0	.8	.8	1.7
160		0	.6	1.2	1.9	2.5
240		.4	.8	1.7	2.5	3.3
340	0	.6	1.2	2.1	2.9	3.5
460	0	.9	1.5	2.4	3.3	3.9
1,000	.4	1.2	2.0	2.9	3.8	4.7

Table 6.5 Determination of Sample Size—
Percentage of Occurrences in Sample

For Populations Between 5,000 and 10,000

Sample Size	If the True Population Rate of Occurrence is:					
	.1%	.2%	.3%	.4%	.5%	.75%
240	22	39	52	62	70	84
300	26	46	60	70	78	90
340	29	50	65	75	82	93
400	34	56	71	81	87	95
460	38	61	76	85	91	97
500	40	64	79	87	92	98
600	46	71	84	92	96	99
700	52	77	89	95	97	99+
800	57	81	92	96	98	99+
900	61	85	94	98	99	99+
1,000	65	88	96	99	99	99+

Note: 99+ indicates a probability of 99.5% or greater

Table 6.6 Probability of Including at Least
One Occurrence in a Sample

Reliability (Confidence Level): 95%

Sample Size	Upper Precision Limit Percentage					
	1	2	3	4	5	6
90				0		1
120			0	1		2
160		0	1	2	3	4
240		1	2	4	6	8
340	0	2	4	7	10	12
460	0	4	7	11	15	18
1,000	4	12	20	29	38	47

Table 6.7 Evaluation of Results—
Number of Occurrences in Sample

Levelland, Inc.
VOUCHER TEST WORKSHEET
Years Ended December 31

| | YEAR 1 | YEAR 2 | | | | | | |
CHARACTERISTICS	Column A Sample Size	Column B Estimated Error Rate	Column C Specified Upper Preci- sion Limit	Column D Reliability (Confidence Level)	Column E Required Sample Size	Column F Assumed Sample Size	Column G Number Errors Found	Column H Upper Precision Limit
Non-critical								
1. Invoice in agreement with purchase order or check request.		1.1%	3	95%		460	4	
2. Invoice in agreement with receiving report.		.4%	2	95%		340	2	
3. Invoice mathematically accurate.								
a. Extensions		1.4%	3	95%		1,000	22	
b. Footings		1.0%	3	95%		460	10	
4. Account distributions correct.		.3%	2	95%		340	2	
5. Voucher correctly entered in voucher register.		.5%	2	95%		340	1	
6. Evidence of Accounting Department checks.								
a. Comparison of invoice with purchase order or check request.		2.0%	4	95%		240	2	
b. Comparison of invoice with receiving report.		1.3%	4	95%		160	2	
c. Proving mathematical accuracy of invoice.		1.5%	3	95%		340	10	
Critical								
7. Voucher and related documents canceled.		At or near 0	.75%	95%		600	5	
8. Vendor and amount on invoice in agreement with payee and amount on check.		At or near 0	.4%	95%		800	0	

Required:

a. Year 1:

   An unrestricted random sample of 300 vouchers is to be drawn for Year 1. Enter in Column A of the worksheet the sample size of each characteristic to be evaluated in the sample.

b. Year 2:

   (1) Given the estimated error rates, specified upper precision limits, and required reliability (confidence level) in Columns B, C, and D, respectively, enter in Column E the required sample size to evaluate each characteristic.

   (2) Disregarding your answers in Column E and considering the assumed sample size and numbers of errors found in each sample as listed for each characteristic in Columns F and G, respectively, enter in Column H the upper precision limit for each characteristic.

   (3) On a separate sheet, identify each characteristic for which the sampling objective was not met and explain what steps the auditor might take to meet his sampling or auditing objectives.

(AICPA adapted)

6–48. During the course of an audit engagement, Mr. Command, the senior, decided to use estimation sampling for attributes on certain tests of compliance. He designed a sample plan and instructed Mr. Critical, one of the assistants, to select the sample size from the appropriate table, implement the tests and evaluate the sample results.

Mr. Critical followed the plan and obtained successful results (the upper precision limit was not exceeded) on all the tests except one. On this particular test, the attained precision was 7% at a reliability level of 95%, but the upper precision limit was set at 6%.

Mr. Critical asked Mr. Command what steps should be taken. Mr. Command, looking rather annoyed, stated that Mr. Critical should do the obvious thing, take another sample. Mr. Command went on to say that he was sure the true error rate in the population was about 3% and that an upper precision limit of 6% was adequate; therefore, a new sample should achieve the desired precision of 6%.

Mr. Critical responded by suggesting that the unacceptable sample results should be left alone and that subsequent dollar tests should be adjusted. He felt that a new sample would probably contain the same number of errors, leaving the auditors where they started except for the expended cost and effort.

Miss Overhear, another assistant, happened to catch the conversation and decided to interject an opinion. It was her feeling that the sample should be expanded rather than a new sample taken. She pointed to the table and showed that an additional sample of 50 would bring the precision under 6% if no additional errors were found. Furthermore, an expanded sample would create a large enough sample size so that the risk of incorrectly rejecting compliance would be brought down significantly.

Mr. Command stated that he would accept an expansion of the sample as a second alternative but that, under no circumstances, would he adjust substantive tests until he had better evidence as to the true error rate. He was trying to keep the sample size for substantive tests as low as possible, knowing that the "odds favored good sample results."

Both Mr. Critical and Miss Overhear suggested that the audit manager be called for his opinion. The audit manager took an entirely different viewpoint. He suggested that the problem be expedited by accepting a 7% precision.

Finally, another assistant pointed to the table for evaluating sample results and showed that 7% precision was attained with 95% reliability and that precision between 5% and 6% was attained with 90% reliability. He suggested that they merely accept less reliability and, by doing so, bring precision under 6%. He pointed out that, after all, a 90% reliability was still a 10% protection against incorrectly accepting compliance.

Required:
a. Evaluate the opinions of Mr. Command, Mr. Critical, Miss Overhear, the audit manager, and the other assistant. Comment on the strength and weaknesses of their viewpoints.
b. Do you have any ideas other than those expressed by the five individuals above? If so, why do you feel that yours may be better?

6–49. Refer to the system of cash receipts illustrated in Chapter 5. On the basis of the description of that system, the type of test, and any other relevant factors, indicate which tests would call for higher or lower reliability and precision percentages relative to other tests. Give your reasons.

6–50. Some auditors maintain that the extension of a sample beyond its original size when the original sample result exceeds the upper precision limit is evidence that the auditor is abandoning objectivity. Give reasons for and against this contention.

6–51. One may notice that the heavy black lines in the body of Table set 6.1 of the attribute-sampling tables curve to the right as one reads from top to bottom. Explain why this is true, and what these curved lines express about statistical sampling concepts.

6–52. Examine the list of procedures shown hereafter and indicate whether statistical sampling could or could not be used with each procedure. Support your reasons.
a. Review of flow charts.
b. Completing internal control questionnaires.

    c. Examining payroll checks for proper date, amount, signature, and endorsement.

    d. Oral inquiries to client personnel.

    e. Testing calculations in sales invoices.

    f. Examining purchase invoices for proper payment authorization.

6–53. Comment on this statement: Estimation sampling for attributes has very limited value because all errors are evaluated the same, regardless of the dollar amount of the error.

6–54. The setting of the upper precision limit for tests using estimation sampling for attributes depends in part on the auditor's assessment of the critical nature of the potential errors. For example, many auditors would probably consider an invoice pricing error to be less critical than a missing invoice. The following are lists of potential errors in the areas of sales, payroll, and disbursements. Rank each set of errors from "most critical" to "least critical" and justify your ranking.

Sales	*Payroll*	*Disbursements*
1. Pricing error in the sales invoice.	1. Improper endorsement on the payroll check.	1. Invoice not approved for payment.
2. No shipping document to match the sales invoice.	2. No dual signature on the payroll check.	2. No cancellation stamp on paid invoice.
3. No sales invoice to match the shipping document.	3. Error in the gross pay calculation.	3. No receiving report to match paid purchase invoice.
4. Error in posting amount from sales invoice to the accounts receivable subsidiary ledger.	4. Amount of the check different from the amount in the payroll register.	4. Posting error from the invoice to the purchases journal.
5. Making sales without prior credit approval.	5. No employee receipt for disbursement of payroll in currency.	5. No purchase requisition to match purchase invoice.

Chapter 7

The Study and Evaluation of Internal Control in an EDP System

---

Learning Objectives *After reading and studying the material in this chapter, the student should*

Understand the similarities and differences between the audit of a manual internal accounting control system and an EDP internal accounting control system.

Know the AICPA position on the study and evaluation of internal control in an EDP system.

Know the computer control classifications discussed in *SAS No. 3* and be able to give several examples of each type of control.

Be able to read documentation of an EDP accounting system.

Understand the similarities between reviewing and testing a manual and an EDP accounting system.

Understand the concept of test data and be able to design such data.

Chapter 5 contains discussions of the general nature of internal control and the manner in which auditors study and evaluate the system for the purpose of determining the nature, timing, and extent of future audit tests. The cash receipts procedure used as an illustration in that chapter is assumed to be manual; therefore, no reference is made to documents that are machine-processed.

In this chapter, the elements of internal control in an EDP (electronic data-processing) system are explored. The similarities and differences in auditors' approaches to studying a manual and an EDP system are explained. The cash receipts procedure started in Chapter 5 and partially continued in Chapter 6 is modified as part of a computerized system.

Also, the concepts of statistical sampling discussed in Chapter 6 are applicable to the testing of computer-generated records; thus, no additional sampling theory is introduced here.

The Impact of the Computer

Similarities and Differences

The use of computers in profit-oriented companies dates back only to the early 1950s. Their accounting use was so limited in those early days that auditors paid little attention to them. In the late 1950s and early 1960s, increased use of computers stimulated the publication of numerous books and articles on new audit tools that should be developed to meet the challenge of different systems. As time passed, however, it also became apparent that much of auditors' traditional work would remain unchanged.

Today, it is generally acknowledged that the audit of financial statements compiled from computer-generated records is in many ways similar to, and different from, the audit of statements compiled from manual documents.

Similarities	*Differences*
1. No new auditing standards are required; the standards of field work are the same.	1. Some new auditing *procedures* are developed.
2. The basic *elements* of a good internal control system remain the same, i.e., proper segregation of duties, etc.	2. There are differences in the *techniques* of maintaining good internal control.
3. The major *purposes* of the study and evaluation of internal control still are to provide evidence for an opinion and to determine the basis for the scope, timing, and extent of future audit tests.	3. There is some difference in the *manner* in which the study and evaluation of internal control are made. A significant difference is that people have been removed from some phases of the internal control process.

AICPA Position

The AICPA's position reflects the foregoing points. AU Section 320.33 of AICPA *Professional Standards,* Volume 1 *(SAS No. 1)* contains the following comments on internal control in an EDP system.

Since the definition and related basic concepts of accounting control are expressed in terms of objectives, they are independent of the method of data processing used; consequently they apply equally to manual, mechanical, and electronic data processing systems. However, the organization and procedures required to accomplish these objectives may be influenced by the method of data processing used.

AU Section 321.02 *(SAS No. 3)* entitled *The Effects of EDP on the Auditor's Study and Evaluation of Internal Control* includes these additional observations on the audit role.

Because the method of data processing used may influence the organization and procedures employed by an entity to accomplish the objectives of accounting control, it may also influence the procedures employed by an auditor in his study and evaluation of accounting control to determine the nature, timing, and extent of audit procedures to be applied in his examination of financial statements.

Computer Controls

Let us examine, then, some specific distinctions between a manual and an EDP accounting system by discussing some computer controls. AU Section 321.06–.08 *(SAS No. 3)* divides EDP accounting controls into two major areas. General controls are defined as:

(a) the plan of organization and operation of the EDP activity, (b) the procedures for documenting, reviewing, testing and approving systems or programs and changes, thereto, (c) controls built into the equipment by the manufacturer (commonly referred to as "hardware" controls), (d) controls over access to equipment and data files, and (e) other data and procedural controls affecting overall EDP operations.

Application controls are more specific. Their function is to:

Provide reasonable assurance that the recording, processing, and reporting of data are properly performed.

For purposes of the chapter discussions, these controls are divided into the following categories:

General Controls:

1. Organization or procedure controls

2. Documentation and file controls

3. Hardware controls

Application Controls:

1. Input controls

2. Processing or programming controls

3. Output controls

In the remainder of this section we will discuss the aforementioned controls. The auditor's review and evaluation techniques will be taken up in the succeeding sections.

Organization or Procedure Controls

Organizational controls consist of the general plan designed to effect a smooth operation, accomplish the assigned tasks, and minimize the opportunity for errors or irregularities in the system. Some examples, such as proper job segregation, periodic rotation of duties, and effective operational supervision, are conceptually the same as those used in a manual system. Others, such as proper scheduling of computer runs and proper control over program changes, are unique to EDP operations.

Different Job Responsibilities

One obvious difference in the organization controls of an EDP system is the function performed by the personnel who run the system. Although there is much diversity in practice, an EDP system is likely to have the following positions.

1. *Systems Analyst*

   a.   Reviews potential applications of data processing and works with users in defining information and control requirements.

   b.   Designs and reviews the documentation for computerized systems.

   c.   Designs the various computerized and manual controls.

   d.   Monitors the program maintenance function and maintains systems documentation.

2. *Programmer*

   a.   Codes (not designs systems for) various computerized applications.

   b.   Codes the computer procedures for programmed controls.

   c.   Maintains the computer programs by coding, testing, and debugging modifications.

3. *Machine Operator*

    a.   *Runs* programs and in general maintains custody of the computer hardware.

    b.   Provides physical security over data and program files that are in an operational mode.

4. *Control Group*

    a.   Monitors manual input that is transmitted from the functional areas, such as payroll, shipping, etc. This position includes the origination or handling of control totals.

    b.   Monitors computer output and reconciles this output with the manual input, including control totals.

5. *Input Data Operator*

    a.   Converts manual input to machine-readable form.

    b.   Verifies the machine-readable data produced from manual input.

6. *Librarian*

Maintains custody and control over magnetic tapes, computer documentation, and other computer-related software. A log is kept of the uses made of the various items.

7. *Internal Auditor*

    a.   Provides assistance in defining system controls.

    b.   Performs audit functions to assure integrity and continuity of system.

No one of the listed group should have direct and complete access to the record-keeping system. The systems analysts and programmers should have no control over the day-to-day computer operations. The machine operator's knowledge of detailed programs and records should be sufficient to enable him to perform his job. Output controls should be maintained by the internal auditors or the control group. The responsibility for the custody of files should be assigned to a librarian who is separate from the other data-processing activities.

Unique EDP Control Problems

In addition to different job responsibilities, unique control problems exist in an EDP accounting system. In order to clearly visualize these problems, we should review some elements of the cash receipts system illustrated in Chapter 5.

Credits are posted to the customers' ledger accounts by individuals in one department, and cash is deposited in the bank by an individual in another department. Thus, a proper segregation of duties exists, because the functions of cash recording and cash custody are separated.

What comparable control features should exist in a computer system? First of all, it is important to understand that the basic objective of segregation of duties is equally applicable here. But the segregation is carried out differently than it is in a manual system. The "posting" of credits to customer ledger accounts is accomplished through computer runs under the control of operators and is based on computer instructions coded by programmers. In many cases, the cash deposits are processed in the same manner as they are in a manual system. Thus, it would seem that a proper segregation of duties requires that neither programmers nor machine operators have access to cash; this is a logical assumption, and such segregation apparently is common.

But access to assets is a different problem in EDP installations, because such access can exist *without* the individuals maintaining *physical* contact with the items. This situation is pointed out in AU Section 321.22 *(SAS No. 3)*, which states:

> EDP personnel have access to assets if the EDP activity includes the preparation or processing of documents that lead to the use or disposition of the assets. EDP personnel have direct access to cash, for example, if the EDP activity includes the preparation and signing of disbursement checks. Sometimes access by EDP personnel to assets may not be readily apparent because the access is indirect. For example, EDP may generate payment orders authorizing issuance of checks, shipping orders authorizing release of inventory, or transfer orders authorizing release of customer-owned securities. . . .

What can be done to improve internal control when EDP personnel have this type of indirect access to assets? Several steps are advisable.

1. The computer programs should be safeguarded so that only authorized changes are made. Normally, the computer programmers should not be allowed to make changes without proper approval (a program change control).

2. The programmers themselves should not run the computers.

3. All programs should be documented with flow charts, listings of the programs, and input, output, and file descriptions. Periodic runs should be made by independent personnel to ensure that the actual programs are processing in the way stated in the listings and flow charts.

4. Most errors or exceptions that are detected by the computer (such as incorrect information on a punched card) are produced on some type of device. Operator errors are normally detected by the system aborting, a

message printed by the computer, or the reconciliation of controls by the control group. Routine errors or exceptions resulting from conditions tested by the computer programs should be reported to the control group for disposition. Errors resulting from hardware or vendor software failures should be reported to EDP management.

5. Records should be kept of all processing actions involving the computer to lower the probability of unauthorized runs and provide a documented log for EDP management.

6. Job rotation among computer operators is desirable. If one operator were running a deliberately altered program, collusion would be necessary to perpetrate the irregularity.

Some of the foregoing safeguards (job rotation, for example) are also applicable to manual systems. The techniques for carrying out these safeguards are different in an EDP system, however.

With regard to another feature of the cash receipts system discussed in Chapter 5, recall that there is no reconciliation between the figures on the deposit slip (handled by the cashier) and the amounts on the receipt listings (posted to the customer ledger accounts by the accounts receivable clerk). The absence of such a reconciliation increases the probability of undetected errors or irregularities on both sets of documents. Ideally, an independent party makes this reconciliation.

The basic objective of comparing recorded accountability with a physical count of assets does not change when an EDP system is used. The major difference is that the recorded accountability is produced by computer runs and thus is more reliable *if* proper control measures are followed. For example, one can assume that the cash receipts tallied in the mail room are added accurately to the customer ledger accounts and posted accurately to the accounts receivable control account if prescribed procedures are followed.

Documentation and File Controls

These controls consist of procedures designed to ensure the integrity of computer storage devices and systems documentation. Some examples follow.

1. Data are stored magnetically on tapes or disks. Since the information on the tape or disk will be machine-read, there is always the possibility that an incorrect file will be run. To detect this, a magnetized internal label corresponding to the external label should be the first record of the file so that it can be properly identified by the computer during initial file processing.

2. The grandfather-father-son principle can be used with magnetic tape so that if one tape becomes unreadable or destroyed, a backup tape will be

available to reconstruct the data. The input tape of a processing run is called the "father," the output tape is called the "son," and the input tape used in the previous updating is called "grandfather."[1]

3. There should be controlled access to computer documentation, with records maintained of each usage and each modification.

4. Files should be stored in a secure, fireproof storage area under the strict control of a person charged with librarian responsibilities—critical files should be stored at off-site locations.

5. Computer programs stored on magnetic devices should be duplicated with a securely stored backup file, and strict control over changes should be followed.

In addition, EDP documentation should consist of the following items.

1. Program listing—A listing of a program that shows steps written by the programmer in the specific computer language.

2. Error listing—A list of all of the types of errors that a program has been designed to detect, the probable causes of such errors, and the most likely necessary corrective actions. This listing is prepared by the programmer when he writes the program.

3. Log—A listing of the detail of all computer runs. The logs include (a) errors detected and printed out by the computer; (b) everything pertinent to a machine run, such as identification of the run, setup actions taken, input-output files used, and actions taken by the operator each time the machine halted; (c) an accounting of all machine time, including productive time, idle time, rerun time, etc.; and (d) a listing of all tapes withdrawn from the library and returned to the library, showing date and time, to whom issued, and computer runs for which issued.

4. Systems flow chart—A graphic overview of the flow of information from one machine, document, or location to another.

5. Program flow chart—A graphic representation of the functioning of the computer program. A program flow chart shows one or more computer operations within the systems flow chart.

---

[1]This concept is understood by referring to the computer update run section of the systems flow chart. The master file is the father, the updated master file is the son, and the file that created the master file in the previous day's update is the grandfather.

6. Record layout—A listing of all of the fields of information in a record and the way in which they are arranged in the record. For example, in a punched card, each set of meaningful columns is a field.

Hardware Controls

These controls are built into the machinery itself, in contrast with processing or programming controls that make use of computer programs. There are a number of such controls.

1. Parity check—This is an internal computer check against malfunction in the machine's movement of data. For example, assume that the internal storage of a computer consists of "core-bits" that are magnetized positively to represent the number one and are magnetized negatively to represent the number zero. This binary digit characterization is shown with an illustration of ones and zeros.

Parity check bit	0	0	1	1
Bits representing	0	1	0	0
numbers, letters,	0	0	1	1
and other characters	0	0	1	0
	1	0	0	1

Notice that the parity check bit is magnetized 0 if all other bits have an odd number of ones, and is magnetized 1 if all other bits have an even number of ones. In effect, this is an odd parity check system (there can also be even parity check systems), so that in normal functioning all combinations of bits have an odd number of magnetized ones. If a malfunction occurs, and any zeros are changed to ones or vice versa, an even number of ones is created, and the machine signals an error.

2. Read-after-write check—This control reads information back just after it has been recorded and compares it with the original information.

3. Echo check—This control echoes a character back from the point of transmission to its source. For example, when information is to be transferred from the computer to magnetic tape, the recording device senses what has been received and a signal is echoed back to the computer from the tape unit. This signal is compared for accuracy.

4. Dual heads—This control is similar to an echo check. A reading device senses recorded information and transmits it instantly back to the source for comparison. Dual heads check recorded information, not just the electronic impulse.

Input, Processing (Programming), and Output Controls

AU Section 321.08 *(SAS No. 3)* defines the three application controls as follows.

> Input controls are designed to provide reasonable assurance that data received for processing by EDP have been properly authorized, converted into machine sensible form and identified, and that data (including data transmitted over communication lines) have not been lost, suppressed, added, duplicated, or otherwise improperly changed. . . .
>
> Processing controls are designed to provide reasonable assurance that electronic data processing has been performed as intended for the particular application. . . .
>
> Output controls are designed to assure the accuracy of the processing result (such as account listings or displays, reports, magnetic files, invoices, or disbursement checks) and to assure that only authorized personnel receive the output.

Because the three application controls are closely related, they will be discussed with the same set of illustrations. To introduce the subject, we will furnish brief definitions of several relevant terms.

Input and Output Control Terms

1. Batch control—This is a technique for ensuring that the data flowing through the system are complete and accurate. At an earlier point in the system, independent totals of documents are accumulated by a unit of the company and are matched against future totals derived from the processing of the same documents by another unit of the company. The three most common types of batch totals are:

    a. Control total—Usually a meaningful dollar total such as total charge sales.

    b. Hash total—A total that need not have any significance other than as a batch total. The total of customer numbers is an example.

    c. Record count—A count of the number of documents or transactions processed.

2. Check digit—An extra digit of an identification number that serves only to detect certain types of data transmission or conversion errors.

3. Keypunch verification—The independent verification of data in a punched card by placing the card in a special machine and "punching" the same source data. Instead of repunching the source data, the machine merely verifies the accuracy or indicates the existence of an error in the original keypunching.

4. Control group—A group of employees that checks the accuracy of data flow to and from the computer and sees that output is distributed properly. For example, the control group checks and reconciles batch total differences.

Processing or Programming Control Terms—Processing or programming controls are program steps placed in the computer so that certain predetermined conditions will be detected and certain types of errors in the input records will be detected when computer processing occurs. The major categories are:

1. Completeness test—A check to ascertain that all information fields are complete.

2. Validation test—A check to ascertain that all information fields contain valid data.

3. Sequence test—A check to ascertain that data being processed are in the correct sequence.

4. Limit or reasonableness test—A matching of certain data in the input records being processed against a predetermined limit built into the computer program. The purpose is to detect data above or below this limit. Examples of limits are gross pay or number of hours worked.

An Example of EDP Control Measures

To illustrate some of the controls we have discussed, there are two diagrams that can be used to compare the information "trail" in the Chapter 5 manual cash receipts system with the EDP batch-processing system used in this chapter. We will make the comparison and show how certain EDP controls can be implemented.

By examining the diagram, we can see that in the EDP batch-processing system the manual remittance advices are given to keypunch operators who punch a card for each receipt. All of the punched cards are processed through the computer (along with data on other punched cards), and the customer receipts are subtracted from the account balances maintained on magnetic tape. The information is copied by human means only once, from the manual remittance advices to the punched cards. Therefore, one can be reasonably sure that a customer's record on magnetic tape is accurate, if (1) the cash receipts information has been transcribed correctly from the manual documents to the punched cards, (2) no punched cards have been lost or substituted before computer processing, (3) the computer program that is used to update the customer's balance on magnetic tape contains no errors, and (4) the hardware did not malfunction.

MANUAL CASH RECEIPTS SYSTEM

Manual
Remittance Advice

Receipt No. 02359

Cust. No.	Name	Amount
0053	Leven Bros.	$590.00

Manual
Receipt Listing

Cust. No.	Name	Amount
0053	Leven Bros.	$ 590.00

Manual
Accounts Receivable Subsidiary Ledger

Cust. No.	Name	Dr.	Cr.	Balance
0053	Leven Bros.		$590.00	

EDP BATCH-PROCESSING CASH RECEIPTS SYSTEM

Manual
Remittance Advice

Receipt No. 02359

Cust. No.	Name	Amount
0053	Leven Bros.	$590.00

Punched Card
Remittance Advice

		Columns		
6–11	12–16	17–47	74–80	
Cust. No.	Receipt No.	Name	Amount	
0053	02359	Leven Bros.	$590.00	

Magnetic Tape
Accounts Receivable Subsidiary Ledger

0053
Leven
Bros.
Record

The system operators can increase the probability that none of these types of errors will occur by use of controls such as the following.

1. An input control can be started by requiring the clerks who process the remittance advices to run some type of manual batch total on the documents before giving them to the keypunch operators. This batch total is held by the clerks (not passed to computer personnel). Later, an output control is initiated by the control group, which compares this batch total with a total generated by the computer. If the figures are the same, no further action is taken; if the figures do not agree, the control group investigates the causes and takes corrective action.[2]

As one can imagine, any number of errors could cause the batch total of manual remittance advices to disagree with the computer-generated amount. For example, a keypunching error could be made, or punched cards could be misplaced by either keypunch operators or computer personnel. The basic point, however, is that controls should be established to provide reasonable assurance that input matches output.

Three types of batch totals (control total, hash total, and record count) are illustrated below by a listing of several receipts from Chapter 5.

If the following receipts were processed on a given day	*Customer No.*	*Name*	*Amount*
	0053	Leven Bros.	$ 590.00
	0007	Alpha	3,525.00
	0056	Wolsey	480.00
	0081	Eastgate	347.00
	0036	Keller-Goodman	480.00
	0012	Gamma	5,210.00
	0043	Bart Bros.	288.00
	0008	Griffin	750.00
	0057	Cardwell	1,020.00
	0063	Mann's	2,820.00
The control total would be			$15,510.00
The hash total would be	416	(The sum of customer numbers.)	
The record count would be		10	(The number of documents.)

2. A programming control can be used to enable the computer itself to detect certain errors in the punched cards or the manual remittance advices. Such a control can be installed by running the punched cards

---

[2]This is only one procedure that might be followed. The control group could generate the batch totals, take them directly from the clerks, etc.

through a computer edit program prior to updating the accounts receivable subsidiary records on magnetic tape.[3] A wide variety of potential errors can be detected and indicated by some type of error message. The complexity of the edit program depends on the type of system used. The following two errors are examples of those an edit program might detect.

a.   The punched cards may contain incomplete or inaccurate information in one or more fields. For example, the customer number may have been left blank, or it may have been punched in the field reserved for the customer name. Edit steps designed to detect these types of mistakes are called completeness or validation controls.

b.   The punched card or the manual remittance advice may contain a cash receipt figure that is obviously wrong (a $5,000 receipt from a customer with an average balance of $200). A computer edit program can contain steps that produce an error message if the dollar receipts exceed a predetermined amount. These edit steps are referred to as limit or reasonableness controls.

The Review of Internal Control

We have discussed computer controls and have compared EDP and manual accounting systems. We will now turn to explanations of the review and evaluation of internal control. The illustration is on pp. 284–285.

To explain the review process, an example of a computer batch-processing system is presented. The flow charts and other documentation are not complete replicas of an actual system with all its complexities. Instead, the model is kept relatively simple for teaching and illustrative purposes.

Description of a Systems Flow Chart

The systems flow chart starts at the point at which the information for the sales invoices, credit memos, and remittance advices is in manual form awaiting conversion to punched cards. The manual sales invoices and credit memos are handled in the Billing Department, whereas the remittance advices are created in the mailroom (see the illustration in Chapter 5). In fact, the sales and accounts receivable systems up to and including the creation of these documents are the same in this comparison of EDP and manual systems.

The sales invoices at the top of the systems flow chart are created in the Order Department and sent to the Billing Department, where they are held awaiting shipment notification. When this notification is received, the pricing on the invoices is checked, extensions are made, and one set of copies is sent to the customers. In the manual system the other copies are sent to the Accounts

---

[3]There are several ways of accomplishing this edit step; the method suggested here is one of them.

Receivable Department for hand posting, but in the EDP system a control total is derived, and the copies are given to the control group.

The credit memos at the top of the flow chart originate in the Billing Department after notification of credit approval is received from the Credit Department. One set of copies is sent to the customers, and in the manual system the other copies are given to the Accounts Receivable Department for hand posting. In the EDP system a control total is calculated and the copies are sent to the control group.

The manual procedure for remittance advices is described in detail in Chapter 5. The manual documents are received by the mail clerk, along with receipts on account. At the same time the money is given to the cashier, the remittance advices are passed to the Accounts Receivable Department for hand posting. The computer system operates in the same manner described in the two preceding paragraphs. A control total is computed, and copies of the documents are given to the control group.

The flow chart shows that after the control group checks the documents and assigns batch numbers (a different number for each day), the keypunch operators create a card for each line on each document. For example, if three items of merchandise are sold to a customer, three cards are punched.

Another example of an input control is the machine verification of the data punched cards. This function typically is performed on a verifier that checks the accuracy of the data from the original manual documents.

At the same time the keypunching and verification are done, a count is made of the number of punched cards handled. This additional input control provides assurance that cards are not lost during computer processing. A programming control is added by instructing the computer to tally the number of cards read and processed. The control group then exercises an output control by checking the computer-generated number with the card count number held in the keypunching area.

In this system it is assumed that two computer runs are made, one for editing the input data on punched cards, and the other for (1) updating the customer ledger account balances, and (2) producing a transaction register and exception reports.

The output of the computer edit program consists of (1) a printed list, and accompanying explanations, of designated input data errors that are detected by the edit program, and (2) control totals of the sales invoices, credit memos, and remittance advices. The control group sees that any errors in the input data are corrected and also checks the computer-generated control totals against the adding-machine tapes kept by the Billing Department and the mail clerk. If the two sets of figures do not reconcile, the differences are checked, and another computer run is made. No final updating of the master files can be done until the edit program clears the input data for errors and until the computer-generated control totals agree with batch totals produced by the adding-machine tapes.[4]

---

[4]This is a conceptual statement indicating that total accuracy of the input data must be achieved at some point before the master files can be assumed to be correct. The particular techniques for achieving this accuracy differ among systems.

A SYSTEMS FLOW CHART FOR A COMPUTER SYSTEM – BATCH PROCESSING

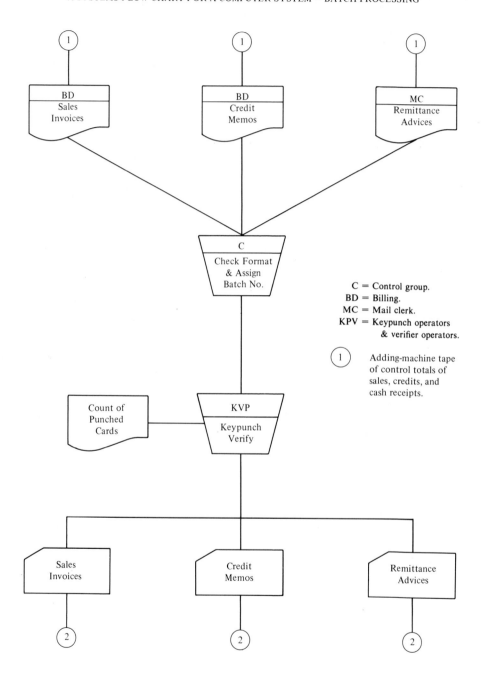

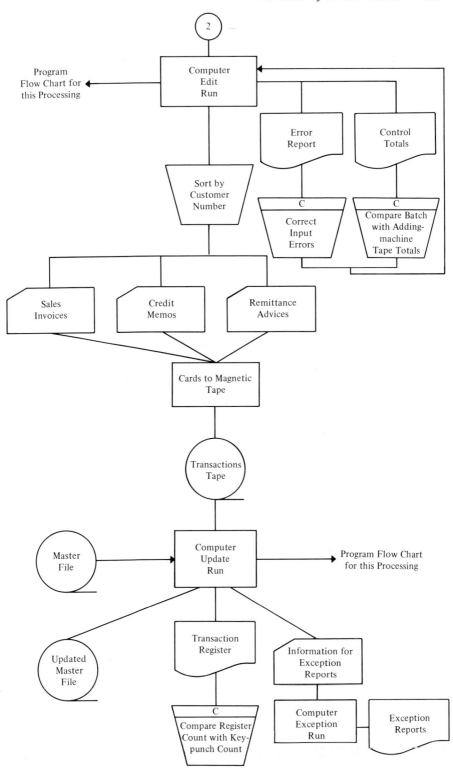

After the input cards clear the first computer run, they are sorted according to customer number. Within the customer number, the cards are arranged into sales invoices, credit memos, and remittance advices (receipts on account). The information on the cards then is transferred to magnetic tape, which is referred to as the transactions or detail tape.

It should be pointed out that the sorting process is mandatory because a sequential record, magnetic tape, is employed. This ordering of records is not necessary if a magnetic disk is used.

The transactions tape and the master file tape are processed with a computer updating run, which results in the following output.

1. An updated master file tape containing the revised information incorporated by the transactions tape.

2. A transaction register that shows:

    a.   The beginning customer account balance.

    b.   The charges to the account.

    c.   The cash receipt credits to the account.

    d.   The non-cash credits to the account.

    e.   The ending customer account balance.

    f.   A record count of the number of cards processed.

The control group implements another output control by comparing this record count with the one produced during the keypunch operation.

3. A deck of punched cards from which a printed exception report is developed. The report shows various items of information that need to be brought to someone's attention. They include:

    a.   Credit balances more than 30 days old.

    b.   Charges to customers with balances more than 60 days old.

    c.   Balances more than 90 days old.

The next page is an illustration of a record layout for sales invoices.

Summary of Controls Shown on Systems Flow Chart

For review, the controls that are apparent or implied from reading the flow chart are listed in the diagram on p. 288, with the applicable sections of the systems flow chart illustrated at the side.

RECORD LAYOUT FOR SALES INVOICES
(INCLUDING EXAMPLES—ONE CARD PER LINE)

								Column(s)	
1 Transac. Code	2–5 Batch No.	6–11 Cust. No.	12–16 Invoice No.	17–47 Name	48–53 Invoice Date	54–65 Descrip- tion	66–68 No. of Items	69–73 Price	74–80 Invoice Amount
1	0501	000007	02102	Alpha Wholesale Co.	X91003	DX1Sofa	012	18000	0216000
1	0501	000007	02102	Alpha Wholesale Co.	X91003	G57Chair	011	14500	0159500
1	0502	000002	02103	Beta Wholesale Co.	X91004	R93Table	003	10600	0031800

The column headings are the fields into which charge sales information will be punched. A 1 punch in Column 1 is read by the computer as a charge sale, which means that the amount in Columns 74–80 is added to the customer's balance. A 2 punch in Column 1 is read as a cash receipt, and a 3 punch in Column 1 is read as a non-cash credit.

Columns 2–5 contain the batch numbers that are assigned to each day's sales.

Note that there is a different batch number for each day. An example of a programming control is to program the computer to read the batch numbers in Columns 2–5 and to accumulate the amounts in Columns 74–80 for all like batch numbers. This computer-generated total is compared by the control group with the adding-machine tape total accumulated in the Billing Department.

Input Control

The generation of batch totals of sales invoices, credit memos, and remittance advices.

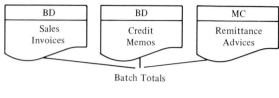

Batch Totals

Input Control

The checking of input and the assigning of batch numbers by the control group.

Input Controls

1. Machine verification of data punched in cards.
2. The count of punched cards processed.

Programming Control

A computer edit run for detecting errors in the punched-card input and for computing control totals.

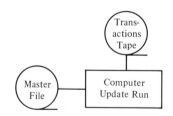

Output Controls

1. Ascertaining that input errors detected by the edit program are corrected.
2. Reconciling computer-generated batch totals with adding-machine batch totals generated in the Billing Department and mailroom.

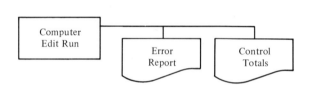

Programming Control

An edit routine in the computer update run that detects certain illogical conditions, such as a missing customer record on the master file.

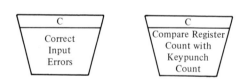

Output Control

Reconciling the computer-generated count of punched cards with the count tallied by the keypunch operators.

Overall Controls

In addition, the EDP system should have this overall control.

	Amounts		File
	Beginning Balance	from	Master file
plus	Today's input	from	Detail file
minus/plus	Errors/rejects	from	Edit file
plus	Internally generated amounts, such as late charge, should equal Ending Balance, which becomes to-morrow's *Beginning* Balance	from	Master file

Note: The master file should also have a computer-generated record count (in and out), as well as a control total(s) over selected dollar amount(s).

Programming Controls in the Computer Edit Run

The Purpose of a Program Flow Chart—One programming control that needs special explanation is the computer edit program that detects and produces messages for various errors in the punched cards. Part of the documentation that backs up this edit program is a program flow chart or a block diagram describing the computer logic. An illustration is on page 290.

Although auditors probably do not need to be capable of reading compli-cated program flow charts (some are very detailed), they should know the prescribed programming controls used by the client company. Reading the program flow charts is one method of gaining this knowledge. Therefore, a step-by-step explanation of computer logic in the edit program follows the illustration of the flow chart.

The Limit and Validation Checks—This particular diagram reads from top to bottom and from left to right. As each card (or record) is read, four preliminary conditions are checked by the computer (as indicated by the first four decision symbols).

1. A check is made to ascertain that the customer number does not contain blanks or zeros. If it does, an error message is printed, and the card must be corrected or completed before it is run through the computer program to update the master files. Note that the arrow branching from the error message symbol comes back into the main part of the program. This indicates that all error conditions are checked on a given card.

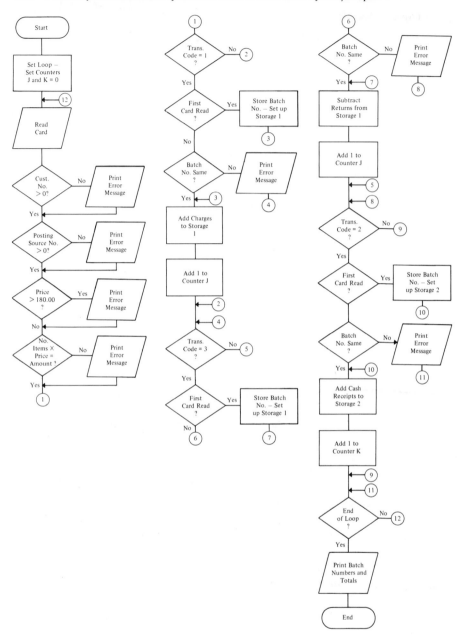

2. The program checks for a posting source number that has blanks or zeros. If any card has this characteristic, the computer follows the same routine as that followed in the customer number check.

3. The program reads the price on the card and compares it with the highest price of the items sold by the company. If the price on the card exceeds this figure, an error message is printed. It should be noted that the program does not produce an error message if the price is wrong but less than $180.00. The edit step is strictly a limit check.

4. The program multiplies the price by the number of items and compares the result with the amount read from Columns 74–80. If the figures do not match, an error message is printed.

It is obvious that these four program checks are but a few of many edit steps that are needed to ensure accuracy, authenticity, and completeness of the punched-card input. This flow chart is limited to four checks for ease of illustration. If the edit program shown in this flow chart actually were used, auditors probably would judge the programming controls to be insufficient.

The Accumulation of Batch Totals—The rest of the program flow chart illustrates the computer accumulation of two sets of batch totals, one set representing the difference between sales invoice charges and sales returns, and the other representing the sum of cash receipts.

At the end of the edit run, the control group compares the first batch total produced by the computer run with the adding-machine total calculated in the Billing Department (the difference between the sales invoice charges and sales returns). The second computer batch total is compared with the adding-machine total derived in the mail room. Any differences are reconciled and, if necessary, errors are corrected.

The computer and adding-machine totals might be different for several reasons, such as incorrect calculations in the Billing Department or mail room, or lost or incorrectly keypunched cards. As has been pointed out, the adding-machine totals should not be seen by the computer operator; otherwise, many of the control features will be lost.

The table on page 292 shows the comparison concept by listing 10 customer cash receipts.

Reading the balance of the flow chart, one notes that card Column 1 is tested for the transaction code. A 1 indicates a charge sale, a 2 denotes a cash receipt, and a 3 signifies a sales return. If the card column does not contain a 1, the computer tests for a 3 (follow connector symbol 2). If a 3 is not punched in Column 1, the computer tests for a 2 (follow connector symbol 5). If none of these numbers is punched in Column 1, the program passes to the end (follow connector symbol 9).

Auditors might observe at this point that no provision is made for numbers other than 1, 2, or 3 that might be punched in the transaction code field. This seems to be a weakness of the edit program and probably would be noted.

If the transaction code is 1, the program tests to determine whether this is the first card read with a transaction code 1. If so, an arithmetic storage section is set up in the computer to accumulate the charge sales amounts read from

Customer No.	Name	Correct Amount	Assume that keypunch transposition errors were made, and these customers had the following amounts in their cards.
0053	Leven Bros.	$ 590.00	$ 590.00
0007	Alpha	3,525.00	3,255.00
0056	Wolsey	480.00	480.00
0081	Eastgate	347.00	347.00
0036	Keller-Goodman	480.00	480.00
0012	Gamma	5,210.00	5,120.00
0043	Bart Bros.	288.00	288.00
0008	Griffin	750.00	750.00
0057	Cardwell	1,020.00	1,020.00
0063	Mann's	2,820.00	2,820.00
The adding-machine batch total is		$15,510.00	
The computer batch total is			$15,150.00

Columns 74–80 of the cards. If this is not the first card read that contains transaction code 1, the card is tested to determine whether it contains the same batch number as previous cards with transaction code 1. If it does, the charge sales are added to the arithmetic storage; if not, an error message is printed.

This logic is shown on page 293 by a section of the flow chart.

Again, the auditor might observe that if the first card to show a transaction code has the wrong batch number, the sales amount for that card will be stored incorrectly. Also, succeeding cards containing that transaction code and having the *correct* batch number will be rejected. This problem can be illustrated by using part of the record layout for sales invoices and simulating several sales.

Column(s)

1 Transaction Code	2–5 Batch No.	48–53 Invoice Date	74–80 Invoice Amount
1	0501	X91003	0216000
1	0501	X91003	0159500

Assume that the first card processed with a transaction code 1 is the record shown above with an amount of $2160.00. *If* 0501 is the correct batch number, the program will correctly place $2160.00 in a storage accumulation. All

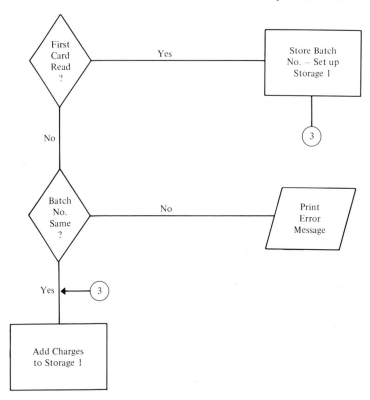

succeeding cards with transaction code 1 *and* batch number 0501 will process correctly, and the amounts in Columns 74–80 will be added to the storage accumulation.

However, assume a situation in which the batch number of the first card is wrong.

<table>
<tr><th colspan="4">*Column(s)*</th></tr>
<tr><td>1</td><td>2–5</td><td>48–53</td><td>74–80</td></tr>
<tr><td>Transaction</td><td>Batch</td><td>Invoice</td><td>Invoice</td></tr>
<tr><td>Code</td><td>No.</td><td>Date</td><td>Amount</td></tr>
<tr><td>1</td><td>0502 (wrong)</td><td>X91003</td><td>0216000</td></tr>
<tr><td>1</td><td>0501</td><td>X91003</td><td>0159500</td></tr>
</table>

Here, the sales amount of $2160.00 would be stored incorrectly, and the sales amount of $1595.00 would be rejected incorrectly, because the batch numbers do not match. If the first card of every transaction code does not contain the correct batch number, this edit program will not produce the correct totals.

Therefore, an auditor observing the discrepancy would evaluate this part of the edit program as a weakness in programming controls.

The rest of the program flow chart contains the same routines for transaction codes 3 and 2. The final output is a listing of the batch numbers and batch totals that should match the adding-machine totals calculated in the Billing Department and mail room.

The Internal Control Questionnaire

The foregoing discussion of flow-chart review contains observations that undoubtedly would be a part of an internal control questionnaire, for example, the comments about controls established over sales invoices, credit memos, and remittance advices. Nevertheless, it is useful at this point to pose some hypothetical questions and point out differences between an EDP and a manual system.

Question	*Reason for the Question*
Is there a separation between the functions of programming and computer operation?	If these functions were combined, the programmer might be able to place some substitute steps in the program, instructing the computer to branch around an important edit check. An example is the total number of hours worked. The program steps could branch around the edit of a certain employee's time records and ignore the excessive hours recorded for this individual. The use of a separate operator would not ensure that an irregularity of this type would not occur, but it would improve the chances, particularly if the programmer had no further access to the program. In addition, if these functions were combined into one job, the excessive workload might impair efficiency.
Are programs periodically tested by a control group or an internal audit group?	Program steps that are run on the computer may not be the same as the steps shown on the program listing. They should be tested occasionally by an outside group for intentional or unintentional changes. Standard procedures should be used for initiating and reviewing program changes.
Are programmers or machine operators allowed to correct record errors and reconcile the control totals?	Allowing the programmer to perform these functions is similar to allowing an accounts receivable clerk to keep the general ledger control account. Because errors or irregularities could be covered up, a separate person or group (preferably a member of a control group) should perform this function.
Is there controlled access to data files and other EDP documentation?	Unauthorized access to programs or files could result in improper alterations. Limited access to ledgers and journals is required in a manual system, but in an EDP system, alteration might be more difficult to detect.

Two points should be noted about these questions. The terms used are unique to EDP systems, which indicates that auditors make different inquiries and observations on these systems in order to answer the questionnaire. But the underlying *philosophy* of good accounting control is the same in both an EDP and a manual system.

For example, if one thinks of computer programming as a type of transaction recording and of computer operation as potential access to assets resulting from transaction, then it is obvious that, ideally, the two functions should be separated. The periodic testing of records by an outside group is designed to lower the probability that errors or irregularities can be perpetuated and is a control device equally applicable to manual systems (although computer programs obviously are not used).

Summary of Procedures for Reviewing EDP Controls

We have discussed an EDP system in detail, defined the relevant terms, examined a flow chart, a narrative description, and part of an internal control questionnaire. Now, we can summarize the review procedures that an auditor should conduct to give himself reasonable assurance that the system contains the necessary controls to prevent or detect material errors and irregularities. The list is not comprehensive but is suggestive of the comprehensive nature of EDP review procedures.

Organization or Procedure Controls

1. Review the organization chart for job descriptions, titles, etc.

2. Determine whether the positions of (a) system analyst, (b) programmer, and (c) computer operator are separated.

3. Review the procedures for changing master files and programs.

4. Determine whether computer operators are rotated periodically among computer runs.

5. Review procedures used by the control group in handling input and output data, reconciling differences in totals, etc.

6. Ascertain whether access to console messages and console logs is limited to proper personnel.

7. Review the internal audit procedures in the EDP areas.

8. Ascertain whether logs are kept on each computer run. Review certain of these logs for error messages and review their ultimate disposition.

9. Review the procedures for handling error messages on the computer console or printer.

10. Determine whether EDP personnel are bonded.

11. Determine whether the internal auditors periodically check program listings against program runs.

12. Ascertain whether internal auditors periodically run programs with test data.

Documentation and File Controls

1. Determine whether the master tapes, program listings, and other files are kept in safe, fireproof storage.

2. Determine whether access to tapes, program listings, and other computer documentation is limited to proper personnel. Determine whether a log is kept of the usage.

3. Review the log of tape and computer program usage.

4. Determine whether changes made to master files are authorized properly and are recorded.

5. Determine whether, where applicable, external and internal labels are used on files.

6. Determine whether file protection rings are used on magnetic tapes.

7. Review the insurance coverage over EDP files.

8. Determine whether master file data are printed out periodically and checked.

9. Ascertain whether grandfather-father-son or some appropriate backup file system is used.

10. Ascertain whether documentation for the EDP system includes the following:

    a.   Systems flow charts

    b.   Program flow charts

    c.   Program listings

    d.   Error listings

  e. Run manuals

  f. Operator instructions

  g. Records of computer runs

  h. Records of changes in computer programs

  i. Listings and explanations of controls

  j. Record layouts

 11. Review the procedures used to modify the documentation.

Hardware Controls—Review the hardware controls in operation, such as echo checks, read-after-write, dual heads, and parity checks.

Input/Output Controls

 1. Determine whether controls are established over input data before computer processing.

 2. Determine whether computer-generated control totals are reconciled with controls established over input data.

 3. Review the policies for distributing computer output.

 4. Determine whether data rejected by the computer are reviewed at the appropriate level, corrected and re-entered.

 5. Review the procedures and the forms used to keep batch totals.

 6. Review the use of batch totals in the system and other controls such as control totals, hash totals, and record counts.

 7. Determine whether input data are sequentially numbered.

Processing or Programming Controls

 1. Determine whether edit programs are used for the applicable computer runs.

 2. Review the edit programs to ascertain whether they contain the proper programming controls. In general, edit programs would include:

  a. Limit or reasonableness tests.

  b. Validity tests.

  c. Sequence checks.

    d.   Accumulation of control totals, hash totals, or record counts (or some combination of the three).

    e.   Tests for unmatched master and detail data.

Preliminary Evaluation

The final step in the review process of the EDP system, like its manual counterpart in Chapter 5, should be a preliminary evaluation to assess its major strengths and weaknesses and to form a judgment on the system's reliability. To facilitate easy comparison with the preliminary evaluation of the manual system, we will use the AICPA model shown in Chapter 5. We should point out, however, that different auditors may have varying opinions of the system. These ideas represent only one point of view.

Errors or Irregularities that Could Occur in the System	Internal Control Procedures (if any) that Should Prevent or Detect These Errors or Irregularities	Internal Control Procedures that are Documented as Part of the System
1. The machine operator could make unauthorized changes to the programs that would result in such things as duplicate payments to vendors, etc.	1. The functions of programmer, machine operator, and librarian should be separated.	1. These functions are separate.
2. Errors could occur when machine-readable documents are created, verified, or processed by the computer.	2. Batch controls should be established over the source documents before they are released for further processing.	2. Batch totals are accumulated where the source documents are created. These totals are matched with the computer outputs by the control group.
3. Keypunch errors could occur when machine-readable documents are created from source document information.	3. There should be verification of the source document data and a system should be used to allow the computer to detect errors missed in the verification.	3. There is machine verification, but a check digit system on the source documents is not used.
4. Blank fields, illogical conditions, and out-of-limit figures could be on the machine-readable documents even after verification takes place.	4. The machine-readable documents should be processed through a computer program that contains programming controls.	4. The machine-readable documents are processed through a computer edit program that tests for certain error conditions.

Like its manual counterpart, the EDP system used as an illustration has some definite strengths.

1. There appears to be an adequate separation of the sensitive jobs of programmer, machine operator, and librarian.

2. The control group reconciles manual input and computer output. This step is very important, because the visible audit trail disappears.

3. Control totals are established at the point at which the information from the source documents is converted into machine-readable form. These totals are retained by the originators.[5] A reconciliation is performed by the control group.

4. There is machine verification of the source document data that are keypunched.

5. *Some* programming controls have been established. Error messages are printed for invalid and out-of-limit data, and control totals are calculated by the computer.

On the negative side, the following areas appear to be weak.

1. No check-digit system is used for the customer number. A check digit is an extra digit added to the number. In itself, the digit is meaningless, serving only as a device for detecting transposition errors in keypunching or miscoded data on the source documents. There are many types of check-digit systems; one method, based on a multiple of 10, works as follows.

To a five-digit customer number, such as 25018, a sixth digit is added at the end. It is calculated by successively multiplying, from left to right, alternate digits by the multipliers 1 and 2. For example, 25018 is multiplied by 12121 as follows.

$$
\begin{array}{ccccc}
2 & 5 & 0 & 1 & 8 \\
1 & 2 & 1 & 2 & 1 \\
\hline
2 & 10 & 0 & 2 & 8
\end{array}
$$

The individual products are added $(2 + 10 + 0 + 2 + 8 = 22)$. This sum is subtracted from the next highest multiple of 10 $(30 - 22 = 8)$; 8 is the check digit. The total customer number is 250188. The digit 8 is the only one that can possibly be used with 25018. The computer is programmed to perform the same arithmetic and calculate a digit. The computer compares its calculated digit with the one it reads in the data card. If the digits do not match, an error message is printed. For instance, if the keypunch operator transposes digits in the customer number and punches 251088, the computer-calculated check digit is 9 $(25108 \times 12121 = 21, 30 - 21 = 9)$, and, obviously, the numbers do not match.

[5]The system could provide that the control group receive all data transmitted from other departments.

2. The edit program has an inadequate variety of tests. Several other input areas should be checked. (Examine the program flow chart.)

3. The computer could be utilized more effectively. For example, the invoice pricing and extensions could be programmed rather than performed manually.

4. There is no check for the arithmetic sign. The edit program treats all numbers as positive.

5. The processing of cash receipt cards could produce an erroneous error message for an invalid price or improper amount. This could occur because cash receipt cards have nothing in the number of items or price columns.

The purpose of this preliminary evaluation is to provide a basis for the future testing of account balances. This evaluation is subject, of course, to the results of the compliance tests, which are discussed in the next section.

Tests of Compliance in an EDP System

Around or Through the Computer

The basic reason for conducting tests of compliance is the same in an EDP system as it is in a manual system; auditors are making a determination as to whether the system is functioning as prescribed. However, some of the methods may differ, depending on the philosophy adopted by the auditors. If auditors choose to ignore the EDP processing and work "around the computer," they perform tests that are almost identical to the steps used in a manual operation. An example of this approach can be illustrated by showing examples of manual input and printed output.

Sales Invoices		Credit Memos		Remittance Advices	
Block Whlsl.	$1,020.00	L Street	$145.00	Block Whlsl.	$940.00
Beta Furn.	696.00	Kellon	205.00	Beta Furn.	509.00
Hogan Display	208.00				

The data are transferred to machine-readable form, run through two computer programs, and (in part) printed out.

TRANSACTION REGISTER

	Beginning Balance	Charges	Sales Returns	Cash Receipts	Ending Balance
Block Whlsl.	$940.00	$1,020.00	$ —	$940.00	$1,020.00
Beta Furn.	509.00	696.00	—	509.00	696.00
Hogan Display	—	208.00	—	—	208.00
L Street	145.00	—	145.00	—	—
Kellon	305.00	218.00	205.00	100.00	218.00

Because the audit trail figuratively disappears, complete tracing cannot be done. But auditors can "go around" by testing the validity of the manual input and the printed output. If satisfied, they assume that the computer programs are processing correctly.

In addition, several tests of compliance can be performed without either going through or going around the computer. A case in point is the check for proper approval on credit memos; another is the observation that machine verification is performed on punched cards.

There is some risk in using an "around-the-computer" approach. Suppose that the program has been altered to understate or ignore charges to certain groups of customers. The printout on the transaction register could have the appearance of correctness even though the balances stored on magnetic tape are understated. There are ways by which this procedure may be detected or discouraged, such as surprise confirmation of magnetic tape balances or a strong control system that requires independent reconciliation of output forms. Another weakness is that auditors see only the results of transactions *actually* processed, which may not reveal potential problem areas. Thus, despite the practical advantages of auditing around the EDP system, auditors can leave gaps when they do so.[6]

Test Data

Recognizing this potential problem, members of the accounting profession have devised a method of going "through the computer" itself. This technique, which involves the use of test data, was designed for the batch systems that have been employed since the early 1960s. The method operates as follows.

1. Simulated transactions are created and entered in the system (punching the data on cards is one way).

[6]Auditing through the computer has been advocated for a number of years. See Wayne S. Boutell, "Auditing Through the Computer," *The Journal of Accountancy,* November 1965, pp. 41–47.

2. These transactions are entered in the auditor's worksheet, along with the predetermined computer results.

3. The simulated transactions are run with the client's computer program, and the computer results are compared with the predetermined results. If the two sets match, the client's program is presumed to be functioning as called for in the system documentation.

The test-data procedure is depicted in the following diagram.

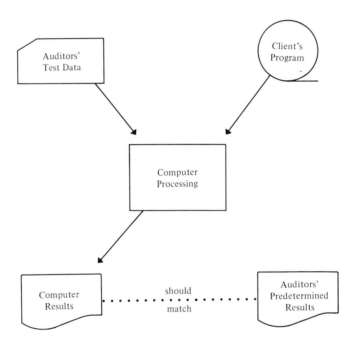

It should be noted that the diagram shows the auditors entering the system by punching the simulated transactions on cards. If they wish, auditors could create the transactions in the source documents, which would be tantamount to testing the keypunching verification techniques.

What assurance do auditors have that they are using the correct program? If a different program is being used to perpetuate an irregularity, the client might furnish the auditors with a program that correctly processes the test data, while he uses the other program to process the real data. There is no assurance that this substitution will not be carried out, but precautionary steps can be taken to lower the probability of such an occurrence.

For instance, the test data could be run on a surprise basis. It is not unusual for audit steps to be performed in this way; counts of cash and negotiable securities are prime examples. However, a more practical approach might be for auditors to obtain a copy of the client's program in advance of the test-data run. Then, at the client's convenience, the test data could be processed, and the program used in this processing would be compared with the program obtained earlier.

An Example of Test Data

The foregoing discussion provides a background for a specific illustration of the test-data procedure. Reference to the systems flow chart shows that there are two computer runs. The first run is an edit program designed to (1) reject and print an error message for certain inaccurate or invalid data on the invoices, credit memos, and remittance advices, and (2) print control totals that match similar totals produced in the Billing Department and the mail room. The second run is an updating program that incorporates current transactions into the file records. Though both computer programs should be tested by the auditor, the illustration can be simplified by confining the discussion of test data to the edit program.

There are some preliminary edit tests incorporated in the program flow chart.

1. A validity test of the customer and posting source numbers. If the computer reads a punched card that has a blank or zeros in either of these fields, an error message is printed.

2. A limit test of the price. The price of the most expensive merchandise sold by the client is $180.00. If the computer reads a punched card that has a price in excess of this amount, an error message is printed.

3. A validity test of the amount. If the computer reads a punched card that contains an extension error on the amount, an error message is printed.

Next, there are three tests on the batch numbers. If this number is incorrect for any or all of the transaction codes, an error message(s) is produced. Finally, totals are accumulated for all cards that contain the same and correct batch number.

Following are illustrations of (1) test data designed by auditors, (2) auditor's predetermined results, and (3) the actual computer output. Observe that auditors enter at least one input card without errors to test the program output under conditions in which there are no inaccuracies. In addition, auditors enter at least one card for every type of error condition tested in the edit program. Because seven errors are possible, eight cards are illustrated.

EXAMPLES OF TEST DATA DESIGNED BY AUDITORS

Type of Error	Column 1 Transaction Code	Columns 2–5 Batch No.	Columns 6–11 Customer No.	Columns 12–16 Source No.	Columns 17–47 Name	Columns 48–53 Date	Columns 54–65 Description	Columns 66–68 No. of Items	Columns 69–73 Price	Columns 74–80 Amounts
No Errors	1	0501	000007	02102	Alpha	X91003	DX1 sofa	012	18000	0216000
Invalid Customer No.	2	2499	000000	02349	Leven	X91003				0059000
Invalid Source No.	3	0501	000010	00000	Kellon	X91003	TX1 sofa	001	11200	0011200
Excess Price	1	0501	000007	02102	Alpha	X91003	G57 chair	011	24500	0269500
Extension Error	1	0501	000007	02102	Alpha	X91003	G57 chair	011	14500	0145500
Billing Batch No. in Error	1	0502	000007	02102	Alpha	X91003	G57 chair	011	14500	0159500
Billing Batch No. in Error	3	0502	000010	00371	Kellon	X91003	X45 table	001	11200	0011200
Mail Room Batch No. in Error	2	2500	000053	02349	Leven	X91003				0059000

EXAMPLES OF PREDETERMINED OUTPUT DESIGNED BY AUDITORS

The first card's processing should not show any errors.

The second card's processing should produce an error message; the Customer No. field contains zeros.

The third card's processing should produce an error message; the Source No. field contains zeros.

The fourth card's processing should produce an error message; the Price field contains $245.00.

The fifth card's processing should produce an error message; the Amount field contains $1,455.00.

The sixth card's processing should produce an error message; the Batch No. field contains 0502.

The seventh card's processing should produce an error message; the Batch No. field contains 0502.

The eighth card's processing should produce an error message; the Batch No. field contains 2500.

The batch totals produced by the computer should be:
(1) $6,198.00—the sum of (Cards 1, 4, and 5 amounts minus Card 3 amount).
(2) $590.00—Card 2 amount.

ACTUAL OUTPUT FOR
COMPUTER EDIT PROGRAM
FOR CUSTOMER CHARGES AND CREDITS

Transac. Code	Batch No.	Cust. No.	Document No.	Date	No. of Items	Price	Document Amount
2	2499	0	2349	X91003	0.	0.00	590.00
Invalid Cust. No.							
3	501	10	0	X91003	1.	112.00	112.00
Invalid Document No.							
1	501	7	2102	X91003	11.	245.00	2,695.00
Price Over Limit							
1	501	7	2102	X91003	11.	145.00	1,455.00
Incorrect Extension·							
1	502	7	2102	X91003	11.	145.00	1,595.00
Incorrect Billing Batch No.							
3	502	10	371	X91003	1.	112.00	112.00
Incorrect Billing Batch No.							
2	2500	53	2349	X91003	0.	0.00	590.00
Incorrect Mail Room Batch No.							

Billing Batch No.	Mail Room Batch No.	No. Cards	Total
501		4	6,198.00
	2499	1	590.00

Critique of the Test-Data Approach

What, then, is the major advantage of using test data, particularly in comparison with an around-the-computer approach to compliance testing? A major benefit is the greater assurance that is gained about the reliability of the client's computer programs. The auditors are sure that the programs they run function as prescribed, even though they may not be completely certain that they are running the right programs.

However, the use of test data does have certain limitations, and it is not the final answer to the audit of EDP records.

1. For instance, a successful test-data run does not necessarily indicate a strong system of internal control, because other types of errors or irregularities could occur outside the computer processing area. A prime example is the failure of a mail clerk to report all cash receipts.

2. Test data determine whether the program that has been *furnished* the auditor is functioning as it should. The auditor cannot be *certain* that this is the same program used in daily operations, although proper precautions can greatly increase that assurance.

3. The test-data approach may be limited to a check of the functions by the *client's* program, and that program may be inadequate to edit the client's machine-readable records. This possibility is demonstrated by the program flow chart analyzed in this chapter.

4. The test-data approach was designed for a batch-processing system. There is some question as to its suitability for the newer online systems, where there is no intermediate processing between the input devices and the central processing unit, and where more continuous processing takes place.

Despite these drawbacks, the use of test data is definitely a step in the direction of going through the computer and, with proper safeguards, can provide auditors with insights into the EDP system that are not gained by ignoring the computer.

Variations of the Test-Data Approach

In recent years, variations of the test-data method have been suggested by members of the accounting profession.[7] One such innovation is the integrated test facility (ITF). By this "parallel simulation" method, both simulated data and actual client data are run simultaneously with the client's program, and computer results are compared with auditors' predetermined amounts. Experience with

---

[7]For more detail, refer to Barry R. Chaiken and William E. Perry, "ITF—A Promising Computer Audit Technique," *The Journal of Accountancy,* February 1973, pp. 74–78.

this technique is still limited. A theoretical advantage of this method over the traditional test-data approach is the ability of auditors both to check the functioning of the client's program and to test the accuracy of the client's output.

Another variation of the test-data approach is for auditors to run the client's program with actual data several times during the year. The purpose of this procedure is to provide relative assurance that the client's records have been processed consistently during the year.

This approach appears to be an EDP counterpart of the traditional test of transactions in a manual system, in which data are traced, for selected time periods, from source documents to journals and ledgers. In both procedures, auditors are attempting to satisfy themselves that record-keeping procedures are employed consistently. Undoubtedly, there are benefits to be gained from monitoring of this sort, but it could prove to be costly.

The Auditors' Program

In recent years, a new and distinctly different approach to compliance testing in an EDP system has been implemented by some CPA firms, particularly larger ones. The *auditors'* computer programs are run with actual client data. The computer results are compared with actual client output to test its validity. If the computer output matches the previously processed client output, it is assumed that the client's computer processing, in this area, is acceptable.

This method is illustrated below by a comparison of its diagram with the diagram for the test-data approach.

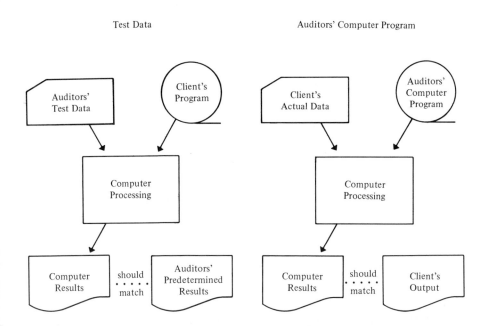

Test Data   Auditors' Computer Program

Several advantages of the newer approach should be apparent. First, auditors can be almost certain that the correct program is being run, because it is designed either by them or by their firm. With the test-data approach, auditors have only relative assurance that the client program being processed is the correct one.

Another distinct advantage of auditors' programs is that tests can be made of the controls that auditors believe are important in the client's system. For example, the edit program illustrated previously in this chapter shows a certain number of programming controls, such as a limit test on the price and a validity test on certain fields. If auditors use their own program, they can test for *any* condition, regardless of whether or not that condition is tested in the client's edit program. For example, auditors may wish to check for alphabetic data in the customer number. A program run using actual client data should detect errors of this sort, even if the client's programming routine does not.[8] Thus, auditors' programs *might* detect control inadequacies not otherwise pinpointed by the use of test data.

To explain this concept more fully, the diagram of the auditors' program technique is reproduced below with an example of sales invoice data that are extended improperly.

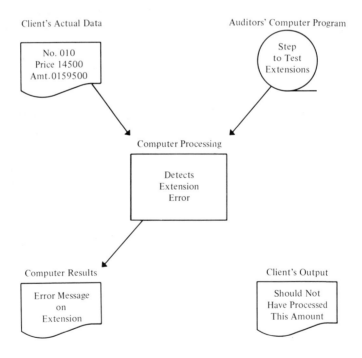

---

[8] It should be pointed out, however, that actual edit programs used in computer installations are much more comprehensive than the one illustrated in this chapter.

There is also a disadvantage to the auditors' use of their own programs, namely the cost of writing and periodically modifying them. In fact, many small firms do not have either the money or the expertise to use this technique. Because the test-data approach does not require program preparation, it is probably the cheaper method.

Finally, it should be mentioned that auditors' programs typically are designed both to test compliance with internal control and to gather audit evidence on year-end financial statement balances. For this reason, additional discussion of this technique is deferred to Chapter 14, which concerns the gathering of audit evidence in an EDP system.

Auditors' Participation in EDP Design

Discussion of audit techniques for compliance testing in an EDP system must include examination of auditors' participation in the design of such systems. As explained in Chapter 1, there is a sharp distinction between accounting and auditing. Accounting is a process of compiling figures and is essentially a record-keeping function. Auditing is an analytical process of reviewing, observing, calculating, etc., for the purpose of ascertaining the fairness of the figures that are recorded by the company. Although the AICPA sanctions, under certain conditions, the audit of one's own bookkeeping, members of the accounting profession generally consider this a practice to be avoided. As pointed out in Chapter 2, the SEC's rules against this practice are strict.

In manual systems, this separation of activities has not presented any technical problem, because auditors, trained extensively in the function of recording manual transactions, are generally able to acquire fairly quickly the tools to audit such systems. Thus, with the proper training and experience, auditors can function adequately without participating in the design of a manual accounting system.

In contrast, as EDP systems become more complicated, auditors may not be able to conduct an examination without extensive knowledge of the system. In the opinion of some practitioners, this knowledge can be gained only through participation in the system's design. Such an opinion was expressed by an AICPA task force on computer auditing and appeared in an article in the January 1975 issue of *The Journal of Accountancy.* The article is entitled "Advanced EDP Systems and the Auditor's Concerns."

> The auditor's participation in the system design will become more critical in the case of an advanced system. This involvement will enable the auditor to provide a valuable contribution, particularly from the control standpoint, to the company's new systems and will, since new audit requirements can be specified during the initial design phase, decrease the chance of an "unauditable" system's being developed.

Although the possibility is not mentioned in the foregoing comments, participation by auditors in EDP systems design might lower the probability of computer frauds, because better accounting controls would be installed at the outset. Participation of this sort also should enable auditors to use better evidence-gathering techniques and should improve the probability of detecting situations similar to that in the *Equity Funding* case (in which computers were used to process a large volume of false data).

But it is also possible that certain ethical considerations eventually will arise concerning independence. Interpretation 101-3 of the AICPA's *Code of Professional Ethics* lists the following conditions under which members may audit their own bookkeeping:

1. The member must not have any conflict of interest which would impair his objectivity.

2. The client must know enough about his activities and financial condition so that he can accept responsibility for the statements.

3. The member must not assume the role of employee or manager, i.e., have custody of assets or exercise authority on behalf of the client.

4. The auditor must still make sufficient tests and conduct the audit in accordance with proper standards.

Whether participation in EDP design creates a violation of any of the above-listed criteria for maintaining independence is a question that can be answered only in individual circumstances. General conjectures probably would be premature. The question is interesting and demonstrates the complexity of the environment in which auditors operate, particularly in the examination of statements generated by EDP systems.

Online Real-Time Systems

Some Differences Between Batch and Online Real-Time Systems

Although the illustrations throughout this chapter are based on a batch-processing system, many of the concepts are equally applicable to online real-time systems. Separation of duties is still important, program changes should be approved, and independent groups should periodically test programs.

However, the different structure of online systems calls for a change in the emphasis placed on some controls. For instance, it would be more important to limit access to terminals in an online system than to restrict use of the keypunch machine in a batch-processing system. An unauthorized punched card has more chances to be detected before the data on the card enter the computer files. Machine verification, visual inspection by control groups, and preliminary edit programs are examples of checkpoints. But with the exception of edit programs designed to reject certain types of improper input, these checkpoints are lacking in an online system. Therefore, it is important that access to computer terminals be considerably restricted.

A hypothetical illustration of how an online system might function is shown below. Each department has a terminal with direct access to the CPU and the magnetic disk storage files, so that inquiries can be made on the status of customer invoices and orders; a teletypewriter with a display screen shows these data. In addition, updates can be made to a customer's account balance directly from the Accounts Receivable Department. Because disk storage equipment is used, random access processing can be employed, making it unnecessary to process records in sequence.

EDP – ONLINE SYSTEMS FLOW CHART

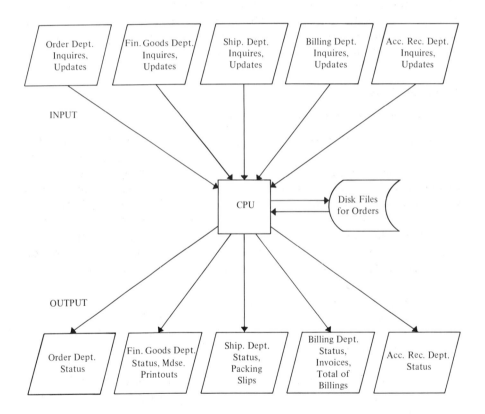

The Auditor's Approach to Batch and Online Real-Time Systems

From the auditor's standpoint, one of the biggest problem areas in an online real-time system is the compliance testing phase.[9] Some practitioners

[9]From an article by William F. Lewis, "Auditing Concepts of On-Line Computer Systems," *The Arthur Young Journal,* Winter/Spring 1971.

maintain that the test-data approach, designed for batch systems, may not be feasible for online real-time applications. The problem is not the physical inability of the auditors to run test data. In a system that is used continually, the intrusion of simulated entries into the disk files may introduce errors in the data base that cannot be reversed easily.

One suggestion has been made for overcoming this problem. Auditors could keep a tally of the simulated entries and could give a list of these entries to the client, so that they could be reversed at a later time. Another suggestion is to establish "audit divisions" that sort the audit transactions out of the client's regular data.

A monitoring process also could be conducted by auditors. For example, the terminal operations could be observed over a period of time, and observers could ascertain whether error messages are produced with inaccurate input, etc. This procedure appears to have some of the characteristics of the "walk-through" described in Chapter 5. Although observation of the client's procedures is an acceptable auditing approach, it is not yet possible to tell whether monitoring is an adequate substitute for the use of test data.[10]

Computer Fraud

The Uniqueness of the Programming Function

Much has been written about computer frauds and the various adversities that have plagued companies in recent years. Contrary to the opinion of many laymen, the use of EDP does not solve all the control problems. Although EDP ensures that data will be transmitted faster, it can create control problems that are more serious than those in manual operations. In the authors' opinion, there is *no* way for management to be certain that computer frauds will not occur.

One of the potential weaknesses in any computer system is the opportunity that programmers or operators (sometimes the same people) have to commit irregularities. The literature is replete with examples, such as the following.

1. A computer programmer changes a program designed to print out a list of bank customer overdrafts so that his particular account is omitted. He writes checks for a total amount greatly in excess of his bank balance, but no overdraft of his account appears on the listing.

2. A computer programmer changes an accounts receivable billing program so that his account or the account of a friend is not billed for a number of sales.

3. A computer programmer changes a program so that double payment of invoices is made to a friend's business establishment, or payment is made to a "dummy" company, one that does not exist.

---

[10]See William F. Lewis, "Auditing On-Line Computer Systems," *The Journal of Accountancy,* October 1971, pp. 47–52.

In a manual system, it is very difficult to find any parallel opportunity for individuals to commit these types of fraud, because the information trail consists of visible, hand-prepared documents. Clever programmers, unchecked by adequate controls, can literally "hide" their fraud in computer programs. The Auditing Standards Executive Committee of the AICPA has recognized the control problems inherent in a computer system in its statement in AU Section 321.12–.13 *(SAS No. 3)*:

> Frequently, functions that would be considered incompatible if performed by a single individual in a manual activity are performed through the use of an EDP program or series of programs. A person having the opportunity to make unapproved changes to any such programs performs incompatible functions in relation to the EDP activity. . . . EDP data files frequently are basic records of an accounting system. They cannot be read or changed without the use of EDP, but they can be changed through the use of EDP without visible evidence that a change has occurred. A person in a position to make unapproved changes in EDP data files performs incompatible functions.

Although there is no way for a company to be certain that programming irregularities will not happen, controls similar to those suggested throughout the chapter will lower the probability of such occurrences.

Distinction Among Computer Frauds

In the wake of irregularities that occurred in the *Equity Funding* case, some members of the accounting profession issued statements about what they believe to be misconceptions concerning computer fraud.[11] Strictly speaking, a computer fraud is similar to the situations described in the preceding section in that one or more individuals defraud a company by a unique application of the computer.

The situation in the *Equity Funding* fraud was somewhat different (although in many ways even more serious). Management and other groups within the company defrauded the public, directly, by falsifying records to support fictitious earnings. The computer was used to keep these records, and thus perpetrated the fraud, but the defalcation itself did not depend on the existence of the computer.

The distinction may appear moot to some members of the accounting profession but, in the authors' opinion, such a situation presents different auditing problems. The ability of a programmer to defraud a company by clever computer applications might signal a serious weakness in internal control, which perhaps can be remedied by tighter control measures. In contrast, massive collusion of the *Equity Funding* type might not be amenable to correction by standard control devices simply because the basic propositions upon which

---

[11]See *Report of the Special Committee on Equity Funding,* issued in 1975 by the American Institute of Certified Public Accountants.

internal control relies (segregation of duties, etc.) apparently break down in such a situation.

In Chapter 14 is an illustration of audit steps that have implications with regard to the *Equity Funding* fraud.

Commentary

When first approaching the study of EDP, one may tend to regard it as something mysterious and apart from the mainstream of auditing. This chapter is intended to show that this is not the case. The concepts underlying good internal control are the same in both a manual and an EDP system, and many of the techniques are similar. The major dissimilarities are the different job responsibilities and the auditors' option of going through the EDP system.

The study and evaluation of internal control in a computer environment are still in an initial stage of development. As time passes and more is learned, techniques undoubtedly will improve, and in the not-too-distant future auditors may feel as "comfortable" with computer installations as they generally do now with manual devices.

At the present time, however, CPA firms appear to be somewhat indefinite about requiring their staff members to perform EDP audits. It is generally agreed that their personnel must understand the basic terminology and should be able to recognize problem areas. But if the system being examined is fairly complicated (advanced online configurations, for instance), auditors with special training must be called in. The situation is analogous to the preparation of income tax returns—each staff member has *some* knowledge of the subject matter but must refer to specialists for complicated problems.

Techniques for gathering audit evidence in EDP systems are presented in Chapter 14.

Chapter 7 References

Boutell, Wayne S., "Auditing Through the Computer," *The Journal of Accountancy,* November 1965, pp. 41–47.

Chaiken, Barry R., and William E. Perry, "ITF—A Promising Computer Audit Technique," *The Journal of Accountancy,* February 1973, pp. 74–78.

Davis, Gordon, "Standards for Computers and Information Processing," *The Journal of Accountancy,* September 1967, pp. 52–57.

Davis, Gordon, *Auditing & EDP,* American Institute of Certified Public Accountants, 1968.

Devlin, Gerald W., "Internal Control is Not Optional," *Management Accounting,* August 1975, pp. 49–51.

Horwitz, Geoffrey B., "EDP Auditing—The Coming of Age," *The Journal of Accountancy,* August 1970, pp. 48–56.

Lewis, William F., "Auditing Concepts of On-Line Computer Systems," *The Arthur*

Lewis, William F., "Auditing On-Line Computer Systems," *The Journal of Accountancy,* October 1971, pp. 47–52.

Porter, W. Thomas, Jr., "A Control Framework for Electronic Systems," *The Journal of Accountancy,* October 1965, pp. 56–63.

Report of the Special Committee on Equity Funding, American Institute of Certified Public Accountants, 1975.

Chapter 7
Questions Taken from the Chapter

7–1.   Name three similarities between the audit of EDP systems and the audit of manual systems. Name three differences.

7–2.   What comments are contained in *SAS No. 1* concerning internal control in an EDP system?

7–3.   Name five job responsibilities usually held in an EDP system but not held in a manual system.

7–4.   What two job responsibilities in an EDP system should be fully separated?

7–5.   How can EDP personnel have access to assets without maintaining physical custody of these assets?

7–6.   Name five steps that can be taken to improve internal control when EDP personnel have indirect access to assets.

7–7.   What intermediate processing steps disappear when an online system is used rather than a batch-processing system?

7–8.   What is the difference between control totals, hash totals, and record counts?

7–9.   What purpose is served by batch totals? What role is played by control groups in making the batch-total process effective?

7–10.   What is a programming control? Name two types of programming controls.

7–11.   Why do programming controls often need to be more complex in an online system than in a batch-processing system?

7–12.   Why are frauds sometimes easier to perpetrate in an EDP system than in a manual system?

7–13.   What is the difference between the auditing problems of a computer fraud and the irregularities that ocurred in the *Equity Funding* case (according to *Report of the Special Committee on Equity Funding*)?

7–14.   List and briefly define six types of computer controls.

7–15.   Name three review procedures, each of which should be used to check the six computer controls.

7–16.   What is the purpose of the transaction code field in the punched card for sales invoices? What is the purpose of the batch number field?

7–17.   What is a special problem of running test data in an online system?

7–18.   Give the reason why each of the following questions probably would appear on an internal control questionnaire designed to evaluate an EDP system.
    a.   Is there a separation between the functions of programming and computer operation?
    b.   Are programs periodically tested by a control group or an internal audit group?

      c. Are programmers or machine operators allowed to correct input errors and to reconcile the control totals?

      d. Is there controlled access to data files and other EDP documentation?

7–19. List four strengths of the EDP system as described in the chapter. List three weaknesses.

7–20. What is meant by an "around-the-computer" approach to testing compliance in an EDP system? Name one risk in using this approach.

7–21. List the three steps in the test-data approach.

7–22. What is the purpose of auditors' predetermined computer output?

7–23. What tentative conclusions can auditors draw when their predetermined output matches the actual computer output in a run of test data?

7–24. What is the major advantage of using the test-data approach rather than an around-the-computer approach?

7–25. Name three limitations of the test-data approach.

7–26. What is ITF? How does it differ from the traditional test-data approach? What is the advantage of ITF?

7–27. What is the advantage of running data several times a year with the client's program? What is the disadvantage?

7–28. Describe the technique of using the auditors' own programs.

7–29. Name two advantages to be gained from the use of an auditors' program rather than test data. Name a disadvantage.

7–30. Why has it been suggested that auditors should participate in the design of EDP systems? What ethical problems are raised?

7–31. Name the four conditions under which members of the AICPA may audit their own bookkeeping.

7–32.   Why is control over access to computer terminals particularly important in an online system?

Chapter 7
Objective Questions Taken from CPA Examinations

In each of the following six items, two independent statements (numbered I and II) are presented. You are to evaluate each statement individually and determine whether it is true. Your answer for each item should be selected from the following responses:

a.   I only is true.
b.   II only is true.
c.   Both I and II are true.
d.   Neither I nor II is true.

7–33.   I.   One of the techniques for controlling batch processing is the use of hash totals.

II.   A hash total is obtained by counting all documents or records to be processed.

7–34.   I.   An important improvement in recent computer hardware is the ability to automatically produce error listings.

II.   Echo-check printouts provide the primary documentation upon which a company relies for an explanation of how a particular program operates.

7–35.   I.   The control of input and output of accounting transactions to and from the data-processing department should be performed by an independent control group.

II.   An internal-audit computer program which continuously monitors computer processing of accounting transactions is a feasible approach for improving internal control over online, real-time systems.

7–36.   I.   Read-after-write, dual-read, parity check, and echo check are all types of hardware controls.

II.   A limit check in a computer program is comparable to a decision that an individual makes in a manual system to judge a transaction's reasonableness.

7–37.   Computer files are usually maintained on magnetic tapes or disks.
  I.   A principal advantage of using magnetic tape files is that data need not be recorded sequentially.
  II.   A major advantage of disk files is the ability to gain random access to data on the disk.

7–38.   Meridian Corporation's EDP department personnel include a manager, a programmer-systems analyst, three machine operators, a librarian, four key-punch operators, and a control clerk.
  I.   Assuming that employees' activities are strictly limited to their assigned responsibilities, internal control would not be strengthened by rotating periodically among machine operators the assignment of individual application runs.
  II.   If work volume expands sufficiently to justify individual positions, it would be desirable for Meridian to separate the duties of the systems analyst and the programmer.

7–39.   A computer programmer has written a program for updating perpetual inventory records. Responsibility for initial testing (debugging) of the program should be assigned to the
  a.   EDP-department control group.
  b.   Internal-audit control group.
  c.   Programmer.
  d.   Machine operator.

7–40.   What is the computer process called when data processing is performed concurrently with a particular activity, and the results are available soon enough to influence the particular course of action being taken or the decision being made?
  a.   Real-time processing.
  b.   Batch processing.
  c.   Random access processing.
  d.   Integrated data processing.

7–41.   A customer inadvertently ordered part No. 12368 rather than part No. 12638. In processing this order, the error would be detected by the vendor with which of the following controls?
  a.   Batch total.
  b.   Key verifying.
  c.   Self-checking digit.
  d.   An internal consistency check.

Items 7–42 and 7–43 are based on the following information:

A sales transaction card was designed to contain the following information:

Card Column	Information
1–10	Customer account number
11–30	Customer name
31–38	Amount of sale
39–44	Sales date
45–46	Store code number
47–49	Sales clerk number
50–59	Invoice number

7–42.   If such a card is rejected during computer processing because the sales clerk whose identification number appears on the record does NOT work at the store indicated by the numbers in card columns 45 & 46, then the error was probably detected by which of the following?
   a.   A self-checking number.
   b.   A combination check.
   c.   A valid-character check.
   d.   A limit check.

7–43.   If the last letter of a customer's name is erroneously entered in card column 31, which of the following is most likely to detect the error during an input edit run?
   a.   A logic check.
   b.   A combination check.
   c.   A valid-character check.
   d.   A self-checking number.

7–44.   A company uses the account code 669 for maintenance expense. However, one of the company's clerks often codes maintenance expense as 996. The highest account code in the system is 750. What would be the best internal control check to build into the company's computer program to detect this error?
   a.   A check for this type of error would have to be made before the information was transmitted to the EDP department.
   b.   Valid-character test.
   c.   Sequence check.
   d.   Valid-code test.

7–45.   Evaluation of the electronic data-processing aspects of a system of accounting control should
   a.   Not be a part of the auditor's evaluation of the system.
   b.   Be a separate part of the auditor's evaluation of the system.
   c.   Be an integral part of the auditor's evaluation of the system.
   d.   Be coordinated with the auditor's evaluation of administrative control.

7–46. Which of the following *best* describes a fundamental control weakness often associated with electronic data-processing systems?

    a. Electronic data-processing equipment is more subject to systems error than manual processing is subject to human error.

    b. Electronic data-processing equipment processes and records similar transactions in a similar manner.

    c. Electronic data-processing procedures for detection of invalid and unusual transactions are less effective than manual control procedures.

    d. Functions that would normally be separated in a manual system are combined in the electronic data-processing system.

7–47. An advantage of manual processing is that human processors may note data errors and irregularities. To replace the human element of error detection associated with manual processing, a well-designed electronic data-processing system should introduce

    a. Programmed limits.

    b. Dual circuitry.

    c. Echo checks.

    d. Read-after-write.

7–48. Which of the following is *not* a problem associated with the use of test decks for computer-audit purposes?

    a. Auditing through the computer is more difficult than auditing around the computer.

    b. It is difficult to design test decks that incorporate all potential variations in transactions.

    c. Test data may be commingled with live data, causing operating problems for the client.

    d. The program with which the test data are processed may differ from the one used in actual operations.

7–49. An auditor's investigation of a company's electronic data-processing control procedures has disclosed the following four circumstances. Indicate which circumstance constitutes a weakness in internal control.

    a. Machine operators do not have access to the complete run manual.

    b. Machine operators are closely supervised by programmers.

    c. Programmers do not have the authorization to operate equipment.

    d. Only one generation of backup files is stored in an off-premises location.

7–50. Which of the following is an example of application controls in electronic data-processing systems?

    a. Input controls.

    b. Hardware controls.

    c. Documentation procedures.

    d. Controls over access to equipment and data files.

7–51.   The grandfather-father-son approach to providing protection for important computer files is a concept that is most often found in
a.   Online, real-time systems.
b.   Punched-card systems.
c.   Magnetic tape systems.
d.   Magnetic drum systems.

7–52.   In its electronic data-processing system a company might use self-checking numbers (check digits) to enable detection of which of the following errors?
a.   Assigning a valid identification code to the wrong customer.
b.   Recording an invalid customer's identification charge account number.
c.   Losing data between processing functions.
d.   Processing data arranged in the wrong sequence.

7–53.   So that the essential accounting control features of a client's electronic data-processing system can be identified and evaluated, the auditor must, at a minimum, have
a.   A basic familiarity with the computer's internal supervisory system.
b.   A sufficient understanding of the entire computer system.
c.   An expertise in computer systems analysis.
d.   A background in programming procedures.

7–54.   Program controls, in an electronic data-processing system, are used as substitutes for human controls in a manual system. Which of the following is an example of a program control?
a.   Dual read.
b.   Echo check.
c.   Validity check.
d.   Limit and reasonableness test.

7–55.   Some electronic data-processing accounting control procedures relate to all electronic data-processing activities (general controls), and some relate to specific tasks (application controls). General controls include
a.   Controls designed to ascertain that all data submitted to electronic data-processing for processing have been properly authorized.
b.   Controls that relate to the correction and resubmission of data that were initially incorrect.
c.   Controls for documenting and approving programs and changes to programs.
d.   Controls designed to assure the accuracy of the processing results.

7–56.   The auditor looks for an indication on punched cards to see if the cards have been verified. This is an example of a

a. Substantive test.
b. Compliance test.
c. Transactions test.
d. Dual-purpose test.

7–57. An internal administrative control that is sometimes used in connection with procedures to detect unauthorized or unexplained computer usage is
a. Maintenance of a computer tape library.
b. Use of file controls.
c. Maintenance of a computer console log.
d. Control over program tapes.

Chapter 7
Discussion/Case Questions

7–58. Mr. Traditional had a modest accounting practice consisting, for the most part, of local clients. None of the firms he serviced had a computer system, although several of them had talked of obtaining one.

In early March, Mr. Traditional received a telephone call from Unlimited Horizons, Inc., one of his client firms that had undergone considerable growth within the last few years. The client indicated that there were tentative plans to acquire computer equipment and to convert part of the accounting system from manual to EDP. The purpose of the call was to solicit Mr. Traditional's help in installing a suitable system. As the client's controller put it, "You will be auditing it later this year, so this will be a good opportunity to familiarize yourself with the characteristics of the new setup."

Mr. Traditional hesitated to accept the engagement, but decided to do so to avoid losing the client to another CPA firm. He consulted an up-to-date auditing text and found that a proper segregation of duties calls for a separation among (1) systems analyst, (2) computer programmer, (3) computer operator, (4) keypunch operator, (5) file maintenance, and (6) control group.

When Mr. Traditional suggested this type of job segregation to the controller of Unlimited Horizons, Inc., it was rejected as being too expensive. The controller decided that one person could design systems, program, and operate the computer, and he also felt that a control group could be omitted. It was left to Mr. Traditional to decide what controls would be built into the system.

Later in the year, Mr. Traditional was asked to conduct the audit of Unlimited Horizons, Inc. But he felt that because he had participated in the setup of the system, a senior in the firm should do the field work. Mr. Traditional would still review the audit and sign the report. This arrangement was carried out, and the audit was completed with an unqualified opinion.

Several months later, Unlimited's programmer confessed to fraud. He had perpetrated a scheme to have the computer print out checks far in excess of the purchase amount to several vendors, who then gave the programmer a "kickback" of part of the excess payment.

Upon learning of this confession, Mr. Traditional called in the senior and asked him about the situation. The senior replied that he had used all the auditing techniques that he considered appropriate, including the running of test data with all the client's programs. The senior did indicate that he had "wondered" about the lack of segregation between the duties of programming, machine operation, and systems design. The senior also had noticed that the programmer was the only person who ever handled computer output, including error messages. However, he was reluctant to say anything to Mr. Traditional, because Mr. Traditional had aided in setting up the system.

Required:
a.   What do the official pronouncements of the AICPA state about auditors' responsibility for fraud detection?
b.   Do you believe that either Mr. Traditional or the senior will be able to use these official pronouncements as a defense in the event of a lawsuit?
c.   Do you think that either Mr. Traditional or the senior violated any auditing standards? If so, which one(s)? Why?
d.   Assume that you were the senior and had reported these control deficiencies to Mr. Traditional. Assume that Mr. Traditional dismissed them as minor. What would you do?

7–59.   Listed below are six jobs or responsibilities that might be performed in an EDP accounting system. Because of a limited budget, however, only three people are available to handle the six jobs. How should these six jobs be divided?
a.   Systems analyst
b.   Computer operator
c.   Programmer
d.   Keypunch operator
e.   Verifier operator
f.   File librarian

7–60.   It is sometimes asserted that a programmer in an EDP system *could* be in a unique position to perpetrate fraud that remains undetected. Indicate three ways that a programmer could take assets from the company through unauthorized programming activities. Describe the lack of controls that might make it possible for the loss of assets to go undetected.

7–61. Examine the batch-processing systems flow chart in the chapter and do the following.
a. List the *input* documents that are being processed on punched cards.
b. List the *output* documents that are being produced on punched cards.
c. List the *input* documents that are being processed on magnetic tape.
d. List the *output* documents that are being produced on magnetic tape.
e. List the *output* documents that are being produced on a printer.
f. Indicate the processing areas that call for *program* flow charts.
g. List three duties that are included here that would not be needed in a manual system.
h. Indicate where batch totals would be made up.
i. Indicate where batch totals would be checked.
j. Indicate which document or documents would be considered exception reports.

7–62. Examine the program flow chart in the chapter and do the following.
a. List the steps that involve a decision.
b. List the steps for data input and for data output.
c. List the processing steps.
d. Briefly describe the purpose of each connector symbol and the route that the program takes if it branches to the steps depicted by the symbol. (Example—connector symbol 2 is followed if the transaction code punched in the card is not a 1. The program branches to the section that tests for transaction code 3.)

7–63. A feature of internal control is considered weak if errors or irregularities could occur as a result of the feature's existence. For each of the following situations, indicate the error or irregularity that could occur.
a. Allowing the computer operator, who has some knowledge of programming, to have unlimited access to programs.
b. Forwarding the control totals to the computer operator.
c. Failing to give magnetic tapes proper labels.
d. Providing only visual verification of the data on the punched cards.
e. The computer program has no programming controls.
f. No log is kept of computer runs.

7–64. Refer to the record layout of sales invoices in the body of the chapter. Exclusive of the edit checks discussed, list ten additional ones that could be placed in a computer program to test for invalid, incomplete, or out-of-limit conditions in the charge sales data.

7–65. Mr. A. O. Steady, a programmer for One Corporation, had been employed in his job for a number of years. He was the only programmer. There was a separate computer operator, keypunch operator, verifier, and librarian.

Part of the computer system calls for data in the purchase invoices to be keypunched into cards. The cards are processed through the computer, and checks for vendors are printed by the computer. Each check is manually compared with the purchase invoice by another individual before mailing.

One day it was discovered that Mr. Steady had been receiving "kickbacks" from a company that had received money from One Corporation without sending them any merchandise. The president of the company was puzzled about how Mr. Steady could have perpetrated this fraud. There seemed to be a proper segregation of duties, and the programmer never handled the checks that were printed by the computer.

Required:
a. Explain how this fraud may have occurred and remained uncovered.
b. Discuss some controls, including non-computer types, that might have prevented or uncovered the fraud.

7–66. George Beemster, CPA, is examining the financial statements of the Louisville Sales Corporation, which recently installed an offline electronic computer. The following comments have been extracted from Mr. Beemster's notes on computer operations and the processing and control of shipping notices and customer invoices:

To minimize inconvenience, Louisville converted without change its existing data-processing system, which utilized tabulating equipment. The computer company supervised the conversion and has provided training to all computer department employees (except keypunch operators) in systems design, operations, and programming.

Each computer run is assigned to a specific employee, who is responsible for making program changes, running the program, and answering questions. This procedure has the advantage of eliminating the need for records of computer operations, because each employee is responsible for his own computer runs.

At least one computer department employee remains in the computer room during office hours, and only computer department employees have keys to the computer room.

System documentation consists of those materials furnished by the computer company—a set of record formats and program listings. These and the tape library are kept in a corner of the computer department.

The Corporation considered the desirability of programmed controls but decided to retain the manual controls from its existing system.

Company products are shipped directly from public warehouses, which forward shipping notices to general accounting. There a billing clerk enters the price of the item and accounts for the numerical sequence of shipping notices from each warehouse. The billing clerk also prepares daily adding-machine tapes ("control tapes") of the units shipped and the unit prices.

Shipping notices and control tapes are forwarded to the computer department for keypunching and processing. Extensions are made on the computer. Output consists of invoices (in six copies) and a daily sales register. The daily sales register shows the aggregate totals of units shipped and unit prices, which the computer operator compares to the control tapes.

Required:
Indicate the weaknesses that you see in this system and your recommendations for correcting these weaknesses.

(AICPA adapted)

7–67. In connection with his examination of the financial statements of the Olympia Manufacturing Company, a CPA is reviewing procedures for accumulating direct labor hours. He learns that all production is by job order and that all employees are paid hourly wages, with time-and-one-half for overtime hours.

Olympia's direct labor hour input process for payroll and job-cost determination is summarized in the flow chart on page 328.

Required:
For each input processing Step A through F:
    a. List the possible errors or discrepancies that may occur.
    b. Cite the corresponding control procedure that should be in effect for each error discrepancy.

Note: Your discussion of Olympia's procedures should be limited to the input process for direct labor hours, as shown in Steps A through F in the flow chart. *Do not discuss* personnel procedures for hiring, promotion, termination, and pay rate authorization. *In Step F, do not discuss* equipment, computer program, and general computer operational controls.

Organize your answer for each input-processing step as follows:

Step	Possible Errors or Discrepancies	Control Procedures

(AICPA adapted)

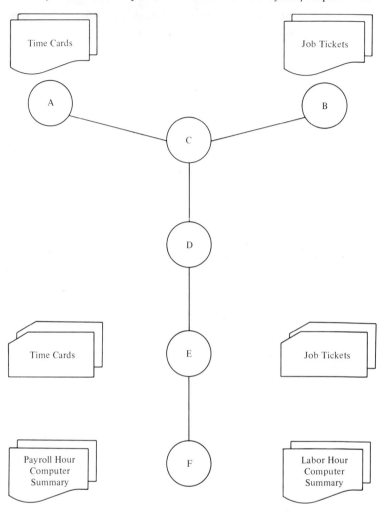

7–68. You are reviewing audit work papers containing a narrative description of the Tenney Corporation's factory payroll system. A portion of that narrative is as follows:

Factory employees punch timeclock cards each day when entering or leaving the shop. At the end of each week, the timekeeping department collects the time-cards and prepares duplicate batch-control slips by department showing total hours and number of employees. The timecards and original batch-control slips are sent to the payroll accounting section. The second copies of the batch-control slips are filed by date.

In the payroll accounting section, payroll transaction cards are keypunched from the information on the timecards, and a batch total card for each batch is keypunched from the batch-control slip. The timecards and batch-control slips are then filed by batch for possible reference. The payroll transaction cards and batch total card are sent to data processing, where they are sorted by employee number within batch. Each batch is edited by a computer program which checks the validity of employee number against a master employee tape file and the total hours and number of employees against the batch total card. A detail printout by batch and employee number indicates batches that do not balance and invalid employee numbers. This printout is returned to payroll accounting to resolve all differences.

In searching for documentation, you found a flow chart of the payroll system which included all appropriate symbols (American National Standards Institute, Inc.) but was only partially labeled. The portion of this flow chart described by the narrative above appears on page 330.

Required:

a. Number your answers 1 through 17. Next to the corresponding number of your answer, supply the appropriate labeling (document name, process description, or file order) applicable to each numbered symbol on the flow chart.

b. Flow charts are one of the aids an auditor may use to determine and evaluate a client's internal control system. List advantages of using flow charts in this context.

(AICPA adapted)

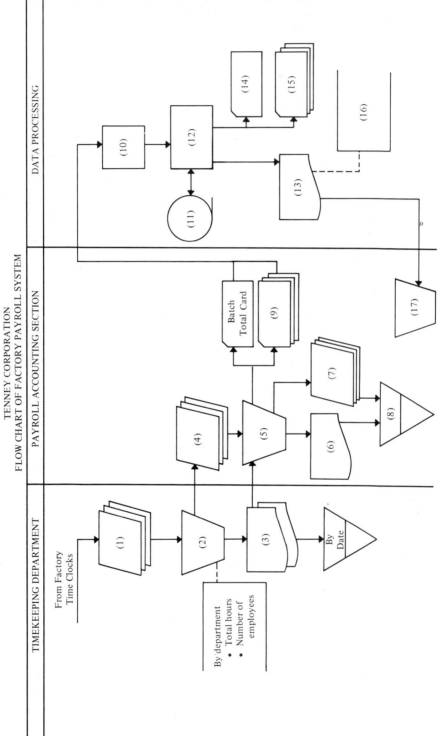

TENNEY CORPORATION
FLOW CHART OF FACTORY PAYROLL SYSTEM

Chapter 8

Reports on Internal Control

Learning Objectives   *After reading and studying the material in this chapter, the student should*

Understand the reason why CPAs issue reports on internal control to their clients.

Know the AICPA's position on the issue of reports on internal control to the potential recipients.

Understand the basic component parts of the AICPA's suggested form for a report on internal control.

Be able to write a letter on internal control, given the deficiencies of the internal control system.

The last three chapters concerning the importance of the study and evaluation of internal control have stressed that the purpose of this process is the same regardless of the accounting system employed. This chapter covers a direct output of the study and evaluation process, namely, a report to the client detailing certain deficiencies that are found during this preliminary phase of the audit. The importance of this report has been made more significant by the issuance of *SAS No. 20*. This pronouncement requires that auditors inform their client's senior management and the board of directors or audit committee of any material weaknesses that come to their attention during the course of an ordinary examination.

The Report as a Service to Clients

Actually, auditors have been giving advice for many years on matters of internal control, because they generally consider it an obligation to help clients improve their operations. It is also mutually beneficial for auditors to render such services, for any increase in the efficiency of accounting controls decreases the amount of testing needed and furnishes a better basis for reliance on the accounting records.

Because it is traditional for the study and evaluation of internal control to be conducted before the end of the client's fiscal year, auditors are in a good position to apprise the client quickly of any deficiencies discovered during this preliminary examination. Usually the client's management is informed orally so that corrective action can be taken as soon as possible. Later, a formal written report is issued to the president or board of directors explaining the weaknesses and the steps suggested to strengthen the system. Sometimes an informal report is issued to the controller or another company officer in addition to, or in place of, the formal report.

Types of Items Included in a Report

It is not difficult to imagine the potentially large variety of topics that could be included in a report on internal control. Even if auditors confine their review and testing to accounting controls (safeguarding assets and insuring accuracy and reliability of the accounting records), a substantial number of areas ordinarily are covered. And any organization that is reasonably complex is likely to have areas in which controls need strengthening and/or in which prescribed procedures are not being followed. Some possible examples are:

1. Inadequate controls over payments made from cash funds.

2. Sales made to customers without proper credit approval.

3. Failure to bill customers properly on a timely basis.

4. Failure to keep perpetual inventory records, or the keeping of unreliable inventory records.

5. Inadequate insurance coverage (this has some effect on administrative controls).

6. Failure to take purchase discounts (this also has some effect on administrative controls).

7. Duplicate payment of purchase invoices.

8. Inadequate follow-up of standard cost variances.

9. Compliance deviations in excess of a tolerable amount (such as pricing, extensions, postings, approvals, etc.).

Specific Examples of Internal Control Deficiencies

Even though control problems are brought promptly to the attention of client personnel, the written report may not be issued until the report on the financial statements is released several months after the end of the client's fiscal year. Ideally, the auditors' recommendations are being applied by that time; nevertheless, the formal report contains descriptions of the deficiencies found during the examination.

A description of auditors' findings might appear in language similar to the following.

> During the course of our work, we noted that unit costs shown on actual production orders were significantly different from the standard costs, and we could not find any evidence that these variances were investigated. Personnel insist that they do not have the time to investigate these variances.

> We recommend that a detailed review be made of the cost accounting system. This review should develop a set of written procedures for: (1) standard cost updating, (2) requirements for job cost accumulation, (3) changes in labor and overhead cost factors, (4) investigation of variances, and (5) product-line costing versus job costing.

This excerpt is fairly typical of comments made by auditors. Basically it consists of three parts: (1) a description of the deficiency (failure to investigate significant variances), (2) the client's reaction when orally apprised of this deficiency (lack of available time to investigate the variances), and (3) the auditors' recommendation (a system of variance investigations should be initiated).

Note that the above-described deficiency appears to be an example of an inadequacy in the system, rather than the client's failure to comply with prescribed procedures. It probably was found by inquiry and observation, without application of statistical sampling techniques.

In contrast, the weakness described below *could* have been brought to light by the application of statistical sampling, although it appears from the language that judgment sampling or general observation was used.

> It also came to our attention that there is generally a lack of clerical accuracy testing on sales invoices. This lack of testing occurs both in the Order Department and the Billing Department and reportedly is due to a lack of personnel. We

recommend that such clerical accuracy testing be performed on all invoices over $500 and that appropriate notation be made on the sales invoice of this clerical accuracy test.

Because the auditors recommend the testing of clerical accuracy only on invoices over $500, it is possible that any future clerical errors would be noted by auditors only if the error exceeded a certain dollar amount (this procedure is mentioned in Chapter 6).

In some situations, auditors may believe that even *one* compliance deviation is important enough to warrant special attention. For example:

> X Co. rents lease and well equipment to certain customers. In one instance we noted that an uncollectible rental (accounts receivable) was charged to sales rather than being charged to the bad-debt reserve. This procedure hinders management in evaluating the effectiveness of the Credit Department and also distorts the results of the company's rental business. Uncollectible amounts should be charged to the reserve for bad debts.

The final example shows the diversity of report recommendations. The three previously listed examples probably do not require the client to make major revision of the accounting system. However, the recommendation below does necessitate a substantial change.

> Under present procedures, the company's sales branches are responsible for initiating the billings to industrial customers. Most branches do their billing in the latter part of the month and, as a result, the cost of sales cannot be accumulated in the home office in time to make a proper accounting closing. . . . The situation referred to above could be eliminated by centralizing the billing function, and it is our understanding that the company has considered this possibility. It is our opinion that centralized billing would prove more efficient.

The Consequences of Internal Control Deficiencies

When auditors find problems of the sort described in the previous section, they probably increase the year-end testing in the areas involved. For example, the failure of client personnel to check the clerical accuracy of sales invoices might result in circulation of more requests asking customers to verify the accuracy of their balances as of the end of the fiscal year. Also, the improper accounting treatment of the uncollectible rental might necessitate the extension of normal audit procedures for the allowance-for-bad-debts account.

However, auditors do not issue qualified or adverse opinions (see the appendix to Chapter 1 for a discussion of the types of opinions) *directly* as a result of deficiencies found in the system of internal control. The study and evaluation of internal control are only the preliminary phase of the audit, and any decision on a qualified or adverse opinion normally would depend on the outcome of the test of year-end account balances.

There may be circumstances in which controls are so weak (or practically non-existent) that special action must be taken by the auditors. In these cases, several alternatives are available.

1. Auditors can extend their testing to cover an extremely large percentage of transactions. Often this procedure is necessary when the client is a small organization and cannot afford the personnel to maintain an adequate control system (charitable institutions are examples). But the volume of records also may be small enough in these cases that auditors can gather sufficient evidence without spending an excessive amount of time.

2. Auditors can simply withdraw, believing that the risk of overlooking material errors or irregularities is too great to be overcome, even with an increased volume of testing.

3. Auditors can withdraw temporarily and ask the client to do the work necessary to make the system "auditable." When auditors encounter poorly constructed and error-filled records, they might be tempted to perform corrective bookkeeping themselves. For instance, if the accounts receivable subsidiary ledger did not balance with the control account, auditors might try, on their own, to reconcile the difference. Though this might be a practical alternative, it involves a risk of violating the AICPA *Code of Professional Ethics* by performing bookkeeping for the audit client (the audit-bookkeeping conflict is discussed in Chapter 2). Actions of this sort by auditors would be even more likely to violate the SEC rule on bookkeeping and auditing.

Reports Issued by Internal Auditors

In Chapter 1, it is pointed out that the role of the internal auditor differs in many respects from that of the external or independent auditor. The fundamental distinction is that the former is an *employee* of the company, whereas the latter is hired as an independent accountant.

Though there is *some* overlap in the respective auditors' purposes for studying internal control, the scope of the internal auditors' work tends to be much broader. The external auditors' function is to render an opinion on the fairness of the firm's financial statements; thus, their examination of internal control is confined, for the most part, to accounting controls (safeguarding assets and insuring accuracy and reliability of the records). The internal auditors' job usually covers a review of administrative controls as well (efficiency of operations and adherence to managerial policy). For this reason, their reports on internal control include areas that usually are not examined in as much detail by external auditors.

For example, external auditors might include in their report a criticism of the accounting treatment of accounts receivable write-offs because this improper treatment has a potentially *direct* effect on the fairness of the financial statements. Contrasted to this are comments that internal auditors might make on the slow pace of accounts receivable collections, because this situation has an effect on the efficiency of operations.

However, there would also be *some* similarity between the contents of the two internal control reports; both the internal and external auditors are concerned with matters of accounting control.

The Recipient of Reports on Internal Control

Type of Recipients

Years ago, reports on internal control generally were considered to be documents intended "only for management's eyes," whereas the report on the financial statements was designed for stockholders, creditors, and other members of the public. But in recent years, the question has arisen as to whether these internal control reports should be distributed to others.

The AICPA considered this question and decided that four groups have some potential use for the report on internal control:

1. Management (directors, officers, etc.).

2. Regulatory agencies (SEC, stock exchanges, etc.).

3. Other independent auditors.

4. The general public (present and prospective investors, creditors, customers, etc.).

Use by Management,
Regulatory Agencies, and Independent Auditors

With regard to the disposition of reports to the first three groups listed above, the AICPA has expressed the following comments in AU Section 640.04 of *AICPA Professional Standards,* Volume 1 *(SAS No. 1).*

Management is responsible for establishing and maintaining internal accounting control. Regulatory agencies may be concerned with such control because it is relevant to their primary regulatory purpose or to the scope of their examination functions. Independent auditors of one entity or organization unit may be concerned with internal accounting control of another because it is relevant to the scope of their examination. It may be presumed that these groups include persons whose training and experience or intimate knowledge of the organization should provide a reasonable basis for understanding the nature and effectiveness of internal accounting control and the auditor's evaluation of it. Consequently, it is evident that reports on internal accounting control can serve a useful purpose for management, regulatory agencies, and other independent auditors.

Use by the General Public

An entirely different sentiment is expressed in AU Section 640.05 *(SAS No. 1)* concerning the distribution of reports to stockholders, creditors, and other members of the general public. The belief is that any possible action taken by the general public as a result of reading the report would be indirect; therefore, the usefulness of such reports to these groups would be limited.

AICPA's Conclusions on Report Distribution

AU Section 640.10 *(SAS No. 1)* also contains a capsule version of the conflicting views. The arguments in favor of releasing reports on internal control to the general public seem to center around the belief that the public needs to be apprised of management's performance. The contrasting view is that the general public does not have the expertise to evaluate these reports properly and that their release might result in a distorted appraisal of management's performance. Therefore, the AICPA's conclusions are that "the decision as to whether reports on an auditor's evaluation of internal accounting control would be useful for some portion or all of the general public in particular cases or classes of cases is the responsibility of management and/or any regulatory agencies having jurisdiction."

Is this opinion justified? The answer depends on one's viewpoint. Some believe that in light of the public's misconception about the relationship between internal control and the auditor's opinion, a strong case can be made for the AICPA's position. A weakness in one or more accounting control areas does not obligate auditors to issue a qualified or adverse opinion, because audit procedures can be extended to overcome the deficiency. However, the weakness should be reported to someone who is in a position to correct it. Consider, for example, the manual system of cash receipts illustrated in Chapter 5. A layman reader of a report that contains a description of deficiencies such as these might wonder how the auditor can justify an unqualified opinion.

A Suggested Form for Reports[1]

It is entirely possible that distribution of internal control reports to the public may become a common practice among management and regulatory agencies, because pressure for more disclosure is being exerted by this same public.

In anticipation of the possible release of reports, the AICPA has placed in AU Section 640.12 *(SAS No. 1)* a suggested form for use by auditors. To relate

---

[1]Copyright © (1972) by the American Institute of Certified Public Accountants, Inc.

this illustration to the material in Chapters 5, 6, and 7, the hypothetical report shown below includes recommendations based on the deficiencies discussed in the Chapter 5 manual system and the Chapter 7 EDP system.

Description of the Report Contents	*The Form of the Report (from AU Section 640.12)*
A distinction between the opinion report and the study and evaluation of internal accounting control.	We have examined the financial statements of ABC Company for the year ended December 31, 19X1, and have issued our report thereon dated February 23, 19X2. As a part of our examination, we reviewed and tested the Company's system of internal accounting control to the extent we considered necessary to evaluate the system as required by generally accepted auditing standards. Under these standards the purpose of such evaluation is to establish a basis for reliance thereon in determining the nature, timing, and extent of other auditing procedures that are necessary for expressing an opinion on the financial statements and to assist the auditor in planning and performing his examination of the financial statements.
The objective of internal accounting control.	The objective of internal accounting control is to provide reasonable, but not absolute, assurance as to the safeguarding of assets against loss from unauthorized use or disposition, and the reliability of financial records for preparing financial statements and maintaining accountability for assets. The concept of reasonable assurance recognizes that the cost of a system of internal accounting control should not exceed the benefits derived and also recognizes that the evaluation of these factors necessarily requires estimates and judgments by management.
The limitations of internal accounting control.	There are inherent limitations that should be recognized in considering the potential effectiveness of any system of internal accounting control. In the performance of most control procedures, errors can result from misunderstanding of instructions, mistakes of judgment, carelessness, or other personal factors. Control procedures whose effectiveness depends upon segregation of duties can be circumvented by collusion. Similarly, control procedures can be circumvented intentionally by management with respect to the execution and recording of transactions or with respect to the estimates and judgments required in the preparation of financial statements. Further, projection of any evaluation of internal accounting control to future periods is subject to the risk that the procedures may become inadequate because of change in conditions and that the degree of compliance with the procedures may deteriorate.
Disclaimer on detecting all weaknesses.	Our examination of the financial statements made in accordance with generally accepted auditing standards, including the study and evaluation of the Company's system of internal accounting control for the year ended December 31, 19X1, that was made for the purposes set forth in the first paragraph of this report, would not necessarily disclose all weaknesses in the system because it was based on selective tests of accounting records and related data. However, such study and evaluation disclosed the following conditions that we believe to be material weaknesses.

[*Authors' note:* If the auditor becomes aware of a material weakness, it should be reported to management. However, if certain immaterial deficiencies cannot be corrected because of impracticality and auditors are willing to omit any mention of these deficiencies, the language in the following paragraph can be used.]

(from AU Section 640.13)

However, such study and evaluation disclosed the following conditions that we believe to be material weaknesses for which corrective action by management may be practicable in the circumstances.

A description of the weakness.[2]	One of your internal accounting control policies requires that cash receipts be deposited intact each day. We noted that this policy was violated a substantial number of times during the year.
A recommendation for improvement, and client reaction.	We suggest that the Company enforce the policy of depositing cash receipts intact at the completion of each day's business. In addition, we suggest that the daily bank deposit be reconciled with daily cash receipts on account and daily cash sales before the deposit is made. After oral discussions with the Controller, he agreed to adhere more closely to the policy in the future.
A description of the weakness.	We noted that the validated bank deposit slip is returned to the Cashier. Good internal accounting control requires that there be a separation of duties so that no one person is in a position both to perpetrate and to conceal errors or irregularities in the normal course of his duties.
A recommendation for improvement, and client reaction.	Therefore, we suggest that the Company adopt a policy of having the bank mail the validated deposit slip to the Company Treasurer, or other person of appropriate authority, other than the Cashier. The Controller agreed to make this change.
A description of the weakness.	During the course of our review, we found that the edit program for accounts receivable does not have an adequate number of tests. In our opinion, several additional tests should be made to reduce the risk of input errors going undetected.
A recommendation for improvement, and client reaction.	We recommend that a thorough review be made of the edit program for the purpose of adding necessary steps to detect all of the various types of input errors that might be expected. This matter was discussed with the Financial Vice-President, who agreed to revise the edit program.
A standard ending.	As indicated above, we have discussed these deficiencies and recommendations for improvement with the appropriate management personnel and have been assured that corrective action will be taken on each matter.

[2]The rest of the illustration is hypothetical recommendations taken from examples in Chapters 5, 6, 7. This is not a direct quote from AU Section 640.12.

We appreciate the cooperation of your personnel. If we can be of any assistance in implementing these recommendations, please let us know.

Sincerely,

Hunt & Find, CPAs

Recurring Deficiencies

The auditors hope that their recommendations for the improvement of internal control will be followed by the client, and that the deficiency will have been corrected when they return for the next year's audit. Unfortunately, the auditors' suggestions are not always followed. As indicated in one of the chapter examples, clients sometimes disagree on the practicality of following auditors' recommendations. A case in point is the client's reluctance to investigate standard cost variances because of lack of time.

What should auditors do when they return the following year and observe that certain recommendations for improvement of internal control have not been implemented? Unless the deficiency seriously impairs record reliability, auditors probably will make the necessary adjustment in the scope of their audit procedures and will repeat the recommendation in the current year's report on internal control.

This position may appear to be a strange and somewhat timid one for independent auditors to take, but one should remember that the study and evaluation of internal control are only a preliminary phase of the audit. The final product is the auditors' opinion on the fairness of the financial statements; the study of internal control forms *part* of the basis for this opinion. Many poor control features can be overcome with increased testing of the evidence that supports financial statement balances.

One also should remember from Chapter 1 that the accuracy and authenticity of the accounting records are the responsibility of management. Auditors cannot *force* their clients to change the system of internal control, although they certainly can exert pressure by placing recommendations for change in the letter that goes to the board of directors.

However, in many cases (perhaps in most cases) the question of recurring deficiencies is moot, because client personnel are anxious to correct embarrassing problem areas and to hold down the cost of the audit. Often, clients seek auditors' aid in strengthening the system.

A Less Formal Report

The report form illustrated in this chapter is designed on the assumption that management *might* release the report to the public, despite the auditors' general reluctance to authorize this course of action. If auditors wish to issue a report for management's internal use only, a less formal style can be used.

In some situations, auditors may believe that an oral suggestion is sufficient, particularly if the problem is very minor or if the client has asked for a suggestion (such as how a particular journal entry should be handled).

The prudent course of action, however, is for auditors to give most recommendations (some auditors might say *all* recommendations) in written form, including those also given in oral form. This course of action results in more "protection" against client lawsuits in the event that serious errors or irregularities show up at a later time.

Reports on Internal Control
for Unaudited Financial Statements

How does an auditor report on internal control for engagements that result in unaudited financial statements? Ever since a successful lawsuit several years ago against a CPA who issued unaudited statements (the *1136 Tenants' Corporation* case discussed in Chapter 3), members of the accounting profession have been more careful about stating their responsibilities in this area.

This extra care is expressed, in part, by the following excerpt from AU Section 640.11 *(SAS No. 1)*.

> In no event, however, should an auditor authorize a report on his evaluation of internal accounting control to be issued to the general public in a document that includes unaudited financial statements.

This pronouncement does not answer the question of what might happen if a report on internal control were issued for client use only, and eventually this report, accompanied by unaudited financial statements, became available to the general public. Because of the potential risks involved, some CPAs might be reluctant to issue any type of report on internal control for engagements that result in unaudited financial statements.

This question, and other similar questions related to unaudited statements, probably will remain unsettled until CPAs' responsibilities in this area become clearer.

Chapter 8
Questions Taken from the Chapter

8-1.   When do auditors traditionally conduct the study and evaluation of internal control?

8–2. Why is the client informed orally of an internal control deficiency before this deficiency appears in a written formal report?

8–3. This chapter contains listings of nine types of deficiencies that might appear in the auditors' letter of internal control. Name five of them.

8–4. Name the three parts to the deficiency excerpts shown in the first part of the chapter.

8–5. When auditors find internal control deficiencies, what adjustment normally is made to year-end audit procedures?

8–6. Why would auditors not issue a qualified or an adverse opinion directly as a result of deficiencies found in the study and evaluation of internal control?

8–7. Name three types of special action that auditors can take if they find the client's controls to be extremely weak or practically non-existent.

8–8. What possible ethical problem is involved if auditors perform the corrective bookkeeping when they find inadequate client records?

8–9. What is the difference between internal and external auditors' relationships to a company?

8–10. To which controls is the external auditors' examination generally confined? What additional controls are included in the scope of the internal auditors' work?

8–11. Name four groups that have some potential use for reports on internal control. Which group has the least need for such reports, according to AU Section 640.05 *(SAS No. 1)*? Why? What position is taken by the AICPA regarding distribution of reports on internal control to this group?

8–12. The AICPA believes that three of these four groups referenced in question 8–11 have need for reports on internal control. Name the special reasons why the AICPA believes each of the three groups has this need.

8–13. The following questions relate to the AICPA's suggested form for reports on internal control.
   a. According to the form, what is the objective of internal accounting control?

b. What statement is contained in the form on cost and benefits of an internal control system?

c. According to the form, there are three limitations of internal accounting control. What are they?

8–14. Unless the deficiency seriously impairs record reliability, what will auditors normally do when they return for the following year's audit and find that same deficiency?

8–15. What is the final product of the auditors' engagement?

8–16. Who has responsibility for the accuracy and authenticity of accounting records?

8–17. Why is it a good idea for auditors to give written recommendations on most (if not all) of their recommendations on internal control deficiencies?

8–18. What statement is contained in AU Section 640.11 *(SAS No. 1)* concerning the issuance of both reports on internal control and unaudited financial statements? Why?

Chapter 8
Objective Questions Taken from CPA Examinations

8–19. Hickory Company, whose financial statements are unaudited, has engaged a CPA to make a special review and report on Hickory's internal accounting control. In general, to which of the following will this report be least useful?
a. Hickory's management.
b. Present and prospective customers.
c. A regulatory agency having jurisdiction over Hickory.
d. The independent auditor of Hickory's parent company.

8–20. George Green, CPA, is preparing a report on internal control. He has already discussed the internal control weaknesses with the appropriate client officials. During these discussions, the client stated that, given its circumstances, there was no practicable corrective action which could be taken for one of the major weaknesses and therefore asked that it not be included in Green's report. In the final analysis, Green concurred that no corrective action by management is practicable. Which of the following is the most appropriate course of action for Green to take?
a. He must include this weakness in his report; otherwise, he will be in violation of generally accepted auditing standards.

  b. He may omit this weakness from his report without any further mention.

  c. He may omit this weakness from his report but should send a confidential memo to the Board of Directors pointing out the nature of the weakness and why it was omitted from his report.

  d. He may omit this weakness from his report but should clearly state that the report is restricted to material weaknesses for which corrective action by management may be practicable in the circumstances.

Items 8–21 through 8–24 are based on the following information:

The Mastermind Security Company has asked you to prepare a report on its internal control. The report is to be based on your study and evaluation of the company's control system made in conjunction with your recently completed annual audit of Mastermind. Mastermind is considering the inclusion of the report on internal control in a document that will be sent to stockholders. Mastermind's stock is widely held and actively traded.

8–21. Under which of the following conditions would the inclusion of the internal-control report in a document to stockholders be prohibited?

  a. If the document contains only the internal control report and unaudited interim financial statements.

  b. If only a select class of stockholders receives the document.

  c. If Mastermind has requested the internal control report because of a ruling by a regulatory agency.

  d. If the internal control report indicates management negligence.

8–22. If the report on internal control is distributed to the general public, it must contain specific language describing several matters. Which of the following must be included in the specific language?

  a. The distinction between internal administrative controls and internal accounting controls.

  b. The various tests and procedures utilized by the auditor during his review of internal control.

  c. The objective of internal accounting controls.

  d. The reason(s) Mastermind's management requested a report on internal control.

8–23. If the report on internal control will be distributed to Mastermind's stockholders, the opening paragraph of the internal control report should indicate its timeliness by including which of the following pairs of dates?

  a. The date on which your review of internal control was completed and the date on which you agreed to prepare the internal control report.

  b. The date of the audit report issued on Mastermind's financial statements and the date on which you agreed to prepare the internal control report.

    c.   The date of Mastermind's financial statements and the date on which your review of internal control was completed.

    d.   The date of Mastermind's financial statements and the date of the audit report issued on those financial statements.

8–24.   How should the report on internal control deal with the possibility that some stockholders might use it for speculation about future adequacy of Mastermind's internal control system?

    a.   The report should contain an opinion as to whether the present internal control system can be relied upon for the next accounting period.

    b.   The report should make no mention of the possibility of making such projections.

    c.   The report should disclose management's opinion about the number of future accounting periods during which the present internal control system can be relied upon.

    d.   The report should discuss the risk involved in making such projections.

A CPA has completed an annual audit. The client has requested a report of internal control which it intends to submit to a regulatory agency. The CPA has drafted the first three paragraphs of the report on internal control; they are presented in Items 8–25, 8–26, and 8–27, respectively. Each paragraph contains a deficiency or inappropriate statement. Each sentence or part thereof of each paragraph corresponds to a response—a, b, c, and d.

Select the response (a, b, c, or d) in each (Items 8–25, 8–26, and 8–27) that corresponds to the sentence or part thereof that contains the *deficiency* or *inappropriate* statement.

8–25.   First paragraph:

    a.   We have examined the financial statements of ABC Company and have issued our report thereon.

    b.   As a part of our examination, we reviewed and tested the Company's system of internal accounting control to the extent we considered necessary to evaluate the system as required by generally accepted auditing standards.

    c.   Under these standards the purpose of such evaluation is to establish a basis for reliance . . .

    d.   . . . in determining the nature, timing, and extent of other auditing procedures that are necessary for expressing an opinion on the financial statements.

8–26.   Second paragraph:

    a.   The objective of internal accounting control is to provide reasonable, but not absolute, assurance as to the safeguarding of assets against loss from unauthorized use or disposition, . . .

    b. ... and the reliability of financial records for preparing financial statements and maintaining accountability for assets.

    c. The concept of reasonable assurance recognizes that the cost of an effective system of internal accounting control may often have to exceed the benefits derived.

    d. ... and also recognizes that the evaluation of these factors necessarily requires estimates and judgments by management.

8–27.   Third paragraph:

    a. There are inherent limitations that should be recognized in considering the potential effectiveness of any system of internal accounting control.

    b. In the performance of most control procedures, errors can result from misunderstanding of instruction, mistakes of judgment, carelessness, and other personal factors. Control procedures whose effectiveness depends upon segregation of duties can be circumvented by collusion.

    c. Similarly, control procedures can be circumvented intentionally by management with respect either to the execution and recording of transactions or with respect to the estimates and judgments required in the preparation of financial statements.

    d. Further, projection of any evaluation of internal accounting control should not extend beyond the next fiscal year because, beyond that period, changed conditions and the degree of compliance with the procedures could materially weaken the overall system of internal control.

8–28.   A CPA should *not* issue a report on internal control if

    a. The report is to be sent to stockholders with unaudited interim financial statements.

    b. The CPA has not audited the company's financial statements.

    c. The report is to be given to creditors.

    d. The report is to be given to prospective investors.

8–29.   Dey, Knight, & Co., CPAs, has issued a qualified opinion on the financial statements of Adams, Inc., because of a scope limitation. Adams, Inc., requested a report on internal control which it intends to give to one of its major creditors. What effect, if any, would the qualified opinion have on the internal control report that Dey, Knight, & Co. intends to prepare based on its audit engagement?

    a. The audit scope limitations should be indicated in the report on internal control.

    b. A report on internal control cannot be issued based on a qualified opinion.

    c. The audit scope limitation has no effect but Dey, Knight, & Co. should not issue the report if it will be given to a creditor.

    d. The audit scope limitation has no effect on a report on internal control.

8–30. Which of the following groups does *not* have the responsibility to decide whether reports on an auditor's evaluation of internal accounting control would be useful to the general public?

a. Regulatory agencies.

b. Directors.

c. Officers.

d. Stockholders.

8–31. An auditor was engaged to study the internal control procedures of a governmental agency, and the agency set forth the criteria for the study in questionnaire format. The auditor then performed a study based on such criteria. The auditor's report should *not*

a. Identify the matters covered by the auditor's study.

b. Express a conclusion, based on the agency criteria, concerning the procedures studied.

c. Exclude any relevant condition that the auditor believes to be a material weakness although *not* covered by the criteria of the agency.

d. Indicate whether the study included tests of compliance with procedures covered by the auditor's study.

Chapter 8
Discussion/Case Questions

8–32. During the course of the study and evaluation of a client's internal control, Mr. Inves, a CPA, discovered that, in a number of instances, the pertinent information had not been posted from sales invoices to the customer ledger accounts. Upon inquiring about this situation, Mr. Inves was told by the controller that the posting had not been done because of a temporary shortage of personnel. The controller suggested that Mr. Inves complete the posting himself, because 90% of it was already done.

Mr. Inves decided to follow this suggestion in order to expedite the audit. He completed the posting and balanced the accounts receivable control account to the total of the customer ledger accounts. Mr. Inves considered the possibility of mentioning the posting deficiency in his formal letter of internal control to the board of directors, but he decided against including it, because he considered it a minor matter and because the client's shortage of personnel was the reason for the deficiency. No informal letter of internal control was issued to the controller.

Required:

a. What are the four conditions under which the AICPA approves bookkeeping services and audit services for the same client? (Refer to ethical rule on independence in Chapter 2.)

    b.   Do you think that Mr. Inves violated the AICPA Code of Ethics? Why or why not?

    c.   Assume that there is no bookkeeping/auditing provision in the AICPA Code of Ethics. Do you think that Mr. Inves violated the spirit of the rule on independence?

    d.   Do you think that Mr. Inves followed the right course of action in omitting the posting inadequacy from the formal letter of internal control? What would your answer be if Mr. Inves also issued an informal letter to the controller?

8–33.   For several years, Mr. Accom performed a variety of accounting services for his clients, including the preparation of unaudited financial statements. This year, one of Mr. Accom's clients asked him to conduct a review of the internal control system and to issue a report on the deficiencies, along with recommendations for improvement. The client indicated that this report would *not* be released outside the firm. Mr. Accom agreed to conduct the review.

However, upon completion of the review, the client gave a copy of both the report on internal control and the unaudited financial statements to a local bank with whom the client had an outstanding loan. The report on internal control omitted an important deficiency which had allowed a substantial inventory shortage to go undetected. Although no one knew it at the time, the inventory on the client's balance sheet was overstated by a material amount.

Required:

    a.   Should Mr. Accom have agreed to conduct the review of internal control and to issue the report?

    b.   Define the term "materiality" as it is commonly used by auditors. What significance does a materially understated inventory figure have as far as the local bank is concerned?

    c.   If the local bank were unable to recover the loan and filed a lawsuit against Mr. Accom for negligence, do you think Mr. Accom would have an adequate defense? Why or why not?

8–34.   During the audit of Clements Manufacturing Co., Bill Gahagan, CPA, noticed that scrap steel from the manufacturing process was piled in an unfenced vacant lot next to the plant. No records were kept of the amount of scrap generated in the manufacturing process or the amount on hand at any time. Scrap sales were recorded when made by the maintenance foreman. Gahagan included these weaknesses in his report on internal control to management, together with his recommendations for strengthening controls by fencing the area and weighing and recording the daily scrap production; however, the management of Clements took no action on the recommendations included in the report.

Six months later, it was discovered that the maintenance foreman reported only a small portion of the actual scrap that was sold and had kept most of the proceeds from the scrap sales for himself. The president of Clements then called Gahagan and asked him why he had not caught the maintenance foreman in his audit and suggested that he might hold Gahagan liable for the loss.

Discuss how Gahagan should handle this situation and how he would reply to the president. What liability do you think Gahagan would have in this situation? Would your answer be different if the letter of internal control was sent to the board of directors?

8–35.　During the course of an audit, Mr. Robin, CPA, observed that one of the clerks in the small toy department consistently was taking money from the customers in areas other than the cash register. At the time, he gave little thought to this practice, because most items were priced in "whole dollars" and the customer did not need any change. He did mention this to the controller, who had the same reaction. The audit was completed and the report was issued.

The following year it was discovered that the clerk had been keeping the money from these "whole dollar purchases" rather than putting it in the cash register.

Required:
a.　Could the auditor be held liable for failure to detect this fraud? Should his comment to the controller make any difference?
b.　Assume that the auditor decided to include his observation in a letter of internal control, along with his recommendation on how to eliminate this weakness and provide relative assurance that currency receipts are placed in the cash register drawer. Draft the relevant part of such a letter.
c.　What audit procedures might have provided evidence that could have resulted in the detection of this fraud?

8–36.　Why should a report on internal control not be issued with unaudited interim financial statements?

8–37.　During an audit of a loan company, the auditor discovered that the recipient of a loan with a principal balance of $2,100 had received only $2,000 when the loan was written. The auditor checked the loan agreement and noted that it called for a $2,000 check and $100 in currency to be given to the customer who borrowed the money.

The auditor checked other loan agreements and found similar wording in each of them. When the controller was asked about this practice, he replied that this "service" was given to the customers so that they could have immediate access to currency. He thought that the $100 inconsistency was an error and could be corrected easily. The auditor decided to drop the matter.

Early the following year, a class-action lawsuit was brought against the loan company by several customers who discovered that they had received less than the principal amount of the loans. It was discovered that the controller had kept the currency. The company's board of directors notified the auditor that they would take legal action against him because of his failure to inform them of what he had learned about this matter.

Required:

a. Could the auditor be held liable for either negligence or gross negligence in this case?

b. Assume that the auditor chose to include this loan agreement practice in a letter of internal control to the board of directors. Draft the relevant portion of such a letter. Include a recommendation of the controls that can be implemented to prevent such an occurrence.

c. What audit procedures might detect this fraud?

8–38. Name three internal control weaknesses that might not be correctible because of impracticalities.

Chapter 9

The General Nature
of Evidence-Gathering

Learning Objectives *After reading and studying the material in this chapter, the student should:*

Understand the relationship between the second and third standards of fieldwork.

Have an appreciation of the auditors' inherent risk in gathering evidence.

Know the AICPA guidelines on competency and sufficiency of audit evidence.

Understand the concept of materiality and how to relate this concept to evidence-gathering.

Know the difference between audit objectives and audit procedures.

Possess an understanding of the purpose, design, and indexing techniques of working papers.

Section III of the text is an extended discussion of the study and evaluation of internal control, including the use of statistical sampling and the techniques of EDP auditing. This section contains a similar series of discourses on the gathering of evidence to establish the fairness of financial statement balances.

In this chapter we will explore the *general* characteristics of evidence-gathering techniques. Comprehensive coverage of selected audit procedures is included in Chapters 10, 11, and 12. Chapters 13 and 14 discuss statistical sampling and EDP techniques for evidence-gathering.

Substantive Tests

The Third Standard of Field Work

AU Section 330.01 of the AICPA *Professional Standards,* Volume I *(SAS No. 1)* describes the third standard of fieldwork as follows.

Sufficient competent evidential matter is to be obtained through inspection, observation, inquiries, and confirmations to afford a reasonable basis for an opinion regarding the financial statements under examination.

Although the term "evidence-gathering" typically is used in a narrow sense to refer to the acquisition of evidence for year-end financial statement balances, in a broader context the term represents most of the acts performed by auditors to form an opinion as to the fairness of financial statements. In the latter sense, then, evidence-gathering includes the study of internal control.

Distinction Between Substantive and Compliance Tests

There is a difference between evidence-gathering for the purpose of testing the company's compliance with prescribed internal control procedures and evidence-gathering to establish the fairness of account balances included in the financial statements upon which the opinion is rendered. The former procedures are called tests of compliance, and the latter are called substantive tests.

The chart on page 353 shows the relationship between these two types of tests, and describes the two categories of substantive tests.

In practice, however, there is some overlap between compliance and substantive tests. For example, when auditors trace a sample of entries from the cash receipts listings to the customer ledger accounts, they actually are performing a "dual-purpose" test.

1. Auditors are testing the compliance with prescribed posting procedures. If too many errors are found (a long lag in dates between the listings and the postings, for example), auditors may conclude that compliance is not satisfactory.

2. Auditors are gathering evidence to help them form a basis for reliance on the sales and accounts receivable balances in the financial statements. If large dollar errors are found when the transactions are traced from the receipt listings to the customer ledger accounts, it may be necessary to request the client to adjust the account balance.

The Philosophy of Evidence-Gathering

The Risk of Insufficient or Invalid Evidence

There is always a risk that evidence gathered during the examination is insufficient or incompetent to provide auditors with an adequate basis for an

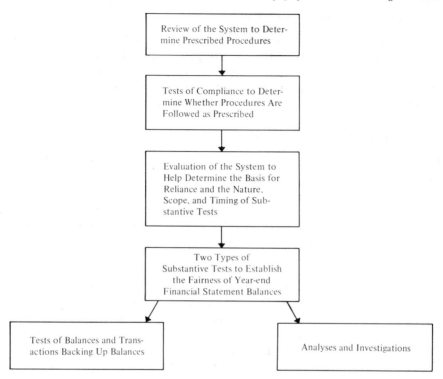

opinion on the fairness of the client's financial statements. Thus, auditors constantly must be aware of the possibility that they may have drawn wrong conclusions about the fairness of the statements.

To *reduce* (not completely eliminate) the probability of their drawing invalid conclusions, auditors exercise judgment in the selection of the evidence they gather to support an opinion. Sometimes this judgment proves to be faulty, and as a result, auditors often suffer the consequences of a lawsuit for negligence. The *Barchris Construction Corporation* case is an illustration. Apparently, the auditors believed that the evidence they gathered on events occurring after the balance sheet date was sufficient to sustain an opinion that the statements were fair. According to the judiciary's comments, the auditors should have detected a deteriorating financial condition that existed after the balance sheet date.

The courts sometimes have taken the position that the auditors' judgment was so faulty that it made it possible for a form of fraud to be committed. An early trend was established for judicial positions of this type when the judge ordered a new trial in the *Ultramares* case on the basis that gross negligence may have occurred. Evidence of the existence and ownership of accounts receivable had been gathered by the auditors, who apparently believed that internally generated documents were sufficient to establish the validity of the accounts. This evidence proved to be insufficient. Several years later, the AICPA established an expanded criterion for sufficiency of evidence when they required direct confirmation with the client's customers.

Guidelines for Evidence-Gathering

It is desirable for auditors to have guidelines to help them decide on the competency and sufficiency of audit evidence. Some *general* guidelines are furnished by the AICPA in AU Section 330 *(SAS No. 1)* but otherwise, auditors must apply their own judgment in selecting specific audit procedures and deciding the amount and types of evidence to be gathered.

The Competency of Audit Evidence

Evidence from Independent Sources—AU Section 330.08 *(SAS No. 1)* states that:

> When evidential matter can be obtained from independent sources outside an enterprise, it provides greater assurance of reliability for the purposes of an independent audit than that secured solely within the enterprise.

Although this statement would seem to be axiomatic in today's audit environment, it should be remembered that two landmark fraud cases *(Ultramares* and *McKesson & Robbins)* might not have occurred, or the frauds might have been detected, if evidence of accounts receivable and inventory had been acquired from independent outside sources. Even today, as a result of the *Equity Funding* case, a suggestion has been made by a special committee of the AICPA that independent confirmations be obtained of an insurance company's insurance in force if the amount is material.

Therefore, there is probably little opposition to the theoretical guideline that "outside" evidence should be obtained, if possible. The auditors' practical problem, however, is determining the specific outside evidence that should be gathered. Discussion of certain accounts shows which ones might qualify for outside verification.

The key, of course, is whether the account in question is susceptible to outside verification. For example, cash and marketable securities probably can be verified independently, because they are held physically by outside agencies (except for cash and/or marketable securities held in the client's office). In contrast, deferred income tax charges and goodwill are created as the result of different types of transactions, and separate confirmation of authenticity cannot be made to auditors.

Accounts receivable and accounts payable usually qualify for independent outside verification, because they represent individual balances owed both by customers and to creditors. Sales and cost of products sold typically do not qualify for the same basic reason that deferred income tax and goodwill do not— these two income statement accounts cannot be authenticated by any outside parties or organizations.

As a general rule, then, auditors will attempt to verify from an independent outside source any account that is composed of (1) tangible commodities held by outside organizations, (2) debts owed to or by the client, and (3) other items for which there is reason to believe that an outside source can verify authenticity.

Evidence and Internal Control—Another guideline of AU Section 330.08 *(SAS No. 1)* states:

> When accounting data and financial statements are developed under satisfactory conditions of internal control, there is more assurance as to their reliability than when they are developed under unsatisfactory conditions of internal control.

The question of how much or what type of audit adjustment to make for the quality of internal control is a matter of judgment for which there are few definite criteria. Some auditors might confine year-end tests of certain accounts to analytical reviews if the system of internal control is exceptionally strong. Other auditors might take a different viewpoint and make only a small change in the amount of substantive testing, regardless of the condition of internal control.

The inventory account is a specific example of the inverse relationship between internal control and substantive testing. The inventory account often constitutes a significant portion of both current and total assets and has an important effect on net income. Thus, it is natural to assume that inventories are given substantial testing by auditing firms.

What change can auditors make in the nature, timing, or scope of inventory testing if internal control is strong? One answer is that the observation of a client's inventory count can be carried out at times other than year-end *if* a good perpetual inventory system is in use. An illustration of this concept follows.

Hypothetical Perpetual Inventory Records (in dollars) from 11-30 through 12-31				*Hypothetical Physical Inventory Count (in dollars) as of 11-30*
	Purchases	Issues	Balance	
11-30	$	$	$20000	$19,500 (inventory count observed and ac-
12- 3	1000*		21000	cepted by auditors)
12- 5		500*	20500	
12-10		400*	20100	
12-15	1000*		21100	
12-20	500*		21600	
12-27		800*	20800	? (not taken by client)

*Transactions reviewed by auditors.

If the auditors' review of the inventory system and tests of the perpetual inventory records give them relative assurance that these records are reliable, a

December 31 balance sheet figure near $20,800 probably would be acceptable. Some review would be made of the intervening transactions in the client's perpetual inventory records from November 30 through December 31. But because the reliability of the inventory records already has been established, this would not be a detailed review.

Evidence from Direct Knowledge—The third AU Section 330.08 *(SAS No. 1)* guideline for judging the competency of audit evidence is:

> Direct personal knowledge of the independent auditor obtained through physical examination, observation, computation, and inspection is more persuasive than information obtained indirectly.

Marketable securities provide a prime example of the relative competency of evidence. Assume that this account consists of bond investments bearing the client's company name and held by a financial institution. A letter from that institution attesting to the validity and ownership of such bonds furnishes auditors with persuasive evidence that the marketable securities do exist and that these securities are owned by the client. However, physical inspection of these bonds gives auditors *more* persuasive evidence of existence and ownership.

It also should be noted that even physical examination of such bonds would not furnish auditors with completely reliable evidence, because the bonds might be forged. Therefore the term "persuasive" is used rather than "convincing."

One can easily ascertain some accounts that are suitable for evidence-gathering through direct personal knowledge:

1. Cash on hand in the client's office can be counted.

2. Marketable securities can be examined.

3. The client's count of physical inventories can be observed.

4. Titles to land and other fixed assets can be inspected.

5. Taxes on income can be computed and checked with the client's computation.

6. Depreciation and amortization can be computed (probably on a test basis) and checked with the client's computation.

The Sufficiency of Audit Evidence

The Persuasiveness of Evidence—AU Section 330.10 *(SAS No. 1)* also contains criteria for judging the sufficiency of audit evidence, i.e., how much and what type should be gathered. One guideline relates to persuasiveness:

The independent auditor's objective is to obtain sufficient competent evidential matter to provide him with a reasonable basis for forming an opinion under the circumstances. In the great majority of cases, the auditor finds it necessary to rely on evidence that is persuasive rather than convincing.

There may be no better illustration of the difficulty of gathering convincing evidence than the process of determining the fairness of the allowance for doubtful accounts. Consider the reason for the account and its composition. The purpose of the account is to reduce the accounts receivable balance to its estimated realizable amount without crediting the receivable itself. The allowance estimate is based on the client's judgment as to what percentage of the accounts receivable balance may be uncollectible.

There is no way to confirm the account balance by an independent source, and certainly, it cannot be verified by inspection. Auditors usually find out what detail method the client used to calculate the allowance percentage (perhaps an aging schedule). Next, the auditors check the accuracy of the client's calculations. Finally, the auditors render a *judgment* as to whether the client's assumptions are sound. In all likelihood, some of the best evidence that auditors can acquire is a tabulation of the collection history and some opinions (often from the client) on the collectibility of slow-paying accounts. This evidence is persuasive, but far from convincing.

Another prime example of the difficulty of gathering convincing evidence is the process of determining contingent liabilities arising from pending lawsuits against the client. Normally, auditors do not have the expertise to make a reasonable prediction of the outcome of unsettled litigation, and typically, they turn to the client's lawyer (or other legal counsel) for an opinion. The lawyer's response is important evidence because his opinion might necessitate the auditors' request for an entry to record a liability on the client's books. The lawyer's response is also important in determining the type of opinion the auditors will render on the company's financial statements. In certain situations, auditors may be forced to give a qualified opinion if the lawyer is unwilling or unable to give an evaluation of the probable outcome of the litigation pending against the client.

Therefore, any evidence obtained from a legal counsel on matters discussed in the preceding paragraph is classified as persuasive but not convincing, because it is entirely possible that the lawyer's opinion is wrong. In addition, any response received from the client's lawyer may contain some bias, because of the lawyer's vested interest in receiving a favorable verdict from the courts.

The Cost of Obtaining Evidence—Another guideline on the sufficiency of evidence is contained in AU Section 330.12–.13 *(SAS No. 1)*.

An auditor typically works within economic limits; his opinion, to be economically useful, must be formulated within a reasonable length of time and at reasonable cost. . . . As a guiding rule, there should be a rational relationship between the cost of obtaining evidence and the usefulness of the information obtained.

Consider the two types of confirmation requests that auditors might send to accounts receivable customers. One type, referred to as positive, contains a request for the customer to respond to the auditors regardless of whether or not the customer agrees with the client's book figure. The other type of confirmation request, referred to as negative, contains a request for the customer to respond to the auditors only if the customer does *not* agree with the client's book figure.

For positive confirmations, it is usually standard practice for auditors to send second and, if necessary, third requests to customers who fail to respond to the initial letter. Though this approach makes the choice of positive confirmations a more costly alternative, it also reduces the amount of audit risk. Recipients of negative requests may fail to respond to the auditors because (1) they agree with the client's book figure, or (2) they simply neglect to return the request to the auditors. At some point, the value of the additional audit evidence gained from circularizing positive confirmation requests ceases to exceed the additional cost of sending out this type of request. Auditors must make a judgment as to when this point is reached.

The following diagram helps to show the cost-benefit relationships of positive and negative confirmation requests.

Positive Requests   *Negative Requests*

Number of positive requests in relation to negative requests (1)   Number of negative requests in relation to positive requests (2)

Value of the audit evidence   Risk of obtaining insufficient evidence

(1) The slope of the line is a matter of judgment. The line is drawn with a sharp upward slope near its end to indicate that, at some point, the marginal value of audit evidence gained from additional positive requests becomes much smaller.

(2) The slope of the line is a matter of judgment. The line is drawn with a flat slope near its end to indicate that, at some point, the marginal risk of obtaining insufficient evidence becomes much greater when negative requests are sent out rather than positive requests.

Another appropriate example of the cost-benefit relationship of evidence-gathering is the audit of property, plant, and equipment accounts. Except, possibly, for initial examinations, auditors generally do not conduct a physical inspection of the fixed assets that constitute the beginning figure in the client's balance sheet. They usually confine their audit procedures to the additions and retirements that are recorded, and should have been recorded, during the year.

Though it is true that an annual inspection of fixed assets would provide auditors with additional evidence of the accounts' validity, the cost of such a procedure in most cases would greatly outweigh the benefit. The turnover rate of fixed assets is usually low, and evidence of the authenticity of these accounts can be gathered during other phases of the audit.

The Objectivity of Evidence—The last criterion listed for judging the sufficiency of audit evidence is contained in AU Section 330.15 and concerns objectivity.

> In developing his opinion, the auditor must give consideration to relevant evidential matter regardless of whether it appears to support or to contradict the representations made in the financial statements.

A sample with a large number of errors is an example of a matter that requires the auditors' objectivity toward evidence. The temptation to look for items that contain no errors is strong. This temptation should be resisted, however. The effect on the financial statements of events that occur after the balance sheet date is another matter that calls for objectivity. At the end of the year, the statements may appear to be fair, but at a later time, certain events may occur or particular conditions may change, so that adjustments and/or disclosures are necessary. One example is destruction of a material portion of the company's property; another is the sudden bankruptcy of a customer with a substantial accounts receivable balance at the end of the fiscal year.

When first considering the question of post-balance-sheet events, one may think that such events have no effect on the audited financial statements. But financial statements are prepared on the assumption that the business enterprise will outlast its assets. In addition, many financial statement estimates are based on future projections, for example, the allowance for doubtful accounts and depreciation expense. Therefore, auditors should take advantage of any "hindsight" available to them. If an event occurs during the period of the fieldwork, and in the auditors' opinion that event has a material effect on the audited financial statements, an adjustment of the statements or disclosure is necessary, even if it has an adverse effect on such statements.

It is the auditors' responsibility to exercise their judgment in deciding (1) whether a subsequent event has a material effect on the audited financial statements, and (2) whether an adjustment of the accounts or disclosure is required. If the evidence gathered by the auditors indicates that the event relates to conditions that existed at the balance sheet date (the bankruptcy of a customer whose balance was considered of doubtful collectibility at the balance sheet date), an adjustment to the statements should be made. If the evidence shows that the event relates to conditions that arose after the balance sheet date (sale of bonds or property destruction), a footnote disclosure is considered sufficient.

Materiality and Evidence-Gathering

A General Definition of Materiality

Throughout the examination, auditors constantly are faced with the questions of how *much* evidence to gather, what *types* of evidence to secure, and what *actions* to take on the basis of the evidence inspected. Generally, the answers to these questions depend on the auditors' feelings about an important but somewhat nebulous concept, materiality.

In accounting, an item is considered material if an error in or omission of this amount would cause prudent individuals who can read financial statements intelligently to change decisions that they might make on the basis of the company's financial statements.[1] Thus, evidence may be gathered on certain account balances because errors in these amounts would be material (positive confirmations on certain accounts receivables). Adjustments should be made to the statements if certain events occur, because failure to make these adjustments would have a material effect on the statements (certain types of subsequent events).

Percentage Criterion for Materiality

One obvious criterion for judging materiality is the dollar size of the amount in comparison with some key figure, such as the asset and equity totals at the balance sheet date. By referring to the following chart, one can isolate the account balances that auditors probably would consider to be material by the criterion of relative size; the amounts are in thousands of dollars.

Account Title	Amount (Dollars in Thousands)	Percentage of Total Asset and Equity Figure of $1,010,349
Accounts Receivable	$183,425	18
Inventories	306,718	30
Buildings and Improvements	130,881	13
Machinery and Equipment	363,752	36
Accounts Payable	315,947	31
Retained Earnings	$178,067	18

The retained earnings account at the bottom of the list might be audited indirectly by obtaining evidence on other accounts. However, a reasonable percentage of the total audit time would be devoted to gathering competent and sufficient evidence on the other accounts listed.

Often, auditors' judgments on materiality are based on the relationship of the item to net income, because this figure is of such importance to many investors. For example, auditors might find evidence to support an allowance for

---

[1] A narrower definition is that prudent investors would change their decision about investing in the company's securities.

doubtful accounts that is $200,000 more than the amount shown on the client's books. If an adjustment would require a charge to the Bad Debt expense account and would decrease Net Earnings from $27,700,000 to $27,500,000, it is uncertain whether the auditors would pursue the matter, if they used the potential change in Net Earnings as a primary guideline for determining the materiality of the proposed adjustment.

Are there any definite percentage guidelines on materiality? Unfortunately, there are not. Some members of the accounting profession might argue that materiality should be left completely to the judgment of auditors. However, there is ample evidence to suggest that if guidelines are not established by auditors, the courts or other outside authorities will apply their own criterion when the circumstances arise. Such application of a materiality criterion occurred in the *Barchris Construction Corporation* case (summarized in Chapter 3). The judiciary ruled that a certain overstatement of earnings was not material, because earnings would have increased substantially without the falsification that caused the overstatement.

Some thought has been given to the problem of materiality guidelines. Leopold A. Bernstein, in the April 1968 issue of the *Lybrand Journal,* suggested "border zones" for general guidelines. One possible set of border zones is illustrated in the chart below.

Percentage by Which the Item Is Expected to Change Net Income[2]	*Decision on Materiality*
Less than 10%	Not material
10 to 15%	Left to auditors' judgment
More than 15%	Material

Other Criteria for Materiality

Dollar amounts and percentages are not the only criteria for forming judgments on materiality. Another important element is the pervasiveness of the item on the financial statements. For instance, the cash account may represent a small percentage of total assets or other key figures. Nevertheless, because of the many transactions affecting cash, its importance usually outweighs its relative dollar size. As a result, auditors usually spend more time gathering evidence on cash than on most accounts of a comparable dollar amount. (Some auditors would state that higher risk is the reason for this emphasis on cash.)

Another example of an item that could be material because of its pervasive nature is a subsequent event that might require footnote disclosure in the financial statements. Auditors would request the client to make a disclosure if

[2]It is possible that many auditors rely on smaller percentages, e.g., less than 5%, not material; 5 to 10%, left to auditors' judgment; more than 10%, material.

there were reason to believe the financial statements would be misleading without such a disclosure; the judgment on materiality would not necessarily depend on a dollar amount. Transactions between key officers and the company might be a case in point.

Other factors to consider in forming a judgment as to materiality are past trends and the relative size of past balances and transactions. For example, if a company's earnings had increased by approximately 10% each year for several years, an adjustment that would reduce the increase in earnings in the current year to 2% might be considered material, even though it might be only 8% of total earnings. Some auditors believe that if a company's earnings have averaged $1,000,000 for the past several years and amount to only $10,000 in the current year, the company has for all practical purposes broken even, and that a $5,000 adjustment would not be material. They reason that even though the adjustment may constitute 50% of current-year earnings, in light of the company's size and past operations, either $5,000 or $10,000 of earnings would still indicate break-even operations. Thus, the auditors would be relating materiality to "normal" operations and conditions. This practice may prove hazardous in abnormal times, but it does illustrate the point that many factors must be considered in making a materiality judgment, and fixed percentages cannot be blindly applied.

Audit Objectives

Before an auditor begins his work, he must know what he is trying to accomplish. In other words, he must know what his objectives are in the audit of each account. Though this statement may sound rather basic, sometimes there is a tendency for auditors to begin their work by mailing confirmations to customers, examining invoices and canceled checks, and performing numerous other tasks without first considering what they are attempting to accomplish for each account, and then determining the best, most efficient way to proceed. Only by knowing what his objectives are can an auditor know whether they have been accomplished. For example, an auditor may decide that one of his objectives in auditing cash is to determine that it is not restricted and is subject to immediate withdrawal; but if this objective is not specified, he may omit the audit steps necessary to accomplish it. Similarly, by establishing objectives, the auditor can avoid excess work that does not contribute to their accomplishment.

The broad objective of auditing a given account is to authenticate its existence, ownership, valuation, and classification.[3] Within this framework, each individual account must be analyzed to determine specific objectives.

In some cases, the auditor must determine the proper accounting treatment for the financial statement item either by reference to official pronouncements or

---

[3]Some authors take a narrower view of audit objectives and include such items as study and evaluate internal control, establish a proper cutoff of transactions, and verify related income statement amounts. In our view, however, these are audit procedures rather than objectives. We have attempted to determine audit objectives of accounts within the context of the financial statements.

by a knowledge of the "generally accepted" accounting method in order to establish his objective. As an example, AC Section 5121.04, .05, and .07 of *AICPA Professional Standards,* Volume 3 *(ARB No. 43)* contains these comments on inventory.[4]

(Cost Principles)

The primary basis of accounting for inventories is cost, which has been defined generally as the price paid or consideration given to acquire an asset. As applied to inventories, cost means in principle the sum of the applicable expenditures and charges directly or indirectly incurred in bringing an article to its existing condition and location.

(Pricing Options)

Cost for inventory purposes may be determined under any one of several assumptions as to the flow of cost factors (such as first-in first-out, average, and last-in first-out); the major objective in selecting a method should be to choose the one which, under the circumstances, most clearly reflects periodic income.

(Lower of Cost or Market Principle)

A departure from the cost basis of pricing the inventory is required when the utility of the goods is no longer as great as its cost. Where there is evidence that the utility of goods, in their disposal in the ordinary course of business, will be less than cost, whether due to physical deterioration, obsolescence, changes in price levels, or other causes, the difference should be recognized as a loss of the current period. This is generally accomplished by stating such goods at a lower level commonly designated as market.

It follows, then, that one specific objective in auditing inventory is to ascertain that the amount of inventory shown on the balance sheet is stated at the lower of cost or market, by use of one of the generally accepted pricing methods. This objective, in turn, relates to the broad objective of determining that the inventory is valued properly.

In other cases, the specific objectives flow naturally from the broad ones. Again with inventory as an example, the broad objective of authenticating existence leads to the specific objective of determining that the inventory amount on the balance sheet is represented by physical items actually on hand, in transit, or on consignment. By this approach, the auditor can arrive at the following specific objectives for the audit of inventories.

1. Determine that the inventory amount on the balance sheet is represented by physical items actually on hand, in transit, or on consignment (existence).

---

[4]All accounting pronouncements are in Volume 3 of the Commerce Clearing House, Inc. book. The pronouncements are divided into Volumes 3 and 4 of the CCH, Inc. loose-leaf service.

2. Determine that the inventory is calculated properly at the lower of cost or market in accordance with generally accepted accounting principles consistently applied (valuation).

3. Determine that the inventory belongs to the company and that any liens on the inventory are disclosed properly (ownership).

4. Determine that any excess, slow-moving, or special-purpose items are properly valued and classified (valuation and classification).

A similar approach can be used to arrive at the objectives for other financial statement amounts.

Audit Procedures

After the audit objectives have been determined, the next step is to define the audit procedures that will accomplish the specified objectives. There is no one set or official list of audit procedures that can be used on each engagement, because the nature of the accounts and their materiality vary. For example, the procedures for the audit of inventory quantities and pricing of a manufacturer of a complex product with several stages of completion and a sophisticated standard cost system would be different from those for a company whose inventory consists of a large pile of coal. Also, if inventory is an immaterial amount in the financial statements, only a few limited procedures may need to be applied to this account. Finally, an important determinant of both the extent and nature of auditing procedures in a particular engagement is the effectiveness of the client's system of accounting procedures and internal controls.

As pointed out in Chapter 5, the weaker the system of internal control, the more reliance the auditor must place on substantive tests of balances and transactions, and vice versa. There is no scale of strength and weakness upon which internal control can be measured; neither is there a related table indicating the percentage of inventory items to be counted or accounts receivable to be confirmed. Some auditors have lamented this fact, and certain academicians have performed mathematical studies attempting to quantify the effectiveness of internal control and the extensiveness of audit procedures. Judgment, however, with a tempering of experience, remains the basis for both the evaluation of internal control and the determination of the extent of audit procedures. The exercise of this judgment is, of course, one trademark of a professional.

Although an all-inclusive list of audit procedures cannot be prepared, the following procedures are representative of those often followed in the audit of inventories. Note that the procedures are keyed to the objectives to which they relate.

1. Review the accounting procedures and internal controls for inventories and update the memorandum describing inventory procedures in the permanent file (existence, ownership, valuation, and classification).[5]

2. At year-end, observe the client's inventory counting and recording procedures to determine whether they are adequate to result in an accurate inventory. Make and record test counts of inventory and compare them with those made by the client. Recount any items for which differences are found (existence).

3. Confirm the existence and ownership of inventory held on consignment or in public warehouses (existence and ownership).

4. Test the propriety of the cutoff of inventory shipments and receipts by recording the numbers and descriptions of the last five shipping and receiving reports during the year-end inventory observation, and at a later date, inspect the related shipping and receiving documents in addition to the first five after the end of the year. Analyze the sales and purchase invoices that correspond with those shipping and receiving documents and determine that (a) items recorded as sales in the current period were excluded from inventory, (b) items recorded as sales in the subsequent period were included in inventory, (c) items recorded as purchases in the current period were included in inventory, and (d) items recorded as purchases in the subsequent period were excluded from inventory (existence and ownership).

5. During the inventory observation, look for and inquire about any excess, slow-moving, obsolete, or unsalable inventory. Indications of such items would be a covering of dust or rust, or prior-year inventory tags (valuation and classification).

6. Account for all prenumbered inventory tags before and after the physical inventory (existence).

7. Obtain a copy of the final inventory listing and compare inventory prices with purchase invoices. Check the calculation of the inventory amount on the basis of the method used (i.e., FIFO, LIFO, average) and compare with market, i.e., the lower of replacement cost or net realizable value (valuation and ownership).

8. Review confirmations received from banks and other creditors and minutes of the board of directors for indications of pledges or assignments of inventories (ownership).

---

[5]Permanent audit files are discussed later in the chapter.

The foregoing procedures are necessarily more general than those in an actual audit engagement, but they demonstrate how audit procedures are designed to accomplish the audit objectives. Normally, several procedures are required to achieve an objective fully and, in many cases, one procedure will contribute toward the fulfillment of more than one objective.

Working Papers

Definition and Purposes of Working Papers

For auditors to conduct their examination properly and provide adequate support for their opinion, they must prepare audit working papers.

AU Section 338.03 of *AICPA Professional Standards,* Volume 1 *(SAS No. 1)* contains the following definition.

> Working papers are the records kept by the independent auditor of the procedures he followed, the tests he performed, the information he obtained and the conclusions he reached pertinent to his examination.

As pointed out in Chapter 1, AU Section 338.05 *(SAS No. 1)* contains *general* guidelines as to what working papers should include or show.

a. Data sufficient to demonstrate that the financial statements or other information upon which the auditor is reporting were in agreement with (or reconciled with) the client's records.

b. That the engagement had been planned, such as by the use of work programs, and that the work of any assistants had been supervised and reviewed, indicating observance of the first standard of field work.

c. That the client's system of internal control had been reviewed and evaluated in determining the extent of the tests to which auditing procedures were restricted, indicating observance of the second standard of field work.

d. The auditing procedures followed and testing performed in obtaining evidential matter, indicating observance of the third standard of field work. . . .

e. How exceptions and unusual matters, if any, disclosed by the independent auditor's procedures were resolved or treated.

f. Appropriate commentaries prepared by the auditor indicating his conclusions concerning significant aspects of the engagement.

In summary, properly prepared working papers are necessary for an auditor to demonstrate compliance with the standards of field work. They should show how the work was planned (primarily by the use of audit programs) and the extent of supervision of assistants (indication of reviews made by the auditor), and should contain specific items of evidence such as internal control checklists (see Chapter 5), confirmations from creditors, bank reconciliations, etc.

Ownership of Working Papers

Though there is some variation in state laws, working papers prepared by an auditor in connection with his examination of a client's financial statements are generally the property of the auditor. The client normally has no claim to the working papers, regardless of the fact that he paid the auditor to perform the audit in which the working papers were prepared. Working papers generally are not "privileged" in the manner of communications between an attorney and his client, and must be surrendered in response to a subpoena or other legal action. This means that information obtained by an auditor during his examination of financial statements could be used against his client in a legal proceeding. Of course, even though the working papers belong to the auditor, he is required to comply with the confidentiality requirements of Rule 301 of the *Code of Professional Ethics,* as discussed in Chapter 2.

Importance of Working Papers

Because the audit working papers constitute the auditor's evidence of the work he performed, they may either help or damage him if problems subsequently arise concerning the audited financial statements. They will become the most important documents involved in any subsequent litigation and, because they are subject to subpoena, they may provide evidence for the plaintiff or prosecution as well as for the auditor's defense.

It is unfortunate that the term "working papers" has evolved to describe the evidence that an auditor accumulates during an examination of financial statements. The term connotes an unfinished product such as an accumulation of preliminary notes and calculations on scratch pads. Though only a very imprudent auditor would prepare working papers in such a manner, many auditors are not as careful as they should be in this regard. For example, an auditor may spend many hours considering a serious accounting or reporting problem of a client, but make only a brief note of his conclusion once it is reached. Without evidence to show the careful consideration given the problem, it may appear later that only superficial thought was given to it. Even if working papers are complete, sloppy preparation with many erasures, misspelled words, and incomplete sentences will cast doubt on them. An adversary attorney might use such working papers to demonstrate a careless approach to the audit. The auditor must keep in mind that he may not be the only one to read the working papers he prepares, and he must consider the impression others may gain from reading them.

Types of Working Papers

Auditors normally maintain two types of working-paper files. One is referred to as a permanent or continuing audit file, and the other often is called the current-year audit file.[6]

Permanent Audit Files—The permanent audit file is used to store documents, schedules, and other data that will be of continuing significance to several years' audits. For example, an auditor must obtain a copy or extract of a client's articles of incorporation to verify the types (common and preferred), par values, and number of authorized shares of stock that the company may issue, as well as restrictions on payment of dividends, purchase of treasury stock, or other matters requiring disclosure in the financial statements. Rather than obtaining a copy or extract of the same document each year, the auditor places one copy in the permanent file, which is a part of each year's audit evidence. Of course, it is necessary to check each year for any amendment to the articles of incorporation and to indicate any changes on the document in the permanent file. Amendments normally would be detected by the auditor during his review of minutes of stockholders' and directors' meetings, because approval generally is required by one or both of these bodies.

Although the organization of permanent files varies, most would contain such sections as the following.

1. Historical information regarding the company—This section usually includes a memorandum describing the company and its operations, major plants and manufacturing processes, and products, distribution facilities, and important customers. An organization chart listing the names and positions of key officers and employees and any recurring audit administrative matters also are shown. This type of information is particularly important to an auditor assigned to a client for the first time. It allows him to learn something of the operations of the company in a brief period of time and makes him aware of any unusual matters concerning the audit, such as timing deadlines, reporting requirements, etc. It saves the client the task of acquainting different members of the auditing firm with basic information about the company. A partial example of such a memorandum follows.

X Co.

Company Operation

and

Audit Administration Memo

---

[6]In addition to audit working-paper files, the auditor may maintain tax working-paper files (for working papers developed in connection with the review or preparation of a client's tax returns), correspondence files (to maintain a record of all communications with each client), billing files (to maintain the current status and historical record of billings to and collections from each client), and client files (for developing a record of the relationship with each client).

X Co. was formed in 1970 by Mr. Fitzgerald and Mr. Hamilton, both of whom remain 50% stockholders. X Co. operates a 40,000 barrel per day refinery near Granville, Arkansas (also the location of its administrative offices), at the intersection of Highways 65 and 71. The President and Plant Manager is Mr. Foote and the Treasurer, with whom we arrange the timing of our work, is Miss Elaine. The company acquires most of its raw materials (crude oil) in the open market and is subject to the regulations of the Department of Energy (a copy of the DOE regulations is filed behind this memo). The crude oil is refined into premium and regular gasoline which is sold to a chain of independent gasoline stations. The company is on an August 31 fiscal year. We are to prepare a long-form audit report for the shareholders which is to be delivered to them by October 10. In the past we have experienced difficulty and delay in obtaining confirmation of the significant receivable from the chain of independent gasoline stations, so we must be sure to mail the confirmation at the earliest possible date and include sufficient details to. . . .

An auditor who had never worked on the X Co. engagement would know, by reading the memorandum, where the company is located and what it does, whom to contact at the company, what some of the important reporting and timing requirements are, and that he would need a knowledge of DOE regulations in the audit.

2. Company accounting procedures and internal controls—This material might consist of narrative descriptions of the client's accounting procedures and internal controls, internal control checklists, flow charts, or any combination of these items. A chart of accounts and samples of any records or forms that would aid in understanding company procedures also would be included. A brief example of a description of accounting procedures relating to notes payable and long-term debt is set forth below.

X Co.
Procedures for Notes Payable
and
Long-Term Debt

X Co. has outstanding a $5 million issue of 6% bonds due in installments of $500,000 per year beginning in 19XX. All refinery property and equipment is pledged to these bonds. The bond indenture restricts payment of dividends to $200,000 per year. The company also borrows on short-term notes for working capital purposes.

All borrowings are authorized by the Board of Directors, and the banks or other creditors are specifically mentioned in the minutes. Both the President and the Treasurer must sign any notes that are issued. The Treasurer maintains a schedule showing the due dates of all bond, note, and interest payments. . . .

Examples of internal control checklists and flow charts are shown in Chapter 5.

3. Corporate documents—In addition to the articles of incorporation, the permanent file normally contains copies or extracts of bylaws, loan agreements, bond indentures, labor contracts, stock option plans, pension plans, important long-term operating agreements or contracts, and other documents. It is clear that all these documents could significantly affect the company's operations and its financial statements; therefore, the auditor must have evidence of the provisions of these documents and of his reviews thereof.

4. Continuing analysis of certain accounts—It is often more efficient to maintain cumulative or carry-forward schedules in the permanent files for certain accounts with little activity, or for which comparisons with several prior years are helpful, than to prepare such schedules each year in the current files. Continuing analysis might be used for capital stock, long-term debt, checklists for compliance with loan agreements, net operating loss carry-forward schedule, equity in earnings of subsidiaries, and gross profit ratios by major product class. An illustration of such an analysis is shown on page 371.

This page 371 analysis should alert the auditor that there has been a deterioration in accounts receivable during the year, and it should raise a question as to the adequacy of the allowance for doubtful accounts.

5. Audit planning—This section could include a master copy of the audit program that could be revised and mechanically copied each year rather than completely rewritten; schedules of plant capacity and volumes of tanks, bins, and other containers (it would be embarrassing for an auditor to discover after he had satisfied himself as to inventory that his client did not have the physical capacity to store the amount of inventory shown in the accounting records); and, if certain procedures are performed on a rotating basis, a record of the accounts (cost centers, bank accounts, etc.) or locations (branch offices, subsidiaries, etc.) tested each year to assure that nothing is overlooked, or that the same account or location is not tested repeatedly year after year, while others are never tested.

The permanent audit file can be a very useful tool of the auditor if it is kept current and used. Occasionally, an auditor, in his haste to complete the current-year audit, will neglect to review and update the permanent file. When this happens, the file becomes less useful and less used each succeeding year, until it becomes a file of obsolete or superseded data. At that point, it becomes less than useless to the auditor; it becomes a threat to him, because it is evidence of negligent and inadequate work.

X Co.
___

Analysis of Uncollectible Accounts

For the year—	*19X6*	*19X7*	*19X8*	*19X9*
Sales	$4,365,000	$5,837,000	$5,679,000	$5,928,000
Bad debt provision	54,000	68,000	63,000	71,000
Bad debt charge-offs	54,000	67,000	63,000	91,000
Balance 12-31				
Current	$372,000	$498,000	$501,000	$530,000
30-60 days	31,000	54,000	53,000	57,000
Over 60 days	2,000	19,000	16,000	52,000
Total	$405,000	$571,000	$570,000	$639,000
Allowance for doubtful accounts	$30,000	$35,000	$35,000	$15,000
Ratios—				
Bad debt charge-offs to sales	1.2%	1.1%	1.1%	1.5%
Allowance to total accounts receivable	7.4%	6.1%	6.1%	2.3%
Allowance to accounts receivable over 60 days	15.0	1.8	2.2	0.3
Days' sales in accounts receivable	33.9	35.7	36.6	39.3

Current Audit Files—The current audit files for each year contain the evidence gathered and the conclusions reached in the audit for that year. The material in the current files includes schedules and analyses of accounts, memoranda of audit work performed in certain areas and audit problems considered and resolved, an audit program, correspondence with third parties (banks, customers, creditors, legal counsel, etc.) confirming balances, transactions and other data, review sheets containing questions and comments regarding the audit work performed which are prepared by supervisory personnel during their review of the working papers, a schedule of time spent on the engagement by individual auditors, and other documents. Examples of the foregoing items are provided in this or following chapters.

Working papers organized in a logical manner improve the efficiency of an audit and the effectiveness of its review. Although an auditor must be aware of the interrelationships among accounts (such as between sales and accounts receivable, or accounts payable, inventory, and cost of sales), and design his audit procedures in recognition thereof, the most logical approach to the organization of working papers is to begin with the financial statements on which the auditor expresses his opinion, which, of course, are the final product of the client's records and accounting system.

Using the financial statements as the apex, the auditor dissects the individual financial statement items to the point at which they are most efficiently and effectively audited. This point varies by company and by account. For example, it would be difficult to perform much effective auditing on the total accounts receivable balance shown in the financial statements. This account first must be broken down into sub-accounts, such as customer, notes, officer and employees, interest, and other receivables on a schedule sometimes referred to as the lead schedule. Then the more significant of the sub-accounts would be analyzed further by customer, note, officer and employee, etc., to obtain amounts which the auditor can subject to such audit tests as confirmation, aging, etc. However, if all salesmen are paid a fixed commission on sales, it may be possible to test sales commission expense by multiplying total sales by the commission rate rather than analyzing commission expense by salesmen. Thus, the organization of working papers can be likened to the following triangle.

FINANCIAL
STATEMENTS

Audit

Trial

Balances

Lead Schedules

Detail Audit Schedules

This concept is illustrated in the discussion of indexing the working papers in another section.

Although the auditor would prefer to use the financial statements as the starting point for his audit, there are often practical reasons why he cannot. Sometimes the financial statements have not been prepared while the audit is in progress. In other cases the auditor processes adjusting entries as a result of his work, which would change any financial statements that have been prepared. Therefore, the auditor normally prepares an audit trial balance which resembles the financial statements (without footnotes), but contains columns for adjustments and reclassifications proposed as a result of the audit. The adjustments columns are for the posting of adjusting entries proposed by the auditor as a result of his work to correct errors in the accounting records. On many audit engagements the auditor has no adjustments, whereas on others, numerous ones may be made. The reclassification columns are for the posting of reclassification entries, for example, entries to change the classification of an item for financial statement presentation purposes. Adjusting entries always must be posted to the accounting records by the client, because they are corrections of errors in those records, but reclassification entries are not posted to the accounting records, because they are only rearrangements of the ledger accounts for financial statement purposes.

The audit trial balance shown in Figure 9.1 contains certain adjusting and reclassifying entries (from Figures 9.2 and 9.3). The column entitled "As adjusted and reclassified" should agree with the amounts shown in the client's financial statements. The proposed adjusting and reclassifying entries must be concurred with by the client company's management, because the responsibility for the financial statements is theirs. If such entries are proposed by the auditor, the documentation and support for the entries must be included in the working papers. Therefore, the proposed entries usually are shown in three places: (1) the working-paper schedule on which the entry was computed, (2) a schedule listing all such entries, and (3) the audit trial balance. Note that the entries in Figures 9.2 and 9.3 include references to working-paper schedules where the support for the entries can be found.

In addition to material errors and misclassifications, auditors often find minor errors and misclassifications, no one of which is sufficiently material to require an adjusting or reclassifying entry. These items should be reviewed with the client company's management to determine whether they want to record them. Clients often wish to correct minor errors so they can begin a new year with a more accurate set of accounting records. The auditor must prepare a schedule summarizing all such uncorrected errors to be sure that, in the aggregate, they do not have a material effect on the financial statements. This schedule often is called a summary of immaterial entries or a summary of entries passed. Figure 9.4 is an example of such a schedule. The net amount of the entries not made is compared with the amounts to be shown in the financial statements to determine whether it is material in the aggregate. In determining the materiality, the auditor must keep in mind all aspects of the audit. For example, if a company's loan agreement provides that it will maintain an excess of current assets over current liabilities of at least $100,000 or be in default, and if current assets total $200,100, and current liabilities total $100,000, a $101 adjustment could be material. Normally, the senior, manager, and partner assigned to the audit state or approve the conclusion that the entries not recorded are not material.

TB-1

X Co.
Audit Trial Balance
Assets
12-31-X8

J. Jones
01-20-X9

	Index	As Adjusted 12-31-X7	Per Books 12-31-X8	Adjustments Dr	Adjustments Cr	As Adjusted 12-31-X8	Reclassifications Dr	Reclassifications Cr	As Adjusted and Reclassified 12-31-X8
Current Assets — Cash	A	73430	65141			65141			65141
Accounts Receivable	B	121189	140767		(2) 6000	134767 (2a)	187197		155564
Inventories	C	216508	243030			243030			243030
Prepayments	D	11343	12255			12255			12255
		482464	461193			455193			455990
Property and Equipment	F	581750	604988			604988			604988
Less allowance for depreciation		89246	108716			108716			108716
		492484	496272			496272			496272
Other Assets	H	17342	16629			16629			16629
		992340	974094			968094			986891
Columns footed		T	T			T			T

Figure 9.1 Audit Trial Balance

	Index	As Adjusted 12-31-X7	Per Books 12-31-X8	Adjustments Dr.	Adjustments Cr.	As Adjusted 12-31-X8	Reclassifications Dr.	Reclassifications Cr.	As Adjusted and Reclassified 12-31-X8
Current Liabilities									
Notes Payable	M	50000	60000			60000		(a) 50000	110000
Accounts Payable	N	110139	103677		(1) 4832	108509		(b2) 18797	127306
Accrued Liabilities	P	30693	81778	(3) 5000		76778			76778
		240769	245455			245587			314084
Long-term Debt	R	350000	300000			300000	(b1) 50000		250000
Stockholders' Equity:									
Common Stock	S	100000	100000			100000			100000
Retained Earnings-	T								
Beginning of Year		282263	301518			301518			301518
Net income of year		68315	77061			71329			71329
Dividends		<50000>	<50000>			<50000>			<50000>
End of year		301578	328839			322807			322807
		401578	428839			429807			429807
		922340	974094			968094			986881

TB-2

X Co.

Audit Trial Balance

Liabilities

12-31-X8

J Jones 01-20-X9

Figure 9.1 (Continued)

X Co.
Audit Trial Balance
Revenue and Expense
12-31-X8

TB-3
J. Jones 1-20-X9

	Index	As Adjusted 12-31-X7	Per Books 12-31-X8	Adjustments Dr	Adjustments Cr	As Adjusted 12-31-X8	Reclassifications Dr	Reclassifications Cr	As Adjusted and Reclassified 12-31-X8
Sales	10	1349906	1417232			1417232			1417232
Cost of goods sold	20	935929	1055838	① 4832		1060670			1060670
Gross profit		311977	361394			356562			356562
Selling expenses	30	83897	106236		② 6000	112236			112236
Administrative expenses	40	59315	76997			76997			76997
Interest expense	40	34850	31100			31100			31100
		182062	214333			220333			220333
Net income before income taxes		129315	147061			136229			136229
Federal and state income taxes		60000	70000		③ 5000	65000			65000
Net income		69315	77061			71229			71229

Figure 9.1 (Continued)

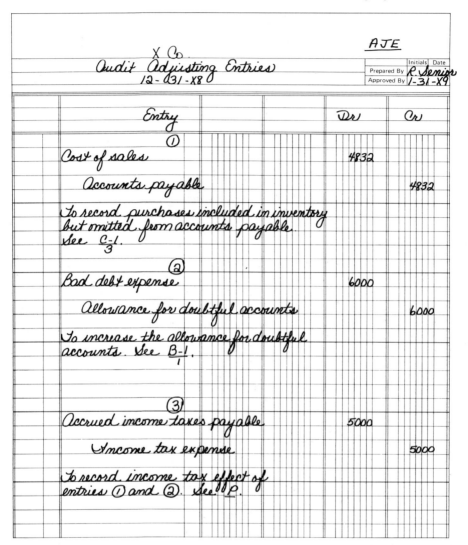

X Co.
Audit Adjusting Entries
12-31-X8

AJE

Prepared By R. Senior
Approved By 1-31-X9

Entry	Dr	Cr
①		
Cost of sales	4832	
Accounts payable		4832
To record purchases included in inventory but omitted from accounts payable. See C-1/3.		
②		
Bad debt expense	6000	
Allowance for doubtful accounts		6000
To increase the allowance for doubtful accounts. See B-1/1.		
③		
Accrued income taxes payable	5000	
Income tax expense		5000
To record income tax effect of entries ① and ②. See P.		

Figure 9.2 Audit Adjusting Entries

An example of the next level in the triangle, the lead schedule, is shown in Figure 9.5. It illustrates the division of the total inventory amount into subgroups by stage of completion. In other cases, the total inventory amount might be divided according to product lines, inventory storage location, or other criteria. The type of breakdown shown on the lead schedule is determined by the classifications used by each client in its accounting records. Thus, in this, as in many other aspects of auditing, no single format is applicable to all engagements; each must be tailored to the particular aspects of each audit.

RJE

X Co.
Audit Reclassifying Entries
12-31-X8

	Initials	Date
Prepared By	R. Senior	1-31-X9
Approved By		

Entry	Dr.	Cr.
(101)		
Long-term debt	50000	
Notes payable		50000
To reclassify as a current liability the portion of long-term debt due within one year. See R.		
(102)		
Accounts Receivable	18797	
Accounts Payable		18797
To reclassify credit balances in accounts receivable to accounts payable. See B.		

Figure 9.3 Audit Reclassifying Entries

The base of the working-paper triangle symbolizes the detail audit schedules—a myriad of schedules and documents, each representing a specific piece of evidence gathered during the examination. Figures 9.6 through 9.8 illustrate a few of the detail audit schedules that might be included in the inventory section.

Figure 9.6 is an example of an inventory listing of raw materials. Note that the total of this schedule agrees with the amount shown as raw materials on the lead schedule in Figure 9.5. On this schedule, part of the total inventory amount (raw materials) has been analyzed to the extent of individual inventory items. Only at this point can many important audit procedures be performed. Physical quantities can be compared with test counts made by the auditor during the inventory observation (Figure 9.7), and prices used to calculate the inventory (Figure 9.8) can be compared with the purchase prices of the items (this is a simplified example which assumes a FIFO method; many inventory pricing tests are more complicated).

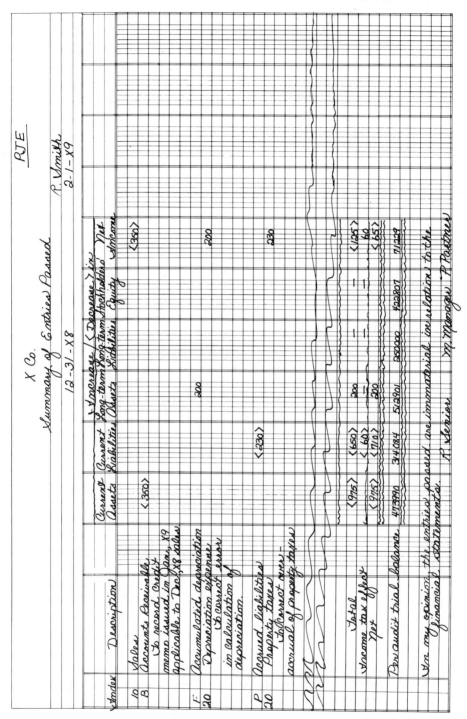

Figure 9.4 Summary of Entries Passed

C

J. Jones 1-20-X9

X Co.
Inventory Lead schedule
12-31-X8

Index	As Adjusted 12-31-X7	Per Books 12-31-X8	Adjustments Dr.	Adjustments Cr.	As Adjusted 12-31-X8
Raw materials	C-1	75425 ✓	51014 ✓		
Work in progress	C-2	29760 ✓	36253 ✓		
Finished goods	C-3	151903 ✓	137692 ✓		
Supplies	C-4	18414 ✓	18011 ✓		
		275502 Ⓣ	243030 ① Ⓣ		
		TB-1	TB-1		

✓ Traced to general ledger.
✓ Traced to prior year working papers.
Ⓣ Column footed.

① The plant manager stated that the decline in inventory was due to a decline in sales orders during the last quarter of this year. This was confirmed by examination of the sales backlog report as of 12-31-X8.

Figure 9.5 Inventory Lead Schedule

At this point, the concept of the working-paper triangle should be clear. From the audit trial balance consisting of three schedules, the accounts are analyzed on the lead schedules of which, in this case, there would be approximately 15, and then each lead schedule is supported by several detail audit schedules. Thus the base expands as each account is analyzed in greater detail.

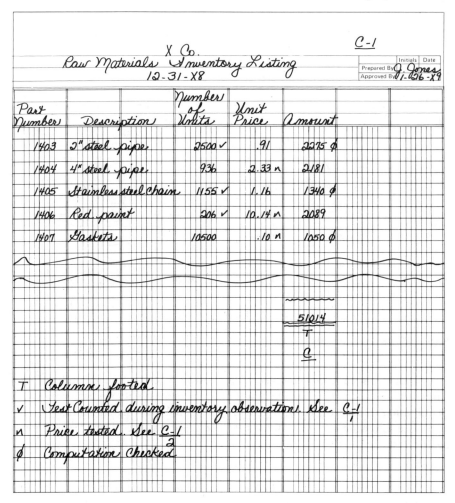

Figure 9.6   Raw Materials Inventory Listing

Management Representations

During an audit made in accordance with generally accepted auditing standards, an auditor is required by *SAS No. 19* to obtain certain representations from management. Some of these representations take the form of letters from management to the auditor, and others are preprinted certificates which the auditor furnishes management for signature. Management representations serve the dual purpose of (1) emphasizing to the client company's management their responsibility for complete and accurate financial statements, and (2) providing audit evidence, particularly in areas not susceptible to normal audit procedures.

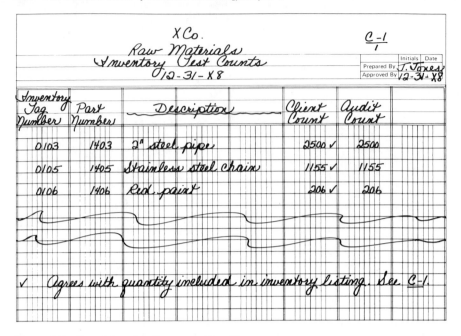

Figure 9.7   Raw Materials Inventory Test Counts

They should not be used, however, as a substitute for customary audit procedures, and they do not relieve the auditor of the responsibility for performing the examination in accordance with generally accepted auditing standards.

The following representations commonly are obtained from management by auditors.

1. A representation that the financial statements being examined by the auditor are the final statements for the year, and that they are presented fairly in accordance with generally accepted accounting principles applied on a basis consistent with that of the preceding year. This representation emphasizes management's responsibility for the fairness of the financial statements and, if an error is discovered after the financial statements are issued, it prevents management from maintaining that they knew of and disclosed the error to the auditor.

2. A representation that there are no material unrecorded assets or liabilities and no contingent assets or liabilities that are not disclosed properly in the financial statements. Examples are unrecorded claims receivable or payable and liabilities from lawsuits or as guarantor of third-party obligations. Because transactions of this nature are difficult to detect with normal audit procedures, additional assurance is sought from management as the party most likely to be aware of such transactions.

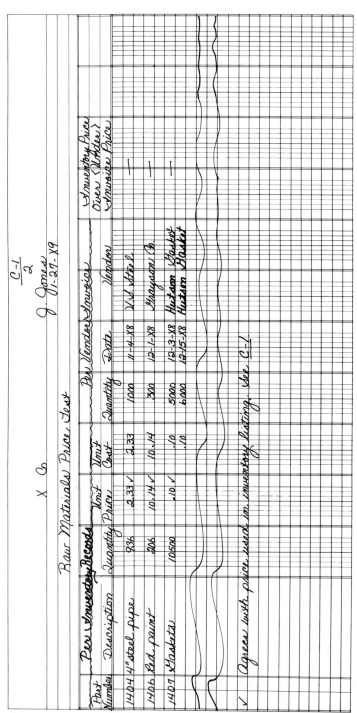

Figure 9.8 Raw Materials Price Test

3. A representation that all agreements relating to capital stock are disclosed properly in the financial statements. This representation concerns the disclosure of such items as stock repurchase agreements and stock option plans. Although items of this nature usually are approved by the board of directors or the stockholders and included in the minutes of their meetings, such approval is not a legal requirement in all states. Therefore, a representation covering these matters is requested from management.

4. A representation that transactions with other entities were based on arm's-length dealings. This representation is one of several audit procedures suggested in AU Section 335.12 *(SAS No. 6)* for determining the existence of related parties. Disclosure of transactions with related parties that were not arm's-length may be necessary for a fair presentation of the financial statements.

5. A representation that there have been no events or transactions subsequent to the audit date that would have a material effect on the financial statements that have not been disclosed therein. The auditor's responsibility extends beyond the audit date for material transactions that affect the financial statements; a representation is obtained to remind management that their responsibility also extends beyond the audit date and to supplement the limited procedures performed by the auditor during this period.

6. A representation that all financial and accounting records were furnished to the auditor and none were withheld. This representation is primarily a reminder to management of their obligation to the auditor.

7. A representation that minutes of all meetings of stockholders and directors (and any subcommittees) held during the period from the beginning of the year being audited to the date the audit is completed have been furnished the auditor. This representation, usually signed by the corporate secretary, constitutes evidence for the auditor that all minutes have been accounted for and that none have been overlooked or withheld.

The representations (other than the minutes representation) usually are signed by the chief executive officer (chairman of the board of directors or president) and the chief financial officer (financial vice-president, treasurer, or controller), because these individuals should be the most knowledgeable about the subjects of the representation. Auditors should not accept the representations of lower-level officers in place of those of the chief executive and financial officers. The dates of the representations should coincide with the last day of audit work in the client's office (which, as discussed in Chapter 15, is the date of the audit report), which would make them effective through the date to which the auditor's responsibility extends.

Minutes

As part of each examination, the auditor must extract or obtain copies of the minutes of all meetings of stockholders (or partners in the case of a partnership), board of directors, and subcommittees of the board of directors (the executive committee, compensation committee, finance committee, etc.). The extracts or copies must be made from or compared with the original signed minutes to ensure their authenticity, and a notation should be made on them to this effect.

Because almost all important actions of a company must be approved in these minutes, they must be subjected to a careful review. Matters that might be found in a review of the minutes include:

1. Authorization for new bond or stock issues.

2. Authorization of dividend payments.

3. Plans for plant expansion (which might involve significant purchase commitments) or the acquisition of other companies.

4. Threatened or pending litigation against the client.

5. Adoption of pension, profit-sharing, or stock option plans.

6. Approval of important contracts and agreements.

7. Approval of the purchase or sale of significant assets.

The auditor next must ascertain that important actions noted in the minutes have been properly reflected or disclosed in the financial statements. Important actions also are cross-referenced to the sections of the working papers in which they are recorded or audited. For example, a resolution of the board of directors declaring a dividend of $.50 a share would be cross-referenced to the retained earnings section of the working papers where a calculation such as the following might be found.

Number of shares of common stock outstanding, per examination of stockbook	100,000
Dividend per share, per board of directors' authorization	$\times$ .50
Total dividend paid per *TB-2*	$50,000

Minutes of prior years also must be reviewed during an auditor's first examination of a particular company. In this review, the auditor is primarily interested in actions having a continuing effect on the company and its financial statements.

Form and Indexing

The examples in Figures 9.5 through 9.8 also illustrate the methods of indicating the performance of audit work and the indexing of working papers.

The actual audit procedures performed can be indicated in several ways, three of which are shown in the examples. First, Figure 9.5 illustrates a narrative documentation of evidence at ①. This inquiry of the plant manager could have been made in connection with an audit procedure to explain unusual variations in inventory balances between years. Note that the auditor confirmed the plant manager's explanation by examination of the sales backlog report. A second method of illustrating the performance of audit procedures is shown in Figure 9.8. In this case, the schedule headings describe the examination of certain records and documents. For example, "Per Vendor Invoice" denotes the examination of purchase invoices from vendors to obtain the information shown on the schedule. The third method, illustrated in Figure 9.6, is the use of the "tick" or "tie" mark. In this case, the tick mark ϕ shows that the calculation of the number of units multiplied by the unit price has been checked. Also, note that the T under the total means that the amount column has been totaled or footed. Other tick marks in the same example (see ✔ and ↗) indicate that certain amounts have been audited on other schedules. It does not matter what type of tick or tie mark is used, as long as the auditor and anyone reviewing the working papers can understand its meaning.

Other aspects of the form of working papers that can be noted from the foregoing illustrations are (1) each schedule has a heading consisting of the client's name, a description of the information shown on the schedule and the audit date, (2) each schedule is indexed, and (3) each schedule is signed and dated by the auditor performing the work.

To facilitate the organization and review of the working papers, auditors generally use some type of indexing or coding system to identify each one. One such indexing scheme, used in Figures 9.4 through 9.8, is the designation of each balance sheet account with a letter (such as "A" for cash, "B" for accounts receivable, etc.) and each income statement account group with a number (such as "10" for sales, "20" for cost of goods sold, etc.). These designations are modified further with numbers to provide the indices of the detail audit schedules.

As shown in the illustrations, each account group listed on the audit trial balance is assigned a single letter or number which is also the index of the lead schedule for that account group. On the inventory lead schedule in the example, each sub-account is assigned an index consisting of the letter "C" (indicating that it is part of and should be filed in the inventory section) and a number (by type of inventory). Similar indexing would appear on other lead schedules. The schedules showing the detail audit work performed on each sub-account would be indexed with the sub-account number (indicating that the work relates to a particular sub-account) and an additional number to designate the schedule on which the work was performed. All schedules would be filed first in alphabetical order and then in numerical order. This type of indexing is simple and can be expanded in many ways.

The working-paper triangle with the indices of the schedules described is shown below.

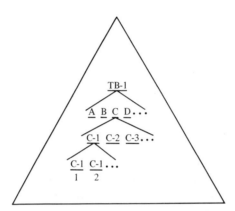

If a supervisor or other person had reason to review the audit work performed in the example on raw material inventory prices, rather than turn through what might be hundreds of pages of working papers, he would look first on the audit trial balance (Figure 9.1) to find the index of inventory (C)[7]; turning to the inventory lead schedule (Figure 9.5), he would locate the index of raw materials (C-1) that prices were tested on $\dfrac{\text{C-1}}{2}$ (Figure 9.8). Considerable time thus can be saved if a logical indexing system is used.

Client Assistance

Clients often assist their auditors by preparing certain working-paper schedules. This practice can be advantageous to both the auditor and the client. From the auditor's standpoint, it reduces the time he must spend on clerical and mechanical tasks and allows him to concentrate on more important and significant aspects of the audit. Of course, he must test the preparation of all such

[7]In practice, most firms use a standard index system, and this first step would not be necessary; inventory always would be indexed as "C" or some other designation.

schedules to the extent necessary to satisfy himself as to their accuracy and credibility, but the time required to do so is normally much less than would be necessary for the initial preparation of the schedules.

From the client's standpoint, any clerical work that can be shifted from the auditor to the client should reduce the total audit fee, which usually is based on the total time required to perform the audit.

As a time-saving device, auditors sometimes incorporate copies of client documents, schedules, and worksheets into their working papers. For example, rather than copy the inventory listing shown in Figure 9.6 from the client's listing, the auditor probably would make a mechanical copy of the client's listing and include it as a schedule in the working papers. This practice not only saves time, but also eliminates the possibility of an error in manual copying. The audit work would be performed on this copy. If the inventory information is maintained on data-processing equipment, the auditor might request a copy of the data-processing inventory listing for his working papers. The fullest use should be made of mechanical aids to reduce hand copying.

According to AU Section 322 *(SAS No. 9)*, auditors also can use the client's internal audit staff to assist in preparing working papers and performing both substantive and compliance tests. The independent auditor should consider the competence and objectivity of the internal auditors and supervise and test their work to the extent he considers necessary. All judgments concerning the audit must be made by the independent auditor. AU Section 322 *(SAS No. 9)* contains discussions of the effect of the internal audit function on the scope of the independent auditor's examination.

Security of Working Papers

Because an auditor's working papers represent the support for his professional opinion and contain confidential information regarding his client's operations, it is imperative that he maintain control of them at all times. On the client's premises, he should use a trunk or briefcase that can be locked to secure them at night, during lunch, or any other time they are not in use. Within his office, an auditor normally maintains a fireproof safe for the protection and security of his working papers.

Systems Evaluation Approach

The Appendix I to the text contains descriptions of a new audit approach to the final phase used by the accounting firm of Peat, Marwick, Mitchell & Co.

Appendix to Chapter 9

General Tools Used for the Preparation of Audit Programs

Audit Tools	*Potential Uses*
1. Confirmation	1. Evidence of existence (or ownership) of cash in bank, accounts receivable, inventory on consignment, insurance coverage, accounts payable, long-term debt, stock outstanding, contingent liabilities, and pension commitments.
2. Observation	2. Evidence of existence of inventories, cash on hand, notes receivable, investment securities, loan collateral, and property and equipment.
3. Inspection	3. Evidence of ownership, existence, and valuation represented by vendor invoices, receiving reports, purchase orders, sales invoices, and shipping documents.
4. Inquiry	4. Evidence of ownership, existence, valuation, and classification represented by oral and written communications with clients.
5. Calculation	5. Evidence of existence and valuation of accrued liabilities, deferred income taxes, depreciation expense, and interest expense.
6. Comparison	6. Evidence of ownership, existence, valuation, and classification represented by changes from the prior year or budgeted amounts, ratio analysis, and regression techniques.
7. Analysis	7. Evidence of ownership, existence, valuation, and classification represented by reviews for reasonableness and predictable relationships among accounts.

Chapter 9 References

Chazen, Charles, and Kenneth I. Solomon, "The Art of Defensive Auditing," *The Journal of Accountancy,* October 1975, pp. 66–71.

Dyer, Jack L., "Toward the Development of Objective Materiality Norms," *The Arthur Andersen Chronicle,* October 1975, pp. 38–49.

Mautz, R. K., "The Nature and Reliability of Audit Evidence," *The Journal of Accountancy,* May 1958, pp. 40–47.

Sauls, Eugene, "Nonsampling Errors in Accounts Receivable Confirmation," *The Accounting Review,* January 1972, pp. 109–115.

Report of the Special Committee on Equity Funding, American Institute of Certified Public Accountants, 1975.

Toba, Yoshihide, "A General Theory of Evidence as the Conceptual Foundation in Auditing Theory," *The Accounting Review,* January 1975, pp. 7–24.

Chapter 9
Questions Taken from the Chapter

9–1.   What is the third standard of fieldwork?

9–2.   What is the difference between a test of compliance and a substantive test?

9–3.   What risk is taken by auditors when they gather evidence during an audit examination?

9–4.   Give an example of a situation in which auditors' faulty judgment has resulted in a lawsuit for negligence.

9–5.   Cite the legal case in which auditors' faulty judgment resulted in a criminal conviction.

9–6.   What does AU Section 330.08 *(SAS No. 1)* state about the competency of audit evidence obtained from outside sources?

9–7.   Name three fraud cases (two of them fairly old and one fairly recent) that were related to the failure of auditors to gather outside evidence.

9–8.   Name four accounts for which the balance can be verified independently. Name four accounts for which the balance cannot be verified independently.

9–9.   As a general rule, auditors will attempt to verify, from an independent outside source, the existence of accounts that have what characteristics?

9–10.   What does AU Section 330.08 *(SAS No. 1)* state about the relationship between reliability of account data and conditions of internal control?

9–11.   What change can be made in the observation of the client's inventory count if internal control is strong?

9–12.   What does AU Section 330.08 *(SAS No. 1)* state about the competency of audit evidence gained through the personal knowledge of the auditor?

9–13.   The chapter contains a list of six accounts suitable for evidence-gathering through direct personal knowledge. Name the six.

9–14.   What does AU Section 330.10 *(SAS No. 1)* state about the persuasiveness of audit evidence?

9–15.   What is the purpose of the allowance for doubtful accounts?

9–16.   What steps are used to collect evidence on the validity of the allowance for doubtful accounts?

9–17.   Where do auditors typically turn for evidence of the possible outcome of litigation pending against the client?

9–18.   What possible consequences might arise if auditors are unable to obtain an evaluation of the probable outcome of litigation pending against the client?

9–19.   What does AU Section 330.12–.13 *(SAS No. 1)* state about the cost of obtaining evidence?

9–20.   Identify and describe the two types of accounts receivable confirmation requests.

9–21.   What are the two reasons why recipients of negative accounts receivable confirmation requests may fail to respond to the auditors?

9–22.   Why do auditors generally not conduct an annual inspection of fixed assets?

9–23.   What does AU Section 330.15 *(SAS No. 1)* state about the objectivity of audit evidence?

9–24.   Define materiality.

9–25.   Name three base figures with which other dollar amounts in the financial statements are compared in order to determine materiality.

9–26.   In what lawsuit did the judiciary apply its own criterion of materiality? What was this criterion?

9–27.   What are the materiality border zones discussed by Bernstein in the *Lybrand Journal* article?

9–28.   Name a suggested criterion for materiality other than dollar amounts or percentages. Give two examples of accounts or events that might be material for reasons other than high dollar amounts or high percentages.

9–29.   Why should the auditor consider the objectives of auditing each account before beginning his work?

9–30.   What are the four broad objectives of auditing a given account?

9–31.   Why is a knowledge of generally accepted accounting principles important to an auditor?

9–32.   Why is there no official list of audit procedures that could be used on every engagement?

9–33.   How is the effectiveness of a client's system of internal control related to the extensiveness of substantive tests?

9–34.   State the AU Section 338.03 *(SAS No. 1)* definition of working papers.

9–35.   What are the general guidelines in AU Section 338.05 *(SAS No. 1)* as to what working papers should include?

9–36.   Illustrate how properly prepared working papers can demonstrate compliance with the standards of fieldwork.

9–37.   To whom do audit working papers generally belong?

9–38.   What is the significance of the lack of privilege between an auditor and his client?

9–39.   Explain the importance of the audit working papers.

9–40.   What are the two types of working-paper files?

9–41.   What is the purpose of the permanent audit file?

9–42.   List five sections that might be included in a permanent audit file, and give two examples of data that might be found in each section.

9–43.   List six types of data found in the current audit files.

9–44.   Explain the general concept of the organization of working papers in the current files.

9–45.   Why does the auditor prepare an audit trial balance?

9–46.   Explain the difference between adjusting and reclassifying entries proposed by the auditor.

9–47.   What is a lead schedule in a current working-paper file?

9–48.   Explain the concept of the working-paper triangle.

9–49.   Who has the final responsibility for determining whether or not a proposed adjusting entry will be recorded?

9–50.   Name the three audit working-paper schedules on which a proposed adjusting entry may be found.

9–51.   What is the purpose of a summary of immaterial entries or summary of entries passed?

9–52.   State three ways of indicating the performance of audit work in the working papers.

9–53.   What are the purposes of management representations?

9–54.   List and explain seven types of representations that may be obtained from management.

9–55.   Which client personnel normally make the representations to the auditor?

9–56.   Why is an auditor concerned with minutes of meetings of stockholders and board of directors?

9–57.　List seven examples of matters of interest to an auditor which may be found in a review of minutes.

9–58.　What is the purpose of an index system for the working papers? Describe the system illustrated in the text.

9–59.　Clients often assist their auditors by preparing certain working-paper schedules. What are the advantages of this practice to the auditor and to the client?

9–60.　List two mechanical aids an auditor might use in preparing working papers.

9–61.　To what extent can an independent auditor use the client's internal audit staff?

9–62.　How does an auditor maintain the security of his working papers?

Chapter 9
Objective Questions Taken from CPA Examinations

9–63.　The following statements were made in a discussion of audit evidence between two CPAs. Which statement is *not* valid concerning evidential matter?
a.　"I am seldom convinced beyond all doubt with respect to all aspects of the statements being examined."
b.　"I would not undertake that procedure because at best the results would be persuasive and I'm looking for convincing evidence."
c.　"I evaluate the degree of risk involved in deciding the kind of evidence I will gather."
d.　"I evaluate the usefulness of the evidence I can obtain against the cost to obtain it."

9–64.　Evidential matter supporting the financial statements consists of the underlying accounting data and all corroborating information available to the auditor. Which of the following is an example of corroborating information?
a.　Minutes of meetings.
b.　General and subsidiary ledgers.
c.　Accounting manuals.
d.　Worksheets supporting cost allocations.

9–65. Although the validity of evidential matter is dependent on the circumstances under which it is obtained, there are three general presumptions which have some usefulness. The situations given below indicate the relative reliability a CPA has placed on two types of evidence obtained in different situations. Which of these is an *exception* to one of the general presumptions?

   a. The CPA places more reliance on the balance in the scrap sales account at plant A, where the CPA has made limited tests of transactions because of good internal control, than at plant B, where the CPA has made extensive tests of transactions because of poor internal control.

   b. The CPA places more reliance on the CPA's computation of interest payable on outstanding bonds than on the amount confirmed by the trustee.

   c. The CPA places more reliance on the report of an expert on an inventory of precious gems than on the CPA's physical observation of the gems.

   d. The CPA places more reliance on a schedule of insurance coverage obtained from the company's insurance agent than on one prepared by the internal audit staff.

9–66. In connection with a lawsuit, a third party attempts to gain access to the auditor's working papers. The client's defense of privileged communication will be successful only to the extent it is protected by the

   a. Auditor's acquiescence in use of this defense.

   b. Common law.

   c. AICPA *Code of Professional Ethics.*

   d. State law.

9–67. A CPA observes his client's physical inventory count on December 31, 1971. There are eight inventory-taking teams, and a tag system is used. The CPA's observation normally may be expected to result in detection of which of the following inventory errors:

   a. The inventory-takers forget to count all of the items in one room of the warehouse.

   b. An error is made in the count of one inventory item.

   c. Some of the items included in the inventory had been received on consignment.

   d. The inventory omits items on consignment to wholesalers.

9–68. When conducting an audit, errors that arouse suspicion of fraud should be given greater attention than other errors. This is an example of applying the criterion of

   a. Reliability of evidence.

   b. Materiality.

   c. Relative risk.

   d. Dual-purpose testing.

9–69.   For what minimum period should audit working papers be retained by the independent CPA?
   a.   For the period during which the entity remains a client of the independent CPA.
   b.   For the period during which an auditor-client relationship exists, but not more than six years.
   c.   For the statutory period within which legal action may be brought against the independent CPA.
   d.   For as long as the CPA is in public practice.

9–70.   As part of an audit, a CPA often requests a representation letter from his client. Which one of the following is *not* a valid purpose of such a letter?
   a.   To provide audit evidence.
   b.   To emphasize to the client his responsibility for the correctness of the financial statements.
   c.   To satisfy himself by means of other auditing procedures when certain customary auditing procedures are not performed.
   d.   To provide possible protection to the CPA against a charge of knowledge in cases where fraud is subsequently discovered to have existed in the accounts.

9–71.   The auditor interviews the plant manager. The auditor is most likely to rely upon this interview as primary support for an audit conclusion on
   a.   Capitalization vs. expensing policy.
   b.   Allocation of fixed and variable costs.
   c.   The necessity to record a provision for deferred maintenance costs.
   d.   The adequacy of the depreciation expense.

9–72.   During an audit engagement, pertinent data are compiled and included in the audit work papers. The work papers primarily are considered to be
   a.   A client-owned record of conclusions reached by the auditors who performed the engagement.
   b.   Evidence supporting financial statements.
   c.   Support for the auditor's representations as to compliance with generally accepted auditing standards.
   d.   A record to be used as a basis for the following year's engagement.

9–73.   In determining validity of accounts receivable, which of the following would the auditor consider *most* reliable?
   a.   Documentary evidence that supports the accounts receivable balance.
   b.   Credits to accounts receivable from the cash receipts book after the close of business at year end.
   c.   Direct telephone communication between auditor and debtor.
   d.   Confirmation replies received directly from customers.

9–74. Audit working papers are used to record the results of the auditor's evidence-gathering procedures. When preparing working papers, the auditor should remember that

a. Working papers should be kept on the client's premises so that the client can have access to them for reference purposes.

b. Working papers should be the primary support for the financial statements being examined.

c. Working papers should be considered as a substitute for the client's accounting records.

d. Working papers should be designed to meet the circumstances and the auditor's needs on each engagement.

9–75. Which of the following *best* describes the element of relative risk which underlies the application of generally accepted auditing standards, particularly the standards of fieldwork and reporting?

a. Cash audit work may have to be carried out in a more conclusive manner than inventory audit work.

b. Intercompany transactions are usually subject to less detailed scrutiny than arm's-length transactions with outside parties.

c. Inventories may require more attention by the auditor on an engagement for a merchandising enterprise than on an engagement for a public utility.

d. The scope of the examination need *not* be expanded if errors that arouse suspicion of fraud are of relatively unsignificant amounts.

9–76. An auditor generally obtains from a client a formal written statement concerning the accuracy of inventory. This particular letter of representation is used by the auditor to

a. Reduce the scope of the auditor's physical inventory work but *not* the other inventory audit work that is normally performed.

b. Confirm in writing the valuation basis used by the client to value the inventory at the lower of cost or market.

c. Lessen the auditor's responsibility for the fair presentation of balance sheet inventories.

d. Remind management that the primary responsibility for the overall fairness of the financial statements rests with management and *not* with the auditor.

Chapter 9
Discussion/Case Questions

9–77. Susan Start, a newly hired staff assistant of a CPA firm, was assigned to the audit team examining the financial statements of Rel-Hep Finance Company. The senior in charge of the engagement decided to assign Susan to the phase of the audit concerning the Allowance for Doubtful Accounts. She was given an aging schedule and told to check its accuracy.

During the course of this work, Miss Start noticed that the Allowance for Doubtful Accounts figure was only 1% of the Accounts Receivable balance, despite the fact that many large accounts were very old. She passed this information on to the senior who, in turn, asked the client about the low allowance percentages.

The client indicated that there was no particular problem, because the company's policy was to refinance slow-paying customers. For example, if Customer A had not made any recent monthly payments on a $500 one-year note, the company would refinance the note over a two- or three-year period, thus lowering the monthly payments. Generally, additional credit was not refused a customer, regardless of the status of payments on an existing account. The client maintained that this policy made possible a low allowance because few accounts had to be written off.

Required:
a. If you were suspicious of the small size of the allowance account, what evidence would you gather to verify or alleviate your suspicions?
b. With accounts that are constantly being refinanced, what evidence can be gathered to develop a reasonable assurance that these accounts are collectible?

9-78.   Com-See-Me, a retail furniture company, made a practice of assigning certain customer account balances to a local bank which, in turn, advanced the furniture company money. When the designated customers paid their bills, these amounts were forwarded to the bank.

As a condition of this financing arrangement, the bank required an annual audit by a CPA firm. During the course of the audit, the senior in charge was informed by the company controller that no circularization of accounts receivable confirmations would be allowed, because the company did not wish their customers to find out about the financing arrangements with the bank.

After consulting with the partner in charge of the audit, the senior informed the controller that no confirmation letters would be sent.

Required:
a. Do you think that the CPA firm lost its independence by agreeing to this restriction on the scope of the audit procedures?
b. (1) Without regard to question a, indicate whether you think this restriction should warrant a qualified opinion or a disclaimer of opinion. AU Section 509.10 of *AICPA Professional Standards,* Volume 1 *(SAS No. 2)* requires one or the other.

For purposes of this decision, assume the following relationships.

Accounts Receivable 12-31	$ 100,000
Total Current Assets 12-31	$ 500,000
Total Assets 12-31	$1,000,000
Net Income for Year ended 12-31	$ 30,000

    (2) What additional factors would you consider, other than the figures above, in making your decision on a qualified opinion or disclaimer of opinion?

  c. What evidence would you attempt to acquire in place of accounts receivable confirmations?

9–79.   One of the clients of Mr. Cain, a practicing CPA, is a local financial institution. As a part of the standard audit procedures, Mr. Cain sent accounts receivable confirmations to 50 of the 200 customers.

Although a return envelope was sent with the confirmations, most of the customers chose to bring the letter to the office. Mr. Cain was interrupted constantly to explain orally the nature of the confirmation letter. He would tell the customers that their signatures were to be placed on the letter without any additional comment only if they agreed that they owed the amount printed on the letter. Otherwise, the customers were asked to indicate the amount they believed they owed.

Few of the customers seemed to understand the oral instructions given by Mr. Cain, and most merely signed the letter and left it. Mr. Cain was perplexed by these types of responses but he did not know whether to send additional letters, look for alternative types of evidence, or simply accept these signatures without additional audit procedures.

Required:
  a. Which of the three courses of action would you recommend? If you reject all three courses of action, what would you recommend?
  b. Comment in general on the validity and limitations of audit evidence gained from confirmation letters sent to the public.

9–80.   One of the audit clients of Brown and Brown, CPAs, is We-Fit Manufacturing Company, makers of shirts, sweaters, and other clothing items. The finished goods inventory consists of merchandise placed in hundreds of sealed boxes on the warehouse floor. The client's inventory count consists of a count of boxes which, for the most part, are supposed to contain a standard number of shirts or sweaters. The only way the auditors can be certain about the contents of the boxes is to break them open and count the pieces of clothing. They do not wish to use this method of verification except on a few selected boxes.

Required:

a. What criterion would you use in deciding how many boxes to open, if any, during the inventory observation?

b. What are the possible legal implications associated with a massive number of empty boxes?

c. Would you seek out extra audit evidence on the validity of inventory simply as a result of the client's count procedure? If so, what types?

9–81. In each of the following cases, rank the various items of evidence on a scale from most persuasive to least persuasive. Furnish support for your rankings.

a. Evidence to support the cash account.
   (1) Bank reconciliation prepared by the client.
   (2) A written confirmation of the bank balance, sent directly to the auditor.
   (3) Written certification by the client that the bank balance is correct.
   (4) Oral assurance by the client that the bank balance is correct.
   (5) The year's canceled checks and validated deposit slips held by the client.

b. Evidence to support Accounts Receivable (not including the Allowance for Doubtful Accounts).
   (1) Sales invoices held by the client.
   (2) Written certification by the client that the bank balance is correct.
   (3) Written confirmations of the balances, sent by the customer directly to the auditor.
   (4) An accounts receivable aging schedule.
   (5) Shipping documents held by the client, showing the dollar amount of merchandise sent to customers.
   (6) Deposit slips held by the client, showing the cash received from customers during the month after year-end.

c. Evidence to support inventory quantities.
   (1) Purchase invoices held by the client.
   (2) Checks issued to vendors during the month following year-end.
   (3) Observation of the client's physical count.
   (4) A written certification from the client that the amount shown as inventory is correct.

9–82. You are the auditor of Star Manufacturing Company.

A trial balance taken from the books of Star one month prior to year-end follows:

	Dr. (Cr.)
Cash in bank	$ 87,000
Trade accounts receivable	345,000
Notes receivable	125,000
Inventories	317,000
Land	66,000
Buildings, net	350,000
Furniture, fixtures and equipment, net	325,000
Trade accounts payable	(235,000)
Mortgages payable	(400,000)
Capital stock	(300,000)
Retained earnings	(510,000)
Sales	(3,130,000)
Cost of sales	2,300,000
General and administrative expenses	622,000
Legal and professional fees	3,000
Interest expense	35,000

There are no inventories consigned either in or out.

All notes receivable are due from outsiders and held by Star.

Required:
a. Which accounts should be confirmed with outside sources?
b. Briefly describe by whom they should be confirmed and the information that should be confirmed.
c. Organize your answer in the following format.

Account Name	By Whom Confirmed	Information to Be Confirmed

(AICPA adapted)

9–83. You have been engaged to examine the financial statements of the Elliott Company for the year ended December 31, 19X9. You performed a similar examination as of December 31, 19X8.

Following is the trial balance for the Company as of December 31, 19X9:

	Dr. (Cr.)
Cash	$ 128,000
Interest receivable	47,450
Dividends receivable	1,750
6½% secured note receivable	730,000
Investments at cost:	
Bowen common stock	322,000
Investments at equity:	
Woods common stock	284,000
Land	185,000
Accounts payable	(31,000)
Interest payable	(6,500)
8% secured note payable to bank	(275,000)
Common stock	(480,000)
Paid-in capital in excess of par	(800,000)
Retained earnings	(100,500)
Dividend revenue	(3,750)
Interest revenue	(47,450)
Equity in earnings of investments carried at equity	(40,000)
Interest expense	26,000
General and administrative expense	60,000

You have obtained the following data concerning certain accounts:

The 6½% note receivable is due from Tysinger Corporation and is secured by a first mortgage on land sold to Tysinger by Elliott on December 21, 19X8. The note was to have been paid in 20 equal quarterly payments beginning March 31, 19X9, plus interest. Tysinger, however, is in very poor financial condition and has not made any principal or interest payments to date.

The Bowen common stock was purchased on September 21, 19X8, for cash in the market where it is actively traded. It is used as security for the note payable and held by the bank. Elliott's investment in Bowen represents approximately 1% of the total outstanding shares of Bowen.

Elliott's investment in Woods represents 40% of the outstanding common stock that is actively traded. Woods is audited by another CPA and has a December 31 year-end.

Elliott neither purchased nor sold any stock investments during the year other than those noted above.

Required:
For the following account balances, discuss (1) the types of evidential matter you should obtain, and (2) the audit procedures you should perform during your examination.

a.   6½% secured note receivable.
b.   Bowen common stock.
c.   Woods common stock.
d.   Dividend revenue.

(AICPA adapted)

9–84.   In his examination of financial statements, an auditor must judge the validity of the audit evidence he obtains. Assume that you have evaluated internal control and found it satisfactory.

Required:
a.   In the course of his examination, the auditor asks many questions of client officers and employees.
   1. Describe the factors that the auditor should consider in evaluating oral evidence provided by client officers and employees.
   2. Discuss the validity and limitations of oral evidence.
b.   An auditor's examination may include computation of various balance sheet and operating ratios for comparison to prior years and industry averages. Discuss the validity and limitations of ratio analysis.
c.   In connection with his examination of the financial statements of a manufacturing company, an auditor is observing the physical inventory of finished goods, which consists of expensive, highly complex electronic equipment. Discuss the validity and limitations of the audit evidence provided by this procedure.

(AICPA adapted)

9–85.   Your Way Co. owns a chain of 27 hamburger restaurants in several states, franchised from Big Burger Co. Although the gross profit ratio varies among restaurant locations, it is fairly constant over time at any individual location. The company's employees belong to the U.H.C.U. (United Hamburger Cookers' Union). The company's primary financing consists of 20-year notes sold to an insurance company, but it also financed some of its restaurants through a sale/lease-back agreement. You decide that you will visit nine stores each year, because only a few limited cash records are maintained at the stores.

This is the first audit of Your Way Co. and you have been assigned to prepare the permanent file. List the documents, schedules, and other material you would include in the file and state how each would be used in your audit.

9–86.   Using the indexing method described in the text, assign indices to the following schedules:
Audit trial balance—assets.
Accounts receivable lead schedules containing trade, officer, and notes receivable.

Aged trial balance of trade accounts receivable.

Schedule of trade accounts receivable confirmed.

Trade accounts receivable confirmations.

Schedule of subsequent collection of trade accounts receivable.

Officer accounts receivable.

Notes receivable.

Allowance for doubtful accounts receivable.

9–87. Harlan, Inc. is a large, highly diversified, organization engaged in feed and flour milling; the manufacture of plastic products; the manufacture of highly specialized machinery; and the operation of poultry hatcheries, farms, and processing plants. The company's facilities are located throughout the United States.

During the past several years, the Company has been expanding the business through acquisitions, mergers, and natural growth of the original operations.

The Company finances poultry-growing operations of feed customers until the flocks are marketed. Poultry prices have been depressed for the past several years; and, as a result, large bad debt losses have been sustained. Many accounts are secured by collateral other than poultry and are not considered current assets.

Your review of the internal controls and accounting procedures currently in effect disclosed the following changes from prior years:

1. The Company organized an internal audit staff during the year.

2. The internal audit staff is reconciling all bank accounts.

3. The Company has instituted a procedure whereby divisions send the Home Office an "invoice apron" which lists the information required for payment of the invoices and the due date. The invoices and other supporting data (receiving reports, purchase orders, etc.) remain at the receiving location.

4. A physical inventory of office furniture and fixtures has been taken at the Home Office.

Discussions with Company personnel and preliminary audit work disclosed the following information:

1. The Company plans to discontinue extending credit to certain customers whose poultry-raising operations have deteriorated to the point that their ability to repay the Company over several years is doubtful. It is anticipated that a large provision for doubtful accounts may be necessary to reduce these accounts to estimated realizable value on a forced realization basis (the Company has second mortgages as collateral on many of the farms).

2. During the physical inventory at a plant that has a significant part of the total inventory, you observed that the inventory was recorded on sheets, by part number (prenumbered tags were not used); and receiving reports were not

being used. You made numerous test counts and obtained extensive data for testing the receiving cutoff.

3. The tool crib inventory consists of a conglomeration of miscellaneous items, most of which are small quantities with very minor unit prices. This inventory totals $42,395.89, which is an insignificant portion of the total inventories.

4. The purchased parts stockroom at another plant is segregated from the production areas by a wire fence. While visiting this plant, you noted that the gate was left open all day, and access to the stockroom (which contains many valuable TV tubes and similar items) was available to any employee. The stockroom's perpetual inventory records were formerly checked by an employee who made test counts on a cycle basis. This employee has been terminated and has not been replaced. As a result, the cycle counting has ceased. We expanded our tests in view of these situations and are satisfied that the perpetual records reasonably reflect the quantities on hand.

These situations did not exist last year. You are currently using last year's audit program as a guide in performing the current audit.

Required:
a. Discuss the relationship of the review of internal control to the audit program. Make any comments on the planning of this engagement that you consider pertinent.
b. Discuss the audit procedures you would use in each of the situations described.

(Used with permission of Ernst & Ernst)

Chapter 10

Auditing the Working Capital Cell

Learning Objectives *After reading and studying the material in this chapter, the student should*

Know the characteristics of the accounts comprising the working capital cell.

Understand the important features of internal control associated with each account in the working capital cell and how they are tested.

Know the audit objectives of each account in the working capital cell.

Be able to determine and apply the audit procedures necessary to accomplish the audit objectives of each account in the working capital cell.

Know how to prepare the audit working papers to document the audit procedures for each account in the working capital cell.

Chapter 9 covers the practical approaches to gathering evidence and preparing working papers. The purpose of this chapter and the following two chapters is to provide examples of audit objectives, procedures, and working papers that illustrate specific evidence-gathering and working-paper preparation techniques. Because the emphasis is on learning techniques, rather than on memorizing procedures or formats, a complete set of working papers is not considered necessary and is not included. This chapter includes a discussion of the objectives, procedures, and selected working-paper examples for the working capital accounts shown on the audit trial balance in Chapter 9.

Accounts in the working capital cell tend to turn over rapidly, so the balance at the end of a period contains few if any elements that were included in the balance at the beginning of the period. Thus, audit emphasis is placed on the ending account balance, rather than on changes in the account during the period being audited.

Current Assets

Although assets may be understated as well as overstated, auditors normally find few cases of unrecorded or understated assets. Management's desire for increased earnings tends to result in more auditor-proposed adjustments to reduce earnings and assets than to increase them (though there may be exceptions motivated by income tax factors). Thus, auditors place primary emphasis on verifying recorded amounts when they audit assets. (The emphasis is different in the audit of liabilities.)

Cash

Although the cash balance often is not one of the larger amounts on the balance sheet, the volume and dollar amount of transactions flowing through the cash account usually are greater than for any other account, because most business transactions ultimately are settled in cash. Also, being the most liquid of all assets, cash is very susceptible to defalcation. For these reasons, the audit procedures applied to cash are more extensive than the dollar size of the balance sheet account would seem to warrant.

Most business transactions are carried out by check, although it is common for businesses to maintain small petty cash funds for minor disbursements. Unless the balance or volume of transactions in a petty cash fund is large, the auditor seldom makes a count of the fund. Cash always is counted if it is material, however, as in the audit of a bank.

Unless otherwise disclosed in the financial statements, the cash balance included in current assets is considered to be unrestricted and available for immediate withdrawal. AC Section 2031.06 of *AICPA Professional Standards, Volume 3 (ARB No. 43),* contains the following comments concerning the classification of cash:[1]

> ... the nature of current assets contemplates the exclusion from that classification of such resources as: (a) cash and claims to cash which are restricted as to withdrawal or use for other than current operations, are designated for expenditure in the acquisition or construction of noncurrent assets, or are segregated for the liquidation of long-term debts.

[1] All accounting pronouncements are in Volume 3 of the Commerce Clearing House (CCH), Inc. book. The pronouncements are divided into Volumes 3 and 4 of the CCH, Inc. loose-leaf service.

Restrictions on cash can take many forms, from an informal agreement with a bank to maintain an average balance above a certain amount to support borrowing and other credit arrangements (referred to as a compensating balance) to formally restricted escrow or other accounts from which immediate withdrawal cannot be made. Compensating balance arrangements should be disclosed in a note to the financial statements, whereas formally restricted accounts may be classified as non-current assets.

The general concept of internal control—the separation of responsibilities for recording transactions affecting an asset from custody of the asset—has special applicability to cash. The implementation is sometimes referred to as a separation of the treasury and accounting functions. In large companies, these functions frequently are separated into different departments, with a treasurer supervising the custody function and a controller supervising the recording function. Even if such a complete separation of functions is not possible, certain specific duties can be separated among the available personnel to achieve the strongest control possible under the circumstances. For example, an employee who receives and deposits cash and checks should not prepare sales invoices, post detail accounts receivable records, reconcile the detail of accounts receivable to the control account, prepare or mail customer statements, approve write-offs of customer accounts, receive or reconcile the bank statements, etc. Similarly, an authorized check-signer should not receive or reconcile the bank statement, open the mail, approve vouchers for payment, etc. All elements of strong internal control will not be used in every company, and it is the auditor's evaluation of the nature and effectiveness of the controls that are in use that serves as the basis for determining the scope of his work.

It has been pointed out in an earlier chapter that no account is audited in a vacuum, i.e., evidence gathered in the examination of one account may be applicable to other accounts as well. Audit procedures applied to other accounts often generate evidence concerning the cash account which supplements the evidence gathered in the cash section. Thus, procedures listed in this section do not generate the only evidence related to cash.

Audit Objectives

The objectives in the audit of cash are:

1. To determine that the amount shown as cash in the financial statements constitutes cash on hand, in banks, or in transit.

2. To determine that any restricted cash is properly classified and disclosed.

Audit Procedures and Working Papers

The first procedure to accomplish these objectives must be a review and test of the accounting procedures and internal controls relating to cash. Internal control related to cash received is often tested in an overall credit sales-cash receipts test, such as that illustrated in Chapter 5. A cash disbursements test (see Appendix to Chapter 5) is used as a test of compliance for cash payments. The "proof of cash" or "block reconciliation" described later in this section is a dual-purpose test (both compliance with procedures and account balances are tested). The result of these tests determines the extensiveness of other cash audit procedures. Noted throughout this section are examples of audit procedures that can be varied in scope on the basis of the auditor's evaluation of internal control.

The next procedure could be to prepare a lead schedule analyzing the amount shown in the audit trial balance by bank account (Figure 10.1). The amounts on the lead schedule should be compared with the general ledger or prior-year working papers, as applicable.

Bank Reconciliations and Cutoff Statements—The auditor then could prepare or obtain a copy of the client's bank reconciliation for each account. The reconciliation for A Bank is shown in Figure 10.2. Similar reconciliations also would be prepared or obtained for B and C Banks. The company's procedures provide that only a limited number of very minor items can be purchased with petty cash. The auditor's review of procedures indicated that this policy was being followed, and therefore, no additional work will be performed on the petty cash balance.

The bank reconciliation provides the auditor with an important means of accomplishing one of his objectives with regard to cash—to determine that the cash is represented by deposits in banks or in transit. If the amount shown as the balance in the bank is confirmed directly with the bank, and if the reconciling items (deposits in transit and outstanding checks) are audited satisfactorily, substantial persuasive evidence will have been accumulated in support of the cash balance. These procedures might appear in the audit program as follows.

1. Foot the bank reconciliation and any supporting details (indicated by T in Figure 10.3).

2. Confirm the balance per bank directly with the bank by means of a standard bank confirmation (indicated by $\cancel{C}$ in Figure 10.3; also see Figure 10.4).

3. Obtain a cutoff bank statement directly from the bank for the period from 1-1-X9 to 1-15-X9.

4. Trace the balance per bank in the reconciliation to the year-end bank statement and the beginning balance of the cutoff bank statement (indicated by $\checkmark$ and $\oslash$ in Figure 10.3).

5. Trace balance per books to the cash book (indicated by $\cancel{N}$ in Figure 10.3).

A

X Co.
Cash Lead Schedule
12-31-X8

J. Jones 1-18-X9

	Index	As Adjusted 12-31-X7	Per Books 12-31-X8	Adjustments Dr	Adjustments Cr	As Adjusted 12-31-X8
A Bank	A-1	51315 ✓	46782 ʌ			46782
B Bank	A-2	12065 ✓	8309 ʌ			8309
C Bank	A-3	10000 ✓	10000 ʌ			10000
Petty Cash		50 ✓	50 ʌ			50
		73430	65141			65141
		T	T			T
		TB-1	TB-1			TB-1

ʌ Traced to general ledger.
✓ Traced to prior year working papers.
T Column footed.

Figure 10.1 Cash Lead Schedule

6. Trace the dates and amounts of deposits in transit to the cutoff bank statement and to authenticated deposit slips obtained directly from the bank (indicated by (ʌ) and ⋉ in Figure 10.3).

7. For canceled checks received in the cutoff bank statement and dated prior to the audit date, perform the following steps (indicated by (✓) in Figure 10.3).

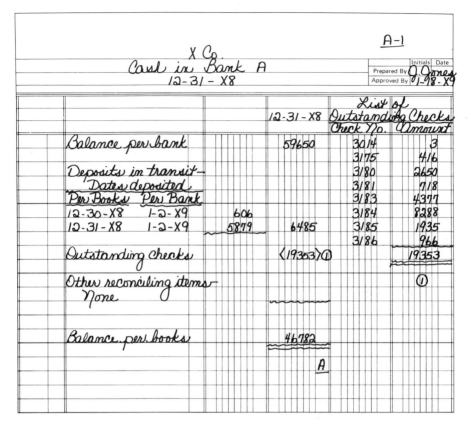

Figure 10.2 Cash in Bank A

    a. Trace to the outstanding check list supporting the bank reconciliation to determine whether they are shown properly.

    b. Compare the signature with the list of authorized check signers in the permanent audit file.

    c. Examine the endorsement to see that the check is endorsed by the payee and that there are no unusual second endorsements.

    d. Compare with cash book as to date, number, payee, and amount.

8. For checks received in the cutoff bank statement and dated after the audit date, examine the first bank endorsement to see that it does not precede the audit date (indicated by Note 1 in Figure 10.3).

9. For checks on the outstanding check list that did not clear with the cutoff bank statement, compare check number and amount with cash book and investigate any that have been outstanding for an unusually long time (indicated by ⊗ in Figure 10.3).

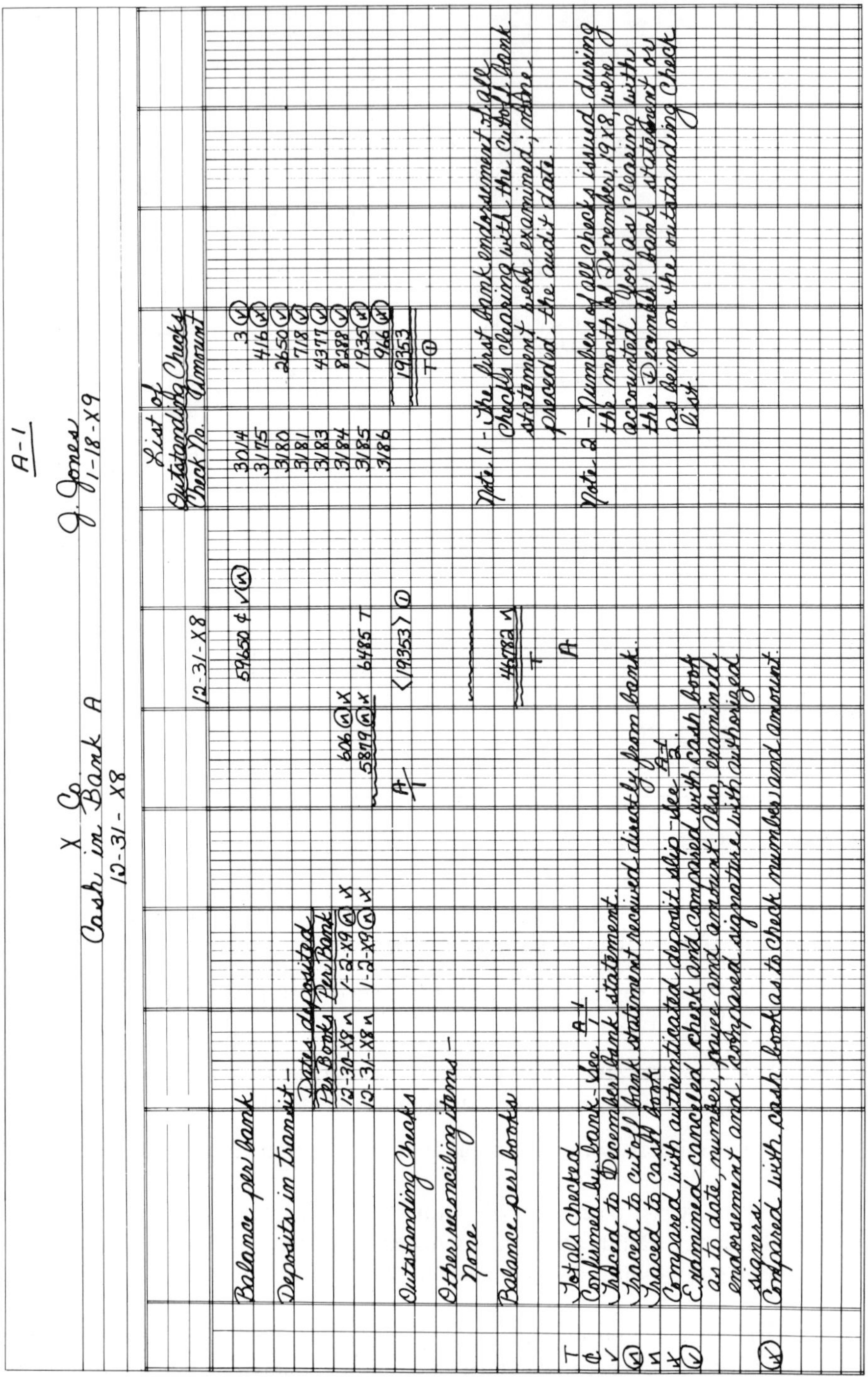

Figure 10.3 Cash in Bank A

STANDARD BANK CONFIRMATION INQUIRY

Approved 1966 by

AMERICAN INSTITUTE OF CERTIFIED PUBLIC ACCOUNTANTS

AND

BANK ADMINISTRATION INSTITUTE (FORMERLY NABAC)

Dear Sirs: January 1   19 X9

Your completion of the following report will be sincerely appreciated. IF THE ANSWER TO ANY ITEM IS "NONE", PLEASE SO STATE. Kindly mail it in the enclosed stamped, addressed envelope <u>direct</u> to the accountant named below.

Report from   Yours truly,

           X Company

           (ACCOUNT NAME PER BANK RECORDS)

(Bank)   Bank A

       Hypothetical, Arkansas   By *Bruce Lee*

                                 Authorized Signature

           Bank customer should check here if confirmation of bank balances only (item 1) is desired. ☐

Partner & Co.
999 Verify Street   NOTE–If the space provided is inadequate,
Hypothetical, Arkansas   70000   please enter totals hereon and attach a statement giving full details as called for by the columnar headings below.

Dear Sirs:

1. At the close of business on December 31   19 X8 our records showed the following balance(s) to the *credit* of the above named customer. In the event that we could readily ascertain whether there were any balances to the credit of the customer not designated in this request, the appropriate information is given below.

AMOUNT	ACCOUNT NAME	ACCOUNT NUMBER	SUBJECT TO WITHDRAWAL BY CHECK?	INTEREST BEARING? GIVE RATE
$ 59,650 A-1	Regular	10 0000	Yes	No

2. The customer was directly liable to us in respect of loans, acceptances, etc., at the close of business on that date in the total amount of $ **360,000** , as follows:

AMOUNT	DATE OF LOAN OR DISCOUNT	DUE DATE	INTEREST RATE	PAID TO	DESCRIPTION OF LIABILITY, COLLATERAL, SECURITY INTERESTS, LIENS, ENDORSERS, ETC.
$ 300,000 B	12-1-19X3	12-1-19X4	8%	12-1-19X8	Land and Building
60,000 M	10-30-19X8	10-30-19X9	8%	12-31-19X8	Accounts receivabl

3. The customer was contingently liable as endorser of notes discounted and/or as guarantor at the close of business on that date in the total amount of $ none , as below:

AMOUNT	NAME OF MAKER	DATE OF NOTE	DUE DATE	REMARKS
$				

4. Other direct or contingent liabilities, open letters of credit, and relative collateral, were

     none

5. Security agreements under the Uniform Commercial Code or any other agreements providing for restrictions, not noted above, were as follows (if officially recorded, indicate date and office in which filed):

     none

Yours truly, (Bank) Bank A, VP and Cashier

Date January 5   19 X9   By *R. Senneff*

                                   Authorized Signature

Additional copies of this form are available from the American Institute of CPAs, 666 Fifth Avenue, New York, N. Y. 10019

Figure 10.4   Bank A Letter

10. Account for all check numbers issued during the month as having cleared with the year-end bank statement or as being on the outstanding check list (indicated by Note 2 in Figure 10.3).

It is pointed out in Chapter 5 that there should be a separation between authorization of cash disbursements, signing and mailing checks, and the bank reconciliation. If such separation existed, audit procedures 7 through 10 might be applied on a test basis. Auditors sometimes use a minimum dollar amount as a basis for selecting the items to be tested in this type of situation to be sure of testing all significant items. Other types of sample selection previously discussed could also be used.

The foregoing procedures include three requests from the bank: a bank confirmation, a cutoff bank statement, and authenticated deposit slips. All three items are necessary, because the information requested in each is different. The bank confirmation includes data about direct and contingent liabilities, restrictions on cash, etc., as well as the account balance. The cutoff bank statement contains canceled checks, debit memos, and credit memos, as well as a listing of transactions. The deposit slips show the individual cash items included in the deposit, whereas the bank statement shows only the total deposit.

Because of the large number of companies being audited each year, the requests impose a considerable burden on the banks, particularly after the end of the calendar year. The standard bank confirmation (Figure 10.4) was designed to facilitate the furnishing of information by banks to auditors, and should be used whenever possible. The auditor sends these forms *directly* to the bank or other financial institution and receives them directly back. It is common practice for the auditor to send confirmation requests to every bank with which the client has an account or has done recent business (to detect unrecorded bank accounts and direct or contingent liabilities). Note that Item 2 on the form is a request for information on outstanding loans that the client has with the bank as of the balance sheet date. Thus, this form can serve as a working paper for both cash and liabilities. Items 3 and 4 are intended to gather information on contingent liabilities that may require either adjusting entries on the client's records or footnotes to the financial statements. Item 5 is placed on the form to acquire information on any cash restrictions that might require a reclassification of certain cash items to a non-current asset category.

The cutoff bank statement is a normal bank statement that has been cut off as of some date during the month. The auditor requests that the cutoff statement be sent directly to him for use in testing the propriety of reconciling items at year-end. Authenticated deposit slips are copies of deposit slips which are mailed to the bank for comparison with copies in its files. The bank indicates on the deposit slip that this has been done (authenticates the deposit slip) and mails it directly to the auditor. The purpose of this procedure is to test for lapping—the substitution of cash items to cover misappropriated funds and the delayed posting of collections to the detail accounts receivable records. The effectiveness of this procedure is limited, however, if details of the deposit are not included on the deposit slip or if the bank does not make an item-by-item comparison of the deposit slip (banks sometimes decline to do so).

Interbank Transfers—The foregoing procedures would provide evidence of the individual cash account balances, but the auditor also must consider the effect of transactions between accounts, often referred to as "interbank transfers." (Intrabank, interdivision, and intercompany transfers also should be tested.) The objective of testing interbank transfers is to determine that for each transfer near the end of a year, the amount added (deposited) to one account was subtracted (withdrawn) from another account during the same year. For example, if a transfer is added to one account on December 31 and subtracted from another on January 2, the cash balance as of December 31 would be overstated by the amount of the transfer (a liability account such as vouchers payable also would be overstated). This practice, sometimes referred as as "kiting," has been used by companies to conceal weak cash positions and to increase their current ratios (if current liabilities exceed current assets, an equal increase in each will improve the current ratio). Auditors test interbank transfers with an interbank transfer schedule such as that shown in Figure 10.5. The time period covered by this schedule depends on the normal time for a check to clear the banks involved. One or two weeks is not unusual.

The first transfer in Figure 10.5 was recorded on the books and cleared both banks in the same year and therefore is not a reconciling item. The second transfer was recorded on the books in one year and cleared both banks in the subsequent year. Therefore, it should appear in Bank A's reconciliation as a deposit in transit (as in Figure 10.3) and in Bank B's reconciliation as an outstanding check.

Bank accounts that are closed during a year and show a zero book balance should be confirmed. The auditor also may ask the bank whether it has been notified formally of the closing of the account (often they are not) and whether there have been any transactions after the date of the last book entry. Unless the necessary formal steps are taken by the client to close such an account, it could be used for the unauthorized deposit and subsequent misappropriation of funds.

Testing Cash Transactions—In addition to applying audit procedures to the year-end cash balance, auditors frequently test the recording of cash transactions for some period during the year. To do this, they prepare what is referred to as a "block reconciliation" or "proof of cash" for the period selected. An example of a block reconciliation is shown in Figure 10.6 (the month ending on the audit date is used to relate it to the previous examples, but any one or more months within the period being audited could have been selected). The reconciliation must balance, i.e., the amounts in each column must sum to the total of the column and the amounts in each row must sum to the balance at the end of the period. This form of reconciliation "blocks in" a period, so that all cash transactions in a particular account for the period can be accounted for and tested.

A/1

X Co.

Interbank Transfer Schedule

For the Period 12-22-X8 to 1-10-X9

12-31-X8

J. Jones 1-9-X9

Check Number	Description	Date of Deposits		Date of Withdrawals		Bank A	Bank B	Bank C
		Per Books	Per Bank	Per Books	Per Bank			
3182	Transfer from A to B	12-30-X8 ✓	12-31-X8 ✓	12-30-X8 ✓	12-30-X8 ✓	<10000>	10000	
0613	Transfer from B to A	12-31-X8 ✓	1-2-X9 ✓	12-31-X8 ✓	1-3-X9 ✓	5879	<5879>	
						A-1	A-2	

✓ Traced to cash book.
✓ Traced to bank statement.

Figure 10.5 Interbank Transfer Schedule

X Co.
Block Reconciliation
Bank A
12-31-X8

A-1
4

	Initials	Date
Prepared By	J. Jones	
Approved By		1-20-X9

	Balance 11-30-X8	Deposits	Withdrawals	Balance 12-31-X8
Balance per bank	63277 (a)	130244 (e)	⟨133871⟩ (h)	59650 (k)
Deposits in transit				
Per books / Per bank				
11-30-X8 / 12-1-X8	5121 (b)	(5121) (b)		
12-30-X8 / 1-2-X9		606 } (g)		606 } (g)
12-31-X8 / 1-2-X9		5879 } (g)		5879 } (g)
Outstanding checks – Prior to 12-1-X8	⟨14603⟩ (c)		14600 (c)	⟨3⟩ (i)
After 11-30-X8			(19350) (i)	⟨19350⟩ (i)
Other - None				
Balance per books	53795 (d)	131608 (f)	⟨138621⟩ (j)	46782 (d)

Figure 10.6 Block Reconciliation

Audit procedures that could be applied to the amounts in the block reconciliation are listed below (the letters correspond to those in Figure 10.6).

a. Confirm with bank and trace to ending balance on November bank statement and beginning balance on December bank statement.

b. Trace to December bank statement and authenticated deposit slip.

c. Obtain or prepare a list of outstanding checks; determine whether each check on list (i) cleared with December bank statement (by examination of canceled check) or (ii) was listed as outstanding at December 31.

d. Compare with general ledger or cash book.

e. Prepare a listing of deposits from bank statement and indicate date of each deposit (see f).

f. Prepare a listing of deposits from cash book and indicate date of each deposit; compare this listing with that prepared in step e and investigate any significant delays between dates deposits were recorded in cash book and by bank.

g. Trace to cutoff bank statement and authenticated deposit slip.

h. Prepare and total a listing of canceled checks included with bank statement.

i. Same as steps 7, 8, and 9 listed heretofore as applicable to year-end bank reconciliation.

j. Prepare and total a listing of disbursements from cash book and
   i. account for serial numbers of all checks issued during the period.
   ii. examine canceled checks clearing during the month for signatures and endorsements and compare with cash book as to date, number, payee, and amount.

k. Confirm with bank and trace to ending balance on December bank statement and beginning balance on cutoff bank statement.

Classification and Disclosures—Most of the procedures discussed previously in this section relate to the objective of determining the existence of the cash balance. To determine that restricted cash is properly classified and disclosed, the auditor must gather additional evidence. This evidence may take the form of answers to inquiries of management (sometimes included in the management representation letter) and answers regarding the restriction of cash on the standard bank confirmation (Figure 10.4). The standard bank confirmation, however, does not define the word "restricted," and in completing the confirmation form most banks have not considered compensating balances and other such informal agreements to be restrictions. Also, the confirmations often are completed by the bank's internal auditor, who may not have knowledge of informal restrictions agreed upon by the bank loan officer and the client. Therefore, when the SEC began requiring additional disclosures of cash restrictions, including compensating balances, most auditors began sending a special confirmation to each bank requesting specific confirmation of compensating balances and other agreements between the bank and the client.[2] Although it is required only in the audits of clients subject to SEC jurisdiction, many auditors have included the specific confirmation of compensating balances as a standard procedure in all audits. Another source for detection of restrictions on cash is the audit of liabilities, particularly the review of the provisions of lease and loan agreements and related documents.

---

[2]See *Accounting Series Release No. 148* issued by the Securities and Exchange Commission on November 13, 1973.

Receivables

Receivables usually represent uncollected revenues, but may include other items such as claims and other credits due from vendors, amounts due from sale of scrap or assets not held for resale, and accrued income such as interest and unbilled construction contract work (accounted for on the percentage-of-completion basis).

Accounts receivable included in current assets are considered to be collectible within one year or during a company's natural operating cycle. Examples of current assets in AC Section 2031.04 *(ARB No. 43)* include:

> ... (c) trade accounts, notes and acceptances receivable; (d) receivables from officers, employees, affiliates and others, if collectible in the ordinary course of business within a year; (e) installment or deferred accounts and notes receivable if they conform generally to normal trade practices and terms within the business; ...

Receivables not meeting these criteria should be classified as long-term assets. AC Section 5111.01 *(ARB No. 43)* provides, "Notes or accounts receivable due from officers, employees, or affiliated companies must be shown separately and not included under a general heading such as notes receivable or accounts receivable," and AU Section 335.17 *(SAS No. 6)* requires disclosure in the financial statements of "amounts due from or to related parties and, if not otherwise apparent, the terms and manner of settlement." Therefore, the segregation of certain receivables is necessary for a proper classification.

Receivables sometimes are pledged as collateral for a loan, and sometimes are discounted or sold (with or without recourse). Disclosure is required of the direct or contingent liabilities that may arise from such transactions.

Receivables, like all other assets, should not be stated at an amount in excess of net realizable value. An allowance or reserve account is used most commonly to reduce the receivable balance to the amount expected to be realized. It is important that the allowance for doubtful accounts be sufficient to cover not only losses from accounts known to be uncollectible, but also an estimate of current receivables that subsequently will become uncollectible. In addition, the account must be adequate to cover discounts and other allowances.

Internal controls relating to accounts receivable are discussed in Chapter 5. The basic concept of segregation of duties requires that such responsibilities as maintaining detail receivable ledgers, receiving cash, maintaining the receivable control account, preparing sales documents, writing off bad debts, and approving credit be separated to the extent possible.

Audit Objectives

The objectives in the audit of receivables are:

1. To determine whether the amount shown as receivables in the financial statements represents bona fide amounts due from others and owned by the client.

2. To determine that receivables are not stated in excess of net realizable value.

3. To determine whether receivables are classified properly.

4. To determine whether all liens on and pledges of receivables have been disclosed properly.

Audit Procedures and Working Papers

The auditor should begin work in this section with a review and test of the client's accounting procedures and internal control relating to sales, shipping, billing, and accounts receivable. An example of tests of some of these procedures is shown in Chapter 5. The evaluation of these procedures and controls should affect directly the work performed in this section and particularly the scope of one of the most important procedures in this section—the direct confirmation of account balances.

A system that has accountability checks and segregation of duties, such as that shown in the cash receipts system in Chapter 5, might allow the auditors to reduce the number of confirmations sent.

The auditor must prepare a lead schedule for receivables to determine types and amounts of major sub-accounts. Figure 10.7 shows that X Co.'s receivables are made up primarily of customer accounts receivable. The officer and employee and other accounts should be reviewed to determine that there were no significant transactions through them during the year. If there were none, no additional work normally would be performed on these accounts, because they are clearly immaterial.

The Aging Schedule—The auditor must obtain a listing of the individual customer trade receivables in order to have sufficiently detailed information to which he can apply audit procedures. Because it is more efficient to use one audit schedule for two tests than to prepare separate schedules, a detail listing of customer trade receivables is combined with an aging schedule in Figure 10.8. Note the following audit procedures which have been performed on this schedule.

1. The listing was totaled and determined to be in agreement with the general ledger (by cross-referencing it to the lead schedule where the balance had been traced to the general ledger).

2. Customer account balances were confirmed on a test basis (relates to bona fide existence of the receivable).

3. Alternate procedures (the examination of invoice, shipping documents, remittance advice from customer evidencing subsequent payment by the

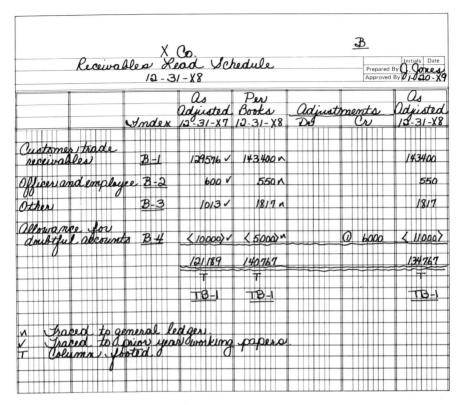

Figure 10.7 Receivables Lead Schedule

customer) were applied to customer balances selected for confirmation but for which no reply was received (relates to bona fide existence of the receivable).

4. The adequacy of the allowance for doubtful accounts was tested by (a) a review of credit files and discussions with the credit manager of the collectibility of all accounts over $2,000, (b) a review of subsequent collection of account balances, and (c) a comparison with the prior year of the percentage of accounts receivable in each of the aging categories (valuation at net realizable value and proper classification).

Confirmations—AU Section 331.01 *(SAS No. 1)* states:

> Confirmation of receivables and ... are generally accepted auditing procedures. The independent auditor who issues an opinion when he has not employed them must bear in mind that he has the burden of justifying the opinion expressed.

Figure 10.9 is an example of a *positive* confirmation request in which the customer is asked to return the confirmation directly to the auditor, indicating

B-1

X Co.
Customer Trade Receivables
Aged Listing
12-31-X8

J. Jones 1-20-X9

Acct. #	Customer Name	Balance 12-31-X8	Current	30-60 days	60-90 days	Over 90 days	Subsequent Collections	Comments
1103	Cunningham, Inc.	5630 ¢	5630				—	
1110	Ritter & Co.	1935	1860		75		1860	
1112	McFlynn & Sons, Ltd.	278				278	—	Bankrupt
1130	March Co.	3811	3101	1110			3811	—
	Total	B 143400	110945	18159	8286	5610	64423	
	Percent of total	100	77	13	6	4		
	Prior year percentages	100	83	11	5	1		

T Columns footed
¢ Confirmed by customer—See B-4
v Second request mailed.
√ Non-reply. Examined invoices, shipping documents and remittance advices for subsequent payments.

Client computes allowance balance as 1/10%. Showed credit files and discussed 6 accounts over $2000 with the credit manager and noted approximately $5000 of accounts in the 30-60 and 60-90 categories that are doubtful. Recommend the following adjustment:

Bad Debt Expense 6000
Allowance for Doubtful Accts 6000

J Jones 1-20-X9

Figure 10.8 Customer Trade Receivables Aged Listing

thereon agreement or disagreement with the amount shown on the client's records. Often the confirmation request is accompanied by a statement showing the detail of the accounts receivable balance, which allows the customer to determine agreement or disagreement more accurately.

A negative confirmation request contains much of the same wording as a positive request, except that the customers are asked to return the confirmation directly to the auditor *only* if they disagree with the amount shown in the client's records.

It should be noted that the letter is addressed from the client to the customer in such a way as to convey the message that a confirmation to the auditor is requested, rather than a payment to the company. This point is particularly important if the customers are members of the general public who are not used to receiving this type of letter and are likely to misconstrue it as a payment request. Occasionally, a note is printed in boldface type at the top of the letter stating that it is *not* a request for payment.

If a positive confirmation request is sent and no reply is received within a reasonable period of time, a second and sometimes third request is sent. If a reply is not received after additional requests are circularized, the auditors will attempt to obtain the information either by oral inquiry or from an inspection of internal records such as sales invoices, shipping documents, and evidence of payment. The important point is that once an account has been selected for verification, some type of evidence must be obtained. The matter may not be dropped simply because a reply is not received to a confirmation request.

The auditor must investigate all exceptions to both positive and negative confirmation requests. Often the difference between the client's records and the customer's reply can be explained by a cash receipt in transit or a late charge by the client. Sometimes the client has made an error, in which case an adjusting entry might be proposed if the difference is material. Not only exceptions but also gratuitous comments should be investigated. For example, if the comment "Paid in full on January 8, 19X9" were added by a customer to a confirmation as of December 31, 19X8, the auditor should trace the payment to cash receipts on or near January 8, 19X9.

The auditor should understand that a confirmation is evidence of the bona fide existence of a receivable, but it is *not* evidence of its collectibility. A customer may agree completely that a balance is due, but may be without funds to make payment. Similarly, the collection of a receivable after the audit date is evidence of its collectibility but not of its existence at the audit date (the receivable and related sale actually may have occurred after the audit date).

Valuation of Receivables—Because the determination of the adequacy of the allowance for doubtful accounts can be very subjective, the exercise of good business judgment by the auditor is crucial in this area. An auditor cannot automatically accept amounts computed by standard client procedures (such as a percentage of sales, a percentage of accounts past due, etc.), even if such procedures produced the proper results in the past. Conditions change, and procedures that produced reliable results at one time may not do so later. Among

X CO.
122 WEST AVE.
HYPOTHETICAL, ARKANSAS 70000

January 19, 19X9

Cunningham, Inc.
423 Best Road
Hypothetical, Arkansas   70000

Gentlemen:

Our auditors, Partner & Co., are now engaged in an examination
of our financial statements. In connection therewith, they
desire to confirm the balance due us on your account as of
December 31, 19X8, which was shown on our records (and the
enclosed statement) as $5,630.00. *B-1*

Please state in the space below whether or not this is in
agreement with your records at that date. If not, please
furnish any information you may have which will assist the
auditors in reconciling the difference. After signing and
dating your reply, please mail it directly to Partner & Co.,
999 Verify Street, Hypothetical, Arkansas   70000. A stamped,
addressed envelope is enclosed for your convenience.

It is very important that Partner & Co. receive your prompt
reply. We sincerely appreciate your assistance and will
return the courtesy extended to us whenever X Co. receives
your corresponding audit request.

Very truly yours,

Bruce Lee

B-1   Bruce Lee
                             Controller

The above balance of $ *5,630⁰⁰* due X Co. agrees with our
records at December 31, 19X8 with the following exceptions
(if any):_____

_____*None*_____
Date___*1-20-X9*_____ Signed *Harold Cunningham*

Figure 10.9 Accounts Receivable Confirmation

the factors an auditor would consider in evaluating the adequacy of the
allowance for doubtful accounts are: (1) a review of the effectiveness of the
client's current credit and collection policies, (2) statistical analyses such as
number of days' sales in accounts receivable, bad debts written off as a

percentage of sales, bad debt expense as a percentage of sales, percentage of accounts receivable in each aging category, etc. (most useful if the volume of accounts is high and the dollar value of individual accounts is low), (3) review of credit files and payment histories of individual accounts (a very effective procedure, but it can be impractical if more than a few large accounts are involved), (4) review of subsequent collections (also an effective procedure, but many collectible accounts may not be paid in the normal course of business before the audit is completed), and (5) consideration of general economic and industry conditions, the business approach (for example, an auditor would intuitively expect the allowance for doubtful accounts of a consumer loan company to be higher during a period of recession and high unemployment, because many customers would be out of work and unable to repay their loans). The auditor must require the client to bear the burden of justifying the adequacy of the allowance for doubtful accounts.

The auditor's earlier evaluation of the system of credit sales might affect the scope of audit work in this area. For example, if compliance with the credit policy had been violated consistently, and customers with questionable credit had been sold merchandise, the auditor might do more investigation of slow-paying accounts.

Matters Requiring Disclosure—To meet the objective of adequate disclosure of liens on or pledges of receivables, the auditor would (1) inquire of management as to any financial arrangements made to assign, discount, or pledge receivables, (2) inspect loan agreements for the existence of and minutes for the approval of such financial arrangements, and (3) request confirmation of the existence of such arrangements from financial institutions with which the client transacts business (see Figure 10.4). The business approach is also useful here. For example, an auditor should know that a client company that is short of cash and already has pledged its property and equipment is much more likely to have pledged its receivables than a client with an adequate cash flow.

Inventories

Inventories is one of the most significant accounts for many industrial and commercial companies, because of its effect on both working capital and gross profit, and the determination of inventory amounts can involve some of the most complex calculations in a company's accounting process. These facts, plus the susceptibility of inventories to misappropriation and misstatement, result in the application of extensive audit procedures to inventories if the amounts are material.

Inventories consist of physical goods to be sold, or consumed in the production of goods to be sold, in the ordinary course of business. In practice, materials and supplies that can be used in construction of property and equipment also are included in inventories.

The valuation of inventories has been the subject of several accounting pronouncements, three of which are set forth in Chapter 9. Basically, these statements provide that inventories must be stated at cost, that cost may be determined under any one of several assumptions as to the flow of cost factors (first-in first-out, last-in first-out, average, etc.), and that inventories should not be valued in excess of market (generally the lower of replacement cost or net realizable value). The relationship between these accounting principles and the determination of audit objectives for inventories is illustrated in Chapter 9.

Though the principles appear to be straightforward, the auditor will encounter many situations that do not fit precisely within the principles. It is in arriving at an informed judgment as to the proper accounting in these circumstances that an auditor earns his professional reputation. Consider the following examples. A plant, because of lack of sales orders, operates below capacity. Is the increased fixed cost per unit a properly inventoriable cost or is it an excess capacity cost that should be charged to expense as incurred? A client markets a new product which fails to gain customer acceptance and is selling very slowly. What is the net realizable value of the 100,000 units on hand when only 100 are sold each week at a retail price of $50 per unit? The client sells a chemical, the price of which fluctuates widely in the open market. Shortly after the end of the year, the price drops sharply. Should the client reduce the value of inventory to reflect the lower market price, or ignore the price drop on the basis that such fluctuations are normal and expected?

Because of the complexities of inventory calculations, the auditor must be particularly alert for inconsistencies in the inventory procedures. A change from the last-in first-out to the average cost method of pricing inventories would be apparent; changes in the method of determining quantities (i.e., the inclusion of goods in the current year that were excluded in the prior year because of obsolescence) and in the application of pricing methods (i.e., the inclusion of the storeroom operating cost in inventory one year and not in the next) can be less obvious, but still have a significant effect on net income.

An auditor normally finds it necessary to review and test the client's cost accounting system in his tests of inventory pricing. To do so, the auditor must be knowledgeable of the most generally used cost accounting systems, including process cost, job order cost, and standard cost systems. The auditor also must understand the physical flow of material and labor through the client's facilities to fully understand the operation of the cost system. Thus, the need to look beyond the numbers is evident.

Internal control over inventories is related closely to purchasing, receiving, and accounts payable. Ordinarily, the responsibilities for ordering, receiving, storing, shipping, and accounting for the goods should be separated to the extent possible. In addition, the client should adopt procedures to ensure an accurate physical inventory (including specific written instructions, prenumbered inventory cards or listings, double counts of inventory items, etc.) and pricing. Because internal controls in several areas affect inventories, the auditor must consider the results of several compliance tests in evaluating the effectiveness of internal control related to inventories. Tests of purchasing (purchasing and receiving), cash disbursements (accounts payable), payroll (labor costs), and sales (shipping) affect inventories and are illustrated in Chapter 5. Dual-purpose tests in the inventory area include physical inventory test counts and inventory pricing tests, both of which are illustrated later in this section. The auditor's evaluation of these controls determines the extent of his procedures.

As an example, it is pointed out in Chapter 5 that the authorization of purchases, the recording of purchases, and the custody of the assets resulting from purchases should be separated. If such segregation of duties exists, the auditors might be willing to adjust the audit procedures that test inventory quantities.

Audit Objectives

The objectives in the audit of inventories are:

1. To determine whether the amount shown as inventories in the financial statements is represented by physical items actually on hand, in transit, or on consignment.

2. To determine whether the inventory is calculated properly at the lower of cost or market in accordance with generally accepted accounting principles consistently applied.

3. To determine whether the inventory belongs to the company and that any liens on the inventory are disclosed properly.

4. To determine whether any excess, slow-moving, or special-purpose items are properly valued and classified.

Audit Procedures and Working Papers

Because inventories are used in Chapter 9 to demonstrate the relation of audit procedures to audit objectives, there is some duplication of those procedures in this section; however, to minimize redundancy, the procedures considered here are for finished goods rather than raw materials.

The first step is to review the accounting procedures and internal controls relating to inventories and to evaluate the results of compliance and procedure tests of the purchasing, receiving, shipping, cash disbursement, and cost accounting systems. As a result of this review, the auditor establishes the scope of work in this section.

An illustration of the inventory lead schedule is shown in Figure 10.10, and the amount of finished goods is analyzed further by item in the summary of finished goods inventory shown in Figure 10.11. This illustration provides an excellent example of how an auditor satisfies himself as to a total amount by systematically analyzing and auditing the components of that amount. By totaling the amount column, the auditor determines that the amount shown as finished goods is correct, *provided that* the individual amounts are correct. The individual amounts are determined by (a) the number of units, (b) the unit cost, and (c) the multiplication of the two. Thus, if the auditor is satisfied as to these three factors, he is satisfied as to the total finished goods inventory. Conversely, if any step is omitted (such as totaling the amount column), a link between the amount shown on the audit trial balance (and ultimately the financial statements) and the detail audit work (such as physical observation of the inventory) is broken. In this case, the auditor has no assurance that the individual amounts audited are represented in the financial statements. If they are not, any audit procedures performed on them are meaningless. Therefore, it is important that the auditor maintain the audit link between the amounts audited and the amounts shown in the financial statements.

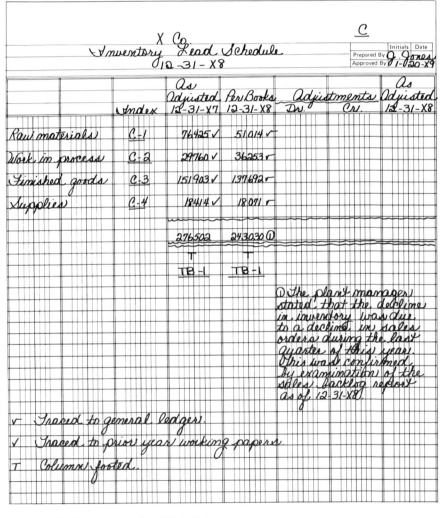

Figure 10.10   Inventory Lead Schedule

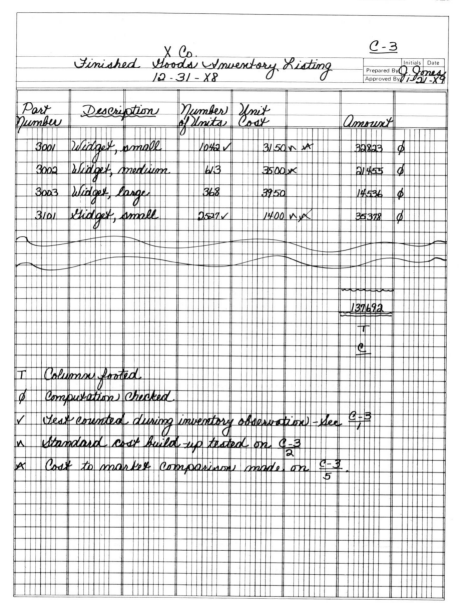

Figure 10.11   Finished Goods Inventory Listing

The Inventory Observation—The number of units is tested in the observation of the physical inventory. This audit procedure is important enough to consider in some detail. AU Section 331.01 *(SAS No. 1)* states, "...and observation of inventories are generally accepted auditing procedures. The independent auditor who issues an opinion when he has not employed them must bear in mind that he has the burden of justifying the opinion expressed."

The client has the responsibility for *taking* an accurate physical inventory. The auditor *observes* the inventory-taking procedures to satisfy himself that an accurate inventory was taken by the client. This distinction is important, because if the auditor takes, rather than observes, the inventory, he assumes the responsibility for a significant item involved in the preparation of the financial statements. His independence with respect to those statements therefore would be in jeopardy.

Before the start of the physical inventory, the auditor should review the client's written inventory instructions, looking for procedures that promote a complete and accurate inventory count. To the extent that he finds controls such as double counts of high-value items, identification procedures to prevent double counting, procedures for accurate recording and collecting of count totals, etc., the auditor adjusts the scope of the observation procedures.

During the actual inventory, the auditor's emphasis should be on observing the counting procedures of the client's inventory teams to determine whether they understand the inventory instructions and are careful and diligent in carrying them out. The auditor also should test-count some of the items (the number depends on the auditor's evaluation of the client's procedures) as an additional test of the effectiveness of the procedures (Figure 10.12). Auditors normally record some, but not all, of their counts for subsequent comparison with the final inventory listing. The purpose of this step is to detect any changes of the counts after they have been completed, as well as to provide a record of specific items counted by the auditor.

Though the recorded test-counts provide a check that counts of actual items are not changed later, they do not prevent the subsequent inclusion of

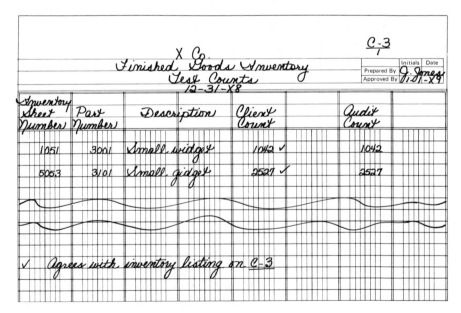

Figure 10.12   Finished Goods Inventory Test Counts

additional count sheets or tags containing fictitious items. To guard against this deception, the auditor should account for the numbers of all inventory sheets or tags as either used or unused. Also, he should record the number of the last line used on several inventory sheets. When the inventory later is compiled, the auditor should determine that it includes only items from inventory sheets he recorded as actually being used and that no items were added to the sheets for which he recorded the last line used.

During the inventory observation, the auditor also obtains and tests the information necessary to check the client's inventory cutoff, i.e., that the physical item and its related cost were treated in a consistent manner. In the finished goods example, the cutoff must be tested to determine whether items included in inventory are excluded from sales, and vice versa. To test the sales cutoff, the auditor prepares a schedule such as the one shown in Figure 10.13. The shipping number, description, quantity, and date shipped are obtained from the shipping department records during the inventory observation. The year in which the shipping number is included in sales is determined at a later date after all accounting transactions have been recorded. The schedule indicates that the sales cutoff was proper, i.e., items shipped prior to the audit date were excluded from inventory and included in sales. Evidence regarding the sales cutoff also is gathered with the confirmation procedures in the accounts receivable section. If an item is recorded as a sale (and account receivable) prior to the end of the year but is not shipped until after the end of the year, the customer should take exception to a confirmation request that includes the unshipped item in the balance. This is one of many examples of the relation of audit procedures between working-paper sections.

				X Co.					$\frac{C-3}{3}$

Finished Goods Sales Cutoff

		Initials	Date
Prepared By	Q. Jones		
Approved By	T.T.	1-7-X9	

\multicolumn{3}{Per Examination of Shipping Records}			Year Shipping Number Included In Sales	
Shipping Number	Description	Quantity	Date Shipped		
A-366	Medium gidget	37	12-30-X8	19X8 ✓
A-367	Large widget	6	12-30-X8	19X8 ✓
A-368	Large widget	118	12-30-X8	19X8 ✓
A-369	Small widget	44	12-31-X8	19X8 ✓
A-370	Small gidget	25	12-31-X8	19X8 ✓
A-371	Unused			19X9 ✓

✓ Per sales journal

Figure 10.13 Finished Goods Sales Cutoff

A final matter with which an auditor must be concerned during the inventory observation is the existence of obsolete, excess, or slow-moving items. Often the plant personnel involved in taking the physical inventory, such as production and shop foremen, storekeepers, and others, are very knowledgeable in this area, and some well-directed inquiries may yield pertinent information. The possibility of obsolescence also can be detected by the alert and inquisitive auditor who notes items with unusual amounts of dust or rust or prior-year inventory tags.

Because many of the audit procedures performed at an inventory observation cannot be quantified in working papers in a reasonable manner (such as observations of count teams, inquiries regarding obsolescence, etc.), it is common practice to summarize the work performed in a memorandum such as that shown in Figure 10.14. The memorandum describes, in reasonable detail, the steps taken to give the auditor relative assurance that a correct count was made. The last paragraph of the memorandum is the auditor's opinion concerning the client's inventory count. The writer of the memorandum noted a possible problem of slow-moving and possibly obsolete items. The note at the bottom of the memorandum indicates that a problem did exist and subsequently was solved by recognition of the obsolescence factor in valuing the inventory. It is important that all problems raised during an audit, both major and minor, be resolved. Unresolved problems in an auditor's working papers will be damaging evidence in any subsequent litigation.

Tests of Inventory Pricing—The inventory pricing method and the type of cost system used by the client largely determine the audit procedures employed in testing inventory pricing. Because pricing methods and cost systems vary widely, so do the related audit procedures. The calculation of LIFO inventory requires not only the pricing of individual items, but also overall calculations of inventory layers based on price index numbers. Compliance with federal income tax regulations is an important aspect of the audit of LIFO inventory calculations. Finished goods prices based on the FIFO or average cost method can be determined by the use of job-order, process, or standard costs. The audit of a job-order cost system usually involves a review of the categories of cost flowing through the control accounts (material, labor, and overhead) and tests of these costs in the jobs in progress at year-end. The audit of a process cost system concentrates on the flow and accumulation of costs by cost center for the major products. In a standard cost system, the audit emphasis is on testing the standard cost buildups for major inventory items and reviewing variances for an approximation of actual costs.

Cost, however, is only one aspect of inventory pricing. The auditor also must determine that cost does not exceed market. As previously noted, the lower of current replacement cost or net realizable value is the proper accounting principle for inventory pricing. The replacement cost test often is made on an overall basis by reviewing the unit cost of production after the end of the year as well as reviewing the unit cost of major raw material items. A decline in either of these costs may indicate that replacement cost is lower than inventory cost and

X CO.

INVENTORY OBSERVATION MEMORANDUM

Prepared by: *Dax Sharp*

Date: *1-1-X9*

Bill Catch and I arrived at the Manufacturing Plant at
7:00 a.m. on 1-1-X9 to begin our inventory observation.
The first thing we did was to obtain a copy of the inventory
instructions and discuss them with the Controller. We took
a plant tour and discussed with the Controller the sequence
in which he wished to have the inventory observed and cleared
since the Company wanted to begin production as soon as
possible in every area and wanted to do so in the production
cycle sequence.

Bill Catch and I began observing the count team procedures
and making test counts in the designated areas. Upon completion of one area, we would clear it and production would
begin. The client was to work only within that area, i.e.,
there was no movement from that area into other areas of
the plant. Then we would go to the next area that was ready.

Bill Catch called me to the raw materials area where he
had been observing procedures and making test counts while
I was clearing the area. He indicated that several dozen
feet of pipe were not on the inventory at all and other
items were recorded at obviously wrong amounts according
to the monthly inventories they have taken; it would have
been impossible to use the quantities that would have been
necessary to get this amount reduced to what the inventory
showed at this time. I discussed this with the Controller
who assigned a new count team to the area with instructions
to recount all items. Subsequent test counts by Bill Catch
indicated that the recounts were accurate. In all other
cases, the inventory teams appeared to understand their
instructions and to be working conscientiously.

We went back to the main plant and continued the inventory
in all areas and cleared them as soon as possible. Bill
Catch and I split up these areas and made selected test
counts on Xerox copies of the client's count sheets which
we maintained for controls and as a record of our test
counts. We inquired of the individuals in charge of the
inventory in each area as to obsolete, excess or slow
moving goods, and they knew of none except for a small
pile of scrap which was excluded from the inventory. We
saw no items during our observation that appeared obsolete,
excess or slow moving.

Figure 10.14 Inventory Observation Memorandum

Inventory Observation Memorandum
Page 2
Date: 1-1-X9

We accumulated the last five receiving documents before
inventory and the first five after the inventory that had
been received on 1-1-19X9. We also had the shipping cut-off
accumulated. This amounted to five shipping documents
before inventory and one shipping document that had been
shipped on 1-1-19X9.

Per the inventory instructions, the client personnel were
to place an X beside each item on their count sheets if it
had not been used at all in the last 6 months. This was
done and there are several throughout the count sheets.
These will have to be followed up at final to see that
they are properly valued in the final inventory listing.

See ① below.

In my opinion, the procedures followed by the client resulted
in a correct and accurate count of all goods on hand at
12-31-19X8, and all slow moving material was properly
indentified as such for later follow up.

Dan Sharp
1-1-X9

① All items indicated by an X were reduced in value by
25% in the final inventory listing to recognize their
slow moving and possibly obsolete nature. Our review
indicates that the 25% is reasonable.

J. Jones
1-21-X9

Figure 10.14 (Continued)

that a detail analysis by product should be made. Net realizable value normally is defined as net selling price less estimated cost to sell. Cost to sell often is estimated as a percentage of the net selling price computed by dividing total selling expense by sales.

Review for Obsolete, Excess, or Slow-Moving Inventory—The auditor's review for obsolete, excess, or slow-moving inventory must extend beyond the inquiries and observations during the inventory observation to include inquiry of top management and a review of perpetual or other inventory usage records. Though top management may not be as familiar with the usage of individual items as a shop foreman, they will be more knowledgeable of major policy decisions that could result in large-scale obsolescence, such as plans to discontinue a product line or to make significant changes in a product. For this reason, the subject of obsolete inventory always is discussed with top management and often is included in the management representation letter. Perpetual or other inventory usage records usually are examined for major items to determine whether the quantity on hand will be used in a reasonable time on the basis of past usage. The business approach can be useful also if the auditor is aware of industry trends in product changes, supplies of raw materials, and manufacturing processes.

Review for Liens or Pledges of Inventory—The inventory audit procedures designed to detect liens and pledges of inventories are similar to those for accounts receivable, including the review of debt instruments, confirmation with financial institutions with which the client does business, and inquiry of management. One unique aspect of inventory subject to lien is that it sometimes is fenced or otherwise segregated from other inventory; financial institutions occasionally place signs in the area stating that the inventory is pledged. Such restrictions on the inventory should be noted during the inventory observation.

Other Inventory Procedures—There is no single approach to the audit of inventories. If inventories are held in public warehouses, confirmation procedures can be used. Certain types of inventories require laboratory analysis (to be sure, for example, that an underground gasoline tank is not filled with water). Analysis and comparison of gross profit ratios are very helpful in evaluating the propriety of inventories of companies in certain industries. The ingenuity of the auditor is a key factor in the examination of inventories.

Prepayments

Prepayments represent costs incurred that are applicable to future periods. Prepayments are of a short-term nature and usually are associated with services to be received within the next year, such as prepaid insurance, rent, and taxes. AC Section 2031.04, .06 *(ARB No. 43)* includes in the definition of current assets:

> ... prepaid expenses such as insurance, interest, rents, taxes, unused royalties, current paid advertising service not yet received, and operating supplies. Prepaid expenses are not current assets in the sense that they will be converted into cash but in the sense that, if not paid in advance, they would require the use of current assets during the operating cycle.

and excludes:

> ... long-term prepayments which are fairly chargeable to the operations of several years, or deferred charges such as unamortized debt issue costs, bonus payments under a long-term lease, and costs of rearrangement of factory layout or removal to a new location.

To establish audit objectives and evaluate the fairness of the financial statements, an auditor must have an extensive knowledge of all current pronouncements comprising generally accepted accounting principles.

Audit Objectives

The objectives in the audit of prepayments are:

1. To determine whether the amounts shown as prepayments and other assets in the financial statements are computed in accordance with generally accepted accounting principles and are applicable to future periods.

2. To determine whether the amounts can reasonably be expected to be realized.

3. To determine whether the amounts are classified properly in the financial statements and disclosure is adequate.

Audit Procedures and Working Papers

The accounting procedures and internal controls to be considered by the auditor in the examination of prepayments include those applicable to cash disbursements, the client's policies regarding the kinds of costs that are subject to deferral, and the systematic procedure for amortizing such costs to expense. Cash disbursement procedures tests are illustrated in the appendix to Chapter 5. Tests of the procedures for proper deferral and amortization are dual-purpose tests; they are described in this section. The evaluation of the effectiveness of these procedures is a significant factor in determining the scope of the auditor's work in this area. For example, a consistently followed policy of deferring the cost of only certain appropriate items might enable the auditor to substantially limit his examination of invoices supporting additions to the prepayments account.

After the auditor's review, test, and evaluation of internal control, the next step normally would be the preparation of a lead schedule. It is assumed in the X Co. example, however, that there is only one prepayment account. A standard lead schedule containing only one account would be pointless, so in this case, the auditor would reference directly from the detail audit schedule for prepaid insurance (Figure 10.15) to the audit trial balance. This schedule (unlike working papers in the cash, accounts receivable, and inventory sections) shows the transactions for the year that affect prepaid insurance and insurance expense (the procedures result in evidence about an expense account as well as an asset account). Examination of the audit work indicated on the schedule shows how it relates to the audit objectives.

Tests of Balance and Amortization—Under generally accepted accounting principles, assets are stated at cost. The auditor has tested this aspect by examination of invoices from insurance companies and the related canceled checks (see ʍ). Additional evidence of the annual premium cost is obtained from examination of the insurance policy (see ✔). The auditor must be aware that the premiums on certain types of policies, such as workman's compensation, sometimes are based on the injury experience of the insured and are subject to retroactive adjustment. In this case, a review of the client's injury experience and consultation with the client's insurance agent may be necessary to arrive at an estimate of the final premium. The applicability of the prepaid portion of the cost to future periods is tested by calculation of the current-year amortization on the basis of the percentage of the total policy period included in the current year (see ✔). An examination of the policy periods shown in the description column indicates that all will expire within one year, and that the classification of the prepaid balance as a current asset is proper. The objective of determining that the amounts can reasonably be expected to be realized is usually self-evident with prepaid insurance, because the balance generally is refunded on cancellation of the policy if not utilized in the future. This is not the case for all prepaid items, however. Consider prepaid rent on a plant that is shut down permanently. Although the cost in this case may apply to some future period, prepayments must generate some future benefit or value at least equal to their cost to be classified properly as assets. If such items do not produce revenues to absorb their costs, they will not be realized.

Review of Insurance Coverage—At the bottom of Figure 10.15 is a notation of a review of the company's insurance coverage with the controller and the auditor's observation that he noted no significant omissions or inadequacies in the coverage. Adequate insurance coverage is not a requirement for a fair presentation of financial statements, although the disclosure of significantly inadequate insurance coverage certainly would be desirable. In addition, the auditor is not an insurance expert with the qualifications to reach an informed opinion about insurance coverage, although he should be familiar with typical business insurance coverage. For example, most businesses maintain fidelity bond coverage to insure against loss from embezzlement and theft, and auditors consider this

D

X Co.

Prepayments

Analysis of Prepaid Insurance and Expense

J. Jones
1-22-X9

12-31-X8

Description		Prepaid Balance 12-31-X7	Current Premiums	Amortization to Expense	Prepaid Balance 12-31-X8
Policy No. REC 12233 for fire and extended coverage on building and contents in the amount of $600,000 for the year from 6-30-X7 to 6-30-X8		5742	—	5742	—
Increase in above policy to $625,000 and extension to 6-30-X9	✓	—	12624 ∩	6312 ✗	6312
Policy No. WC 22244 for workman's compensation coverage of $100,000 from 1-1-X8 to 12-31-X8	✓	—	3291 ∩	3291	—
Policy No. CC 3719 - Blanket fidelity bond coverage in amount of $200,000 from 1-1-X8 to 12-31-X8	✓	—	1873 ✓	1873	—
		11343	37898	36986	12255
		T	T	T	T
		TB-1			TB-1
				20 29704	
				40 7282	
				36986	
				T	

T — Column footed.

✓ — Examined insurance policy noting provisions of the policy and annual premium.

∩ — Examined insurance company invoice and canceled check.

✗ — Checked calculation of amortization.

I discussed the company's insurance coverage with Mr. Lee, Controller. He is satisfied that insurance coverage is adequate to cover the replacement cost of the assets and anticipated liability claims. I noted no significant omissions or inadequacies of insurance coverage.

J. Jones
1-22-X9

Figure 10.15   Prepayments

coverage in their overall evaluation of internal control. The auditor should recommend that the client consult with an insurance representative about any obvious deficiencies in coverage. Important audit information sometimes can be obtained from a review of insurance coverage. For example, a reduction in insurance coverage of property and equipment may lead the auditor to an unrecorded property retirement, and the existence of a loss-payable clause in a policy could result in the detection of an unrecorded liability.

The audit procedures for other types of prepayments are similar to those for prepaid insurance and are not illustrated in detail. A common problem in auditing prepayments is a tendency of the auditor to perform excessive work on relatively immaterial balances. This problem is due in part to the usually uncomplicated and easily verifiable nature of the accounts. An auditor must resist the temptation to dwell on immaterial and uncomplicated accounts and concentrate on the more difficult areas.

Current Liabilities

In examining assets, an auditor places primary emphasis on verifying the amounts recorded as assets. In examining liabilities, however, the auditor must place primary emphasis not on what is recorded, but on what is *not* recorded. The auditor seldom finds amounts recorded as liabilities which are not liabilities, but unrecorded liabilities are not unusual. In fact, they are inherent in an accounting process with periodic reporting. At some point in time, a decision must be made to terminate the accounting process and to issue financial statements with the understanding that revisions will be made only if material.

Although controls normally are established to detect unrecorded liabilities in a major area such as the purchase of raw materials, many minor liabilities such as the late reimbursement of an expense account or small utility bills do not become known until after the accounting process has been terminated. In addition, few companies attempt to record the exact amount of every existing liability at each periodic closing of the accounting records (although the amount unrecorded should be immaterial), because some liabilities such as taxes, insurance. and royalties cannot be determined precisely until some later time. For these reasons, the business approach to the audit must be used if the auditor is to make a sufficient investigation of both recorded and unrecorded liabilities.

Current liabilities include both amounts currently due (such as accounts payable) and amounts that have been incurred but are not yet due (such as accrued interest payable). Normally, the amounts not due are expected to be due within one year. AC Section 2031.07 *(ARB No. 43)* provides the following definition of current liabilities.

> The term *current liabilities* is used principally to designate obligations whose liquidation is reasonably expected to require the use of existing resources properly classified as current assets, or the creation of other current liabilities. As a balance-sheet category, the classification is intended to include obligations for items which

have entered into the operating cycle, such as payables incurred in the acquisition of materials and supplies to be used in the production of goods or in providing services to be offered for sale; collections received in advance of the delivery of goods or performance of services; and debts which arise from operations directly related to the operating cycle, such as accruals for wages, salaries, commissions, rentals, royalties, and income and other taxes. Other liabilities whose regular and ordinary liquidation is expected to occur within a relatively short period of time, usually twelve months, are also intended for inclusion, such as short-term debts arising from the acquisition of capital assets, serial maturities of long-term obligations, amounts required to be expended within one year under sinking fund provisions, and agency obligations arising from the collection or acceptance of cash or other assets for the account of third persons.

Accounts payable and accrued liabilities usually are set forth separately in the financial statements, and accrued liabilities sometimes are analyzed into major categories.[3] Short-term notes payable and current maturities of long-term debt also are shown separately.

Although accounts payable and accrued liabilities normally are unsecured, there could be circumstances in which assets are held by others or are pledged to secure their payment. Also, officers, stockholders, or other companies may guarantee payment. Short-term notes payable and the current portion of long-term debt often are collateralized. All of these conditions must be disclosed in the financial statements. Any other factors regarding current liabilities that are important in evaluating the financial statements should be disclosed, such as amounts due officers and stockholders or significant amounts of past-due accounts payable.

The accounting records in support of current liabilities usually are much less extensive than those supporting current assets. Many companies merely list, and record by journal entry, all unpaid invoices applicable to one year which are received by some specified cutoff date (say the 20th of the following month) in the next year. Accrued liabilities usually are recorded by journal entry and supported by worksheet computations.

Internal control over accounts payable is related closely to controls over purchasing, receiving, cash disbursements, and inventory, whereas control over accruals normally is exercised by controlling the related expense account.

Audit Objectives

The objectives in the audit of current liabilities are:

1. To determine whether all current liabilities existing or incurred as of the audit date are reflected properly in the financial statements.

---

[3]*SEC Regulation S-X* requires that accrued liabilities for payrolls, tax liabilities, interest, rents and royalties, and any other significant item be stated separately.

2. To determine whether current liabilities consist of amounts due within one year or the natural operating cycle of the business.

3. To determine whether disclosures concerning current liabilities are adequate.

Audit Procedures and Working Papers

The audit procedures and working papers applicable to accounts payable and accrued liabilities are considered in this section; the current portion of long-term debt and notes payable is considered in the long-term liability section.

Accounts Payable—As part of the first step of evaluating the accounting procedures and internal control for accounts payable, the auditor must consider, on the basis of the procedures and controls used, the potential for the existence of material unrecorded liabilities. The potential for unrecorded liabilities is evaluated in part on the results of tests of the purchasing and cash disbursements functions shown in the appendix to Chapter 5 and in part on dual-purpose tests of procedures for recording accrued liabilities and the resulting account balances. The magnitude of this potential determines the scope of subsequent procedures.

A review of the purchases section of the appendix to Chapter 5 shows that the vendor's invoice and the company's receiving report are compared in the Accounts Payable Department. The existence of this procedure should lower the probability that received merchandise is not accounted for and that some accounts payable invoices might be unrecorded. On this basis, the auditor might adjust the tests that search for unrecorded liabilities.

Because X Co. in Figure 10.16 has only trade accounts payable, a separate audit lead schedule is unnecessary. The auditor can reference directly from the detail listing of accounts payable to the audit trial balance. This detail listing of accounts payable generally is tested by confirmation of balances with vendors and a review of disbursements subsequent to the audit date.

Confirmation of accounts payable is not designated specifically as a generally accepted auditing procedure in *SAS No. 1,* as is the confirmation of accounts receivable; nevertheless, it is a procedure used by most auditors. One must keep in mind the difference in approaches to auditing assets and liabilities to understand the method of selecting the accounts to which confirmation requests will be sent. Recall that in the audit of liabilities the auditor is more concerned with what is not recorded but should be, than with what is recorded. Therefore, confirmation requests generally are sent to a client's principal vendors and suppliers, regardless of their account balance at the audit date. The auditor's objective is to obtain confirmations from the accounts *most likely* to have large accounts payable balances, not necessarily from the accounts that have large *recorded* balances. For this reason, the auditor often sends confirmation requests

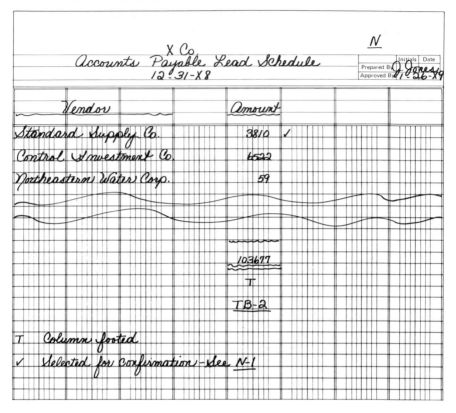

Figure 10.16 Accounts Payable Lead Schedule

to accounts with zero balances. The auditor is not attempting to obtain a high dollar coverage of account balances, but is searching for unrecorded liabilities.[4] Consider the following example.

Vendor	Purchases from Vendor During the Year	Payable Balance at End of Year
A Company	$ 35,000	$ 11,000
B Company	1,150,000	110,000
C Company	1,200,000	—
D Company	246,000	30,000

[4]The requirements for the confirmation of accounts receivable (and the physical observation of inventories) arose as a result of the *McKesson-Robbins* case, which was discussed in Chapter 3. The fraud was largely perpetrated by the use of fictitious accounts receivable and inventories. Had significant unrecorded liabilities been a major issue, confirmation of accounts payable might also have been made a requirement in *SAS No. 1*.

In this case, the auditor probably would select B and C for confirmation, even though both A and D have larger year-end balances than C.

Accounts selected for confirmation can be listed on a confirmation control schedule such as Figure 10.17. This schedule serves the dual purpose of recording the amount confirmed by the vendor and reconciling it to the amount shown in the client's detail accounts payable listing. All differences between the client's accounting records and the confirmation from the vendor must be not only reconciled, but also audited. In Figure 10.17, the payment in transit from the client to the vendor has been traced to the outstanding checklist in the bank reconciliation (indicated by ✔; also see Figure 10.3). The auditor must understand the relationships of the various sections of the audit well enough to realize instinctively that if a vendor had not received a payment from the client which was made before or on the audit date, the check could not have cleared the bank by the audit date and, therefore, must be listed as outstanding. In Figure 10.17, the unrecorded liability was traced to the vendor's invoice for a December purchase that was recorded by the client in January (indicated by ∅). Because it is an immaterial amount, it has been posted to the summary of entries passed for consideration in the aggregate. If the amount had been material, the auditor would have proposed an adjusting entry.

An illustration of an accounts payable confirmation is shown in Figure 10.18. It is similar to an accounts receivable confirmation in that it requests an outside party to provide information about the client directly to the auditor; however, an important difference is that whereas the accounts receivable confirmation contains the customer's balance according to the client's records, the accounts payable confirmation contains a request for the vendor to state the balance due from the client. The difference is due to the different emphasis in auditing assets and liabilities. With the accounts receivable confirmation, the auditor attempts to verify the recorded balance; with the accounts payable confirmation, he attempts to learn of all amounts due to a vendor whether recorded or not.

The second procedure used in the audit of accounts payable is the review of disbursements subsequent to the audit date (usually performed as part of the review of subsequent events and transactions, which is discussed in another section). In making this review, which normally covers the period from the audit date to the date of completion of work in the client's office, the auditor examines unpaid invoices and the invoices or other support for disbursements to determine the period to which they are applicable (often a minimum amount is established to avoid examining minor items). If a disbursement is found that is applicable to a period prior to the audit date, the auditor reviews the accounts payable (and accrued liabilities) listing to determine whether the amount owed was recorded properly as a liability at the audit date. If it was not, it represents an unrecorded liability.

The advantage of this procedure is that it provides much broader account coverage than would be practicable with confirmations; the disadvantage is that any invoice representing an unrecorded liability not paid or received by the client prior to the end of the auditor's work in the client's office would not be detected.

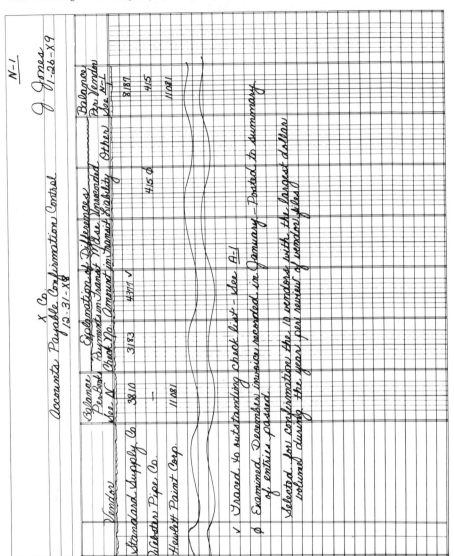

Figure 10.17 Accounts Payable Confirmation Control

Such amounts would be detected by confirmation. Thus, a combination of accounts payable confirmation and a review of subsequent disbursements, together with work performed in other sections such as cash, inventories, and cost of sales, normally provides the audit evidence necessary to satisfy the auditor with regard to accounts payable.

Accrued Liability—The approaches to auditing accrued liabilities are as varied as the types of accrued liability accounts. Some can be tested by reference to the subsequent payment of the liability (accrued payroll and payroll taxes), whereas

N-1
1

X CO.
122 WEST AVE.
HYPOTHETICAL, ARKANSAS 70000

January 2, 19X9

Standard Supply Co.
222 Elm Street
Hypothetical, Arkansas   70000

Gentlemen:

   Our auditors, Partner & Co., are now engaged in an
examination of our financial statements. In connection
therewith, please advise them in the space provided below
whether or not there is a balance due you by this company
as of December 31, 19X8. If there is a balance due, please
attach a statement of the items making up such balance.

   After signing and dating your reply, please mail it
directly to Partner & Co., 999 Verify Street, Hypothetical,
AR 70000. A stamped, addressed envelope is enclosed for
your convenience.

                          Very truly yours,

                          Bruce Lee

                          Bruce Lee
                          Controller

Partner & Co.:

Our records indicate that a balance of $ 8187.00 *N-1* was due
from X Co. at December 31, 19X8, as itemized in the
attached statement.

Dated: __1-11-X9__   Signed: *Von Graham, Treas.*

Figure 10.18   Accounts Payable Confirmation

others must be estimated or calculated on the basis of transactions in other
accounts (accrued interest on the basis of interest-bearing debt outstanding and
accrued royalties on the basis of sales). Figure 10.19 illustrates an audit lead
schedule for accrued liabilities.

X Co.
Accrued Liabilities Lead Schedule
12-31-X8

P
J. Jones
1-20-X9

Index	As Adjusted 12-31-X7	Per Books 12-31-X8	Adjustments Dr.	Adjustments Cr.	As Adjusted 12-31-X8
Accrued payroll	Ⓐ	13079 ✓	10390 ✓	(3) 5000		10390
Accrued interest	R	2333 ✓	2000 ✓			2000
Accrued property taxes	P-1	39199 ✓	22529 ✓			22529
Accrued income taxes	P-2	43103 ✓	44859 ✓			44859
		80663	8ııı8	5000		76778
		TB-2	TB-2			TB-2

Ⓐ Overall test of accrued payroll —
Payroll for the 2-week period ended 1-3-X9   13,177 X × 11/14
Portion applicable to 19X8   10,353
Per books   10,390
Difference not material   37

✓ Traced to general ledger
√ Traced to (prior) year working paper
┬ Column footed
X Traced to payroll register

Figure 10.19   Accrued Liabilities Lead Schedule

In Figure 10.19, accrued payroll has been tested on an overall basis by reference to the total payroll paid in the subsequent period. The auditor has determined the total payroll for the two-week period ended January 3, 19X9, by examination of the payroll register (indicated by ⋊). This amount was multiplied by the fraction representing the payroll period in 19X8 (11 of the 14 days). The computed amount does not agree exactly with the amount recorded on the books because it was computed on an overall basis, whereas the book amount was computed by payroll group or individual employee. The difference, however, is not material and, furthermore, should not be considered an error to be posted to the summary of entries passed. Variations such as this are to be expected when overall tests are used, and they do not imply that the client's records are incorrect.

Accrued interest is cross-referenced to *R,* which is the schedule reference for long-term debt. The audit of accrued interest is discussed in that section.

Accrued property taxes is cross-referenced to a detail audit schedule, *P-1,* which is shown in Figure 10.20. This schedule summarizes the transactions in accrued property taxes for the year. The provision for property tax expense for the current year has been related to the expense section of the working papers and reviewed for reasonableness. This provision is an example of a situation in which an estimate must be made by the client and reviewed by the auditor, because it is assumed that the actual tax for the year will not be known until the tax bills are received at some date after the financial statements are issued. In practice, the period covered by the tax and the date the tax bills are rendered vary by taxing authority within each state. The auditor should be familiar with the practices of the taxing authorities within whose states his clients operate. In the example, the auditor was aware that there had been no tax rate increases and, from the audit of property, knew that there had been no significant variations in property balances. Thus, the auditor found that property tax expense in an amount approximately equal to the amount paid for the prior year was reasonable (indicated by ①). Note that merely comparing the provision for the current year with the amount paid for the prior year, without consideration of the factors that could make them different, would not be effective auditing. The auditor also has examined *receipted* tax bills and canceled checks to substantiate the tax payments (indicated by ✔).

Because of the significance of income tax expense and the related accrued tax liability, it is important that the auditor have a basic understanding of federal income tax laws and regulations, as well as those of the states and any foreign countries in which his clients operate. In addition, the auditor must comprehend the differences between pretax accounting income and taxable income, timing differences and permanent differences, income taxes and deferred taxes, and the myriad of other concepts in AC Section 4091 *(APB Opinion No. 11)* and other pronouncements covering accounting for income taxes. With this knowledge, the auditor would be prepared to begin an audit of income taxes. The procedure would consist generally of an analysis of the accrual account for the year and the testing of payments during the year by reference to canceled checks and prior-year tax returns. The auditor also must check the calculation of income tax expense for the current year and the amount of any deferred taxes arising from timing differences, in all cases considering the propriety of the amounts used in the calculations. In addition, the auditor should inquire as to the status of all unsettled prior-year returns and any revenue-agent examinations in progress. In many accounting firms, the auditor prepares the income tax returns for the clients he audits, unless the client prepares his own return. In some cases, particularly in the larger accounting firms, a separate tax department prepares all tax returns.

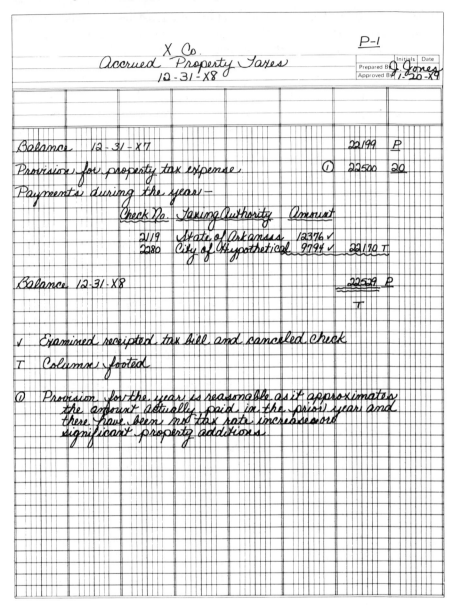

Figure 10.20 Accrued Property Taxes

In auditing accrued liabilities, the auditor must not become so involved with testing calculations and reviewing the reasonableness of estimates that he forgets where the emphasis should be placed in the audit of liabilities. The auditor must consider what accrued liabilities should be recorded that are not recorded. Because accrued liabilities actually are accrued expenses, a good starting point in a search for unrecorded accrued liabilities is the client's expense accounts. It is often necessary to accrue such expenses as payroll, payroll taxes, vacation pay, sick pay, commissions, insurance, income, property and excise taxes, pensions, bonuses, interest, profit-sharing, and royalties. The auditor should consider the necessity of these as well as other accruals in each audit.

Chapter 10
Questions Taken from the Chapter

10–1.　Why are assets more likely to be overstated than understated, and what effect does this have on the auditor's emphasis in the audit of assets?

10–2.　Explain why the audit procedures applied to cash are usually more extensive than might seem warranted by the size of the cash balance.

10–3.　Give an example and explain the importance of restrictions on cash.

10–4.　Explain how a separation of the treasury and accounting functions can be accomplished.

10–5.　State the objectives in the audit of cash.

10–6.　Why is the review of accounting procedures and internal controls one of the first audit procedures performed?

10–7.　In an examination of a bank reconciliation, the auditor gathers evidence regarding the amounts shown therein. Indicate the evidence he would gather or examine in support of:
　　a.　the bank balance.
　　b.　the book balance.
　　c.　deposits in transit.
　　d.　outstanding checks.

10–8.　List three requests an auditor will make of a bank in connection with his audit of cash.

10–9.　The standard bank confirmation provides the auditor with evidence relating to accounts other than cash. State how this is done.

10–10. What is the purpose of obtaining authenticated deposit slips from the bank? Explain the limitations on the effectiveness of this procedure.

10–11. What is the objective of testing interbank transfers?

10–12. Explain what is meant by "kiting" and what it accomplishes.

10–13. What is the purpose of a block reconciliation or proof of cash?

10–14. Why do auditors often send banks special letters confirming compensating balance arrangements rather than rely on bank confirmations?

10–15. State the objectives in the audit of receivables.

10–16. Indicate how the audit objective of proper classification of receivables is related to generally accepted accounting principles.

10–17. What are "alternate procedures" and when are they performed?

10–18. To which audit objectives are the procedures for the test of the allowance for doubtful accounts related?

10–19. To which audit objective is the confirmation of accounts receivable related?

10–20. List the factors an auditor would consider in evaluating the adequacy of the allowance for doubtful accounts.

10–21. Discuss the audit procedures an auditor would employ to detect liens on or pledges of receivables.

10–22. Why would an auditor apply extensive audit procedures to inventories if they are material?

10–23. State the objectives in the audit of inventories.

10–24. Discuss the auditor's responsibility for an accurate physical inventory.

10–25. What is the purpose of the auditor's review of the client's written physical inventory instructions?

10–26. Comment on the following statement. "During the physical inventory, the auditor should make and record as many test counts as possible because his primary concern is to count as large a percentage of the dollar value of the inventory as possible."

10–27. How does the auditor guard against inclusion in the final inventory listing of count sheets or tags containing fictitious inventory items?

10–28. Why does an auditor record some of his inventory test counts?

10–29. What is the purpose of the inventory cutoff test?

10–30. What procedures can the auditor perform during the inventory observation to test for obsolete, excess, or slow-moving items?

10–31. State the general approaches to auditing job-order, process, and standard cost systems.

10–32. Define "market" in the context of "lower of cost or market."

10–33. Why is the subject of obsolete inventory discussed with top management and included in the management representation letter?

10–34. State the objectives in the audit of prepayments.

10–35. Explain how the internal controls relating to prepayments could affect the scope of the auditor's work in that section.

10–36. Explain the form of the lead schedule for prepayments.

10–37. What evidence would an auditor examine in support of additions to prepaid insurance?

10–38. Why does an auditor review a client's insurance coverage?

10–39. What should an auditor do if he believes that a client's insurance coverage is not adequate?

10–40. How do the auditor's approach and emphasis in auditing assets differ from his approach and emphasis in auditing liabilities?

10–41. State the objectives in the audit of current liabilities.

10–42. The auditor normally uses two approaches to the audit of accounts payable. What are they and why are both used?

10–43. How does the auditor select the accounts to which accounts payable confirmations will be sent? Why is this method of selection used?

10–44. What is the purpose of a confirmation control schedule?

10–45. What are the similarities and differences in the form of accounts receivable and payable confirmation letters?

10–46. How and for what period is the review of subsequent disbursements made?

10–47. Name two approaches to auditing accrued liabilities.

10–48. What background knowledge must an auditor possess in order to effectively audit income tax liability and expense?

10–49. What is a good starting point in an auditor's search for unrecorded accrued liabilities?

Chapter 10
Objective Questions Taken from CPA Examinations

10–50. Listed below are four interbank cash transfers, indicated by the letters a, b, c, and d, of a client for late December, 19X8 and early January, 19X9. Your answer should be selected from this list.

	Bank Account 1 Disbursing Date (Month/Day)		Bank Account 2 Receiving Date (Month/Day)	
	Per Bank	*Per Books*	*Per Bank*	*Per Books*
a.	12/31	12/30	12/31	12/30
b.	1/2	12/30	12/31	12/31
c.	1/3	12/31	1/2	1/2
d.	1/3	12/31	1/2	12/31

Which of the cash transfers indicates an error in cash cutoff at December 31, 19X8?

Which of the cash transfers would appear as a deposit in transit on the December 31, 19X8, bank reconciliation?

Which of the cash transfers would *not* appear as an outstanding check on the December 31, 19X8, bank reconciliation?

10–51. The standard bank confirmation form requests all of the following *except*
   a. Maturity date of a direct liability.
   b. The principal amount paid on a direct liability.
   c. Description of collateral for a direct liability.
   d. The interest rate of a direct liability.

10–52. The cashier of Safir Company covered a shortage in the cash working fund with cash obtained on December 31 from a local bank by cashing, but not recording, a check drawn on the company's out-of-town bank. How would the auditor discover this manipulation?
   a. Confirming all December 31 bank balances.
   b. Counting the cash working fund at the close of business on December 31.
   c. Preparing independent bank reconciliations as of December 31.
   d. Investigating items returned with the bank cutoff statements.

10–53. One of the better ways for an auditor to detect kiting is to
   a. Request a cutoff bank statement.
   b. Send a bank confirmation.
   c. Prepare a bank transfer working paper.
   d. Prepare a bank reconciliation at year-end.

10–54. During his examination of a January 19, 19X9, cutoff bank statement, an auditor noticed that the majority of checks listed as outstanding at December 31, 19X8, had not cleared the bank. This would indicate
   a. A high probability of lapping.
   b. A high probability of kiting.
   c. That the cash disbursements journal had been held open past December 31, 19X8.
   d. That the cash disbursements journal has been closed prior to December 31, 19X8.

10–55. Approximately 95% of returned positive accounts receivable confirmations indicated that the customer owed a smaller balance than the amount confirmed. This might be explained by the fact that
   a. The cash receipts journal was held open after year-end.
   b. There is a large number of unrecorded liabilities.

    c. The sales journal was closed prior to year-end.

    d. The sales journal was held open after year-end.

10–56. One of the major audit procedures for determining whether the allowance for doubtful receivables is adequate is

    a. The preparation of a list of aged accounts receivable.

    b. Confirming any account receivable written off during the year.

    c. Vouching the collection of any account receivable written off in prior periods.

    d. Confirming any account receivable with a credit balance.

10–57. The return of a positive account receivable confirmation without an exception attests to the

    a. Collectibility of the receivable balance.

    b. Accuracy of the receivable balance.

    c. Accuracy of the aging of accounts receivable.

    d. Accuracy of the allowance for bad debts.

10–58. Lapping would most likely be detected by

    a. Examination of canceled checks clearing in the bank-cutoff period.

    b. Confirming year-end bank balances.

    c. Preparing a schedule of interbank transfers.

    d. Investigating responses to accounts receivable confirmations.

10–59. Returns of positive confirmation requests for accounts receivable were very poor. As an alternative procedure, the auditor decided to check subsequent collections. The auditor had satisfied himself that the client satisfactorily listed the customer name next to each check listed on the deposit slip; hence, he decided that for each customer for which a confirmation was not received, he would add all amounts shown for that customer on each validated deposit slip for the two months following the balance sheet date. The major fallacy in the auditor's procedure is that

    a. Checking of subsequent collections is not an acceptable alternative auditing procedure for confirmation of accounts receivable.

    b. By looking only at the deposit slip, the auditor would not know if the payment was for the receivable at the balance sheet date or a subsequent transaction.

    c. The deposit slip would not be received directly by the auditor, as a confirmation would be.

    d. A customer may not have made a payment during the two-month period.

10–60. Of the following, the most common argument *against* the use of negative accounts receivable confirmations is that
a. The cost-per-response is excessively high.
b. Statistical sampling techniques cannot be applied to selection of the sample.
c. Recipients are more likely to feel that the confirmation is a request for payment.
d. The implicit assumption that no response indicates agreement with the balance may not be warranted.

10–61. A CPA is engaged in the annual audit of a client for the year ended December 31, 19X9. The client took a complete physical inventory under the CPA's observation on December 15 and adjusted its inventory control account and detail perpetual inventory records to agree with the physical inventory. The client considers a sale to be made in the period that goods are shipped. Listed below are four items taken from the CPA's sales cutoff worksheet. Which item does *not* require an adjusting entry on the client's books?

Date (Month/Day)

	Shipped	Recorded as a Sale	Credited to Inventory Control
a.	12/10	12/19	12/12
b.	12/14	12/16	12/16
c.	12/31	1/2	12/31
d.	1/2	12/31	12/31

10–62. Which one of the following procedures would *not* be appropriate for an auditor in discharging his responsibilities concerning the client's physical inventories?
a. Confirmation of goods in the hands of public warehouses.
b. Supervising the taking of the annual physical inventory.
c. Obtaining written representation from the client as to the existence, quality, and dollar amount of the inventory.
d. Carrying out physical inventory procedures at an interim date.

10–63. The Smith Corporation uses prenumbered receiving reports which are released in numerical order from a locked box. For two days before the physical count, all receiving reports are stamped "before inventory," and for two days after the physical count all receiving reports are stamped "after inventory." The receiving department continues to receive goods after the cutoff time while the physical count is in process. The *least* efficient method for checking the accuracy of the cutoff is to
a. List the number of the last receiving report for items included in the physical inventory count.

    b. Observe that the receiving clerk is stamping the receiving reports properly.

    c. Test-trace receiving reports issued before the last receiving report to the physical items to see that they have been included in the physical count.

    d. Test-trace receiving reports issued after the last receiving report to the physical items to see that they have not been included in the physical count.

10–64. In connection with a review of the prepaid insurance account, which of the following procedures would generally *not* be performed by the auditor?

    a. Recompute the portion of the premium that expired during the year.

    b. Prepare excerpts of insurance policies for audit working papers.

    c. Examine support for premium payments.

    d. Confirm premium rates with an independent insurance broker.

10–65. Only one of the following four statements, which compare confirmation of accounts payable with suppliers and confirmation of accounts receivable with debtors, is true. The true statement is that

    a. Confirmation of accounts payable with suppliers is a more widely accepted auditing procedure than is confirmation of accounts receivable with debtors.

    b. Statistical sampling techniques are more widely accepted in the confirmation of accounts payable than in the confirmation of accounts receivable.

    c. As compared to the confirmation of accounts payable, the confirmation of accounts receivable will tend to emphasize accounts with zero balances at the balance sheet date.

    d. It is less likely that the confirmation request sent to the supplier will show the amount owed him than that the request sent to the debtor will show the amount due from him.

10–66. As part of his search for unrecorded liabilities, a CPA examines invoices and accounts payable vouchers. In general, this examination may be limited to

    a. Unpaid accounts payable vouchers and unvouchered invoices on hand at the balance sheet date.

    b. Accounts payable vouchers prepared during the subsequent period and unvouchered invoices received through the last day of fieldwork whose dollar values exceed reasonable amounts.

    c. Invoices received through the last day of fieldwork (whether or not accounts payable vouchers have been prepared) but must include all invoices of any amount received during this period.

    d. A reasonable period following the balance sheet date, normally the same period used for the cutoff bank statement.

10–67. If a client is using a voucher system, the auditor who is examining accounts payable records should obtain a schedule of all unpaid vouchers at the balance sheet date and

    a. Retrace voucher register items to the source indicated in the reference column of the register.

b. Vouch items in the voucher register and examine related canceled checks.
c. Confirm items on the schedule of unpaid vouchers and obtain satisfaction for all confirmation exceptions.
d. Compare the items on the schedule with open vouchers and uncanceled entries in the voucher register and account for unmatched items.

10–68. The audit procedures used to verify accrued liabilities differ from those employed for the verification of accounts payable because
a. Accrued liabilities usually pertain to services of a continuing nature, whereas accounts payable are the result of completed transactions.
b. Accrued liability balances are less material than accounts payable balances.
c. Evidence supporting accrued liabilities is non-existent, whereas evidence supporting accounts payable is readily available.
d. Accrued liabilities at year-end will become accounts payable during the following year.

10–69. When scheduling the audit work to be performed on an engagement, the auditor should consider confirming accounts receivable balances at an interim date if
a. Subsequent collections are to be reviewed.
b. Internal control over receivables is good.
c. Negative confirmations are to be used.
d. There is a simultaneous examination of cash and accounts receivable.

10–70. From which of the following evidence-gathering audit procedures would an auditor obtain most assurance concerning the existence of inventories?
a. Observation of physical inventory counts.
b. Written inventory representations from management.
c. Confirmation of inventories in a public warehouse.
d. Auditor's recomputation of inventory extensions.

10–71. On the last day of the fiscal year, the cash disbursements clerk drew a company check on Bank A and deposited the check in the company account in Bank B to cover a previous theft of cash. The disbursement has not been recorded. The auditor will best detect this form of kiting by
a. Comparing the detail of cash receipts as shown by the cash receipts records with the detail on the confirmed duplicate deposit tickets for three days prior to and subsquent to year-end.
b. Preparing from the cash disbursements book a summary of bank transfers for one week prior to and subsequent to year-end.
c. Examining the composition of deposits in both Bank A and Bank B subsequent to year-end.
d. Examining paid checks returned with the bank statement of the next accounting period after year-end.

10–72. The auditor obtains corroborating evidential matter for accounts receivable by using positive or negative confirmation requests. Under which of the following circumstances might the negative form of the accounts receivable confirmation be useful?

    a. A substantial number of accounts are in dispute.

    b. Internal control over accounts receivable is ineffective.

    c. Client records include a large number of relatively small balances.

    d. The auditor believes that recipients of the requests are unlikely to give them consideration.

Chapter 10
Discussion/Case Questions and Problems

10–73. In your audit of Ryan Co., for the year ended December 31, 19X8, you note that the bank reconciliation for the Third National Bank account contains a large unlocated difference, as shown below.

Balance per bank statement	$142,267
Deposit in Transit	3,864
Outstanding Checks	(40,793)
Unlocated Difference	10,846
Balance per general ledger	$116,184

From the bank statements (including the cutoff statement you received directly) and cash records, you determine the following:

    a. A deposit in the amount of $3,678 of Rain Co. was credited against the company's account in error in December.

    b. A check in payment of an advertising invoice cleared the bank in December in the amount of $10,318 that was recorded in the cash book at $1,318.

    c. Unrecorded bank service charges for December amounted to $25.00.

    d. Proceeds of a bank loan on December 1, 19X8, discounted for three months at 8%, had not been recorded by the company in the amount of $9,800.

    e. No entry had been made to record the return for NSF of a customer's check of $7,898.

    f. A deposit for the collection of accounts receivable was recorded as $21,079 while the actual deposit in the bank was $13,678.

g. A check for a salesman's expenses recorded in the cash disbursements books and shown on the outstanding checklist as $612 cleared with the cutoff bank statement and was noted to be in the amount of $216.

h. The company is required by an informal agreement with the bank to maintain a compensating balance of $10,000.

Required:

1. State the objectives for the audit of cash.

2. State the procedures you would consider using to accomplish the objectives (the final determination would depend on your evaluation of internal control, although it appears to be weak from some of the items noted above).

3. Prepare the adjusting entries and footnote disclosures necessary for a fair presentation of cash.

10–74. Assume that a CPA's client proposes to have an independent firm that specializes in inventory-taking count the merchandise rather than using their own employees. Under these conditions, would it be acceptable for the CPA to forego the inventory observation? Support your answer with reasons.

10–75. You are in charge of your second yearly examination of the financial statements of Hillsboro Equipment Corporation, a distributor of construction equipment. Hillsboro's equipment sales are either outright cash sales or a combination of substantial cash payment and one or two 60- or 90-day non-renewable interest-bearing notes for the balance. Title to the equipment passes to the customer when the initial cash payment is made. The notes, some of which are secured by the customer, are dated when the cash payment is made (the day the equipment is delivered). If the customer prefers to purchase the equipment under an installment payment plan, Hillsboro arranges for the customer to obtain such financing from a local bank.

You begin your fieldwork to examine the December 31 financial statements on January 5, knowing that you must leave temporarily for another engagement on January 7 after outlining the audit program for your assistant. Before leaving, you inquire about the assistant's progress in his examination of notes receivable. Among other things, he shows you a working paper listing the makers' names, the due dates, the interest rates, and amounts of 17 outstanding notes receivable totaling $100,000. The working paper contains the following notations:

1. Reviewed system of internal control and found it to be satisfactory.

2. Total of $100,000 agrees with general ledger control account.

3. Traced listing of notes to sales journal.

The assistant also informs you that he is preparing to request positive confirmation of the amounts of all outstanding notes receivable and that no audit work has been performed in the examination of notes receivable and interest arising from equipment sales. There were no outstanding accounts receivable for equipment sales at the end of the year.

Required:

a. List the additional audit procedures that the assistant should apply in his audit of the account for notes receivable arising from equipment sales (Hillsboro has no other notes). No subsidiary ledger is maintained.

b. You ask your assistant to examine all notes receivable on hand before you leave. He returns in 30 minutes from the office safe, where the notes are kept, and reports that notes on hand total only $75,000.

List the possible explanations that you would expect from the client for the $25,000 difference. (Eliminate fraud or misappropriation from your consideration.) Indicate beside each explanation the audit procedures you would apply to determine if each explanation is correct.

(AICPA adapted)

10–76. Renken Company cans two food commodities which it stores at various warehouses. The Company employs a perpetual inventory accounting system under which the finished goods inventory is charged with production and credited for sales at standard cost. The detail of the finished goods inventory is maintained on punched cards by the Tabulating Department in units and dollars for the various warehouses.

Company procedures call for the Accounting Department to receive copies of daily production reports and sales invoices. Units are then extended at standard cost and a summary of the day's activity is posted to the Finished Goods Inventory general ledger control account. Next, the sales invoices and production reports are sent to the Tabulating Department for processing. Every month the control account and detailed tab records are reconciled and adjustments recorded. The last reconciliation and adjustments were made at November 30, 19X9.

Your CPA firm observed the taking of the physical inventory at all locations on December 31, 19X9. The inventory count began at 4:00 P.M. and was completed at 8:00 P.M. The company's figure for the physical inventory is $331,400. The general ledger control account balance at December 31 was $373,900, and the final "tab" run of the inventory punched cards showed a total of $392,300.

Unit cost data for the company's two products are as follows:

Product	Standard Cost
A	$2.00
B	3.00

A review of December transactions disclosed the following:

1. Sales invoice #1301, 12/2/X9, was priced at standard cost for $11,700, but was listed on the Accounting Department's daily summary at $11,200.

2. A production report for $23,900, 12/15/X9, was processed twice in error by the Tabulating Department.

3. Sales invoice #1423, 12/9/X9, for 1,200 units of Product A, was priced at a standard cost of $1.50 per unit by the Accounting Department. The Tabulating Department noticed and corrected the error but did not notify the Accounting Department of the error.

4. A shipment of 3,400 units of Product A was invoiced by the Billing Department as 3,000 units on sales invoice #1504, 12/27/X9. The error was discovered by your review of transactions.

5. On December 27 the Memphis warehouse notified the Tabulating Department to remove 2,200 unsalable units of Product A from the finished goods inventory, which it did without receiving a special invoice from the Accounting Department. The Accounting Department received a copy of the Memphis warehouse notification on December 29 and made up a special invoice which was processed in the normal manner. The units were not included in the physical inventory.

6. A production report for the production on January 3 of 2,500 units of Product B was processed for the Omaha plant as of December 31.

7. A shipment of 300 units of Product B was made from the Portland warehouse of Ken's Markets, Inc. at 8:30 P.M. on December 31 as an emergency service. The sales invoice was processed as of December 31. The client prefers to treat the transaction as a sale in 19X9.

8. The working papers of the auditor observing the physical count at the Chicago warehouse revealed that 700 units of Product B were omitted from the client's physical count. The client concurred that the units were omitted in error.

9. A sales invoice for 600 units of Product A shipped from the Newark warehouse was mislaid and was not processed until January 5. The units involved were shipped on December 30.

10. The physical inventory of the St. Louis warehouse excluded 350 units of Product A that was marked "reserved." Upon investigation, it was ascertained that this merchandise was being stored as a convenience for Steve's Markets, Inc., a customer. This merchandise, which has not been recorded as a sale, is billed as it is shipped.

11. A shipment of 10,000 units of Product B was made on December 27 from the Newark warehouse to the Chicago warehouse. The shipment arrived on January 6 but had been excluded from the physical inventories.

Required:

Prepare a worksheet to reconcile the balances for the physical inventory, Finished Goods Inventory general ledger control account and Tabulating Department's detail of finished goods inventory ("tab run").

The following format is suggested for the worksheet.

	Physical Inventory	General Ledger Control Account	Tabulating Department's Detail of Inventory
Balance per client	$331,400	$373,900	$392,300

In addition, assume that you will audit all reconciling items. Indicate by a tick mark the audit work you performed for each item.

(AICPA adapted)

10–77. Late in December 19X9, your CPA firm accepted an audit engagement at Fine Jewelers, Inc., a corporation that deals largely in diamonds. The corporation has retail jewelry stores in several Eastern cities and a diamond wholesale store in New York City. The wholesale store also sets the diamonds in rings and in other quality jewelry.

The retail stores place orders for diamond jewelry with the wholesale store in New York City. A buyer employed by the wholesale store purchases diamonds in the New York diamond market, and the wholesale store then fills the orders from the retail stores and from independent customers and maintains a substantial inventory of diamonds. The Corporation values its inventory by the specific identification cost method.

Required:

Assume that at the inventory date you are satisfied that Fine Jewelers, Inc., has no items left by customers for repair or sale on consignment and that no inventory owned by the corporation is in the possession of outsiders.

a. Discuss the problems the auditor should anticipate in planning for the observation of the physical inventory on this engagement because of the
   (1) Different locations of inventories.
   (2) Nature of the inventory.

b. (1) Explain how your audit program for this inventory would be different from that used for most other inventories.
   (2) Prepare an audit program for the verification of the Corporation's diamond and diamond jewelry inventories, identifying any steps that you would apply only to the retail stores or to the wholesale store.

c. Assume that a shipment of diamond rings was in transit by corporation messenger from the wholesale store to a retail store on the inventory date. What additional audit steps would you take to satisfy yourself as to the gems that were in transit from the wholesale store on the inventory date?

(AICPA adapted)

10–78. Your audit client, Household Appliances, Inc., operates a retail store in the center of town. Because of lack of storage space, Household keeps inventory that is not on display in a public warehouse outside of town. The warehouseman receives inventory from suppliers and, on request from your client by a shipping advice or telephone call, delivers merchandise to customers or to the retail outlet.

The accounts are maintained at the retail store by a bookkeeper. Each month the warehouseman sends to the bookkeeper a quantity report indicating opening balance, receipts, deliveries and ending balance. The bookkeeper compares book quantities on hand at month-end with the warehouseman's report and adjusts his books to agree with the report. No physical counts of the merchandise at the warehouse were made by your client during the year.

You are now preparing for your examination of the current year's financial statements in this recurring engagement. Last year you rendered an unqualified opinion.

Required:
a. Prepare an audit program for the observation of the physical inventory of Household Appliances, Inc. (1) at the retail outlet, and (2) at the warehouse.
b. As part of your examination, would you verify inventory quantities at the warehouse by means of
   1. A warehouse confirmation? Why?
   2. Test counts of inventory at the warehouse? Why?
c. Since the bookkeeper adjusts books to quantities shown on the warehouseman's report each month, what significance would you attach to the year-end adjustments if they were substantial? Discuss.

(AICPA adapted)

10–79. You have just commenced your examination of the financial statements of Vickey Corporation for the year ended December 31, 19X9. Analyses of the company's Unexpired Insurance and Insurance Expense accounts follow.

Vickey Corporation
WORKSHEET FOR DISTRIBUTION OF INSURANCE
For Year Ended December 31, 19X9

Date (19X9)			Folio	Amount Debit	Amount Credit
		Unexpired Insurance			
January	1	Balance forward		$ 5,550	
	10	Premium on president's policy	CD	1,240	
	14	Deposit on workman's compensation policy for 19X9	CD	2,750	
	31	Monthly amortization	JE		$410
April	1	Down payment on fire policy (April 1, 19X9 to April 1, 19Y4)	CD	1,000	
		Totals		$10,540	$410
		Insurance Expense			
January	10	Trip insurance on officers (Inspection tour of dealers in December 19X8)	CD	$ 170	
	31	Monthly amortization	JE	410	
February	21	Balance on workman's compensation policy (Per payroll audit for policy year ending December 31, 19X8) .	CD	250	
April	10	Automobile collision policy (Policy year April 1, 19X9 to April 1, 19Y0)	CD	2,500	
June	10	Increase in fire policy (May 1, 19X9 to April 1, 19Y4)	CD	590	
August	10	Fleet public liability and property damage policy (September 1, 19X9 to September 1, 19Y0)	CD	3,780	
	17	Check from insurance company for reduction in auto collision rate for entire policy year	CR		$120
October	1	Fire policy payment	CD	1,300	
	19	Cost of repair to automobile damaged in a collision	CD	400	
		Totals		$ 9,400	$120

Your examination also disclosed the following information:

1. Only one policy of those prepaid at January 1, 19X9 remained in force on December 31, and it will expire on March 31, 19Y0. The policy was a 24-month policy, and the total premium was $600.

2. Cash value of the life insurance policy on the life of the president increased from $1,110 to $1,660 during 19X9. The Corporation is the beneficiary on the policy.

3. The Corporation signed a note payable to an insurance company for the balance due on the fire insurance policy which was effective as of April 1. The note called for nine additional $1,000 semi-annual payments plus interest at 6% per annum on the unpaid balance (also paid semi-annually).

4. An accrual dated December 31, 19X8 for $170 for insurance payable was included among accrued liabilities.

5. Included in miscellaneous income was a credit dated April 10 for $100 for a 4% dividend on the renewal of the automobile collision insurance policy. The insurance company is a mutual company. Also included in miscellaneous income was a credit dated November 2 for $350 for a check from the same insurance company for a claim filed October 19.

6. An invoice dated November 15 for $1,560 for employee fidelity bonds from November 15, 19X9 to November 15, 19Y0 was not paid or recorded.

7. An invoice dated January 13, 19Y0 for $2,800 for the 19Y0 workman's compensation policy was not recorded. The net amount of the invoice was $2,660 after a credit of $140 from the payroll audit for the year ended December 31, 19X9.

Required:
Prepare a worksheet to properly distribute all amounts related to insurance for 19X9. The books have not been closed for the year. The worksheet should provide columns to show the distribution to unexpired insurance, to insurance expense, and to other accounts. The names of other accounts affected should be indicated. Formal journal entries are not required.

(AICPA adapted)

10–80. Mincin, CPA, is the auditor of the Raleigh Corporation. Mincin is considering the audit work to be performed in the accounts payable area for the current year's engagement.

The prior year's working papers show that confirmation requests were mailed to 100 of Raleigh's 1,000 suppliers. The selected suppliers were based on Mincin's sample, which was designed to select accounts with large dollar balances. A substantial number of hours were spent by Raleigh and Mincin resolving relatively minor differences between the confirmation replies and Raleigh's accounting records. Alternate audit procedures were used for those suppliers who did not respond to the confirmation requests.

Required:
a. Identify the accounts payable audit objectives that Mincin must consider in determining the audit procedures to be followed.

b. Identify situations when Mincin should use accounts payable confirmations and discuss whether Mincin is required to use them.

c. Discuss why the use of large dollar balances as the basis for selecting accounts payable for confirmation might not be the most efficient approach and indicate what more efficient procedures could be followed when selecting accounts payable for confirmation.

(AICPA adapted)

10–81. Arthur, CPA, is auditing the RCT Manufacturing Company as of February 28, 19X9. One of Arthur's initial procedures is to make overall checks of the client's financial data by reviewing significant ratios and trends so that he has a better understanding of the business and can determine where to concentrate his audit efforts.

The financial statements prepared by the client with audited 19X8 figures and preliminary 19X9 figures are presented below in condensed form.

RCT Manufacturing Company
CONDENSED BALANCE SHEETS
February 28, 19X9 and 19X8

Assets	19X9	19X8
Cash	$ 12,000	$ 15,000
Accounts receivable, net	93,000	50,000
Inventory	72,000	67,000
Other current assets	5,000	6,000
Plant and equipment, net of depreciation	60,000	80,000
	$242,000	$218,000

Equities		
Accounts payable	$ 38,000	$ 41,000
Federal income tax payable	30,000	14,400
Long-term liabilities	20,000	40,000
Common stock	70,000	70,000
Retained earnings	84,000	52,600
	$242,000	$218,000

RCT Manufacturing Company
CONDENSED INCOME STATEMENTS
Years Ended February 28, 19X9 and 19X8

	19X9	19X8
Net sales	$1,684,000	$1,250,000
Cost of goods sold	927,000	710,000
Gross margin on sales	$ 757,000	$ 540,000
Selling and administrative expenses	682,000	504,000
Income before federal income taxes	$ 75,000	$ 36,000
Income tax expense	30,000	14,400
Net income	$ 45,000	$ 21,600

Additional information:

The company has only an insignificant amount of cash sales.

The end-of-year figures are comparable to the average for each respective year.

Required:

a. For each year compute the current ratio and a turnover ratio for accounts receivable.

b. On the basis of these ratios, identify and discuss audit procedures that should be included in Arthur's audit of (1) accounts receivable, and (2) accounts payable.

(AICPA adapted)

10–82. During the initial engagement to examine the financial statements of the Graham Company, a medium-sized manufacturing company, you are provided with the following:

The client maintains five bank accounts: three for payrolls on an imprest basis and two operating bank accounts, one of which is for supplies and the other for miscellaneous transactions.

All payments to suppliers are made through the suppliers' operating account. All other transactions are handled through the miscellaneous operating account.

One bank account was closed during the year. A new account, to be used for the same purpose, was opened in another bank more conveniently located to the client's premises.

The bank reconciliation was prepared by the client's staff as of December 31, 19X9, together with listings of the deposits in transit and outstanding checks, which have pertinent details noted thereon.

Suppliers' operating account

Balance per bank statement	$ 6,516.12
Add: outstanding deposits	248,868.19
	$255,384.31
Less: outstanding checks	356,079.27
Overdraft per books	($100,694.96)

Included among the outstanding checks were the following payable to other than the regular trade suppliers:

Date				
11/30/X8	#2122	City of Newark	$ 7,500.00	(Deposit)
1/ 5/X9	#2455	B. F. Goodrich	16,172.00	(1)
6/30/X9	#7798	Mr. Napoleon Deschamps	267.00	(Employee travel advance)
12/15/X9	#17140	The Grand Trust Company	11,866.20	(Dividend disbursing agent)
12/29/X9	#17418	The Graham Company	26,646.09	(Payroll)

(1)Claimed to be a duplicate payment, on which a stop payment had been placed at the bank.

Upon arrival at the client's office, January 6, 19Y0, it was determined that checks totaling $73,515.25 had not been mailed to the respective suppliers.

The bank confirmed the following information upon your request:

Bank balance 12/31/X9:	
Suppliers' operating account	$6,516.12
Miscellaneous operating account	$ 333.74

Deposits between 12/28/X9 and 1/5/Y0:

		Suppliers' Operating	Misc. Operating
12/28/X9		$ 549.81	$330.59
		8,854.84	
		19,330.14	
		65,830.17	
12/29/X9		652.26	
		148,862.49	
1/ 2/Y0	CM	25,000.00	
1/ 3/Y0		211.36	
		32,680.69	
		216,187.50	
1/ 4/Y0		75,000.00	

Signing officers—six persons with no identification as to position. (The client had informed you that there were seven authorized signing officers.)

Bank loan of $3,100,000 guaranteed by parent company, 7¼% interest.

> *Required:*
> a. What procedures will you follow to audit the client's bank reconciliation of the suppliers' operating account?
> b. What additional procedures would be necessary to audit the total cash in banks at December 31, 19X9?
> c. List any recommendations that you feel should be brought to the attention of management.

> *(Used with permission of Ernst & Ernst)*

10–83. Your firm has been engaged to examine the financial statements of RST Inc. for the year ending December 31. RST Inc. is a medium-sized manufacturing company. The Company has approximately 400 open trade accounts receivable and does not prepare monthly statements. The manager assigned to the engagement has decided to circularize the trade accounts receivable as of September 30 (three months before year-end). The senior on the job asks you to be at the company on Wednesday morning, October 1, to mail requests for confirmation. He tells you to ask the company's personnel to prepare 25 positive confirmation requests and 100 negative confirmation requests. He further asks you to obtain an aged trial balance as of September 30, to trace the balances of the open accounts to the trial balance from the subsidiary ledgers, to test the aging, to foot the trial balance, and to compare the total of the trial balance with the accounts receivable control account in the general ledger. The senior also informs you that detailed tests of the sales and credit journals will be made for the month of September.

Required:

a. Enumerate the types of accounts you would want to include in your selection of accounts to be circularized by the positive method.

b. Enumerate the types of accounts you would want to include in your selection of accounts to be circularized by the negative method.

c. Outline a plan for maintaining adequate control over confirmation requests.

d. Outline the additional audit steps that should be undertaken at December 31, in support of the amounts shown as accounts receivable; the Company is preparing for your use an aged trial balance of accounts receivable as of that date.

(Used with permission of Ernst & Ernst)

10–84. You have been assigned to the audit of a medium-sized manufacturer of machine parts whose fiscal year ends October 31. You and the senior arrive at the same time on Monday, November 13, to start the fieldwork for the completion of the audit. The senior introduces you to the controller. The controller gives the senior a package of working papers which he has had prepared for your use in connection with the audit. The senior gives you a copy of the accounts receivable aging schedule prepared by the company, the audit program, the internal control questionnaire, and a folder containing all papers in connection with confirmation of receivables mailed on November 2. The folder contains the following information:

Adding-machine tape of accounts receivable at October 31.

A list showing the name, address, and balance of twelve customers to whom positive requests for confirmation were mailed.

Four customers' statements marked across the face "Do not mail."

Five positive requests that have been returned by customers confirming the balances as being correct.

Eight negative confirmations that have been returned with notations made thereon by customers.

Two positive and one negative requests returned by the post office marked "unknown" or with a similar designation.

The senior introduces you to the credit manager, the accounts receivable bookkeeper, and the billing clerk. He then instructs you to proceed with the tests of the aging schedule and completion of the audit work on accounts receivable, as outlined in the audit program, and advises that he will return the next day to answer any questions you have with respect to the accounts receivable and to review any items you think should be discussed with him.

AGED TRIAL BALANCE—ACCOUNTS RECEIVABLE
October 31

| | Balance | | | | | May– | |
	Dr.	Cr.	Oct.	Sept.	Aug.	June–July	Prior
Allied Products Co.	$ 12,618.32				$12,618.32		
American Manufacturing Co.		$ 612.00					
B & D Machinery, Inc.	57,538.79		$ 35,123.76	$10,078.12	6,312.45	$ 6,024.46	
Best Equipment, Inc.	1,098.45		198.45				$ 900.00
Cooper, Frank M. (Employee)	5,000.00					5,000.00	
Chalmers Motors, Inc.	7,445.83			1,263.17	2,376.28	3,806.38	
Davidson Engineering Company	3,573.35					3,573.35	
Drake Press Division	78,396.21		78,396.21				
Erie Machine Works, Inc.	6,215.63						6,215.63
Evans and Co. (Deposit)	10,000.00		10,000.00				
Franklin Motors, Inc.	17,624.91		11,784.16	5,840.75			
Franklin Motors, Inc. (Note)	25,000.00				25,000.00		
Globe Machinery, Inc.	30,248.65		25,932.40				4,316.25
Globe Machinery, Inc. (Consignment)	103,487.98		50,116.73	28,745.27	12,678.93	8,319.00	3,628.05
Hornblower & Weeks—Hemphill Noyes	14,750.00		14,750.00				
Watkins Company	4,728.16		4,728.16				
Whitman, Inc.	16,512.54		16,512.54				
Young Machinery Co.	8,378.05		6,873.00	1,505.05			
	$495,444.00	$5,612.00	$328,577.92	$53,932.36	$66,768.91	$31,104.88	$15,059.93

The senior informs you that the adding-machine tape was prepared by the accounts receivable bookkeeper for your use in sending out the confirmations. Your representative checked the customers' statements against the tape and the accounts receivable ledger cards, but did not check the total by re-adding the tape. However, you note that the total at the bottom of the tape does agree with the total shown on the aging schedule prepared by the company. You are told that either positive or negative requests were sent to all of the 50 accounts.

The aging schedule is shown on page 471. You are to assume that the internal control is good. You should study the aging schedule and answer the questions which follow. The normal credit terms are net 30.

Required:

a. What auditing procedures should the audit program call for with respect to the aging schedule?
b. Which items on the aging schedule would you select for additional auditing procedures, why would you select them, and what procedures would you use?
c. What would you do with the adding-machine tape prepared at the time the confirmations were mailed?
d. What would you do with the statements marked "Do not mail?"
e. Which items would you select for discussion with the senior?
f. What would you do with the five positive requests that were returned indicating no exceptions?
g. In examining these positive requests, what would you look for?
h. What would you do with the requests returned by the post office marked "Unknown" or with a similar designation?
i. What would you do with the negative requests returned with notations made thereon by customers?

(Used with permission of Ernst & Ernst)

10–85. In order to check a company's handling of the cutoff sales and purchases at the close of the fiscal year ended December 31, 19X8, you have compiled the data listed on the following schedule. All sales and purchases of significant amount from December 26, 19X8, to January 4, 19X9, inclusive, are included.
The company realized a gross profit of 40% on each sale, and all sales and purchases were recorded as of the invoice dates. Items marked "B" were F.O.B. destination; all other items were F.O.B. shipping point.

The physical inventory taken by the company included only those items actually on hand as of the close of business December 31, 19X8. All items on hand were included except a special machine—see "A." This special machine was made to order by the company's supplier for one of the company's customers, and was in the shipping room ready for shipment. It was excluded from the physical inventory.

INVENTORY CUTOFF TEST
ACS COMPANY
12-31-71

Prepared _____ Initials _____ Date _____

Approved _____

Description	Selling Price	Invoice Date	Shipping Date	Receiving Date	Accounts Receivable Dr/<Cr>	Sales Dr/<Cr>	Inventory Dr/<Cr>	Cost of Sales Dr/<Cr>	Accounts Payable Dr/<Cr>
Sales:									
a	6000 –	12-28-71	12-27-71						
b	8000 –	12-28-71	12-29-71						
c	5000 –A	12-29-71	1-3-72						
d	6000 –	12-31-71	1-3-72						
e	7000 –	12-31-71	12-31-71						
f	9000 –B	12-31-71	12-31-71						
g	4000 –	1-3-72	1-4-72						
h	5000 –	1-3-72	12-31-71						
i	6000 –	1-4-72	1-5-72						
Purchases:									
a	5000 –	12-27-71	12-27-71	12-31-71					
b	7000 –B	12-28-71	12-31-71	1-3-72					
c	8000 –	12-31-71	12-31-71	1-3-72					
d	7000 –	12-31-71	1-3-72	1-4-72					
e	6000 –	1-3-72	12-31-71	12-31-71					
f	9000 –	1-3-72	12-31-71	1-3-72					
g	7000 –B	1-4-72	12-31-71	1-3-72					
h	4000 –	1-4-72	1-5-72	1-6-72					

The company maintains a perpetual inventory system. The inventory account and the subsidiary records have been adjusted to the physical inventory.

Complete the schedule on page 473 by showing for each item the required adjustment, if any.

(Used with permission of Ernst & Ernst)

10–86. The following are situations or questions pertaining to the audit of accounts payable:

a. With regard to statements requested from vendors, the auditor's memo states, "We requested statements from all vendors with balances of over $2,000 as shown by the trial balance." Do you feel this procedure is satisfactory? Give reasons for your answer.

b. Why should an auditor be particularly careful to investigate past due accounts payable?

c. Discuss some of the sources from which the auditor can prepare a list of vendors from whom statements should be requested.

d. All vendors' statements on hand and received by the auditor have been reconciled by the company at the auditor's request. How much checking of the reconcilements should be done?

e. In connection with the year-end audit of trade accounts payable, the following are noted; describe briefly what each could indicate and what procedures you would follow in the circumstances:

(1) Several long-standing credit balances.

(2) A number of debit balances included in accounts payable.

(3) A substantial dollar amount of purchase commitments in a situation of declining market prices.

(Used with permission of Ernst & Ernst)

10–87. Your firm has been engaged to examine the financial statements of Brown Appliances, Inc., for the year ended December 31. The company manufactures major appliances sold to the general public through dealers and distributors. Significant financial information of the company, as of December 31, is as follows:

Trade receivables	$12,000,000
Inventories	14,000,000
Property—net	6,000,000
Other assets	3,000,000
	$35,000,000
Trade accounts payable	$ 4,000,000
Other liabilities	6,000,000
Stockholders' equity	25,000,000
	$35,000,000
Net sales for the year	$74,000,000
Net income for the year	3,000,000

You are to audit the trade accounts payable of a division of Brown Appliances, Inc. The trade accounts payable aggregate $2,500,000.

Excerpts from the internal control memorandum follow:

Invoices from suppliers are received in the purchasing department, where they are matched with receiving reports and checked to the applicable purchase order for quantities and pricing. Invoices and receiving reports are then forwarded to the accounting department for clerical checking and final approval for payment.

On the payment date, invoices with attached receiving reports are separated into two groups: one group of invoices with prior-month distribution charges, the other group with current-month distribution charges. The check register is then prepared, with each group having a separate total and check number sequence. The accounts payable for monthly financial statement purposes is the total of the check register for invoices with prior-month distribution charges. A voucher register is not maintained.

The purchasing department holds unmatched receiving reports and unmatched invoices.

Cutoff procedures as established by the company appear adequate; however, the company makes it a practice not to record inventory in transit.

Vendors' statements received by the company are forwarded to a clerk in the accounting department. He does not check all charges appearing on the vendors' statements, but does reconcile all old outstanding charges appearing thereon.

An accounts payable listing has been prepared by the company for the auditors. As explained above, this listing was prepared from the check register of December charges paid in January, and shows vendor, check number, invoice date, date paid, and amount. A quick review of the listing reveals the following:

January-dated invoices amounting to $200,000 appear on the listing payable to Talley and Parks Advertising Agency, for advertising to appear in *Better Homes and Gardens* magazine in February and March. This was included in the year-end accounts payable listing at the request of the vice-president of advertising because he wanted to more closely match advertising department budgeted expense with actual expenditures for the year. The distribution was made to advertising expense.

Amounts appear on the listing as payments for payrolls, payroll taxes, other taxes, profit-sharing plans, etc.

No amounts appear on the listing for legal or accounting services.

Required:
Discuss the problems and procedures involved in auditing this company's accounts payable. Specifically discuss the auditing procedures you recommend be used in your examination and the adjustments you recommend be made to the accounts payable listing.

(Used with permission of Ernst & Ernst)

Chapter 11

Auditing the Capital Asset and Financing Base Cell

Learning Objectives   *After reading and studying the material in this chapter, the student should*

Know the characteristics of the accounts comprising the capital asset and financing base cell.

Understand the important features of internal control associated with each account in the capital asset and financing base cell and how they are tested.

Know the audit objectives of each account in the capital asset and financing base cell.

Be able to determine and apply the audit procedures necessary to accomplish the audit objectives for each account in the capital asset and financing base cell.

Know how to prepare the audit working papers to document the audit procedures for each account in the capital asset and financing base cell.

The approaches to capital asset and financing base accounts are similar in that the beginning balance is established first (normally from the prior-year audit), and then the current-year transactions are audited. The audit of these accounts differs from the audit of working capital accounts, where the auditor anticipates an almost complete turnover of the amounts included in the balance at the end of the year as compared with those at the beginning of the year. Items such as property and equipment, long-term debt, and equity tend to have long lives, and once they have been established as properly recorded amounts, the auditor's concern is with the accounting for the ultimate realization or disposition of these items. Thus, if the auditor is satisfied with the beginning balance and current transactions, he generally is satisfied with the ending balance.

Property and Equipment and Accumulated Depreciation

The property and equipment accounts represent the cost of a company's productive facilities, which may be depreciable (costs of buildings and machinery), depletable (costs of natural resources such as oil and gas), amortizable (cost of leasehold improvements), or fixed (such as land). Leased property and equipment also are included in the balance sheet if they are capitalizable under generally accepted accounting principles. In certain rare situations, such as in a quasi-reorganization, property and equipment are stated on a basis other than cost. AC Section 4072.01 *(APB Opinion No. 6)*, states, however, ". . . that property, plant and equipment should not be written up by an entity to reflect appraisal, market or current values which are above cost to the entity."

Cost is represented by cash or the value of other consideration given in the event of purchase. The determination of purchase cost can be complicated if stock or other securities with no ready market are used. Cost also may be represented by the sum of the direct costs of labor and material in the event of construction (an allocation of overhead may be applicable in some cases). Certain industries such as public utilities that are permitted by regulatory agencies to earn a fixed rate of return include as a cost of property and equipment an allowance for the use of funds during construction.

Even after a unit of property is acquired by purchase or construction, a continuing determination of its cost is necessary, because of the need to distinguish subsequent expenditures that represent maintenance or repair of it. Additions, betterments, and improvements to a unit of property which are capitalized should result either in an extension of its original life (the conversion of a furnace to coal when gas becomes unavailable) or in new or improved operations (the re-gearing of an electric motor to allow it to operate at high speeds), and often result in the retirement (removal from the account) of the converted or replaced part. Maintenance and repairs should be charged to expense as incurred, and include the costs of painting, cleaning, and overhauling, as well as the costs of repairing breakdowns. Although painting and cleaning prolong the life of an asset, such costs usually are taken into consideration in the establishment of its original life and, therefore, should not be capitalized.

Depreciation (which includes depletion and amortization in this discussion) also is required by generally accepted accounting principles (except for certain non-profit organizations such as governmental units). AC Section 4073.05 *(ARB No. 43)*, provides:

The cost of a productive facility is one of the costs of the services it renders during its useful economic life. Generally accepted accounting principles require that this cost be spread over the expected useful life of the facility in such a way as to allocate it as equitably as possible to the periods during which services are obtained from the use of the facility. This procedure is known as depreciation accounting, a system of accounting which aims to distribute the cost or other basic value of tangible assets, less salvage (if any), over the estimated useful life of the unit (which may be a group of assets) in a systematic and rational manner. It is a process of allocation, not of valuation.

Depreciation can be computed by any method that is "systematic and rational." The straight-line, declining-balance, sum-of-the-years'-digits, and unit-of-production methods are the most commonly used.

Because of the low volume of activity and the long-term nature of the assets, a company's detail accounting records for property and equipment often are not as elaborate or as carefully maintained as those for other accounts such as cash, accounts receivable, or inventories. Property and equipment records normally should include a detail property ledger and, if construction activity is significant, a construction work order system. These records are important to the company's determination and the auditor's examination of the cost of retired or abandoned assets. Formal policies regarding approval of capital expenditures, distinction between capital and maintenance expenditures, and the reporting and recording of asset retirements are significant aspects of internal control in the property section.

Property and equipment often constitute the most significant asset in the balance sheet, and depreciation is usually a material amount in the income statement. Therefore, certain minimum disclosures are required with regard to these accounts. AC Section 2043.02 *(APB Opinion No. 12)* provides:

> Because of the significant effects on financial position and results of operations of the depreciation method or methods used, the following disclosures should be made in the financial statements or in notes thereto:
>
> a. Depreciation expense for the period,
>
> b. Balances of major classes of depreciable assets, by nature or function, at the balance-sheet date,
>
> c. Accumulated depreciation, either by major classes of depreciable assets or in total, at the balance-sheet date, and
>
> d. A general description of the method or methods used in computing depreciation with respect to major classes of depreciable assets.

Audit Objectives

The objectives in the audit of property and equipment and related allowances for depreciation are:

1. To determine that the amount shown as property and equipment in the financial statements represents physical facilities owned by the company and associated with the productive process, and that all retired or abandoned property is removed properly from the accounts.

2. To determine that property and equipment are stated at cost that is properly capitalizable and that depreciation expense is adequate and is computed in accordance with generally accepted accounting principles.

3. To determine that the accumulated depreciation is reasonable in relation to the expected useful life of the property.

4. To determine that the disclosures concerning property are adequate and in accordance with generally accepted accounting principles.

Audit Procedures and Working Papers

The procedures necessary to satisfy the auditor depend on his evaluation of the client's accounting procedures and internal controls in the property section, as well as in related areas such as purchasing and cash disbursements (illustrated in Chapter 5). The cash disbursements section of the appendix to Chapter 5 indicates that supporting documents are examined at the time accounts payable vouchers are prepared. Given this procedure, therefore, the auditors might adjust the tests of property purchases.

Tests of procedures for distinguishing between capital and expense transactions and computing depreciation are dual-purpose tests; they will be explained later in this section.

The lead schedule for property and equipment and allowance for depreciation normally summarizes the activity in the major accounts for the year and relates the accumulated depreciation to the applicable asset account (Figure 11.1).

Audit of Property Additions—The audit procedures for property additions are designed to determine that the additions represent physical facilities, are stated at cost, and are properly capitalizable. The auditor must analyze additions in greater detail than that shown on the lead schedule. An example of such an analysis, often referred to as a vouching schedule, is shown in Figure 11.2. In this example, the auditor has chosen as his sample all additions to equipment in excess of $2,000, although other sampling plans also could be used. The audit procedure indicated by ✔ includes (1) examination of vendor invoice and canceled check, indicating that the item was owned by the company and recorded at cost, (2) examination of receiving report, indicating that a physical item was received, and (3) review of the capitalization decision based on the description of the item. Additional evidence of the physical existence of the item is gained from an inspection of it during the inventory observation. The auditor also has noted by ① and ② whether each addition was a new or replacement item. This information is very useful in auditing property retirements.

F1

J. Jones
1-24-X9

X Co.
Property and Equipment and Allowance for Depreciation Lead Schedule
12-31-X8

Depr. Rate	Description	Property and Equipment				Allowance for Depreciation			
		Balance 12-31-X7	Additions	Retirements	Balance 12-31-X8	Balance 12-31-X7	Provision	Retirements	Balance 12-31-X8
—	Land	55041			55041 T				
5%	Buildings	287869	F1 19013	<10854>	296028 T	21451	13037	<10319>	24169 T
10%	Equipment	226526	F2 24757	<9678>	241605 T	61703	21065	<8314>	74393 T
33⅓%	Automobiles	78314		<18314>	18314 T	58313	4101		9914 T
		581150	43190	<20532>	604988 T	89266	38203	<18153>	108716 I
		TB-1	TB-1	E-3	TB-1	TB-1	TB-1		TB-1

T   Columns as now footed.

Figure 11.1 Property and Equipment and Allowance for Depreciation Lead Schedule

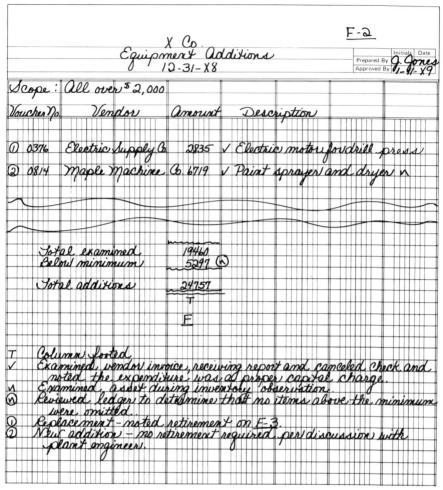

Figure 11.2 Equipment Additions

As a test for property units that may have been charged to expense, an analysis similar to that described above is made of the maintenance expense account, and the work indicated by ✔ is performed. Because the review of the capitalization/expense decision is a significant aspect of this work, it often is performed simultaneously with the work on property additions.

Audit of Property Retirements—In the audit of property retirements, the auditor wants to determine that the recorded retirements are shown properly and, what is even more important, that there are no significant unrecorded property retirements. An illustration of some audit procedures that might be applied to property retirements is shown in Figure 11.3.

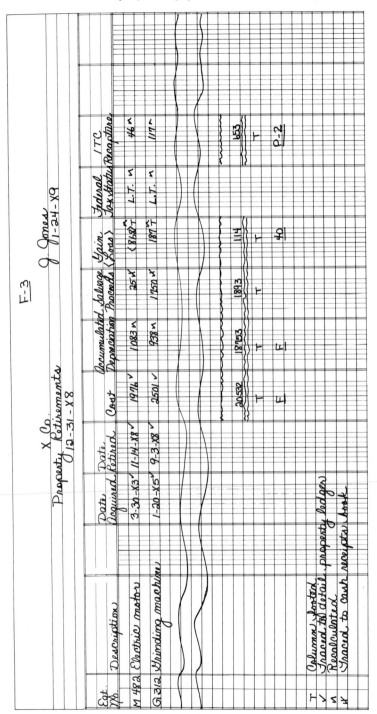

Figure 11.3 Property Retirements

The recorded retirements are tested by tracing the original cost of the item, together with the dates acquired and retired, to the detail property ledger (indicated by ✔). Accumulated depreciation, if not recorded separately for each item, must be recomputed on the basis of the depreciation rate applicable to such assets and the periods they were held (indicated by η). Salvage proceeds can be traced to the cash receipts book or deposit slip (indicated by ✕). With this information, the resulting net gain or loss from property retirements can be recomputed and related to an income or expense account (see the reference to the *40* schedule). The remaining data relate to the federal income tax effect of the transactions and are used by the auditor in his audit of the income tax liability and expense accounts.

The effectiveness of the auditor's search for unrecorded property retirements depends on his business knowledge and vigilance. As noted in the prepayments section, a reduction of property insurance coverage may be due to property retirements. The auditor should investigate any significant reduction in coverage to determine whether it was due to property retirements and, if so, whether they were recorded properly. This example shows how an auditor must be aware of the impact of transactions in one section of the audit upon balances and transactions in other sections. Also, in the work on property additions, the auditor noted whether major additions represented new or replacement items (see ① and ② in Figure 11.2). Because a replacement implies that an old asset was retired when the new asset was placed in service, the auditor can use this information to determine that the retirement was recorded in the accounting records.

Other procedures used in the search for unrecorded retirements include a review of the miscellaneous or other income account for salvage credits and scrap sales, inquiry of operating and management personnel, and a plant tour. Proceeds from the sale of scrap or used equipment may indicate unrecorded retirements and should be investigated if material. In this connection, the auditor should consider the controls over the accumulation and sale of scrap. Lack of controls in this area (which is not unusual) can result not only in unrecorded property retirements, but also in misappropriation of proceeds from the sale of scrap and salvaged items.

Inquiry of operating and management personnel should include discussion not only of actual retirements made during the year, but also of assets on hand which are no longer being used. For example, a decision to discontinue a product line may result in the disposal of many pieces of equipment used to produce that product line. Because the equipment is still on hand, it should not be recorded as a retirement in the accounting records, but it should be reduced in value (after considering the related allowance for depreciation) to the estimated amount to be realized on its disposition. It also may be appropriate to reclassify the equipment to other assets, because it no longer will be used in production.

The plant tour normally should be made early in the audit, because it provides the auditor with a knowledge of the production process that will be important in other sections of the audit, as well as in the search for unrecorded retirements. It sometimes is coordinated with the physical inventory observation.

To be an effective procedure in the search for unrecorded retirements, the plant tour must be planned properly. To stroll through a plant inspecting certain items of equipment provides no evidence about unrecorded retirements (although it may be useful in verifying the physical existence of additions). During the search for unrecorded retirements, the auditor is interested in the pieces of equipment that are *not* in the plant, but still are recorded in the accounting records. To make the test effective, the auditor must select certain items of equipment from the accounting records and then locate and identify them during the plant tour. Any items that cannot be located would represent unrecorded retirements and should be removed from the accounting records. Because of the nature of the audit work performed in the search for unrecorded property retirements, the auditor normally documents it in a memorandum.

Audit of Depreciation—An auditor can become so involved in the detail calculations of depreciation expense that he loses overall perspective of the account. He must remember that depreciation expense results from the allocation of the cost of assets over an *estimated* time period by an *arbitrary* method, and that calculation and verification of this allocation to the nearest dollar are unrealistic. Usually, groups of assets with similar lives and the same depreciation method, such as buildings, equipment, etc., can be tested on an overall basis. For example, if equipment is depreciable on a straight-line basis over 10 years with a 10% salvage value, the auditor would average the beginning and ending balances of the asset account, reduce it by 10%, and multiply the remaining balance by 10%. This figure seldom is exactly equal to the client's depreciation expense because of the uneven rate of additions and retirements during the year, fully depreciated assets, and other factors, but it gives the auditor an idea of the reasonableness of the calculation. If the overall calculation is materially different from the client's amount, detail tests of individual items may be necessary.

The auditor also must be satisfied that all depreciation calculations were made on a basis consistent with the calculation of the prior year. He must review not only the obvious factors, such as the depreciation rates and methods, but also other factors, such as estimated salvage values and methods of considering current-year property additions and retirements. For example, a change in computing depreciation on current-year additions from a standard one-half year to actual number of months in service might not be immediately obvious, but could have a material effect on depreciation expense if large additions were made near the beginning or end of a year.

In addition to tests of depreciation expense for the year, the auditor must evaluate the adequacy of the balance of accumulated depreciation. The auditor is not an appraiser, but depreciation is not a valuation process. It is a process of allocation, and the auditor can use his business judgment to determine the reasonableness of the allocation of the remaining cost of an asset over its remaining useful life. In considering the reasonableness of the remaining useful life of an asset, the auditor must take into account not only its physical condition, but technological innovations and other factors as well. For example, a remaining estimated life of 15 years for a black and white TV camera owned by a TV station would be unreasonable if the station had switched to or planned to switch to full color. Also, an estimated life of 30 years for a natural gas pipeline might be unreasonable if the natural gas reserves amounted to only a 10-year supply (consideration would be given to possible future discoveries). Such examples show the importance of the business approach in the audit.

Property Disclosures—Certain disclosures required by generally accepted accounting principles as they relate to property and equipment and accumulated depreciation have been discussed. In addition to these, disclosure must be made of any mortgages or liens. Many, although not all, mortgages arise in the initial acquisition of an item of property. In the audit of property additions, the auditor must be alert for mortgages associated with any non-cash acquisitions of property. Mortgages and liens also can be detected by review of notes, bonds, and loan agreements, confirmation with financial institutions, review of minutes, and inquiry of management, all of which have been discussed in other sections.

Long-Term Investments and Intangibles

Long-term investments can take many forms. They may be debt or equity, marketable or non-marketable, affiliated or non-affiliated, etc. The auditor must depend on his knowledge of generally accepted accounting principles to determine the appropriate valuations and disclosures in each case. Long-term investments may not be significant for a manufacturing company, but may constitute the most significant asset of an investment company.

The physical evidence of long-term investments usually takes the form of stock or bond certificates (other forms include joint venture agreements, real estate deeds, and mineral leases) with evidence of ownership indicated either on the certificate (registered form) or by possession (bearer forms). Certificates may be held by the company owning them, or by a bank or other financial institution as custodian for the company. If held by the company, the certificates should be stored in a bank safe deposit box or similar location. Internal control is improved if dual access (the requirement that two individuals be present for access) is required for entry to the box.

Long-term investments may be valued on the basis of cost, underlying equity, lower of cost or market, or market value. For example, AC Section 5132 *(FASB Statement No. 12)* provides, among other things:

> The carrying amount for a marketable equity securities portfolio shall be the lower of its aggregate cost or market value, determined at the balance sheet date. . . . Marketable equity securities owned by an entity shall, in the case of a classified balance sheet, be grouped into separate portfolios according to the current classification of the securities for the purpose of comparing aggregate cost and market value to determine carrying amount.

General disclosures relating to long-term investments include the type of investment and basis of valuation; additional disclosures apply in many cases.

Intangibles include goodwill, franchise fees, patents, organizational costs, etc. The existence and ownership of goodwill subsequent to the transaction in which it is established are subjective and difficult to establish. However, other intangibles are represented by legal documents such as franchise agreements, patents, corporate articles of incorporation, etc. Acquisition cost, less amortization, is generally the valuation basis for intangibles. Costs of developing, maintaining, or restoring intangible assets which are not specifically identifiable, have indeterminate lives, or are inherent in a continuing business and related to an enterprise as a whole—such as goodwill—should be deducted from income when incurred.[1] Amortization is usually computed on the straight-line method over the periods estimated to be benefited but not to exceed forty years.

Audit Objectives

The objectives in the audit of long-term investments and intangibles are:

1. To determine whether long-term investments are represented by certificates of ownership owned by and held by or for the company, and that intangibles are represented by contractual rights, privileges, or earning power (goodwill) owned by the company.

2. To determine whether long-term investments are properly stated on the basis of accounting principles applicable in the circumstances, and that intangibles are stated at cost less amortization.

3. To determine whether disclosures concerning long-term investments and intangibles are adequate and in accordance with generally accepted accounting principles.

Audit Procedures and Working Papers

The auditor must consider the accounting procedures and internal controls for the physical security or evidence of ownership of long-term investments, cash disbursements for long-term investments and intangibles, and amortization policies for intangibles.

Security Count—One means of determining that long-term investments exist and are owned by the company is physical inspection of the certificates, often referred to as a security count.

Security counts must be made as of the audit date, unless provisions can be made to verify that there has been no access to the certificates between the audit date and the date they are inspected. This verification can often be provided by examination of banks' records of access to safe deposit boxes (usually confirmed to the auditor by the bank in writing) or the use of seals by the auditor. This

---

[1]AC Section 5141.24 *(APB Opinion No. 17)*

precaution is necessary to prevent the removal and sale or pledge of the investments as of the audit date, and their subsequent replacement without the knowledge of the auditor.

The working-paper schedule evidencing the security count should include such details as company name (unless in bearer form), certificate number, number of shares, par value, face value, interest rate, due date, issue date, etc. An example of such a working paper is shown in Figure 11.4. Note that there is evidence on the working paper that the auditor inspected each certificate, that the inspection was performed in the presence of company officers, and that the certificates were returned intact to the company officers. The security count sheet is later compared with the description of the investments in the accounting records.

Confirmation—Certificates evidencing long-term investments may be held for the company by financial institutions or others. If certificates are held by a reputable bank or other financial institution, the auditor will normally confirm their existence with the holder. If, however, the certificates are held by an institution that is unknown to the auditor, he should not accept a confirmation without some inquiry into the nature of the institution. In addition, in the case of bonds, debentures, loans, and other debt obligations, the amount due should be confirmed directly with the debtor.

Valuation—Once the auditor has established the proper valuation basis for an investment under generally accepted accounting principles, he must apply audit procedures to that basis. If cost is the valuation basis, he should examine documentation (broker's advice, canceled check, etc.) for the acquisition of the investment. If underlying equity is the valuation basis, audited financial statements of the investee constitute sufficient evidential matter as to the equity in the underlying net assets, but unaudited financial statements do not. Therefore, if an investment accounted for by the equity method is material, the scope of an auditor's examination may be limited if the financial statements of the investee company are not audited. Finally, if market is the valuation basis, market quotations from such sources as *The Wall Street Journal* or *Barrons* may be used if the prices are based on a reasonably broad and active market and there are no restrictions on the transfer of the investment.

As intangible assets are generally valued on the basis of cost less amortization, the auditor should examine documentation of their acquisition cost (franchise and merger agreements, invoices, canceled checks, etc.) and test the accuracy of the amortization. He should also evaluate the amortization period to determine whether any current events or circumstances warrant revision.

Investment-Related Income—It is often convenient and efficient to audit dividend and interest income at the time investments are audited, and the audit procedures are usually recorded on the same audit working papers. Dividends can be tested by multiplying the number of shares (from the security count schedule if there were no changes during the year) times the dividend rate (per

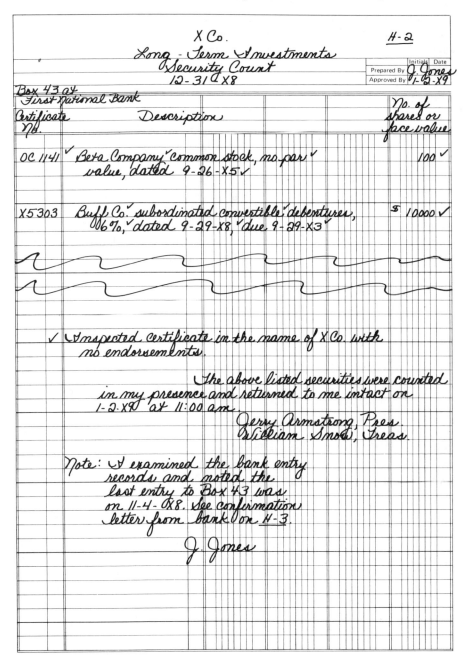

X Co.
H-2

Long - Term Investments
Security Count
12-31-X8

	Initials	Date
Prepared By	J. Jones	
Approved By	V-12-X9	

Box 43 at
First National Bank

Certificate No.	Description	No. of shares or face value
OC 1141 ✓	Beta Company common stock, no par value, dated 9-26-X5 ✓	100 ✓
X5303	Buff Co. subordinated convertible debentures, 6%, dated 9-29-X8, due 9-29-X3	$ 10000 ✓

✓ Inspected certificate in the name of X Co. with no endorsements.

   The above listed securities were counted in my presence and returned to me intact on 1-2-X9 at 11:00 am.
   Jerry Armstrong, Pres.
   William Snow, Treas.

Note: I examined the bank entry records and noted the last entry to Box 43 was on 11-4-X8. See confirmation letter from bank on H-3.

   J. Jones

Figure 11.4 Long-Term Investments—Security Count

Standard & Poor's Dividend Record, audited financial statements of the investee, etc.). Interest income can be recomputed on the basis of face amount, interest rate, and period held (from the security count schedule or confirmation).

Long-Term Liabilities

Long-term liabilities include loans, bonds, and notes payable that generally are due after one year from the balance sheet date, although certain amounts due within one year can be classified as long-term if both the intention and ability to refinance on a long-term basis are demonstrated. AC Section 2033.02 *(FASB Statement No. 6)* includes the following comment: "*Long-term obligations* are those scheduled to mature beyond one year (or the operating cycle, if applicable) from the date of an enterprise's balance sheet." AC Section 2033.10–.11 *(FASB Statement No. 6)* also provides that certain short-term obligations can be excluded from current liabilities if the following conditions are met.

> The enterprise intends to *refinance the obligation on a long-term basis.*
>
> The enterprise's intent to refinance the short-term obligation on a long-term basis is supported by an ability to consummate the refinancing demonstrated in either of the following ways:
>
> a) *Post-balance-sheet-date issuance of a long-term obligation or equity securities . . .*
> b) *Financing agreement. . . .*

Other accounts, often referred to as deferred credits, that also are classified as long-term liabilities include deferred income taxes (attributable to timing differences related to long-term assets or liabilities), deferred investment tax credits, reserves for retirement plans, and minority interests in consolidated subsidiaries.

Long-term liabilities, such as loans, bonds, and notes, have several characteristics that require particular attention from the auditor. First, they often are collateralized or secured by certain of the company's assets. The result is restriction of the free transferability of such assets, and in some cases, restriction of the manner in which they can be used.

Second, the debt instrument may be subject to the provisions of a separate but related debt agreement that also may place restraints on the company's operations. Provisions commonly found in debt agreements include requirements that the company maintain specified working capital and debt-to-equity ratios, as well as restrictions on payment of dividends, purchase of treasury stock, and merger or sale of the company or a significant portion of its assets. Violation of any of these provisions generally constitutes an event of default which allows the debtholder to require immediate payment. If a provision is violated inadvertently and the debtholder has no reason to believe the loan is in danger, he will usually, on request from the company, waive his right to immediate payment. If a waiver is obtained, the financial statements must disclose that a violation occurred and was waived by the debtholder. If a waiver is refused, the debt should be classified as a current liability because payment can be demanded at any time by the debtholder. Obviously, such an event could have a catastrophic effect on a company.

Readers of financial statements normally are interested in many aspects of long-term debt. For example, the impending maturity of a large debt can create cash-flow problems for a company, or the replacement at maturity of low-interest-rate debt with debt bearing a high interest rate can have an adverse effect on earnings. Therefore, disclosures concerning long-term debt should include a complete description of the terms of the debt, as well as the collateral and any restrictions imposed by a debt agreement.

In addition to the controls over cash receipts and disbursements, other internal control features specifically applicable to long-term debt include the requirements that all borrowings be approved by the board of directors and that a checklist be maintained of requirements and restrictions in debt agreements to prevent inadvertent noncompliance.

Audit Objectives

The objectives in the audit of long-term liabilities are:

1. To determine whether all long-term liabilities existing or incurred as of the audit date are reflected properly in the financial statements.

2. To determine whether long-term liabilities consist of amounts due after one year from the audit date (or operating cycle), or short-term obligations for which the intention and ability to refinance on a long-term basis have been demonstrated.

3. To determine whether disclosures regarding long-term liabilities are adequate.

Audit Procedures and Working Papers

The first step is to review and test internal controls applicable to long-term debt. Tests of internal controls over cash receipts and cash disbursements are included in Chapter 5. Dual-purpose tests discussed in this section are used to check other controls such as proper authorization of borrowings, etc. After reviewing and testing the accounting procedures and internal controls, the auditor should prepare an audit lead schedule similar to the one in Figure 11.5.

Tests of Borrowings, Repayments, and Accrued Interest—Because long-term debt transactions often are material, it is not unusual for the auditor to examine all transactions in the account for the year, although samples may be used if the volume of transactions is large and the size of individual transactions is small.

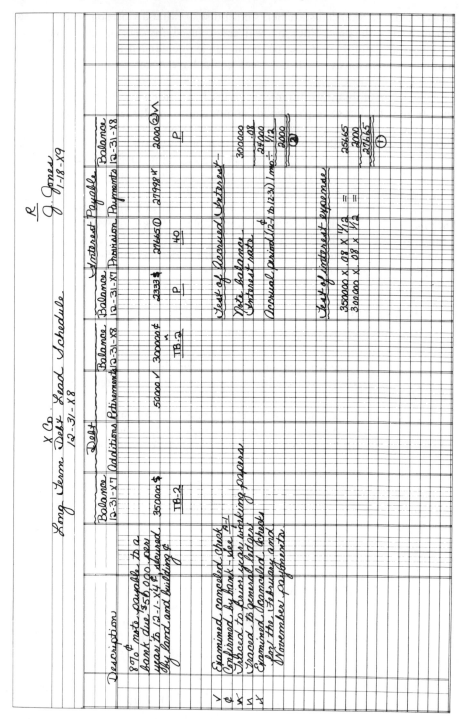

Figure 11.5 Long-Term Debt Lead Schedule

The auditor normally traces the proceeds from any borrowings to the cash receipts records, deposit slips, and bank statement, and audits payments by examination of canceled checks (indicated by ✔ in Figure 11.5) and canceled notes, if paid in full. The authorization of all significant borrowings and any repayments not made in accordance with the terms of the debt instrument should be traced to minutes of meetings of the board of directors.

Interest payable is tested on the same schedule as long-term debt so it can be related visually to the long-term debt balance. The simple example with only one note presents no problem, but if 30 or 40 notes were outstanding, this format would allow the auditor to determine quickly whether interest had been computed for each note. The fact that interest has been computed for each note, however, does not guarantee that it is computed correctly. The auditor tested interest expense at ① in Figure 11.5 by multiplying the outstanding balance by the interest rate and the fraction of the year it was outstanding. The accrued balance at the end of the year was tested in a similar manner at ② with the further step of relating the date to which interest was paid (the beginning of the accrual period) to the bank confirmation filed in the cash section (Figure 10.4). In addition, canceled checks were examined for two months' interest payments (indicated by ✕ in Figure 11.5), and the provision for interest expense was cross-referenced to the related expense account.

Confirmation—The approach to confirming long-term debt is similar to that for accounts payable because the objectives are similar. The confirmation letter ordinarily should not indicate the balance or other details of the debt, but should request that the debtholder furnish that information directly to the auditor. By using this procedure, the auditor anticipates that the debtholder will furnish the details of all notes held, whether or not they are recorded by the client. If the debtholder is a bank, the standard bank confirmation form (Figure 10.4) should be used; otherwise, a separate letter requesting similar information must be prepared. Note that in Figure 11.5 the pertinent data have been indicated as confirmed by the bank and cross-referenced to the bank confirmation (see ¢). Confirmations normally are sent to all financial institutions with which the client has had dealings during the year, regardless of whether or not there was a balance at the end of the year. Although auditors anticipate that they will not receive replies to all of their requests for confirmation of accounts receivable and accounts payable, they usually require confirmation from all significant debtholders before releasing their opinion on the financial statements. Thus, personal followup by the auditor and the client may be needed in addition to second requests.

Review of Debt Agreements—The importance of debt agreements has been stressed. The auditor's review of these agreements must be performed carefully and thoroughly. If the meaning or intent of a provision in an agreement is unclear, the auditor should ask the client to request an interpretation from the debtholder. All violations of a debt agreement, no matter how minor, should be referred to the client so that waivers can be requested. In this area materiality is inherent; if a debtholder believes a provision is significant enough to include in a debt agreement, the auditor must assume that a violation of that provision is material.

The review of debt agreements often is documented with the use of a checklist of the debt agreement provisions in the permanent audit file. This checklist must be reviewed each year and an indication made by each provision as to whether or not it was complied with. Because this procedure requires a knowledge of all aspects of the audit, it generally is performed by the audit senior and is reviewed closely by the audit manager and partner.

Classification and Disclosure—Although the stated maturity date of a note often determines its classification as either a long- or a short-term liability, other factors, such as the intent and ability to refinance or an unwaived violation of a debt agreement, may change that classification. In this phase of the examination, as in others, the auditor must not accept anything at face value, but must consider the possibility that unusual facts or circumstances may result in a different interpretation.

The information about long-term liabilities that is to be disclosed in the financial statements should be requested in the confirmation from the debtholder. Occasionally, the auditor receives a reply directly from a debtholder that is complete except for a description of any liens or collateral, and the auditor is unable to tell whether there are none or whether they were omitted by oversight. The only recourse is to return the confirmation (or preferably a copy of it) to the debtholder with a request that the omitted information be supplied or, if there are no liens or collateral, that this be indicated. It would be imprudent to accept an oral addition or correction to a written confirmation.

Equity Accounts

Equity accounts include common and preferred stock, additional paid-in capital, retained earnings, and treasury stock (partnership and proprietorship capital also would be included, but are not discussed in this text). Transactions in these accounts are relatively infrequent, but can be important. AC Section 2042.02 *(APB Opinion No. 12)* contains the following requirement for disclosure of capital changes.

> When both financial position and results of operations are presented, disclosure of changes in the separate accounts comprising stockholders' equity (in addition to retained earnings) and of the changes in the number of shares of equity securities during at least the most recent annual fiscal period and any subsequent interim period presented is required to make the financial statements sufficiently informative.

The articles of incorporation of a company state the classes of stock it is authorized to issue, as well as the rights and preferences of each class, the number of shares of each class that is authorized, the par value, if any, of such stock, and any restrictions that may attach to it. The articles of incorporation of some companies provide for the restriction of dividends on common stock in the event that all preferred dividends are not paid; those of other companies, particularly companies that are closely held, require stock to be offered to the company at a certain price (such as book value) before it can be sold to another party. Thus, the articles of incorporation are the source of many financial statement disclosures.

The corporate secretary is normally responsible for the records relating to the number of shares of issued and outstanding stock. The procedures used by a company to account for and control the number of shares of stock issued and outstanding are determined largely by the number of its stockholders. If the number of stockholders is small, a company usually maintains a stock certificate book or stockbook. This book is similar to a checkbook in that it consists of a detachable stock certificate attached to a permanently bound stub on which can be entered the number of shares, date the certificate was issued, and name of the stockholder. When stock is sold or transferred, the stock certificate of the original stockholder is surrendered to the company so that it can be marked "canceled" and reattached to the stub from which it was issued. A new certificate is issued to the new stockholder. The number of shares issued at any time is determined by adding the numbers of shares shown on all stock certificate stubs that do not have canceled certificates attached to them. This procedure is adequate if the number of stockholders and stock transfers is small, but obviously would be impractical if hundreds or thousands of shares were transferred daily.

Public companies whose stock is traded actively usually employ banks or other financial institutions to act as transfer agents and registrars. As transfer agents, these institutions receive and issue certificates and maintain lists of stockholders (usually on a computer) for use in mailing stockholder reports, paying dividends, etc. As registrars, they maintain records of the number of shares issued and canceled to check on the transactions of the transfer agent and to guard against mistakes that could result in an overissue of stock.

Treasury stock should be recorded at cost, and it is generally shown as a deduction from the total equity section of the balance sheet, although AC Section 5541.01 *(ARB No. 43)* also provides for the following classification: ". . . it is perhaps in some circumstances permissible to show stock of a corporation held in its own treasury as an asset. . . ." Neither dividends attributable to treasury stock nor gains or losses on its disposition can be recorded as income.

Treasury stock certificates should be stored in a safe deposit box and safeguarded in the same manner as investment securities.

The retained earnings account should contain few transactions other than net income or loss and dividend distributions. *FASB Statement No. 16* requires, with certain exceptions, that all items of profit and loss recognized during a period, including accruals of estimated losses from loss contingencies, be included in earnings for that period. Only two items can be treated as prior period adjustments—corrections of errors in the financial statements of a prior period, and tax adjustments resulting from realization of income tax benefits of pre-acquisition operating loss carry-forwards of purchased subsidiaries.

Many companies have stock option plans for officers and key employees. If these plans are used, records must be maintained of the number of options authorized, granted, exercised, and expired, as well as option prices and periods covered.

Audit Objectives

The objectives in the audit of the equity accounts are:

1. To determine whether the equity accounts and the transactions therein as shown in the financial statements are presented in accordance with generally accepted accounting principles.

2. To determine whether adequate disclosure is made of all restrictions, rights, options, and other matters important to an understanding of the financial statements.

Audit Procedures and Working Papers

In examining equity accounts, an auditor must keep in mind that the audit is being made of the company and *not* of its stockholders. Transfers of shares between stockholders and the number of shares owned by individual stockholders are theoretically of no concern to the auditor, because they have no effect on the financial statements being audited. The auditor is concerned only with the total issued or outstanding stock and not with its ownership. There is no reason to confirm stock ownership with individual stockholders and, in fact, this would be impossible to do if stock were held by nominees or were sold from one individual to another without being sent to the company for transfer. (Such a sale would not affect the ownership of the buyer, but would be imprudent, because dividends and other distributions would continue to be sent to the seller.)

The auditor should review and test the internal control procedures applicable to the equity accounts, including tests of controls over cash receipts and cash disbursements. These controls are discussed in Chapter 5. Because of the limited number of transactions and their importance, the auditor often examines all transactions in the accounts. The audit of the equity accounts follows an approach similar to that used for property and equipment in that a beginning balance is established (usually in the prior-year audit), and then current-year transactions are verified. In addition, certain procedures are applied to the year-end balance.

All equity accounts should be analyzed (often on a schedule maintained in the permanent audit file), and transactions during the year should be audited by examination of supporting documents and approvals. For example, if additional stock were sold during the year, the proceeds should be traced to the cash receipts book and bank statement, and the approval of the sale should be traced to minutes of meetings of stockholders or directors. Particular attention should be paid to the valuation attributed to stock issued for non-cash consideration, the treatment of gains or losses from sale or retirement of treasury stock, the propriety of entries to retained earnings other than net income or loss and dividends, and the accounting treatment of stock dividends and splits.

Tests of Dividends—Total dividends declared during a year can be tested on an overall basis by multiplying the number of shares outstanding at the dividend record date by the dividend rate per share (as previously illustrated). This procedure provides evidence that the proper amount was calculated, but the auditor also must obtain evidence that it was paid to the stockholders.

Some companies engage a bank or trust company as a dividend-paying agent, often the same one that acts as transfer agent. In these cases, the company pays the total dividend to its dividend-paying agent, which makes the dividend distributions to the individual stockholders. The auditor examines the canceled check for the total dividend and the notice of receipt from the agent. The agreement between the company and the dividend-paying agent usually places the responsibility for the proper distribution of the dividend, and liability for incorrect dividend payments, on the agent. If this is the case, the auditor normally does not examine the agent's records of dividend payments to individual stockholders.

If a company pays its own dividends, the auditor should test the propriety of the payments to the individual stockholders. An example of such a test, which can be done on a test or sample basis, is shown in Figure 11.6. The auditor obtained the stockholder's name and number of shares owned of record as of the dividend date from the stockbook (indicated by ✔). He also totaled the number of shares owned by all stockholders and related it to the total outstanding as shown in the account analysis in the permanent audit file (*I-4* is an illustration of a method of indexing the permanent audit file). The auditor then compared the dividend rate with that shown in the minutes of the specific meeting at which the dividend was declared and checked the calculation of the individual dividend payments. Finally, he examined the canceled check in payment of the dividend, noting particularly that it was endorsed properly by the stockholder.

Examination of Stockbook and Treasury Stock—If a company maintains its own stock records, the auditor must examine the stockbook to determine whether the proper amount is shown as being issued in the financial statements. Because many of the same stockbook stubs and canceled certificates are examined each year, the auditor commonly maintains a permanent audit file schedule that can be used to document the examination for several years. An illustration of such a schedule is shown in Figure 11.7. Note that the schedule was used in both the 19X7 audit, when the work was performed by R. Parker, and the 19X8 audit, when it was performed by J. Jones. The auditor must account for all certificate numbers by examination of the stockbook stubs for outstanding shares (indicated by ↗), canceled certificates for retired shares (indicated by ✔), and unissued certificates for the balance. The total of the shares outstanding has been

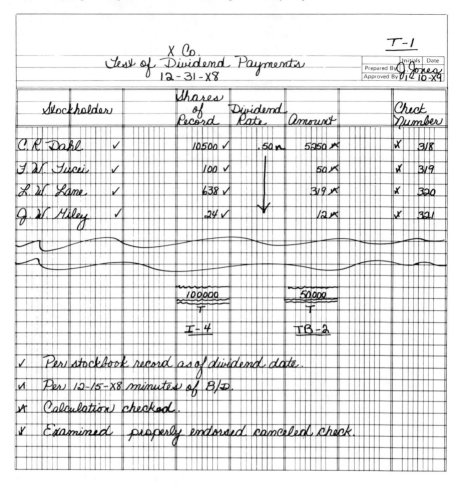

Figure 11.6   Test of Dividend Payments

multiplied by the par value to produce the dollar amount of capital stock shown in the financial statements. A step not indicated in Figure 11.7, but used by some auditors because of the importance of the stockbook, is to ask the client to acknowledge its return in writing after it has been examined.

If the client possesses treasury stock, it should be inspected. The schedule evidencing this inspection should include the certificate number, date, and number of shares. All treasury stock certificates should be in the name of the company. The inspection should be made in a manner similar to that of a security count, described earlier in this chapter.

Confirmation—If the client employs a transfer agent and registrar, there will be no stock certificate book to examine. Instead, the number of shares issued and

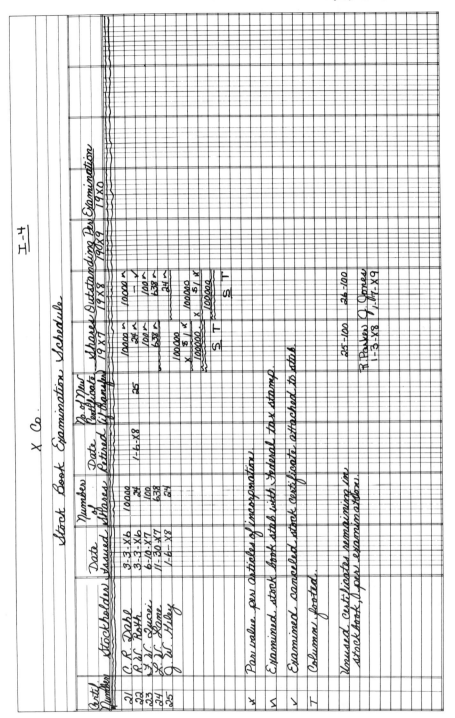

Figure 11.7 Stock Book Examination Schedule

outstanding should be confirmed in writing directly from the transfer agent and registrar to the auditor.

Matters Requiring Disclosure—The general disclosures required of all companies under generally accepted accounting principles, such as capital changes, stock options, and liquidation value of preferred stock, usually can be related to corporate documents such as the articles of incorporation or minutes of meetings of stockholders or directors. Normally, they present no special problems.

Disclosures of restrictions such as limitations or prohibitions on payment of cash dividends and purchase of treasury stock are more difficult to detect. Most restrictions of this nature are due to provisions in debt and lease agreements and should be found by use of the procedures discussed in the long-term debt section, such as review of debt (and lease) agreements and minutes, and confirmations with debtholders. Other restrictions may arise from the laws of the state in which a company is incorporated. For example, several states have laws prohibiting the payment of dividends to the extent of the cost of treasury stock, thereby necessitating the disclosure of a restriction of retained earnings if treasury stock is held. The ability to discover these disclosures requires that the auditor be familiar with the corporate codes of the states in which clients operate.

Appendix to Chapter 11

Illustration of the Property and Equipment and
Accumulated Depreciation Section of an Audit Program

X Co.
Property and Equipment
12-31-X8

Account Description—

This group of accounts represents the land, building, and equipment used in the company's manufacturing process, as well as the salesmen's automobiles. Annual straight-line depreciation rates are 5% for buildings, 10% for equipment, and 33⅓% for automobiles.

Evaluation of Internal Control—

A detailed ledger is maintained showing the individual items of property and equipment, and it is balanced with the control account monthly. Capital expenditures in excess of $5,000 require the approval of the board of directors in the capital expenditures budget. Formal policies have been established to distinguish between capital and maintenance charges. Retirements of property are reported by the shop foremen, but no periodic review is made for unreported retirements and individual pieces of equipment are not tagged or otherwise specifically identified as belonging to the company.

Overall, internal controls over property additions are strong, although controls over property retirements are relatively weak.

Audit Procedures

Time Required				
Estimated	*Actual*	*Audit Step*	*Assigned To*	*Performed By*
Preliminary—				
2		1. Review, test, and evaluate the accounting procedures and internal controls relating to property and equipment and accumulated depreciation (see separate programs).	Staff 1	
1		2. Analyze property and equipment additions through the preliminary audit date listing all additions in excess of $2,000.	Staff 2	
2		3. For all additions in excess of $2,000, examine vendor invoice, canceled check, and receiving report and determine if classification as a capital item is proper. Also indicate on schedule if the addition is a new or replacement item.	Staff 2	
1		4. Analyze maintenance expense through the preliminary audit date listing all charges in excess of $2,000.	Staff 2	
2		5. For all maintenance charges in excess of $2,000, examine vendor invoice, canceled check, and receiving report and determine if classification as an expense item is proper.	Staff 2	
1		6. Trace all additions in excess of $5,000 to capital budget approved by the board of directors.	Staff 2	
1		7. Analyze retirements of property and equipment through the preliminary audit date.	Staff 1	
2		8. Trace the original cost of items retired, together with	Staff 1	

Time Required

Estimated	Actual	Audit Step	Assigned To	Performed By
		dates acquired and retired, to the detailed property ledger. Test calculation of accumulated depreciation to date of retirement and trace any salvage proceeds to cash receipts book. Investigate any significant retirements for which there is no salvage. Recompute gain or loss on the retirement and relate to other income or expense account.		
2		9. Review and supervision.	Senior	
Final—				
1		1. Review the accounting procedures and internal controls relating to property and equipment and accumulated depreciation for any material changes from preliminary review.	Staff 1	
1		2. Prepare or obtain client-prepared lead schedule of property and equipment and accumulated depreciation.	Staff 2	
		3. Trace beginning balances on lead schedule to prior-year audit working papers.	Staff 2	
1		4. Total the detailed property ledger and determine if it balances with the control account.	Staff 2	
1		5. Analyze property and equipment additions from preliminary to year-end, listing all additions in excess of $2,000. Cross-reference the total of this schedule to the lead schedule.	Staff 2	
1		6. For all additions from preliminary date to year-end in excess of $2,000, examine vendor invoice, canceled check, and receiving report and determine if classification as a capital item is proper. Also, indicate on schedule if the addition is a	Staff 2	

Time Required

Estimated	Actual	Audit Step	Assigned To	Performed By
		new or replacement item.		
1		7. Analyze maintenance expense from preliminary date to year-end, listing all charges in excess of $2,000. Cross-reference the total of this schedule to the operating expense lead schedule.	Staff 2	
1		8. For all maintenance charges from preliminary date to year-end in excess of $2,000, examine vendor invoice, canceled check, and receiving report and determine if classification as an expense item is proper.	Staff 2	
1		9. Trace all additions from preliminary date to year-end in excess of $5,000 to capital budget approved by the board of directors.	Staff 2	
1		10. Analyze retirements of property and equipment from preliminary date to year-end. Cross-reference the total of this schedule to the lead schedule.	Staff 1	
1		11. Trace the original cost of items retired after preliminary date, together with dates acquired and retired, to the detailed property ledger. Test calculation of accumulated depreciation to date of retirement and trace any salvage proceeds to cash receipts book. Investigate any significant retirements for which there is no salvage. Recompute gain or loss on the retirement and relate to other income or expense account.	Staff 1	
		12. Perform a search for unrecorded retirements including:		
1		a) Determine whether a retirement was recorded for each addition identified as a replacement.	Staff 1	
1		b) Investigate any significant reduction in property insur-	Staff 1	

Time Required

Estimated	Actual	Audit Step	Assigned To	Performed By
		ance coverage or property taxes to determine if it resulted from property retirements.		
1		c) Review miscellaneous or other income accounts for salvage credits or scrap sales that may indicate the disposal of retired property and equipment.	Staff 1	
1		d) Discuss property and equipment retirements with the shop foremen and the plant manager.	Staff 1	
2		e) Select ten items of equipment from the detailed property and equipment ledger and locate them in the plant.	Staff 1	
1		f) Write a memorandum outlining the work performed in the search for unrecorded retirements.	Staff 1	
		13. Test the current-year provision for depreciation—		
1		a) Compare depreciation methods, estimated lives of assets, and estimated salvage values to prior year for consistency.	Staff 1	
2		b) Make an overall test of depreciation expense by major asset category and investigate any significant variations.	Staff 1	
1		14. Review the balances of accumulated depreciation at year-end to determine the reasonableness of the undepreciated cost of the major assets in relation to their remaining depreciable lives. Consider such factors as obsolescence and technological changes, as well as physical characteristics.	Senior	
2		15. Review and supervision.	Senior	

Conclusion—

Chapter 11
Questions Taken from the Chapter

11–1.   How does an auditor differentiate between (a) an improvement to, and (b) maintenance of, a unit of property?

11–2.   State the objectives in the audit of property and equipment and related allowances for depreciation.

11–3.   Compare the approach to the audit of property and equipment with that of prepayments and inventories.

11–4.   What is a vouching schedule?

11–5.   State the audit procedures applied to property and equipment to determine whether additions are
          a.   recorded at cost.
          b.   represented by actual physical items.
          c.   properly capitalizable.

11–6.   What test does the auditor make to determine if any property units have been charged to expense?

11–7.   On what aspects of property retirements does the auditor place primary emphasis?

11–8.   Explain the procedures for testing recorded property retirements.

11–9.   List five procedures the auditor could use to detect unrecorded property retirements.

11–10.   Explain how and why an auditor usually tests depreciation expense on an overall basis.

11–11.   What factors should an auditor consider in evaluating the adequacy of the allowance for depreciation?

11–12.   What procedures are used by the auditor in his search for mortgages and liens on property and equipment?

11–13. What controls should be exercised over the physical custody of long-term investments?

11–14. What are some physical evidences of ownership of intangible assets?

11–15. State the objectives in the audit of long-term investments and intangibles.

11–16. Why are security counts often made on the audit date? Under what circumstances can they be made at other dates?

11–17. What evidence would an auditor examine for investments valued at cost? Underlying equity? Market value?

11–18. How does the CPA audit dividend and interest income?

11–19. List three characteristics of long-term debt that require special attention of the auditor.

11–20. Name five provisions of accounting significance commonly found in debt agreements.

11–21. Explain the financial statement effects of
a. obtaining a waiver.
b. not obtaining a waiver for a violation of a provision of a debt agreement.

11–22. What is the significance of the violation of a provision of a debt agreement?

11–23. State the objectives in the audit of long-term liabilities.

11–24. What audit procedures does an auditor apply to borrowings and repayments of long-term debt?

11–25. Why is interest payable tested on the same schedule as long-term debt?

11–26. How is the audit of accrued interest payable related to an audit working paper in the cash section?

11–27. Discuss the procedures for confirming long-term debt.

11–28. Why doesn't the concept of materiality apply to violations of a debt agreement?

11–29. Why is the review of debt agreements generally performed by the audit senior?

11–30. Name two factors that can determine the financial statement classification of long-term debt, other than its maturity date.

11–31. What should an auditor do when a confirmation is returned to him with incomplete information?

11–32. List six financial statement disclosures that may be found in a company's articles of incorporation.

11–33. Describe the two methods that may be used to control the number of shares of stock issued and outstanding.

11–34. State the objectives in the audit of the equity accounts.

11–35. Why does the auditor normally not confirm stock ownership with individual stockholders?

11–36. Explain how and why the audit approach to testing dividend payments would differ for companies paying their own dividends and those employing dividend-paying agents.

11–37. Why does the auditor examine the client's stockbook?

11–38. List the information that an auditor should note in his examination of treasury stock certificates.

11–39. Name two sources of restrictions on payment of cash dividends.

Chapter 11
Objective Questions Taken from CPA Examinations

11–40. Tennessee Company violated company policy by erroneously capitalizing the cost of painting its warehouse. The CPA examining Tennessee's financial statements would most likely learn of this error by

    a. Discussing Tennessee's capitalization policies with its controller.

    b. Reviewing the titles and descriptions for all construction work orders issued during the year.

    c. Observing, during the physical inventory observation, that the warehouse has been painted.

    d. Examining in detail a sample of construction work orders.

11–41. In connection with his review of plant additions, the CPA ordinarily would take exception to the capitalization of the cost of the

    a. Major reconditioning of a recently acquired second-hand lift truck.

    b. Machine operator's wages during a period of testing and adjusting new machinery.

    c. Room partitions installed at the request of a new long-term lessee in Pyzi's office building.

    d. Maintenance of an unused stand-by plant.

11–42. An auditor's client has violated a minor requirement of its bond indenture which could result in the trustee requiring immediate payment of the principal amount due. The client refuses to seek a waiver from the bond trustee. Request for immediate payment is *not* considered likely. Under these circumstances, the auditor must

    a. Require classification of bonds payable as a current liability.

    b. Contact the bond trustee directly.

    c. Disclose the situation in the auditor's report.

    d. Obtain an opinion from the company's attorney as to the likelihood of the trustee's enforcement of the requirement.

11–43. When a company has treasury stock certificates on hand, a year-end count of the certificates by the auditor is

    a. Required when the company classifies treasury stock with other assets.

    b. Not required if treasury stock is a deduction from stockholders' equity.

    c. Required when the company had treasury stock transactions during the year.

    d. Always required.

11–44. Mars Company has a separate outside transfer agent and outside registrar for its common stock. A confirmation request sent to the transfer agent should ask for

    a. A list of all stockholders and the number of shares issued to each.

    b. A statement from the agent that all surrendered certificates have been effectively canceled.

    c. Total shares issued, shares issued in the name of the client, and unbilled fees.

    d. Total shares authorized.

11–45. Braginetz Corporation acts as its own registrar and transfer agent and has assigned these responsibilities to the Corporation secretary. The CPA primarily will rely upon his

a. Confirmation of shares outstanding at year-end with the Corporation secretary.

b. Review of the corporate minutes for data as to shares outstanding.

c. Confirmation of the number of shares outstanding at year-end with the appropriate state official.

d. Inspection of the stockbook at year-end and accounting for all certificate numbers.

11–46. In connection with the examination of bonds payable, an auditor would expect to find in a trust indenture

a. The issue date and maturity date of the bond.

b. The names of the original subscribers to the bond issue.

c. The yield to maturity of the bonds issued.

d. The company's debt-to-equity ratio at the time of issuance.

11–47. If a company employs a capital stock registrar and/or transfer agent, the registrar or agent, or both, should be requested to confirm directly to the auditor the number of shares of each class of stock

a. Surrendered and canceled during the year.

b. Authorized at the balance sheet date.

c. Issued and outstanding at the balance sheet date.

d. Authorized, issued, and outstanding during the year.

11–48. Florida Corporation declared a 100% stock dividend during 1975. In connection with the examination of Florida's financial statements, Florida's auditor should determine that

a. The additional shares issued do not exceed the number of authorized but previously unissued shares.

b. Stockholders received their additional shares by confirming year-end holdings with them.

c. The stock dividend was properly recorded at fair market value.

d. Florida's stockholders have authorized the issuance of 100% stock dividends.

11–49. The CPA's examination normally would not include

a. Determining that dividend declarations have been in compliance with debt agreements.

b. Tracing the authorization for the dividend from the directors' minutes.

c. Detail checking from the dividend payment list to the capital stock records.

d. Reviewing the bank reconciliation for the imprest dividend account.

11–50. Which of the following is a customary audit procedure for the verification of the legal ownership of real property?

    a. Examination of correspondence with the corporate counsel concerning acquisition matters.

    b. Examination of ownership documents registered and on file at a public hall of records.

    c. Examination of corporate minutes and resolutions concerning the approval to acquire property, plant, and equipment.

    d. Examination of deeds and title guaranty policies on hand.

Chapter 11
Discussion/Case Questions and Problems

11–51. You are engaged in examining the financial statements of the Ute Corp. for the year ended December 31, 19X9. The following schedules for the property, plant and equipment, and related allowance for depreciation accounts have been prepared by the client. You have checked the opening balances to your prior year's audit work papers.

Ute Corp.
Analysis of Property, Plant and Equipment, and
Related Allowance for Depreciation Accounts
Year Ended December 31, 19X9

Assets

Description	Final 12/31/X8	Additions	Retirements	Per Books 12/31/X9
Land	$ 22,500	$ 5,000		$ 27,500
Buildings	120,000	17,500		137,500
Machinery and equipment	385,000	40,400	$26,000	399,400
	$527,500	$62,900	$26,000	$564,400

Allowance for Depreciation

Description	Final 12/31/X8	Additions*	Retirements	Per Books 12/31/X9
Building	$ 60,000	$ 5,150		$ 65,150
Machinery and equipment	173,250	39,220		212,470
	$233,250	$44,370		$277,620

*Depreciation expense for the year.

Your examination reveals the following information:

1. All equipment is depreciated on the straight-line basis (no salvage value taken into consideration) based on the following estimated lives: buildings, 25 years, all other items, 10 years. The Corporation's policy is to take one-half year's depreciation on all asset acquisitions and disposals during the year.

2. On April 1, the Corporation entered into a 10-year lease contract for a die-casting machine with annual rentals of $5,000, payable in advance every April 1. The lease is cancelable by either party (60 days written notice is required), and there is no option to renew the lease or buy the equipment at the end of the lease. The estimated useful life of the machine is 10 years with no salvage value. The Corporation recorded the die-casting machine in the Machinery and Equipment account at $40,400, the present discounted value at the date of the lease, and $2,020, applicable to the machine, has been included in depreciation expense for the year.

3. The Corporation completed the construction of a wing on the plant building on June 30. The useful life of the building was not extended by this addition. The lowest construction bid received was $17,500, the amount recorded in the Buildings account. Company personnel were used to construct the addition at a cost of $16,000 (materials, $7,500; labor, $5,500, and overhead, $3,000).

4. On August 18, $5,000 was paid for paving and fencing a portion of land owned by the Corporation and used as a parking lot for employees. The expenditure was charged to the Land account.

5. The amount shown in the machinery and equipment asset retirement column represents cash received on September 5 upon disposal of a machine purchased in July 19X5 for $48,000. The bookkeeper recorded depreciation expense of $3,500 on this machine in 19X9.

6. Crux City donated land and building appraised at $10,000 and $40,000, respectively, to the Ute Corporation for a plant. On September 1, the Corporation began operating the plant. Since no costs were involved, the bookkeeper made no entry for the foregoing transaction.

Required:
Prepare the formal adjusting journal entries with supporting computations that you would suggest at December 31, 19X9 to adjust the accounts for the above transactions. Disregard income tax implications. The books have not been closed. Computations should be rounded-off to the nearest dollar.

(AICPA adapted)

11–52. In connection with a recurring examination of the financial statements of the Louis Manufacturing Company for the year ended December 31, 19X9, you have been assigned the audit of the Manufacturing Equipment, Manufacturing Equipment—Accumulated Depreciation, and Repairs to Manufacturing Equipment accounts. Your review of Louis's policies and procedures has disclosed the following pertinent information:

1. The Manufacturing Equipment account includes the net invoice price plus related freight and installation costs for all of the equipment in Louis's manufacturing plant.

2. The Manufacturing Equipment and Accumulated Depreciation accounts are supported by a subsidiary ledger which shows the cost and accumulated depreciation for each piece of equipment.

3. An annual budget for capital expenditures of $1,000 or more is prepared by the budget committee and approved by the board of directors. Capital expenditures over $1,000 which are not included in this budget must be approved by the board of directors, and variations of 20% or more must be explained to the board. Approval by the supervisor of production is required for capital expenditures under $1,000.

4. Company employees handle installation, removal, repair, and rebuilding of the machinery. Work orders are prepared for these activities and are subject to the same budgetary control as other expenditures. Work orders are not required for external expenditures.

Required:
a. Cite the major objectives of your audit of the Manufacturing Equipment, Manufacturing Equipment—Accumulated Depreciation, and Repairs of Manufacturing Equipment accounts. Do not include in this listing the auditing procedures designed to accomplish these objectives.
b. Prepare the portion of your audit program applicable to the review of 19X9 additions to the Manufacturing Equipment account.

(AICPA adapted)

11–53.   A company issued bonds for cash during the year under audit. To ascertain that this transaction was properly recorded, the auditor might perform some or all of the following procedures.
a. Request a statement from the bond trustee as to the amount of the bonds issued and outstanding.
b. Confirm the results of the issuance with the underwriter or investment banker.
c. Trace the cash received from the issuance to the accounting records.
d. Verify that the net cash received is credited to an account entitled "Bonds Payable."

Discuss the audit objectives accomplished by each of the procedures above. Select the procedure you consider the "best," and state why you selected it.

(AICPA adapted)

11–54. In connection with the annual examination of Johnson Corp., a manufacturer of janitorial supplies, you have been assigned to audit the fixed assets. Johnson Corp. maintains a detailed property ledger for all fixed assets. You prepared an audit program for the balances of property, plant, and equipment but have yet to prepare one for accumulated depreciation and depreciation expense.

Required:
Prepare a separate comprehensive audit program for the accumulated depreciation and depreciation expense accounts.

(AICPA adapted)

11–55. One procedure for determining the existence of property and equipment is physical observation, although other methods may be as effective in some cases. Discuss means of gathering evidence as to the physical existence of the following property and equipment items, other than direct observation.
a. A producing oil well.
b. An apartment building.
c. An automobile.
d. A mineral lease.

11–56. During the audit of notes payable of Jones Tractor Co., a CPA was reviewing the terms of a related loan agreement. He noted that the note matured within 12 months of the audit date, but that the loan agreement provided that the term of the note could be extended an additional 12 months at the option of the company, provided there had been "no adverse changes in its financial or operating conditions."

How can the CPA audit the classification of notes payable as a long-term liability in this case?

11–57. Discuss the audit procedures that a CPA might use in gathering evidence of the following transactions and balances
a. The refinancing (cancellation of old note and issuance of a new note) of a note payable to a bank.
b. Acquisition of treasury stock.
c. Year-end balance in additional paid-in capital account (assume that no transactions are shown for the current year).
d. Issuance of common stock to acquire another company.
e. Exercise of employee stock options.
f. Issuance of employee stock options.
g. Sale of treasury stock.

11–58. During your audit of property and equipment, you review the following construction work order listing:

Work Order No.	Description	Amount Authorized	Amount Expended
3103	Construct branch sales office	$38,000	$40,500
3104	Replace fence around plant	7,600	6,950
3106	Resurface parking lot	5,900	6,030
3107	Install additional boiler	18,000	38,100
3108	Construct addition to President's residence	16,000	26,500
3109	Install drapes and carpets in Treasurer's office	2,700	2,800

List the items you would select for additional audit follow-up and give your reasons for listing them.

11–59. Companies that have stock option plans normally include a footnote in their financial statements which describes the plan and states the number of options for shares authorized, granted, exercised, and expired, the option prices, and the market prices of the stock on the grant and exercise dates.

As this information is not normally considered to be a part of the accounting records, what is the auditor's responsibility for it? If you feel audit procedures should be employed, list those that you think should be applied.

11–60. Your new assistant auditor informed you that he has completed the audit of depreciation, and on review of his work you find that his work consisted of checking the clerical accuracy of the multiplication of the depreciation rate times the asset balance. What additional factors should he have considered in the audit of depreciation?

Chapter 12

Auditing the Operations, Contingencies, and Subsequent Events Cell

---

Learning Objectives   *After reading and studying the material in this chapter, the student should*

Know the characteristics of the accounts comprising the operations, contingencies, and subsequent events cell.

Understand the important features of internal control associated with each account in the operations, contingencies, and subsequent events cell.

Know the audit objectives of each account in the operations, contingencies, and subsequent events cell.

Be able to determine and apply the audit procedures necessary to accomplish the audit objectives for each account in the operations, contingencies, and subsequent events cell.

Know how to prepare the audit working papers to document the audit procedures for each account in the operations, contingencies, and subsequent events cell.

This chapter will include examples of the audit objectives, procedures, and working papers involved in the audit of revenue, expense, contingencies, and subsequent events. Important relationships exist among these elements, particularly revenue and expense; therefore, they should be examined together.

515

Revenue and Expense

Investors and other users of financial statements rely on the statement of income as an indication of a company's performance for a given period. They often use information from this statement as a guide in projecting expected future performance. In determining the objectives for the audit of revenue and expense, the auditor should keep in mind the use made of the statement of income.

For net income to be a relevant amount, it should include all revenues and expenses applicable to the audit period and no significant amounts applicable to other periods. In other words, there should be no overstatement or understatement of net income.

FASB Statement No. 16 requires, with certain limited exceptions, such as corrections of errors and realization of income tax benefits of pre-acquisition operating loss carry-forwards of purchased subsidiaries, that all items of profit and loss recognized during a period, including accruals of estimated losses from loss contingencies, be included in the determination of net income for that period.

One useful means of determining whether a particular item of income or expense is applicable to a period is the matching concept. Basically, it requires that all items of expense needed to generate a certain amount of revenue be recognized in the same period that the revenue is recognized. The concept appears in the professional literature in numerous places, including AC Section 5121.03 *(ARB No. 43)*, which states: "A major objective of accounting . . . is the proper determination of income through the process of matching appropriate costs against revenues."

If net income is to be an effective criterion for gauging the performance of the company and estimating its future performance, unusual and infrequent items that are not the result of normal operations and may not recur should be segregated. Otherwise, the reader of the financial statements may be misled into anticipating the continuation of performance that was actually the result of items of a non-recurring nature. The criterion for extraordinary items is set forth in AC Section 2012.19–.20 *(APB Opinion No. 30)*.

> Judgment is required to segregate in the income statement the effects of events or transactions that are extraordinary items. . . . The Board concludes that an event or transaction should be presumed to be an ordinary and usual activity of the reporting entity, the effects of which should be included in income from operations, unless the evidence clearly supports its classification as an extraordinary item as defined in this section.
>
> Extraordinary items are events and transactions that are distinguished by their unusual nature *and* by the infrequency of their occurrence.

Except for the requirement to segregate extraordinary items, companies have considerable leeway in the classification of revenues and expenses within the income statement, provided that the classification is consistent between periods. Most companies that sell a physical product subtract the cost of the product from revenue to arrive at an amount designated as gross profit. Because the profitability of many companies depends on maintaining or increasing the gross profit ratio (the ratio of gross profit to revenue), this ratio is watched closely by most investors. Other companies, particularly those selling services, do not compute gross profit, but subtract all costs and expenses except income taxes from revenue to arrive at net income before income taxes; income tax expense then is subtracted from this amount to arrive at net income.

Several disclosures concerning revenues and expenses are required by generally accepted accounting principles (the amount of research and development expenditures, the segregation of current and deferred income taxes, etc.). Earnings per share is probably one of the most important disclosures. AC Section 2011.12 *(APB Opinion No. 15)* includes the following provisions for its presentation.

> The Board believes that the significance attached by investors and others to earnings per share data, together with the importance of evaluating the data in conjunction with the financial statements, requires that such data be presented prominently in the financial statements. The Board has therefore concluded that earnings per share or net loss per share data should be shown on the face of the income statement.

Revenue and expense accounts are subject to many of the same internal controls (such as those relating to cash receipts and disbursements, accounts receivable, and inventory) as are the asset and liability accounts. In addition, they are often subject to budgetary controls. These controls can be very effective if variations between actual and budgeted amounts are investigated carefully by management.

Audit Objectives

The objectives in the audit of revenue and expense are:

1. To determine whether revenues and expenses are applicable to the audit period and are matched properly in accordance with generally accepted accounting principles.

2. To determine whether all material unusual and infrequent items are segregated properly in the income statement.

3. To determine whether revenues and expenses are classified properly and consistently.

4. To determine whether disclosures concerning revenue and expense are adequate and in accordance with generally accepted accounting principles.

Audit Procedures and Working Papers

Although some detail auditing is performed, the auditor relies heavily on tests of accounting procedures, internal controls, and overall tests in the revenue and expense area. Review and tests of internal controls are discussed in Chapter 5, and shipping, billing, cash receipts, and accounts receivable controls are described. The same controls would apply to revenue. In addition, similar reviews and tests of controls would be made in the areas of (1) receiving, cash disbursements, and accounts payable, and (2) payroll (see appendix to Chapter 5). These controls would apply to expenses. The auditor also may perform other tests of procedures and controls such as tests of material issues and of the cost accounting system. Also, as described in the two previous chapters, some revenue and expense accounts are audited in connection with related asset and liability accounts (e.g., depreciation expense with property and equipment, interest expense with notes payable, allowance for doubtful accounts with accounts receivable, etc.) The tests of procedures plus the audit work performed on the related asset and liability accounts provide the auditor with evidence regarding the revenue and expense accounts. The auditor supplements this evidence with the procedures described in the following sections.

Review of Operations—In its simplest form, the review of operations consists of comparing the income statement for the year being audited with that for the prior year and determining the underlying reasons for significant variations or the lack thereof between years. Determining the variations is simply a clerical function, but obtaining explanations of the variations and evaluating their reasonableness require an ability to work with people, as well as persistence and judgment.

Explanations for significant variations are obtained first from client personnel. The auditor should discuss a variation with the individual who is most knowledgeable in the particular area. For example, the sales manager might be the best person to explain an increase in sales, whereas the plant or maintenance foreman should be asked to explain a decrease in maintenance expense. As discussed in Chapter 9, the explanation should come from someone outside the accounting department, if possible.

Although the auditor prefers information from outside the accounting department, he must be prepared for the inevitable misunderstandings that arise in discussions with individuals who lack a financial background. The auditor can limit such misunderstandings by evaluating carefully the reasonableness of the explanations in light of his overall knowledge of the company's operations.

After explanations have been obtained from client personnel, they must be verified by examination of supporting evidence. Although client explanations can save the auditor considerable time in searching for the reasons for variations, they cannot be accepted as audit evidence without verification. The client is, after all, the subject of the audit. As an illustration, if the auditor were informed that selling expense declined because three market researchers were dismissed, this explanation could be verified by an examination of the payroll records.

The review of operations usually is performed at an early stage of the audit, and it serves two purposes in addition to providing evidence of an analytical review of the income and expense accounts. First, it is an excellent means of familiarizing the auditor with the client's business operations. This knowledge is needed if the auditor is to determine whether the financial statements properly reflect the client's operations. Second, it is a means of spotting, at an early point in the audit, areas that may present problems or require special attention. For example, a decline in maintenance expense when the maintenance foreman says that there has been no reduction in maintenance work may indicate a change in the accounting capitalization policy and may necessitate more detail audit work in the maintenance and property areas; or a decline in interest expense when outstanding debt has increased and interest rates have not changed may indicate that interest expense was charged to an incorrect account or was underaccrued. Thus, the review of operations is an important audit procedure and should be performed by an experienced staff member.

A partial example of one form of a review of operations is shown in Figure 12.1. In this example, sales are analyzed by market area (they could also be analyzed by product, type of service, etc.) and compared with the budgeted and prior-year amounts. Variations were computed (percentage as well as absolute variations may also be helpful), and significant variations were explained. Although the explanations were obtained from the president, the auditor examined evidence to support the explanations.

A review of this type would be important in complying with *SAS No. 21* on segment information, which requires an analysis and inquiry regarding the bases for determining segment information and accounting for intersegment transfers and allocation of common assets and expenses. The auditor must also analytically review the segment information and determine whether it was presented consistently from period to period.

Ratio Tests—The ratio tests that can be applied in each audit vary by industry and by individual company. Many of these ratios have been discussed, such as the gross profit ratio, the ratio of bad debt write-offs to revenues, and the ratio of selling expense to revenue. Another ratio usually computed in audits of corporations is income taxes as a percentage of net income before income taxes. An investigation should be made of any significant difference between this ratio and 50% (48% federal corporate rate plus approximately 2% effective state rate). The auditor should recognize important relationships between financial statement amounts and should review them to see whether any appear out of line or inconsistent.

Test of Reasonableness—Though detail audit procedures are important, the auditor should not become so engrossed with the details that he ignores obvious indications of problems with the financial statements. There are many overall tests of reasonableness (sometimes called predictive auditing) that an auditor can and should apply. For example, a quick overall test of revenue for a particular product can be made by calculating an average price of the product

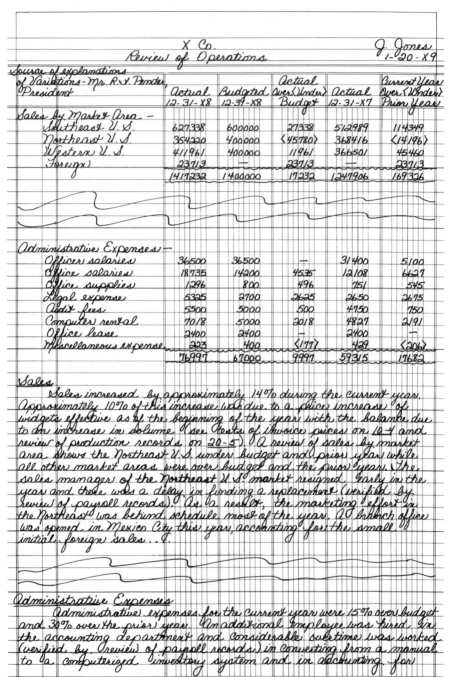

X Co.
Review of Operations — J. Jones 1-20-X9

Source of explanations of Variations—Mr. R. S. Pender, President

	Actual 12-31-X8	Budgeted 12-31-X8	Actual Over (Under) Budget	Actual 12-31-X7	Current Year Over (Under) Prior Year
Sales by Market Area —					
Southeast U.S.	627338	600000	27338	512989	114349
Northeast U.S.	354220	400000	(45780)	368416	(14196)
Western U.S.	411961	400000	11961	366501	45460
Foreign	23713	—	23713	—	23713
	1417232	1400000	17232	1247906	169326
Administrative Expenses —					
Officer salaries	36500	36500	—	31400	5100
Office salaries	18735	14200	4535	12108	6627
Office supplies	1296	800	496	751	545
Legal expense	5325	2700	2625	2650	2675
Audit fees	5500	5000	500	4750	750
Computer rental	7018	5000	2018	4827	2191
Office lease	2400	2400	—	2400	—
Miscellaneous expense	223	400	(177)	429	(206)
	76997	67000	9997	59315	17682

Sales

Sales increased by approximately 14% during the current year. Approximately 10% of this increase was due to a price increase of widgets effective as of the beginning of the year with the balance due to an increase in volume (see Tests of invoice prices on 10-4 and review of production records on 20-5). A review of sales by market area shows the Northeast U.S. under budget and prior year while all other market areas were over budget and the prior year. The sales manager of the Northeast U.S. market resigned early in the year and there was a delay in finding a replacement (verified by review of payroll records). As a result, the marketing effort in the Northeast was behind schedule most of the year. A branch office was opened in Mexico City this year, accounting for the small initial foreign sales. J.

Administrative Expenses

Administrative expenses for the current year were 15% over budget and 30% over the prior year. An additional employee was hired in the accounting department and considerable overtime was worked (verified by review of payroll records) in converting from a manual to a computerized inventory system and in accounting for

Figure 12.1 Review of Operations

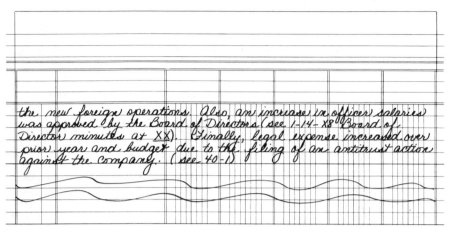

the new foreign operations. Also, an increase in officer salaries
was approved by the Board of Directors (see 1-14-X8 Board of
Directors minutes at XX). Finally, legal expense increased over
prior year and budget due to the filing of an antitrust action
against the company. (see 40-1)

Figure 12.1   (Continued)

(from a review of several sales invoices throughout the year) and multiplying it by the number of units of the product sold (from sales statistics or inventory reports; also, compare sales with plant capacity). The resulting answer seldom exactly equals the recorded revenue for that product, but it should approximate it. Most companies maintain elaborate sales statistics in the sales department which can be used to determine the reasonableness of the sales amount. In addition, budget reports can be very helpful in reviewing the reasonableness of most items of expense. If any overall tests yield unreasonable results, the auditor will expand the detail tests.

Account Analysis—Certain expense accounts usually are analyzed by the auditor because of large unexplained variations from budgeted or prior-year amounts or to gain information about the expenditures that are included in them, as well as to document the authenticity of the expenses. Legal expense is an example of such an account that could contain significant information concerning contingent liabilities. Figure 12.2 illustrates a working-paper analysis of legal expense. Note that the auditor has examined the vouchers in payment of the legal expense as well as the invoices from the attorney and the canceled check (indicated by ✓). The first voucher was for a legal retainer which was of little interest to the auditor. The second voucher, however, was for legal representation in an anti-trust complaint. The anti-trust complaint, which could result in damages being assessed against the company, represents a contingent liability, which (per ①) has been evaluated by the attorney in a letter obtained in connection with the subsequent review and has been noted for disclosure in the financial statements.

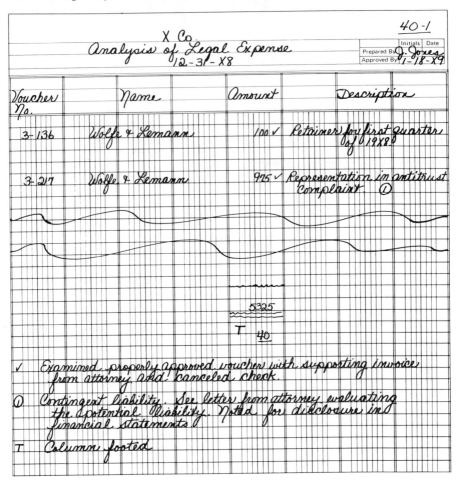

Figure 12.2 Analysis of Legal Expense

Other expense accounts that usually are analyzed include repairs and maintenance (for capital items charged to expense), rents (for identification of lease liabilities), and miscellaneous expense (because it is often the repository of unusual charges and expenses).

Commitments and Contingencies

Although commitments and contingencies are not recorded in the financial statements, generally accepted accounting principles require their disclosure, generally in footnoted form. Because these disclosures are a basic part of the financial statements, they must be subjected to audit procedures just as though they were a recorded amount.

Commitments are important to readers of financial statements because they represent a future cash-flow requirement. The more common commitments are those related to leases, pension plans, and construction expenditures.

Statement of Financial Accounting Standards No. 13 specifies the accounting required for leases. From the lessee's standpoint, leases are classified as either capital or operating leases. Capital leases are reported as assets and liabilities in the balance sheet, while operating leases are disclosed as commitments.[1] Operating lease disclosures include minimum rental commitments in the aggregate and for each of the next five years, total rental expense, and a general description of the leasing arrangements.

The following required disclosures for pension plans are set forth in AC Section 4063.46 *(APB Opinion No. 8)*.

1. A statement that such plans exist, identifying or describing the employee groups covered.

2. A statement of the company's accounting and funding policies.

3. The provision for pension cost for the period.

4. The excess, if any, of the actuarially computed value of vested benefits over the total of the pension fund and any balance-sheet pension accruals, less any pension prepayments or deferred charges.

5. Nature and effect of significant matters affecting comparability for all periods presented, such as changes in accounting methods (actuarial cost method, amortization of past and prior service cost, treatment of actuarial gains and losses, etc.), changes in circumstances (actuarial assumptions, etc.), or adoption or an amendment of a plan.

Disclosure of the amount of construction expenditures anticipated in the succeeding year assists the reader of the financial statements in projecting future cash-flow requirements and usually is included in a footnote.

AC Section 4311.01 *(FASB Statement No. 5)* discusses accounting for contingencies and includes the following definition.

... a contingency is defined as an existing condition, situation or set of circumstances involving uncertainty as to possible gain ... or loss ... to an enterprise that will ultimately be resolved when one or more future events occur or fail to occur. ...

AC Section 4311.10 of the same statement provides guidance as to whether an estimated loss from a contingency should be recorded or disclosed. The requirement for disclosure, if applicable, is specified as follows.

[1]Generally, capital leases are those that (1) transfer ownership of the property to the lessee by the end of the lease term, (2) contain bargain purchase options, (3) have terms equal to 75% or more of the estimated economic life of the property, or (4) have a present value of rentals and other minimum lease payments equal to 90% or more of the fair value of the leased property less any related investment tax credit retained by the lessor.

The disclosure shall indicate the nature of the contingency and shall give an estimate of the possible loss or range of loss or state that such an estimate cannot be made.

Examples of contingencies commonly disclosed in financial statements are pending or threatened litigation, guarantees of indebtedness, threat of expropriation of assets, and proposed assessment of additional taxes.

Audit Objectives

The objectives in the audit of commitments and contingencies are:

1. To determine whether all significant commitments and contingencies as of the balance sheet date are shown properly in the financial statements.

2. To determine whether the disclosures of commitments and contingencies are made in accordance with generally accepted accounting principles.

Audit Procedures and Working Papers

The auditor often finds that clients with strong controls in the conventional areas, such as cash receipts and disbursements and inventories, have surprisingly weak procedures and controls in the area of commitments and contingencies. Perhaps the reason is that internal control is usually the responsibility of an accountant who loses interest in an area rapidly if it does not contain amounts that balance with some predetermined total. At any rate, this is an area in which the auditor sometimes finds it difficult to distinguish between accumulating and auditing the information.

Pensions—Pension plan disclosure data are audited by examining the terms of the contract between the client and the insurance company (a copy should be maintained in the permanent audit file and updated annually) and by confirming these terms on a blind basis with the insurance company.[2] Disclosure of the amount of past service cost, although not required by *APB Opinion No. 8,* is required by the SEC, and the amount usually is included in the confirmation request.[3] Using the information in the confirmation, the auditor normally recomputes minimum and maximum pension expense under *APB Opinion No. 8* and determines whether pension expense recorded by the company falls within this range. He must also review the pension footnote to see whether the required disclosures are made.

---

[2]The term "blind basis" is used to describe confirmation letters that request the person receiving them to furnish information rather than to confirm information furnished by the client. The same type of confirmation letter is used for accounts and notes payable.

[3]See Rule 3-16(g) (4) of Regulation S-X.

Leases—The accumulation of information for the disclosure of lease commitments can be very time-consuming if adequate records are not maintained by the client. Figure 12.3 is an example of a schedule on which lease disclosure information is accumulated and audited. The pertinent lease information has been traced to an *executed* lease agreement (indicated by ✔) and the spreading of the lease payments into time periods has been recomputed (indicated by ↰). The classification of the leases as operating leases under *FASB No. 13* has also been checked. Total rental expense for the year has been cross-referenced to an expense account. The auditor must also review the applicable footnote for proper disclosure.

Other Disclosures—The auditor must be alert in every section of his work for contingent liabilities. Some of the procedures specifically designed to detect contingent liabilities include:

1. Questions about the existence of contingent liabilities on confirmations with banks and other financial institutions (see questions 3, 4, and 5 in Figure 10.4).

2. Review of minutes of meetings of directors and stockholders for actions indicating the existence of contingent liabilities (hiring of special legal counsel, discussion of possible asset expropriation, approval of the guarantee of third-party obligations, etc.).

3. Analysis of the legal expense account for information about litigation and claims for which legal counsel has been engaged.

4. Confirmation directly with legal counsel of the details of litigation and claims being handled for the client. This procedure is discussed further in the subsequent events section.

5. Inquiry of management. The existence of contingent liabilities is a matter that normally is included in the representation letter obtained from management.

Subsequent Events

Generally, an auditor's examination is not completed and the financial statements are not issued until a date subsequent to the audit date. The period between these two dates usually can be divided into three segments: (1) time required for the client to complete, total, and balance the accounting records and otherwise prepare them for audit, (2) time required for the auditor to perform the examination in the client's office (referred to as field work), and (3) time after leaving the client's office required for the auditor to perform quality-control reviews and reproduce and deliver the report.

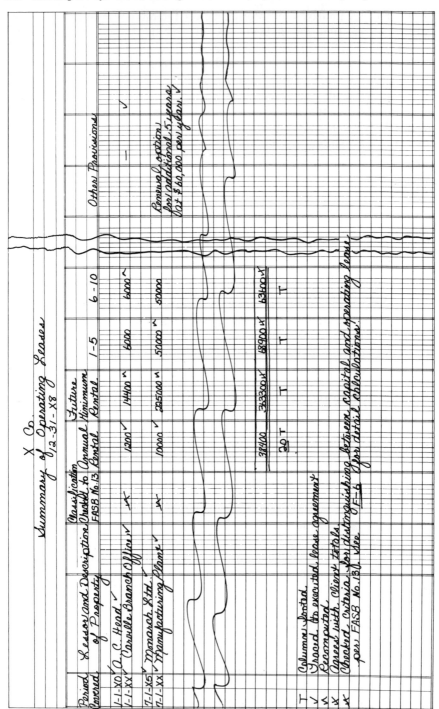

Figure 12.3 Summary of Operating Leases

Because the auditor is in the client's office and has access to the accounting records and management through the end of field work, he must be concerned with material events through that date that might require adjustment of, or disclosure in, the financial statements (the auditor's report also is dated as of the end of field work). Such events can be classified into two types:

1. Events that provide additional information about conditions that existed as of the audit date.

2. Events that provide information about conditions arising after the audit date.

Information provided by the first type of event should be used to evaluate further the accounting estimates reflected in the financial statements, and the financial statements should be adjusted if changes in the estimates are necessary as a result of the additional information. An example would be the final settlement before the end of field work of a major lawsuit that arose prior to the audit date. The settlement would provide additional evidence as to the amount of the liability as of the audit date.

Information provided by the second type of event should not result in adjustment of the financial statements (except that stock dividends and splits can be reflected retroactively), but may require disclosure. An example of this type of event would be an occurrence after the audit date that results in the filing of a major lawsuit against the client. Neither the event nor the lawsuit would affect the amounts shown in the financial statements as of the audit date, and, therefore, the financial statements should not be adjusted. The event should be disclosed, however, because it could have a material future effect on the client.

It is the detection and evaluation of the foregoing types of events that concern the auditor in the audit of subsequent events.

Audit Objectives

The objectives in the audit of subsequent events are:

1. To determine whether the financial statements are adjusted where necessary for subsequent events that provide additional information about accounting estimates.

2. To determine whether adequate disclosure is made where necessary of subsequent events reflecting new conditions that may have a material future effect on the company.

Audit Procedures and Working Papers

The audit procedures applied to the period subsequent to the audit date can be divided into three general groups, each of which is discussed hereafter.

Review of Subsequent Transactions—To the extent that transactions subsequent to the audit date have been summarized in interim financial statements, these statements should be read and compared with the financial statements being audited. Any unusual or unexpected variations or trends (a loss when operations previously had been profitable, significant decreases in assets or increases in liabilities, etc.) should be investigated. For the period from the date of the latest interim financial statements to the last day of field work, the auditor should review the basic accounting records for unusual and significant transactions that may affect the financial statements being audited. Such records include the general ledger, general journal, cash receipts and disbursements records (the review of the cash disbursements records should be coordinated with the review of subsequent transactions as part of the accounts payable work), and sales and expense journals. The review of minutes also should be updated through the last day of field work.

Inquiries of Management—Discussions should be held with the chief executive and financial officers as well as other company officials who may be knowledgeable about the subjects in question. Among the matters discussed would be the following.

1. The existence of any material commitments or contingencies as of the audit date or the last day of field work.

2. Material changes in the equity accounts, long-term debt, or working capital subsequent to the audit date.

3. Unusual adjustments made after the audit date or subsequent changes in accounting estimates made as of the audit date.

4. Changes in trends of sales, expenses, and profit after the audit date.

5. Changes in raw material prices after the audit date and effect on replacement cost of inventories.

6. Cancellation of sales orders or losses of important customers after the audit date.

7. Catastrophes after the audit date (expropriations, fires, explosions, etc.).

8. New contracts or agreements or renegotiation of old ones subsequent to the audit date when sales of products, wages, leases, etc. are affected.

9. Cash-flow requirements for the coming year and sources of financing.

The listing is not all-inclusive, but gives examples of the types of matters that should be covered. The inquiries should be tailored to the operations of each client. The discussions also serve as bases for management's written representations that there were no undisclosed subsequent events that would have a material effect on the financial statements.

Letters from Legal Counsel—The auditor should ask management to request the company's legal counsel to confirm directly to the auditor a description and evaluation of pending or threatened litigation, claims, or other contingent liabilities as of the audit date and the end of field work for which legal counsel has been engaged. For many years, attorneys routinely replied to these requests in various forms, generally without reservation. In the mid-1970's, however, after the SEC charged two prominent law firms with violations of the Securities Exchange Act, auditors began to have difficulty in obtaining adequate responses to such requests. Although the entire matter was complex and involved numerous issues, the principal area of disagreement was unasserted claims. The auditors maintained that unasserted claims should be included in the attorney's response, and attorneys maintained that they could not be responsible for unasserted claims. Furthermore, the attorneys argued that such matters are usually unimportant unless disclosed to the auditor and published in the financial statements, in which case the appropriate party would be likely to assert the claims.

After two years of discussion between committees of the AICPA and the American Bar Association and considerable internal dissent within each group, *SAS No. 12* was issued.[4] In this statement the attorney is asked to comment only on unasserted claims listed by management and to acknowledge professional responsibility to advise management with regard to the possible disclosure of unasserted claims.[5] The following example of an audit request to legal counsel is included in AU Section 337A.01 *(SAS No. 12)*.

In connection with an examination of our financial statements at (balance sheet date) and for the (period) then ended, management of the Company has prepared, and furnished to our auditors (name and address of auditors), a description and evaluation of certain contingencies, including those set forth below involving matters with respect to which you have been engaged and to which you have devoted substantial attention on behalf of the Company in the form of legal consultation or representation. These contingencies are regarded by management of the Company as material for this purpose (management may indicate a materiality limit if an understanding has been reached with the auditor). Your response should include matters that existed at (balance sheet date) and during the period from that date to the date of your response.

Pending or Threatened Litigation (excluding unasserted claims)

[Ordinarily, the information would include the following: (1) the nature of the litigation, (2) the progress of the case to date, (3) how management is responding or

---

[4] Compiled as AU Section 337.

[5] Some auditors consider this accommodation to be unsatisfactory, and have suggested that their responsibilities in the area of unasserted claims be modified unless attorneys become more responsive. See Benjamin Benson, "Lawyers' Responses to Audit Inquiries—A Continuing Controversy," *The Journal of Accountancy,* July, 1977, pp. 72–78.

intends to respond to the litigation (for example, by contesting the case vigorously or by seeking an out-of-court settlement), and (4) an evaluation of the likelihood of an unfavorable outcome and an estimate, if one can be made, of the amount or range of potential loss.] Please furnish to our auditors such explanation, if any, that you consider necessary to supplement the foregoing information, including an explanation of those matters as to which your views may differ from those stated and an identification of the omission of any pending or threatened litigation, claims, and assessments or a statement that the list of such matters is complete.

Unasserted Claims and Assessments (considered by management to be probable of assertion, and that, if asserted, would have at least a reasonable possibility of an unfavorable outcome).

[Ordinarily, management's information would include the following: (1) the nature of the matter, (2) how management intends to respond if the claim is asserted, and (3) an evaluation of the likelihood of an unfavorable outcome and an estimate, if one can be made, of the amount or range of potential loss.] Please furnish to our auditors such explanation, if any, that you consider necessary to supplement the foregoing information, including an explanation of those matters as to which your views may differ from those stated.

We understand that whenever, in the course of performing legal services for us with respect to a matter recognized to involve an unasserted possible claim or assessment that may call for financial statement disclosure, if you have formed a professional conclusion that we should disclose or consider disclosure concerning such possible claim or assessment, as a matter of professional responsibility to us, you will so advise us and will consult with us concerning the question of such disclosure and the applicable requirements of the *Statement of Financial Accounting Standards No. 5*. Please specifically confirm to our auditors that our understanding is correct.

Please specifically identify the nature of any reasons for any limitation on your response.

[The auditor may request the client to inquire about additional matters, for example, unpaid or unbilled charges or specified information on certain contractually assumed obligations of the company, such as guarantees of indebtedness of others.]

The auditor's reporting problems resulting from limitations on a lawyer's response are discussed in Chapter 16.

Chapter 12 Reference

Benson, Benjamin, "Lawyers' Responses to Audit Inquiries—A Continuing Controversy," *The Journal of Accountancy,* July, 1977, pp. 72–78.

Chapter 12
Questions Taken from the Chapter

12–1.   Explain the importance to investors and others of the statement of income.

12–2.   List three disclosures regarding revenues and expenses that are required by generally accepted accounting principles.

12–3.   State the objectives in the audit of revenue and expense.

12–4.   On what procedures does the auditor place heavy reliance in the audit of revenue and expense?

12–5.   Describe the review of operations.

12–6.   What are the purposes of the review of operations?

12–7.   Name four ratio tests that may be applied in an audit.

12–8.   Describe an overall test of revenue for reasonableness.

12–9.   Give two reasons for analyzing an expense account.

12–10. List four expense accounts that would usually be analyzed and indicate the reasons why.

12–11. Why should the auditor be concerned with commitments and contingencies inasmuch as no amounts are shown for them in the financial statements?

12–12. Name three common commitments that exist for many companies.

12–13. How do the lease and pension disclosure requirements differ for the AICPA and SEC?

12–14. Give four examples of contingencies frequently disclosed in financial statements.

12–15. State the objectives in the audit of commitments and contingencies.

12–16.   How are pension plan disclosure data audited?

12–17.   How are lease disclosure data audited?

12–18.   List five audit procedures that are specifically designed to detect contingencies.

12–19.   Describe the time period during which field work is performed.

12–20.   What are the two types of subsequent events that may require adjustment to, or disclosure in, the financial statements? Give an example of each.

12–21.   State the objectives in the audit of subsequent events.

12–22.   List the three general groups of audit procedures applied to subsequent events.

12–23.   Name eight documents or records that would be examined in the review of subsequent transactions. For what period would the review be conducted?

12–24.   List nine matters that would be discussed with management in connection with the audit of subsequent events.

12–25.   What information is a company's legal counsel requested to confirm to the auditor?

12–26.   Discuss the developments that led to the issuance of *SAS No. 12.*

Chapter 12
Objective Questions Taken from CPA Examinations

12–27.   An auditor is reviewing changes in sales for two products. Sales volume (quantity) declined 10% for Product A and 2% for Product B. Sales prices were increased by 25% for both products. Prior-year sales were $75,000 for A and $25,000 for B. The auditor would expect this year's total sales for the two products to be approximately
  a. $112,500.
  b. $115,000.

    c. $117,000.

    d. $120,000.

12–28. The CPA compares current-year revenues and expenses with those of the prior year and investigates all changes exceeding 10%. Choose the error or questionable practice that has the best chance of being detected.

    a. The cashier began lapping accounts receivable in the current year.

    b. Because of worsening economic conditions, the current-year provision for uncollectible accounts was inadequate.

    c. The company changed its capitalization policy for small tools in the current-year.

    d. An increase in property tax rates has not been recognized in the company's current year accrual.

12–29. The audit step most likely to reveal the existence of contingent liabilities is

    a. A review of vouchers paid during the month following the year-end.

    b. Account payable confirmations.

    c. An inquiry directed to legal counsel.

    d. Mortgage-note confirmation.

12–30. The auditor's formal review of subsequent events normally should be extended through the date of the

    a. Auditor's report.

    b. Next formal interim financial statements.

    c. Delivery of the audit report to the client.

    d. Mailing of the financial statements to the stockholders.

12–31. A client has a calendar year-end. Listed below are four events that occurred after December 31. Which one of these subsequent events might result in adjustment of the December 31 financial statements?

    a. Adoption of accelerated depreciation methods.

    b. Write-off of a substantial portion of inventory as obsolete.

    c. Collection of 90% of the accounts receivable.

    d. Sale of a major subsidiary.

12–32. A company guarantees the debt of an affiliate. Which of the following best describes the audit procedure that would make the auditor aware of the guarantee?

    a. Review minutes and resolutions of the board of directors.

    b. Review prior year's working papers with respect to such guarantees.

    c. Review the possibility of such guarantees with the chief accountant.

    d. Review the legal letter returned by the company's outside legal counsel.

12–33. In connection with the annual audit, which of the following is not a "subsequent events" procedure?
   a. Review available interim financial statements.
   b. Read available minutes of meetings of stockholders, directors, and committees, and as to meetings for which minutes are not available, inquire about matters dealt with at such meetings.
   c. Make inquiries with respect to the financial statements covered by the auditor's previously issued report if new information has become available during the current examination that might affect that report.
   d. Discuss with officers the current status of items in the financial statements that were accounted for on the basis of tentative, preliminary, or inconclusive data.

12–34. Which of the following material events occurring subsequent to the balance sheet date would require an adjustment to the financial statements before they could be issued?
   a. Sale of long-term debt or capital stock.
   b. Loss of a plant as a result of a flood.
   c. Major purchase of a business which is expected to double the sales volume.
   d. Settlement of litigation, in excess of the recorded liability.

12–35. A CPA reviews transactions in the repairs and maintenance account for the year and examines supporting documents on a test basis. Which of the following errors or questionable practices has the best chance of being detected by this procedure?
   a. Certain necessary maintenance was not performed during the year because of a shortage of workmen.
   b. The cost of erecting a roof over the storage yard was considered to be maintenance.
   c. The annual painting of the company's delivery truck was capitalized.
   d. Material issue slips were not prepared for some of the maintenance supplies used.

12–36. Accompanied by the production manager, a CPA tours a client's plant. Which of the following errors or questionable practices has the best chance of being detected by this procedure?
   a. Depreciation expense was recognized during the year for a machine which was fully depreciated.
   b. Overhead has been underapplied.
   c. Necessary plant maintenance was not performed during the year.
   d. Insurance coverage on the plant has been allowed to lapse.

12-37.   A CPA examines all unrecorded invoices on hand as of February 29, 19X9, the last day of field work. Which of the following errors or questionable practices has the best chance of being detected by this procedure?

a.   Accounts payable are overstated at December 31, 19X8.

b.   Accounts payable are understated at December 31, 19X8.

c.   Operating expenses are overstated for the 12 months ended December 31, 19X8.

d.   Operating expenses are understated for the two months ended February 29, 19X9.

12-38.   In connection with his review of key ratios, a CPA notes that accounts receivable were equal to 30 days' sales at the end of the prior year and 45 days' sales at the end of the current year. Assuming that there had been no change in economic conditions, clientele, or sales mix, this change most likely would indicate

a.   A steady increase in sales in the current year.

b.   An easing of credit policies in the current year.

c.   A decrease in accounts receivable relative to sales in the current year.

d.   A steady decrease in sales in the current year.

12-39.   A not-for-profit organization publishes a monthly magazine that had 15,000 subscribers on January 1, 19X9. The number of subscribers increased steadily throughout the year and at December 31, 19X9, there were 16,200 subscribers. The annual magazine subscription cost was $10 on January 1, 19X9, and was increased to $12 for new members on April 1, 19X9. An auditor would expect that the receipts from subscriptions for the year ended December 31, 19X9, would be approximately

a. $179,400.

b. $171,600.

c. $164,400.

d. $163,800.

Chapter 12
Discussion/Case Questions and Problems

12-40.   In connection with his examination of Flowmeter, Inc., for the year ended December 31, 19X8, Hirsch, CPA, is aware that certain events and transactions that took place after December 31, 19X8, but before he issues his report dated February 28, 19X9, may affect the company's financial statements.

The following material events or transactions have come to his attention.

1.   On January 3, 19X9, Flowmeter, Inc. received a shipment of raw materials from Canada. The materials had been ordered in October 19X8 and shipped F.O.B. shipping point in November 19X8.

2. On January 15, 19X9, the Company settled and paid a personal injury claim of a former employee as the result of an accident that occurred in March 19X8. The Company had not previously recorded a liability for the claim.

3. On January 25, 19X9, the Company agreed to purchase for cash the outstanding stock of Porter Electrical Co. The acquisition is likely to double the sales volume of Flowmeter, Inc.

4. On February 1, 19X9, a plant owned by Flowmeter, Inc., was damaged by a flood and an uninsured loss of inventory resulted.

5. On February 5, 19X9, Flowmeter, Inc. issued and sold to the general public $2,000,000 in convertible bonds.

Required:

For each of the events or transactions above, indicate the audit procedures that should have brought the item to the attention of the auditor, and the form of disclosure in the financial statements, including the reasons for such disclosures. Arrange your answer in the following format.

Item No.	Audit Procedures	Disclosure and Reasons

(AICPA adapted)

12–41. A CPA evaluating possible audit procedures to test credit sales for understatement considered the following.

a. Age accounts receivable.

b. Confirm accounts receivable.

c. Trace sample of initial sales slips through summaries to recorded general ledger sales.

d. Trace a sample of recorded sales from general ledger to initial sales slip.

Discuss the evidence that each of the procedures above would generate to support the recording of credit sales. Select the best procedure and give the reasons for your selection.

(AICPA adapted)

12–42. Gawdy Fabrics leases store facilities in 15 locations. Each year during the annual audit the CPA reads any new leases (which may consist of 50 pages each) entered into during the year, and records in his audit working papers the lessor, expiration date, and rental amount.

Should the CPA also read or review existing (as opposed to new) leases each year? If so, why? If not, why not? Be as specific as you can.

12–43.   During his review of subsequent events in connection with the audit of Jordon Match Co., Ron Gray, CPA, was informed by the corporate secretary that although there had been two meetings of the board of directors subsequent to the audit date, no minutes had been prepared. The corporate secretary stated that he had been too busy to prepare the minutes, but that the meetings had been routine and no significant matters had been discussed which would require disclosure in the financial statements. He offered to give Gray a letter to this effect.

Discuss each of the following courses of action open to Gray.
a.   Accept the letter from the corporate secretary (who will be the one who subsequently prepares and signs the minutes) and issue the financial statements.
b.   Refuse to accept the letter of the corporate secretary and refuse to issue the financial statements.
c.   Attend the next board of directors' meeting and obtain oral confirmation that no significant matters had been discussed which would require disclosure in the financial statements.

What action should Gray take?

12–44.   Barge Construction Co. constructs large oil barges, each requiring from 6 to 9 months to build. The Company follows the percentage-of-completion method of recording construction revenue for financial statement purposes. At the audit date, two barges were under construction (one estimated as 96% complete and one estimated as 41% complete), and there was a backlog of one barge under firm contract (although informal understandings had been reached to construct five other barges, for which materials had been ordered).

All barges were built under fixed price contracts, including one constructed for the U.S. government during the current year. Warranty costs are recorded as incurred, and have not been significant. The completed-contract method of recording revenue is used for federal income tax purposes; the company's federal income tax returns have never been examined by the Internal Revenue Service.

The CPA performing the audit of Barge Construction Co. is preparing to discuss subsequent events with management. He has a standard checklist of inquiries, but he is wondering if there are not some specific additional questions he should ask in this case. What specific inquiries would you make on the basis of the information above?

12–45.   Commitments and contingencies can take many forms, including the following.
a.   Guarantee of the debt of an affiliated company.
b.   Contract with a builder to construct a plant.
c.   Maintenance of an unrecorded bank account for illegal political contributions.
d.   Discriminatory hiring practices.

    e.   Violation of anti-trust laws.

    f.   Infringement on patent rights.

    g.   Disputes with taxing authorities.

State the audit procedures that could reasonably be employed to detect each of the commitments and contingencies above.

12–46. As auditor of Royal Radio Broadcasting Co., you have learned from past audit experience, familiarity with broadcasting industry practice, and a review of Royal's contracts and agreements the following information regarding Royal's operations.

    a.   Local advertising accounts for about 80% and national advertising about 20% of Royal's revenue.

    b.   Salesmen receive a 15% commission on sales.

    c.   Music license fees average 10% of sales.

    d.   Fixed expenses consist of salaries of about $900,000 and depreciation of $100,000 per year.

    e.   Variable operating expenses such as power, supplies, etc. approximated 10% of sales.

If, upon beginning the audit of Royal for the current year, you learn that total revenue is $2,000,000, what would you expect net income to be under normal operations?

12–47. In general, a lease will be considered a capital lease if it meets any one of the following criteria:

    a.   The lease transfers ownership of the property to the lessee by the end of the lease term.

    b.   The lease contains an option to purchase the property at a bargain price.

    c.   The lease term is equal to 75% or more of the estimated economic life of the property.

    d.   The present value of the rentals and other minimum lease payments is equal to 90% or more of the fair value of the leased property less any related investment tax credit retained by the lessor.

Explain how you would audit a lease classification as capital or operating on the basis of the criteria above. Indicate the specific documentation you would examine, how you would satisfy yourself as to fair value of rentals, estimated economic life, etc.

12–48. During an audit of Miller Carpet Co., Kelley, CPA, learned that the company had been charged with discriminatory hiring policies and that lawsuits totaling $10 million had been brought against it because of this alleged practice. Kelley discussed the matter with the company's legal counsel, who pointed out that the company had a very strong defense and that few successful cases had been brought under the particular statutes involved. He was optimistic that the company would incur no liability.

When Kelley received the legal counsel's letter replying to the request for a description and evaluation of pending or threatened litigation, it contained the same optimistic evaluation of the discrimination suit, but concluded by stating that no opinion could be expressed regarding the potential outcome of the suit. When Kelley contacted the legal counsel by phone to discuss the letter, he was told that the law firm had a policy against giving opinions on the outcome of future litigation, but that the auditor could read between the lines and see that he considered the possibility of an adverse outcome to be very remote.

Discuss the quality of evidence represented by the letter from legal counsel and any other actions or options open to Kelley.

12–49.   When examining documentation supporting costs and expenses, an auditor must consider its availability and its adequacy. Discuss the availability and adequacy of documentation you would expect for the following expenses.

a.   Purchase of stationery and supplies.

b.   Officer travel expenses.

c.   Rental expense on office space.

d.   Cost of taxi to send sick employee home.

e.   Purchase of hams for Christmas gifts to customers.

f.   Contribution (purchase of tickets to Policemen's Ball).

g.   Payment of income taxes.

h.   Payment of directors' fees ($500 per meeting).

i.   Employee salary expense.

j.   Payment of FICA taxes withheld from employees.

k.   Payment of electric bill.

Chapter 13

The Relationship of Statistical Sampling to Evidence-Gathering

Learning Objectives *After reading and studying the material in this chapter, the student should*

Know the difference between the purposes of estimation sampling for attributes and estimation sampling for variables.

Be able to relate the advantages of statistical sampling to the concepts of evidence-gathering.

Understand the limitations of estimation sampling for variables.

Know the difference between mean estimation, ratio estimation, and difference estimation.

Be able to apply the mean estimation, ratio estimation, and difference estimation techniques to audit situations.

Know the key statistical terms used in discussions of estimation sampling for variables.

Several years ago, auditors began to make use of certain statistical concepts to establish a more scientific basis for testing dollar balances on the client's financial statements. These concepts do not alter the specific audit procedures designed to gather evidence on account balances; rather, they refine the scope of the auditors' examination and provide mathematical evaluations of the evidence obtained.

In this chapter we explain how variable sampling can be used in a limited but valuable way to aid auditors in making more scientifically defensible judgments in their audit of financial statement account balances.

It would be helpful to read the appendix before studying the material in this chapter, because it is a primer on estimation sampling and tests of hypotheses.

The Rationale for Using Variable Sampling

The Audit Risk

As explained in Chapter 9, auditors encounter a risk that the evidence they gather will be incompetent or insufficient to give them a valid basis for an opinion on the financial statements. To minimize this risk, auditors exercise judgment on the type and amount of evidence they acquire. The use of variable sampling does not eliminate any of these judgments; it refines *some* of them and leaves the others almost intact.

An example of accounts receivable can be used to illustrate this point. The auditors' task is to decide whether, in their opinion, this figure is a fair representation. To support their viewpoint, they make a variety of judgments on the acquired evidence.

Relationship to Evidence-Gathering Criteria

Independent Outside Sources—Although the evidence-gathering criteria included in AU Section 330 of *AICPA Professional Standards,* Volume 1 *(SAS No. 1)* are not the sole guidelines, they do represent suggestions that are generally accepted by members of the accounting profession. Therefore, these criteria can be used as a convenient reference. For example, evidence gained from independent outside sources is considered to be superior to that acquired from within the client's company. The use of variable sampling has no effect on the auditors' decision to use confirmation requests in obtaining evidence from outside sources. However, the use of these sampling devices will determine the number of confirmations that auditors circulate, and will provide auditors with certain numerical conclusions about the correctness or incorrectness of the client's book balance.

Quality of Internal Control—Another evidence-gathering criterion is that accounting data developed under satisfactory conditions of internal control are more likely to be reliable than the same data developed under unsatisfactory conditions of internal control. To adjust for the effects of good and poor internal control, auditors traditionally vary the size of the sample inversely with the quality of such control. For example, 100 accounts receivable confirmation requests might be circularized if internal control over accounts receivable and sales were adequate; a lower quality of control in the same areas could dictate a larger sample size.

How would the use of variable sampling affect the decision on sample size for substantive tests? First, the auditors would have to convert their qualitative judgments into numerical terms (needed reliability is 95% because internal control is only fair). Then, by the use of statistical concepts, a sample size of accounts receivable confirmations could be obtained that *should* provide auditors with the needed evidence for attesting to the fairness of the account. Variable sampling does not supply auditors with judgments; it furnishes numerical results based on refinements of judgment.

Persuasiveness of Evidence—Auditors often are forced to rely on persuasive rather than convincing evidence. In view of the fact that variable sampling produces numerical results based on a conversion of auditing judgments into quantitative terms, these sampling concepts appear to be of somewhat limited value here. The auditors' judgment as to the persuasiveness of evidence is essentially non-quantitative, although the expression of that persuasiveness sometimes can be shown in numbers. For instance, if auditors are using variable sampling, they might be satisfied as to the fairness of an accounts receivable book balance of $155,078 if the evidence they gather shows that, at 90% reliability, this figure is within ± $10,000 of their estimate of the balance.

Such a situation involves a tacit admission by the auditors that their evidence is persuasive (within ±$10,000), not convincing. Therefore, variable sampling can be an aid in that it expresses this admission of persuasiveness in definite numerical form.

Cost of Evidence—In deciding on the sufficiency of evidence, auditors also can use the guideline that there should be a rational relationship between the cost and benefits of obtaining evidence. If auditors are able and willing to convert some of their judgments into numerical form (e.g., ± $10,000 precision and 90% reliability), it appears that variable sampling can be useful in estimating the cost of obtaining needed additional evidence.

As a case in point, assume that the auditors decide to use variable sampling in the selection of the number of confirmation requests to be circularized. On the basis of their judgment of the needed precision and reliability, the auditors calculate a sample size of 200 to obtain 90% reliability that the true accounts receivable balance is within a certain precision range of their estimate.

But as one knows from studying statistical concepts, a 90% reliability is also a 10% risk (the alpha risk) that the auditors will reject the client's book balance as being materially incorrect when, in reality, it is correct. Recognizing the magnitude of this risk, the auditors may wish to lower it by raising the desired reliability level. An increase in the reliability level means a larger number of confirmation requests and an additional cost of obtaining evidence. If variable sampling is being used, the additional cost of acquiring this evidence can be measured more precisely by calculating the extra number of confirmation requests that must be circularized to increase the reliability from 90% to its desired level.

For example, assume that 200 and 250 confirmation requests are needed to obtain reliability levels of 90% and 95%, respectively. The auditors would need to decide whether the marginal cost of the extra sample of 50 is justified by the benefits to be obtained from the additional reliability of 5%. Though the auditors must still make a cost/benefit decision, they have a more refined basis for doing so.

Objectivity of Evidence—The last listed guideline for deciding on the sufficiency of audit evidence is that the auditors should be willing to give consideration to and be objective about any evidence, regardless of whether this evidence appears to support or contradict the financial statement representations. Auditors cannot depend on variable sampling to help them decide whether a given piece of evidence supports or contradicts the client's figures. But variable sampling is a good device for expressing, in mathematical terms, the extent to which the *total* evidence does or does not support these representations.

The collection of responses from accounts receivable confirmation requests can serve as an illustration. Assume that the auditors are willing to accept the fairness of the client's book figure if the responses from their sample of confirmations show an estimate of the accounts receivable balance that is within $20,000 of the client's book figure. As the auditors evaluate the responses, they make decisions as to whether these responses represent agreement or disagreement with the book figure. Variable sampling is of no use to the auditors in making these types of judgments. If a customer's response shows a receivable balance different from the client's book figure, the auditors investigate this difference and make a judgment as to whether the client's book figure is supported or contradicted by the response.

However, a statistical analysis can be made of the *total* dollar amount of the responses, and this analysis will show whether the evidence collected from accounts receivable supports or contradicts the client's book balance. If the auditors' estimate of the accounts receivable figure is not within $20,000 of the book amount, it would appear that the evidence fails to support the financial statement representation. By using variable sampling techniques, the auditors are able to make more quantitatively definitive statements about whether the evidence does or does not support the financial statement representations.

Techniques of Variable Sampling

The Transition of Judgments to Numerical Terms

If one accepts the proposition that variable sampling has a limited but useful role in the gathering of evidence on the fairness of financial statement balances, it is desirable to delve into the techniques of applying this method to audit situations.

If auditors are to use variable sampling in some of their tests, it is essential that they be prepared to express their judgments in numerical form. It is not suitable for auditors to state that they will accept the fairness of an account balance if their estimate of that balance is *fairly* close to the figure shown on the client's books. The auditors should be specific about the precision they require. Precision is defined in AU Section 320B.09 *(SAS No. 1)* as ". . . the range or limits within which the sample result is expected to be accurate."

Likewise, auditors can no longer express in general terms their certainty about the accuracy of the sample result. They must indicate their desired reliability, which is defined by AU Section 320B.09 *(SAS No. 1)* as ". . . the mathematical probability of achieving that degree of accuracy."

It is not enough, however, simply to state that auditors must convert their judgments into expressions of precision and reliability. Guidelines are needed to make this decision.[1] Fortunately, some general criteria have been issued by the AICPA through AU Section 320A.10 *(SAS No. 1)*.

> Although "precision" and "reliability" are statistically inseparable, the committee believes that one of the ways in which these measurements can be usefully adapted to the auditor's purposes is by relating precision to materiality and reliability to the reasonableness of the basis for his opinion.

Precision

One possible starting point for determining the needed precision is to select the largest dollar difference that could be tolerated between the sample estimate and the true figure. Any difference greater than this figure is considered a material error. One-half of this maximum difference is taken as the precision. For instance, the auditors might consider the client's accounts receivable book figure to be fair if it is within a $20,000 range of their estimate of the book figure. If their estimate is $140,000, the book figure could be as low as $130,000 or as high as $150,000 and it would be acceptable. Thus the precision is expressed as ± $10,000.[2] This particular concept, which is referred to as letting precision represent one-half of materiality, is followed by some CPA firms that use variable sampling.

A variation of the concept was advocated by Elliott and Rogers in the July 1972 issue of the *Journal of Accountancy*. The authors suggest a flexible approach to the relationship between materiality and precision whereby the percentage relationship between the two varies according to the beta risk that the auditors are willing to take. This approach is discussed later in the chapter.

Auditors must answer a fundamental question before precision can be derived: how is materiality itself determined for the audit of a given account? As

---

[1]It is possible that auditors' reluctance to make this transition is partly responsible for the failure of variable sampling to gain more widespread support in the accounting profession.

[2]In this text, variable sampling precision is expressed as a single dollar amount (± $10,000), and precision range is expressed as low and high amounts (from $130,000 to $150,000).

pointed out in Chapter 9, an item is considered material if it would cause prudent individuals to change their minds about a decision concerning the firm's financial statements.

Beyond this statement, however, there is little guidance to help auditors convert their ideas about materiality into dollar precision figures. Some auditors might compare the financial statement figure with the balance sheet total, the current asset total, and the net income for the period. It also might be appropriate to consider the other evidence that is being gathered on the same account. For example, if confirmation requests were the only evidence gathered on the accounts receivable balance of $155,078, the auditors might consider ± $10,000 to be an appropriate precision. However, if confirmations were only a *part* of the total evidence to be acquired on this account (a more likely circumstance), they might be willing to be more lenient and allow ± $20,000 as precision.

Reliability

The Beta Risk—The question of how to derive reliability depends on whether the auditors wish to relate the reliability percentage to the beta or to the alpha risk.[3] The beta risk is the probability that the auditors will accept the account balance as correct when this balance is materially in error. The alpha risk is the chance that the balance will be rejected as being materially in error when such balance is correct.

Consider the first option. It is not necessary to study variable sampling to understand that the sample size of evidence gathered to evaluate the fairness of an account balance should bear an inverse relationship to the quality of internal control and to the nature of other evidence acquired on that account. As an example, assume that the auditors are reasonably concerned about the possibility that the evidence they gather on accounts receivable will lead them to a false impression that the balance is stated fairly. Without any criteria to guide them, the auditors might be inclined to circularize 300 confirmation requests (corresponding to a reliability of 95%).

However, the study and evaluation of internal control in the area of sales and accounts receivable reveal a fairly strong system. In addition, other evidence acquired on accounts receivable affirms the fairness of the account. In this case, the auditors might be willing to adjust their desired reliability to 90% or 80%, thus reducing the number of necessary confirmation requests and increasing the beta risk. The reasons for the auditors' willingness to reduce reliability might be that (1) this particular beta risk pertains only to the risk that the evidence gained from *confirmation requests* will incorrectly affirm the fairness of the account, and (2) the internal control and *other* evidence are positive toward the view that the account balance is fairly stated.

[3]The beta and alpha risks are discussed in the appendix. It is pointed out that when materiality is twice the precision amount, the beta risk is one-half the alpha risk if the two-tailed test of the null hypothesis is conducted.

AU Section 320B.35 *(SAS No. 1)* contains a formula designed to aid auditors in deriving a final reliability figure:

$$S = 1 - \frac{(1 - R)}{(1 - C)}$$

where S = The final reliability level for substantive tests.

R = The reliability level without considering reliance on internal control or other relevant factors. (This is the starting point, which might be 90% or 95%.)

C = The reliance assigned to internal control and other relevant factors.

As can be seen, the higher the value of C, the lower the value of S. For example, if a significant amount of reliance is placed on a good system of internal control, and if other audit evidence gathered in this area is favorable, the value of C might be 90%. If the initial factor is 95%, the final reliability factor is 50%. In other words, a strong system of internal control and other positive factors could allow the auditor to lower the sample size and raise the beta risk on substantive tests. If control were moderately strong and the reliance factor were placed at 30%, an initial reliability of 95% would be lowered to 93%. If control were weak and no reliance placed on it, the final reliability percentage would remain at 95%.[4]

Some auditors might question the lowering of the reliability level to 50%. If the statistical sampling represents the *only* evidence gathered on the account, this question probably is justified. But if this particular test results in only one piece of the evidence obtained on the balance, there may be some justification for a low reliability level, because any incorrect acceptance of the account balance on the basis of one item of evidence is "buffered" by the other evidence on the same account.

The Alpha Risk—What if the reliability level is associated with the alpha risk? In this case there would seem to be a question of whether reliability should be a function of internal control or of the cost of acquiring evidence.

Assume that the auditors start with a 95% reliability (a 5% alpha risk) and decide to adjust the figure downward for a good system of internal control. The existence of such a control system would seem to imply that the records are reliable and that the substantive tests of balances conducted by the auditor should result in a correct conclusion that the account balance is stated fairly.

Yet a reduction of the reliability from 95% to, say, 80% is a substantial increase in the risk that a correct amount will be rejected incorrectly as being materially in error. It would seem that the auditors are in a dilemma if their sampling results show a materially misstated account balance in spite of strong

---

[4]There is an explanation of this *SAS No. 1* formula in the article "A Statistical Interpretation of SAP No. 54," by Donald M. Roberts, and published by *The Journal of Accountancy,* March, 1974, pp. 47–53.

internal control. If the auditors increase the sample size to obtain results that show a correct balance, they do so at an added cost, a cost they might be inclined to consider unnecessary.

Some practitioners raise the alpha risk on substantive tests if the internal control in the same area is strong. But there is a tendency in practice to start out with a very low alpha risk of, say, 2–5% (corresponding with reliability levels of 95–98%). Depending on the auditors' opinion of the quality of internal control, the risk might be raised as high as 10% (a 90% reliability level). By following this pattern, these practitioners retain a fairly low risk of incorrectly rejecting a correct balance; at the same time, they have the benefits of a reduced sample size if internal control is strong.

There seems to be an emerging viewpoint that the alpha risk should vary with the cost of obtaining evidence. Such an opinion was expressed in the Elliott and Rogers article referred to previously. As an illustration, take the case of the auditors' circularization of negative accounts receivable confirmation requests. In comparison with positive confirmations, this type of evidence is relatively inexpensive to obtain, because a non-reply does not require followup.[5] Considering this, the auditors might be inclined to use a fairly high reliability because of the low cost of sending requests.

In contrast, assume that the circularization of confirmation requests is impracticable and that the auditors are forced to gather evidence from alternative sources, such as shipping notices and sales invoices. Assume also that the inspection of these documents in the client company is a costly and time-consuming process. In these circumstances, a fairly low reliability might be appropriate so as to hold down the cost of audit evidence.

A wide variety of approaches are being advocated for the use of variable sampling to gather evidence on year-end account balances. Perhaps as more research is done on the matter, a consistent framework will begin to emerge.[6]

Application of Variable Sampling—Mean Estimation

The foregoing discussion provides background for an application of mean estimation techniques, with accounts receivable balances as a model. As explained more fully later, mean estimation is the method of summing the audited values and using the sample mean of such values as an estimate of the population total. Ratio and difference estimation are discussed in later sections.

Steps in Applying Mean Estimation

Mean estimation, as well as other statistical sampling techniques, can be applied more efficiently by following a sequence of logical steps. These steps are

---

[5]This is not to suggest that the acquisition of evidence by circularization of negative accounts receivable confirmation requests is cheap in comparison with *all* forms of evidence-gathering.

[6]For a more detailed look at the relationship between internal control and reliability, see "Statistical Sampling," John C. Broderick, *The Arthur Young Journal,* Autumn/Winter, 1972–73, pp. 35–44.

outlined in the next paragraph and discussed more fully in the remainder of the section.

1. Determine the distribution of the account balances in order to assess the need for stratification.

2. Stratify the accounts if high variability is a problem.

3. Determine the appropriate sample size by:
   A. Determining the needed reliability.
   B. Determining the desired precision.
   C. Estimating the population standard deviation.

4. Select the sample and perform the necessary audit procedures.

5. Quantitatively evaluate the sample results by estimating the population total and calculating a precision range.

6. Consider a quantitative evaluation of the sample results.

Estimation of an Accounts Receivable Total

Distribution of the Account Balances—Assume that the auditors wish to apply mean estimation methods to accounts receivable confirmations to estimate the October 31 balance. The client's figure is $155,078. One of the first steps logically taken by the auditors is to review the numerical distribution of the accounts. Such a distribution is shown in the following table.

Range of Account Balances	Number of Accounts	Total Dollar Amount of Accounts
Credit Balance	1	$ (920)
$ 0– 999	60	23,705
1,000– 1,999	13	16,793
2,000– 2,999	10	24,742
3,000– 3,999	7	23,620
4,000– 4,999	2	9,466
5,000– 5,999	4	21,160
6,000– 6,999	0	0
7,000– 7,999	1	7,049
8,000– 8,999	0	0
9,000– 9,999	0	0
10,000–10,999	0	0
11,000–11,999	0	0
12,000–12,999	0	0
13,000–13,999	0	0
14,000–14,999	2	29,463
	100	$155,078

It is obvious that the distribution of 100 accounts (99 after taking out the credit balance) is skewed heavily, and also that it has a relatively high standard deviation. The following is some information about the population of accounts receivable (although the auditors may not know this).

1. The mean is $1,576.

2. The total is $155,998 (after adding the credit balance back).

3. The standard deviation is $2,413.

4. The coefficient of variation is 1.53.

5. The skewness is 3.54.

An even more graphic illustration of the nature of this distribution is shown in the following chart (the credit balance is omitted because it probably would be classified as a current liability).

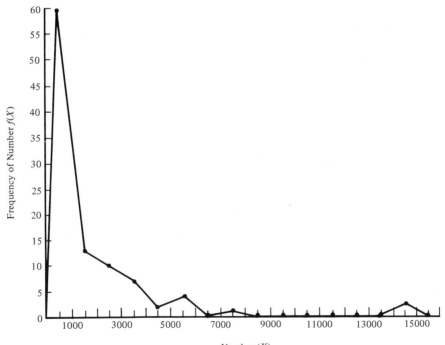

Number (*X*)

Any attempt to determine a statistically derived sample size for this accounts receivable distribution would be ineffective because the size of such a sample probably would be in excess of 99.

Stratification—A Solution to High Variability—How can statistical sampling methods be applied to a distribution of accounts receivable balances with such a large skewness and variability? One answer is to stratify the accounts in two or more sub-groups and then treat each sub-group as a separate population. Review of the distribution of accounts shows the strata into which each dollar group can be placed.

One first notices two very large account balances totaling $29,463 (out of a population total of $155,998, after reclassifying the credit balance as a current liability). Clearly, these two accounts could and should be sampled separately with positive confirmation requests. By so doing, one reduces the variability in the remaining population of 97 and covers approximately 20% of the total dollar balance in accounts receivable.

A further examination of the account balance distribution shows that an additional 14 accounts are listed in the dollar categories starting at $3,000 and ending at $7,999. Positive confirmation requests also might be appropriate for these balances.

Among the 60 accounts listed in the $0–999 classification are 14 with zero balances. Extracting these accounts will lower the variability of the remaining distribution. In addition, the auditors may wish to sample separately accounts with zero balances on the books if there has been unusual or excessive activity.

After separation of the accounts according to the criteria explained in the foregoing paragraphs, the following groups emerge.

Type of Account	Number of Accounts	Type of Evidence
Zero balances	14	Positive confirmation on some accounts with excessive activity
Balances ≥ $3,000	16	Positive confirmation on some or all accounts
Balances between $1 and $2,999	69	Negative confirmations on a sample of accounts—statistical sampling techniques applied

The new population to which variable sampling methods can be applied has a different and more suitable set of statistical measures.

1. The mean is $946.

2. The total is $65,240.

3. The standard deviation is $754.

4. The coefficient of variation is .80.

5. The skewness is 1.14.

Note the distribution of the new population as shown in the following table and chart.

Range of the 69 Remaining Account Balances after Stratification	Number of Accounts	Total Dollar Amount of Accounts
$ 1– 499	21	$ 5,170
500– 999	25	18,535
1,000–1,499	10	11,306
1,500–1,999	3	5,487
2,000–2,499	4	8,660
2,500–2,999	6	16,082
		$65,240

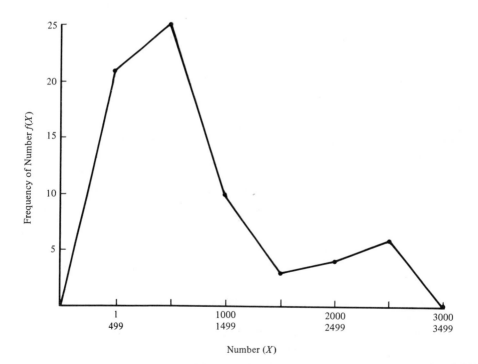

Number (X)

Determination of Needed Reliability—As has been noted, reliability could be related to the beta risk and the quality of internal control, or associated with the alpha risk and the cost of obtaining evidence. The latter association is chosen for the following reasons.

1. Statistically, the reliability or confidence-level percentage is the complement of the alpha risk percentage or the probability of committing a Type I error.

2. It seems logical to allow the alpha risk to vary with the cost of obtaining evidence. This viewpoint has been expressed in the literature and might have a strong appeal to auditing practitioners.

Therefore, consider the desired reliability on the basis of the relative cost of circularizing negative accounts receivable confirmation requests. As pointed out in Chapter 10, the client's customer normally replies to negative confirmation requests only if there is a disagreement between the amounts on the client's and customer's records. In most cases, then, the auditors would not expect to spend any more time on a confirmation once the request is sent to the customer.

Sometimes negative requests can be expedited by adding a short statement to the statement of account mailed monthly to the customers. In view of the relatively inexpensive nature of this evidence, choose a 95% reliability and a 5% alpha risk of incorrectly rejecting the "revised" accounts receivable population total of $65,240.

Determination of Desired Precision—Ascertaining the desired or target precision is a more difficult task because there are wider ranges of figures from which to choose (in auditing practice, a 90–98% reliability is not untypical). The auditors would prefer as narrow a precision range as possible, considering the materiality and importance of the account under examination.

One method of assigning a desired precision is to consider the *largest* amount by which the true accounts receivable total could differ from the client's representation of that total and still be acceptable. In other words, if the true figure is as low as $50,000 or as high as $80,480, would this materially affect the financial statement representation? Would the auditors modify their opinion if a $15,240 ($65,240 − $50,000 or $80,480 − $65,240) adjustment were proposed and not booked by the client? This maximum dollar difference is established ($30,480), and then one-half of the figure is used as the desired or target precision.

Actually, if a two-tailed test is conducted, the precision *range* is the *same* as the maximum allowable difference or materiality. A $10,000 precision means that the true figure is within a $20,000 range (the population estimate plus $10,000 and the population estimate minus $10,000).

In determining the proper materiality for the client's accounts receivable figure of $65,240, two items should be considered.

1. A significant percentage of the original accounts receivable balance of $155,998 will be sampled separately with positive confirmation requests.

2. Negative accounts receivable confirmations represent only part of the evidence gathered to ascertain the fairness of the account total. Other forms of evidence include examination of supporting documents, study of the applicable internal control system, and various inquiries of the client.

Taking into consideration the factors explained in the foregoing paragraphs, assign a $15,000 precision figure, a $30,000 precision range, and a materiality of $30,000.

Estimation of the Population Standard Deviation—The auditors also need a reasonably good estimate of the population standard deviation to determine an optimum sample size. The standard deviation is a measure of variability in the numerical distribution; the higher this variability, the larger the sample size necessary to increase the probability of drawing samples whose characteristics are close to those of the population.

This concept of the effect of variability on the sample size is demonstrated in the following chart and accompanying explanation.

Population No. 1—Mean of 52.4 and
Standard Deviation of 11.27

				50					
			45	50	55	60			
		40	45	50	55	60	65		
	35	40	45	50	55	60	65	70	
30	35	40	45	50	55	60	65	70	75

If a random sample of three is drawn without replacement from this population of 29, the "least representative" or "worst" sample mean is 33.3 [(30 + 35 + 35) ÷ 3] or 71.6 (70 + 70 + 75) ÷ 3. These means are taken from the two extreme ends of the distribution.

If the sample size is increased to four, the "worst" sample means are 35 [(30 + 35 + 35 + 40) ÷ 4] and 70 [(65 + 70 + 70 + 75) ÷ 4]. The use of a sample size of four gives some improvement over the use of three, but the improvement appears to be small, because the variability in the population is not very large.

Population No. 2—Mean of 54.8 and
Standard Deviation of 22.53

				50					
			40	50	60	70			
		30	40	50	60	70	80		
	20	30	40	50	60	70	80	90	
10	20	30	40	50	60	70	80	90	100

Consider Population No. 1 in relation to Population No. 2. A sample size of three from Population No. 2 could produce sample means of 16.6 [(10 + 20 + 20) ÷ 3] and 93.3 [(90 + 90 + 100) ÷ 3], each of which is farther from the true mean than the sample mean in Population No. 1. Because Population No. 2 has a larger standard deviation, a larger increase in the sample size is needed to bring the "worst" sample means fairly close to the true mean. For instance, in a sample of 12, the "worst" sample means are 33.3 [(10 + 20 + 20 + 30 + 30 + 30 + 40 + 40 + 40 + 40 + 50 + 50) ÷ 12] and 76.6 [(60 + 60 + 70 + 70 + 70 + 70 + 80 + 80 + 80 + 90 + 90 + 100) ÷ 12].

What these two charts and sets of explanations show is that in Population No. 2, a sample of 12 is needed to bring the "worst" sample means within about 20 or 25 of the true mean. Yet in Population No. 1, a sample size of only four is needed to achieve comparable results. The reason for this difference is the larger variability in Population No. 2, and this variability is measured by the standard deviation.

What devices can be used to make an estimate of the standard deviation, if ascertaining the actual population figure is impractical? One solution is for the auditors to make an estimate based on past experience. Another possibility is to select a preliminary sample of 30 accounts receivable balances and to take the standard deviation of this sample. As indicated in the appendix, a sample of 30 is considered by statisticians to be large and should provide the auditors with a reasonably good estimate of the population standard deviation. This is true with sets of book values or when audited values (customer replies on accounts receivable requests) are close to book values. If there are large dollar differences between the book and audited amounts, a sample of 30 might be unreliable.

The following list of accounts receivable balances represents the remaining 69 accounts from which a statistically derived sample of negative confirmation requests are circularized.

Customer Number	Balance	Customer Number	Balance	Customer Number	Balance
1	$1,020	32	$ 742	70	$ 660
2	696	33	480	71	200
3	208	34	240	72	2,455
4	1,847	35	480	75	906
6	1,240	37	845	76	100
8	1,190	38	345	77	1,041
9	2,554	39	1,030	78	1,190
10	218	40	2,035	79	576
13	2,106	41	988	81	10
14	402	42	306	83	885
15	1,335	45	2,652	84	104
17	2,507	46	195	86	2,940
18	2,824	47	950	87	576
19	650	48	306	88	1,020
21	144	50	2,605	90	873
22	669	52	850	92	119
23	543	58	582	93	720
24	730	60	1,780	95	975
25	770	64	1,080	96	1,160
26	145	65	425	97	540
28	640	66	569	98	145
29	890	67	2,064	99	710
31	238	69	1,860	100	360
					$65,240

The value of a sample of 30 as an estimator of the standard deviation can be observed by calculating several such estimates and comparing them with standard deviation estimates derived from much smaller samples.

Actual Population Standard Deviation	Standard Deviation of 5 Samples of 30		Standard Deviation of 5 Samples of 10	
$754	$821	746	$927	561
	838	659	910	889
	790		354	

As can be observed, the "worst" 30-sample estimate of the population standard deviation is closer than the "best" 10-sample estimate. Undoubtedly, a sample of 40 would provide an even better estimate, but 30 is the suggested size for the purpose of determining the sample size necessary to achieve the desired precision and reliability.

Use of the Variables to Determine Sample Size—The desired reliability of 95%, the desired precision of $15,000, a $790 estimate of the population standard deviation, and the population size of 69 can be used to derive a sample size. If sampling is conducted with replacement, the formula is:

Sample size =

$$\left(\frac{\text{Estimate of the population standard deviation} \times \text{Reliability coefficient} \times \text{Population size}}{\text{Desired precision}}\right)^2$$

or the statistical notation is:

$$n = \left(\frac{s \times U_R \times N}{A}\right)^2$$

$$n = \left(\frac{790 \times 1.96 \times 69}{15,000}\right)^2$$

$$n = 51.$$

As applied to negative confirmation requests, sampling with replacement does not mean that two letters are mailed to the same customer if the same number is selected in the random sample. It simply means that a larger size must be selected to overcome the possible obstacle of choosing the same number more than once when circularizing only one letter.

A more practical approach may be to sample without replacement and thus derive n separate customer numbers. If this method is adopted, the following correction factor is applied to the sample of 51.

$$n' \text{ (without replacement)} = \frac{n \text{ (with replacement)}}{1 + \dfrac{n \text{ (with replacement)}}{N}}$$

$$n' = \frac{51}{1 + \dfrac{51}{69}}$$

$$n' = 30 \text{ (without replacement).}$$

What the auditors now have is a sample size which, if random drawing is used, should provide an accurate estimate of the population total, within the desired precision range and at the desired reliability level. It is presumed that the estimate of the population standard deviation produced by this final sample of 30 will be close to the original estimate of $790.

Selecting the Sample and Performing the Audit Procedures—Once the sample size is derived, the sample itself is selected. It should be emphasized that the audit procedures are *no different* when statistical sampling is used than they are when judgmental sampling is employed. In this illustration, the auditors mail 30 negative confirmation requests to randomly selected customers.

There are two fairly common techniques for deriving a set of random numbers. One method is to draw samples from a random number table by the techniques described in Chapter 6. Another is to list the numbers from a computer-generated random number program. Assume that the latter method is used. Because accounts receivable customer numbers are available, they can be identified easily with the numbers produced by the computerized random number generator.

Once the 30 customer numbers are selected, negative confirmation requests are sent and the auditors wait for replies. If none is received on a given account, the balance on the client's books is presumed to be correct. If a letter is forwarded by a customer and contains an indication of a disagreement with the book balance, the auditors seek to find the reason for this difference. A satisfactory resolution of this discrepancy is tantamount to a satisfactory reply from the customer.

But what if the auditors are unable to reconcile the difference between the balance listed on the client's records and the amount confirmed by the customer? If it is apparent that the client's figure is wrong, it seems reasonable to use the customer amount to estimate the population total.

Assume that the sample is drawn, the confirmation letters are mailed, and the customer replies are received. On page 558 is a list of the 30 customers receiving confirmation requests. If no reply is received, the account balance on the books is assumed implicitly to be correct, and this assumption is noted on the list. Presume that unresolved discrepancies are errors in the records; therefore, to estimate the population total, the figure confirmed by the customer is used as the correct sample amount.

Estimate of the Population Total—Precision Range—The next step for the auditors is to use the tally of confirmation results to make an estimate of the population total. A precision range is developed around this estimate, and the auditors then may state with 95% reliability that the true population total is within this range.

These sets of calculations are carried out by taking the $29,448 total of the confirmation results and developing the sample mean.

$$\$29,448 \div 30 = \$982$$

Then an estimate of the population total is made.

$$\$982 \times 69 = \$67,758$$

It should be pointed out that $67,758 is only one of *many* population estimates that could have been derived by the auditors. How many could be selected if, from a population of 69, a random sample of 30 is selected without replacement? The answer is

$$\frac{69 \text{ (Factorial)}}{30 \text{ (Factorial) } (69 - 30) \text{ (Factorial)}}$$

Random Number	Customer Number	Client's Book Balance	No Customer Reply	Amount of Customer Reply (error in records)	Amount Used in Population Estimate
1	1	$1,020	x		$ 1,020
2	2	696		$1,534	1,534
10	14	402	x		402
11	15	1,335	x		1,335
14	19	650	x		650
15	21	144	x		144
17	23	543	x		543
22	29	890	x		890
34	45	2,652		2,016	2,016
37	48	306	x		306
38	50	2,605	x		2,605
43	65	425	x		425
44	66	569	x		569
45	67	2,064	x		2,064
46	69	1,860	x		1,860
49	72	2,455	x		2,455
50	75	906	x		906
51	76	100	x		100
52	77	1,041	x		1,041
53	78	1,190	x		1,190
54	79	576	x		576
57	84	104	x		104
58	86	2,940	x		2,940
59	87	576	x		576
61	90	873	x		873
62	92	119	x		119
65	96	1,160	x		1,160
66	97	540	x		540
67	98	145	x		145
69	100	360	x		360
					$29,448

(31,627,280,000,000,000,000). The reason statisticians consider 30 to be a large sample is that there are so many population estimates (and sample means) that the distribution is expected to be fairly close to normal, even if the population from which the sample is taken is skewed.

Because of the large number of possible population estimates, the auditors are not concerned with the fact that the client's book balance is $2,518 removed from their estimate. However, they *are* concerned with whether the amount on the records is within the precision range calculated for their population estimate.

This precision is developed by use of the following formulas.

$$A \text{ (attained population precision)} = s_{\bar{X}} \times U_R \times N$$

$$s_{\bar{X}} = \frac{s}{\sqrt{n}}$$

$$s_{\bar{X}} = \frac{783^7}{\sqrt{30}} = 143$$

$A = 143 \times 1.96 \times 69 = 19,339$ (attained precision with replacement)

This precision is multiplied by a finite correction factor because sampling is done without replacement.

$$19,339 \times \sqrt{\frac{N-n}{N-1}}$$

$$19,339 \times \sqrt{\frac{69-30}{69-1}} = 14,658$$

(attained precision without replacement—rounded figure)

The auditors then are able to make the following statements. At 95% reliability, the true population total of accounts receivable (whatever that total may be) is between

$53,100 ($67,758, the population estimate, minus $14,658)

and

$82,416 ($67,758, the population estimate, plus $14,658).

It should be emphasized that this population estimate and precision range are only single examples of the many that can be generated from this population using a sample of 30. Ninety-five percent of such estimates and ranges include the true population total. However, 5% of the estimates and ranges do not. Thus, there is a 5% probability (the alpha risk) that a sample of 30 could result in an estimate that causes the auditors to reject a true population total.

Is a $53,100-to-$82,416 precision range acceptable for purposes of the audit of the client's $65,240 book balance? The answer is affirmative, because (1) this $65,240 is within the precision range, and (2) the achieved precision of $14,658 is lower than the desired precision of $15,000.

Perhaps a short explanation of the achieved and desired precision figures should be made. The figure $15,000 is used to calculate a sample size. However,

---

[7]The standard deviation of the preliminary sample of 30 used to calculate the sample size is based on the book values. The sample of 30 used to calculate attained precision contains one or more different audit amounts.

the achieved precision based on the standard deviation of the sample is different (and hoped to be narrower) than the desired or target precision. To provide greater assurance that the achieved precision will be narrower, it is suggested that the final sample size be increased by a few numbers.

The following chart is a description of the normal distribution into which the sample means of the final sample of 30 are assumed to fall. The population estimate of the sample is $67,758, and the sample mean is $982. The precision range is $53,100 to $82,416.

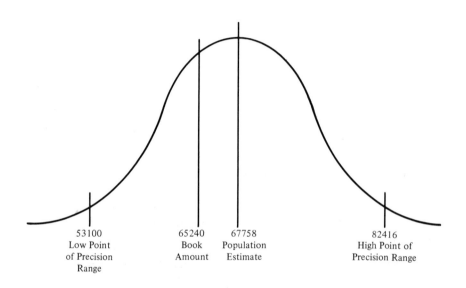

53100	65240	67758	82416
Low Point of Precision Range	Book Amount	Population Estimate	High Point of Precision Range

An Unacceptable Book Value—What if the client's book figure is less than $53,100 or higher than $82,416? Or assume that the difference between the auditors' population estimate and the book value is greater than the attained precision.

Recall that essentially the same question is posed in Chapter 6 in relation to a client's unacceptable error rate. The applicable observations are similar to those in that chapter, which covers statistical sampling for tests of internal control compliance. However, in this chapter we are discussing sampling for dollar balances.

1. The sample size can be expanded in an attempt to derive a population estimate and a precision range that make the book value acceptable. The theoretical advantage of such a decision is that better information is provided as to whether the client's book figure is materially in error or whether a Type I error was committed by the auditors when the initial sample was taken. The practical disadvantage is the additional, and

perhaps unwarranted, cost of the expanded sample, particularly considering that evidence other than confirmation is gathered on the $65,240 accounts receivable total.

2. Another alternative for the auditor is to accept the book figure if it is close to the precision range. As pointed out in Chapter 6, this choice is difficult to defend statistically, but it may be a practical way to expedite the problem. However, the statistical test illustrated in Chapter 6 is a one-tailed test in which the auditors are concerned only with the upper limit of the precision range.

3. The auditor could propose an adjusting entry.

The test illustrated in this chapter is two-tailed, and a book value is rejected if it is too low *or* too high. An implicit assumption is that the auditors are concerned equally with understatement and overstatement of the client's accounts receivable total. However, is this assumption necessarily valid? It may not be if one considers that auditors traditionally view overstatement of assets to be of paramount importance.

It appears, then, that auditors might be more willing to accept a book value slightly under the low point of the precision range than one that is barely above the high point of the precision range. No empirical evidence is known that either supports or contradicts this statement.

Tests of Hypotheses of the Accounts Receivable Total

Setting the Hypothesis—Alpha Risk—Instead of or in addition to developing a precision range around the population estimate, the auditors can ascertain the acceptance or non-acceptance of the accounts receivable book balance by posing tests of hypotheses.[8] One advantage of using this particular technique is that the decision model begins with the auditors' ideas about the acceptable level of the alpha and beta risks.

For illustration, assume that the auditors are gathering the same type of evidence (negative confirmation requests) on the same population of 69 accounts analyzed in the previous section of this chapter. The mean of these accounts is $946, the total is $65,240, and the actual standard deviation is $754.

Therefore, the null hypothesis is that the accounts receivable book value of $65,240 is correct. The auditors will conduct a two-tailed test of this hypothesis because of their concern that incorrect rejection may occur because the population estimate is either too high *or* too low. The total alpha risk is 5%, a 2.5% risk that the book value will be rejected because the population estimate is too low, and a 2.5% risk that it will be rejected because the population estimate is too high.

---

[8]These concepts are discussed in detail in the chapter appendix.

Setting the Hypothesis—Beta Risk—As explained in the appendix, the one-tailed beta risk of incorrectly accepting the client's book figure is one-half of the amount of the alpha risk if precision is one-half of the materiality and a two-tailed alpha test is conducted. However, is it necessary for the auditors to establish a rigid 50% relationship between precision and materiality? The answer is no. By allowing the precision/materiality relationship to vary, the auditors can assign various beta risks, given the same alpha risk.

The concept is illustrated vividly in the following set of charts. The normal curves represent distributions of population estimates around the assumed population total of $65,240. Materiality is placed at $30,000. In the first chart, precision is $15,000, which means that the beta risk is 2.5%, or one-half of the alpha risk of 5%.

Note, however, that precision is only $14,000 in Chart No. 2. This lower precision has the effect of tightening or narrowing the distribution of population estimates. Thus, the normal curve shrinks or collapses, resulting in a smaller beta risk. The determination of the beta risk percentages is explained in a later section on testing the hypotheses.

The opposite result is shown in Chart No. 3. Precision is placed at $16,000, widening the distribution of population estimates. This expands the normal curve and creates a larger beta risk.

CHART NO. 1 – PRECISION ONE-HALF OF MATERIALITY

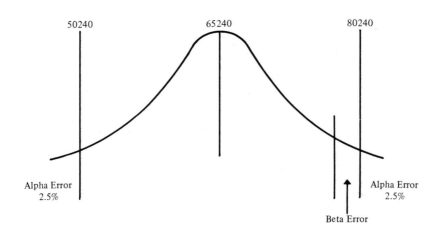

Chart No. 1

CHART NO. 2 – PRECISION LESS THAN ONE-HALF OF MATERIALITY

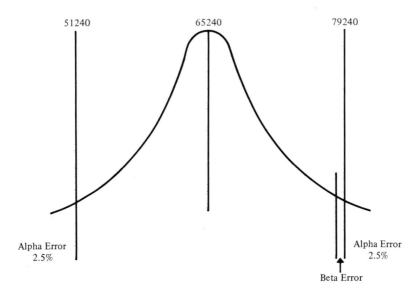

Chart No. 2

CHART NO. 3 – PRECISION MORE THAN ONE-HALF OF MATERIALITY

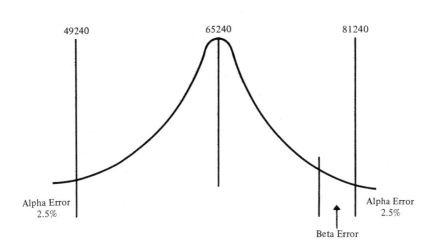

Chart No. 3

The charts clearly demonstrate that the narrower the precision in relation to the materiality, the lower the beta risk, and vice versa. A narrower precision also calls for a larger sample size.

Thus, the auditors can employ a set of decision models in which the following criteria are used to arrive at desired reliability and precision levels.

	Higher Cost of Obtaining Evidence	*Lower Cost of Obtaining Evidence*
Reliability Level—(the alpha risk is the complement of this level)	Lower	Higher

	Better Internal Control	*Poorer Internal Control*
Precision/Materiality Percentage (higher beta risks with higher percentages, vice versa)	Higher	Lower

If the precision/materiality percentage is allowed to vary rather than being fixed at 50%, the beta risk can be assigned independently of the alpha risk, and an optimum sample size can be used to satisfy both criteria.

In the previous section (estimation of the population total) precision was assumed to be one-half of materiality. In this section on tests of hypotheses, varying precision/materiality percentages are used in the analysis.

Materiality is placed at $30,000 for the reasons explained in the previous section of this chapter. Statistically, the alternate hypothesis is that the true population total is $>$65,240$, at an assumed total of $95,240, or $<$65,240$, at an assumed total of $35,240. For our discussion purposes, the concentration is on the "one-tailed" beta risk; therefore, only the $95,240 is studied. Precision may be placed at any percentage of materiality that the auditor chooses. If the percentage is 50%, the one-tailed beta risk is one-half the two-tailed alpha risk; a higher percentage calls for a relatively higher beta risk and vice versa.

In setting the size of the beta risk, the auditors would take several factors into consideration.

1. The system of internal control over sales and accounts receivable, as evidenced in Chapter 5, is reasonably sound, although weak in places.

2. The confirmations are only *one* type of evidence that will be gathered to ascertain the fairness of the book balance of $65,240. Also, positive confirmations will be sent on many of the accounts composing the total accounts receivable balance of $155,998.

Therefore, it seems reasonable to set the beta risk higher than 2.5% and calculate the required precision. If 10% is used, the precision is $18,150 or 60.5% of the materiality of $30,000. The formula used to derive these precisions is as follows. Precision = Materiality/$(1 + Z_\beta/Z_{\alpha/2})$, where Z_β is the normal table value for the beta risk percent and $Z_{\alpha/2}$ is the normal table value for one-half of the two-tailed alpha risk percent.[9] Thus $18,150 = $30,000/$(1 + 1.28/1.96)$.

The alpha risk/beta risk and precision/materiality relationships are illustrated in the following chart.

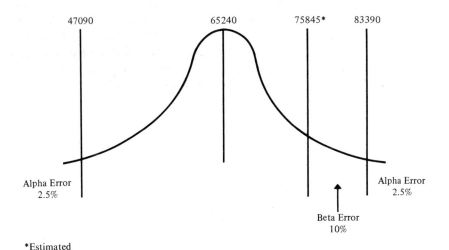

Note that in the chart a normal distribution is drawn around the assumed population total of $65,240, implying that if this figure *is* the true total, the distribution of population estimates will be normal. The $18,150 precision is added and subtracted from the book value of $65,240 to form the range of acceptance of the null hypothesis. If the population estimate is between $47,090 and $83,390, the null hypothesis will be accepted (this will occur in 95% of the estimates if the null hypothesis is correct). Any population estimate below

---

[9]Robert K. Elliott and John R. Rogers, "Relating Statistical Sampling to Audit Objectives," *The Journal of Accountancy,* July, 1972, pp. 46–55.

$47,090 or above $83,390 will cause rejection of the null hypothesis (this will occur in 5% of the estimates if the null hypothesis is correct).

What about the section of the normal curve labeled "beta error"? Because precision has been set at $18,150 rather than $15,000, there is more than a 2.5% beta risk. In fact, there is now a 10% risk that the upper side of the alternative hypothesis (the true total is equal to or larger than $95,240) will be rejected incorrectly and the null hypothesis will be accepted incorrectly.

From the auditors' standpoint the beta risk is particularly important if the error results in a materially misstated accounts receivable book figure. Recall, though, that the auditors are willing to accept this risk, because of the quality of internal control and other factors relevant to this phase of the engagement.

Testing the Hypotheses—The null and alternate hypotheses now can be tested by calculating an estimate of the population total. Essentially, the method for making such an estimate is the same as that described in the previous section of the chapter, except that the population standard deviation is used. The sample size is smaller than 30 because the precision is greater than $15,000.

The calculations for estimating the population total follow.

$$n = \left(\frac{\sigma \times U_R \times N}{A} \right)^2$$

$$n = \left(\frac{754 \times 1.96 \times 69}{18,150} \right)^2$$

$$n = 32 \text{ (with replacement)}$$

$$n' = \frac{32}{1 + \dfrac{n}{N}}$$

$$n' = \frac{32}{1.46}$$

$$n' = 22 \text{ (without replacement)}$$

A random sample of 22 from the 69 balances produces the following amounts.

Customer Number	Correct Balance	Customer Number	Correct Balance	Customer Number	Correct Balance
3	$ 208	39	$1,030	77	$ 1,041
6	1,240	42	306	79	576
17	2,507	45	2,016	81	10
28	640	46	195	83	885
29	890	64	1,080	87	576
31	238	75	906	88	1,020
38	345	76	100	90	873
				92	119
					$16,801

The sample mean, based on the total dollar amount of the sample of 22, is $16,801 ÷ 22 = $764.

The estimate of the population total, based on the derived sample mean, is $764 × 69 = $52,716.

Thus, the auditors would accept the null hypothesis that the true population total is $65,240. If the hypothesis is correct, the auditors would have made the right decision. The alternative hypothesis would be rejected. If the alternative hypothesis is false, a correct rejection was made.

Other Population Estimates—The population estimate derived in the previous section is one of many such estimates that could have emerged from the calculation process. It is important to understand the sampling risk under which the auditors operate, because other population estimates could have resulted in alpha or beta errors, depending on the actual total of the 69 accounts receivable balances.

To illustrate this point, a series of random samples of 22 accounts is listed in the following chart. If the null hypothesis is true, a population estimate based on a sample of 22 will result either in correct acceptance or incorrect rejection of the hypothesis. Likewise, if the alternative hypothesis is true, an estimate will result either in correct acceptance or incorrect rejection of the alternative hypothesis. As can be seen, several conditions can be produced by the various population estimates, and some decisions dictated by these conditions would represent the committing of Type I or Type II errors (same as alpha and beta errors). The following set of samples shows these possible conditions.

Total of the Sample of 22	Sample Mean	Population Estimate	If the Null Hypothesis is True		If the Upper Side of the Alternative Hypothesis is True	
			Correct Acceptance	Incorrect Rejection[a]	Correct Acceptance	Incorrect Rejection[b]
$11,880	$ 540	$37,260		x		
24,610	1,119	77,211	x			x
21,174	962	66,378	x			
20,435	929	64,101	x			
20,986	954	65,826	x			
17,940	815	56,235	x			
19,509	887	61,203	x			
16,694	759	52,371	x			
22,782	1,036	71,484	x			
17,861	812	56,028	x			
18,796	854	58,926	x			
23,713	1,078	74,382	x			
20,941	952	65,688	x			
14,152	644	44,436		x		
20,979	954	65,826	x			
24,443	1,111	76,659	x			x
15,707	714	49,266	x			
20,137	915	63,135	x			
17,267	785	54,165	x			
29,253	1,330	91,770		x	x	

a A Type I error—higher than $83,390 or lower than $47,090.
b A Type II error—between $75,845 and $83,390. Because a discrete rather than a continuous frequency distribution is illustrated, it is assumed that if the alternative hypothesis is correct, no population estimate below $75,845 could be made.

Sampling for Ratio and Difference Estimation

Reasons for Use of Ratio Estimation

During the course of an examination, auditors have many occasions to study accounts in which there may be differences between the client's book figure and the amount calculated or obtained through the audit process.

In such situations, it may be feasible and advantageous to use a variation of variable sampling called "ratio estimation."

This type of sampling is appropriate when the auditors want to know the probable difference between a client-furnished figure (a book figure) and their calculated audit figure. Though the method is suitable for such procedures as footing of perpetual inventory accounts and confirmation of accounts receivable, it is not compatible with such procedures as physical inventory counts (or any procedure in which a figure is being "built up" rather than compared with another figure).

For an oversimplified example of ratio estimation, assume that a client-furnished figure is $100,000. The auditors sample $20,000 of this amount, and their calculated audit figure is $21,000 or 105% of the sampled book figure. They might conclude tentatively that the probable correct population figure is $105,000 (105% of $100,000).

Ratio estimation can be used most effectively when the calculated audit figures are approximately proportional to the client-furnished book figures. For example, the following situation is appropriate for ratio estimation.

Book Figures	Audited Figures	Difference
$ 8,000	$ 7,800	$(200)
10,000	10,250	250
12,000	11,700	(300)
9,000	9,300	300
$39,000	$39,050	$ 50

Conversely, the following situation would *not* be suitable for ratio estimation because the relationship between book and audited figures is not proportional.

Book Figures	Audited Figures	Difference
$ 8,000	$ 8,050	$ 50
10,000	9,000	(1,000)
12,000	11,800	(200)
9,000	10,500	1,500
$39,000	$39,350	$ 350

An Example of Ratio Estimation

Why would it be advantageous to estimate a population total based on ratios rather than arithmetic means (which is the procedure illustrated in the earlier portion of the chapter)? With the same set of data, the standard deviation of ratios is smaller than the standard deviation of the mean. Thus, a smaller sample size can be designed to achieve the same reliability and precision objectives.

This advantage can be illustrated by use of the 99 accounts receivable balances with a book balance of $155,998.

The following list of these 99 accounts includes a column for the amount that would have been confirmed to the auditor had that account been selected in the sample. Also, there is a column for the difference, if any, between the book and audit amounts. Account No. 30 in the original population of 100 has a $920 credit balance. It has been omitted and the accounts 31–100 moved up by one.

Account Number	Client Book Amount	Audit Confirmation Amount	Difference	Account Number	Client Book Amount	Audit Confirmation Amount	Difference
1	$ 1,020	$ 1,020	$ —	26	$ 145	$ 145	$ —
2	696	1,534	838	27	7,049	7,193	144
3	208	218	10	28	640	640	—
4	1,847	1,907	60	29	890	890	—
5	—	—	—	30	238	238	—
6	1,240	1,240	—	31	742	724	(18)
7	5,391	5,418	27	32	480	—	(480)
8	1,190	1,190	—	33	240	240	—
9	2,554	2,554	—	34	480	480	—
10	218	218	—	35	—	774	774
11	5,000	4,027	(973)	36	845	845	—
12	4,606	5,360	754	37	345	345	—
13	2,106	2,106	—	38	1,030	1,030	—
14	402	402	—	39	2,035	2,035	—
15	1,335	1,335	—	40	988	988	—
16	3,240	3,240	—	41	306	306	—
17	2,507	2,507	—	42	—	—	—
18	2,824	2,794	(30)	43	5,551	5,551	—
19	650	650	—	44	2,652	2,016	(636)
20	—	—	—	45	195	195	—
21	144	144	—	46	950	950	—
22	669	669	—	47	306	306	—
23	543	543	—	48	—	—	—
24	730	730	—	49	2,605	2,605	—
25	770	385	(385)	50	3,196	2,836	(360)

Account Number	Client Book Amount	Audit Confirmation Amount	Difference
51	$ 850	$ 850	$ —
52	—	—	—
53	3,722	3,722	—
54	14,522	14,522	—
55	—	—	—
56	—	—	—
57	582	582	—
58	14,941	14,977	36
59	1,780	890	(890)
60	3,008	3,008	—
61	3,329	3,329	—
62	3,175	2,975	(200)
63	1,080	1,080	—
64	425	425	—
65	569	569	—
66	2,064	2,064	—
67	—	—	—
68	1,860	1,860	—
69	660	760	100
70	200	200	—
71	2,455	2,455	—
72	3,950	4,050	100
73	—	—	—
74	906	906	—
75	100	100	—

Account Number	Client Book Amount	Audit Confirmation Amount	Difference
76	$1,041	$1,041	$ —
77	1,190	1,190	—
78	576	576	—
79	—	—	—
80	10	10	—
81	4,860	4,860	—
82	885	885	—
83	104	104	—
84	—	—	—
85	2,940	2,940	—
86	576	576	—
87	1,020	1,020	—
88	—	—	—
89	873	873	—
90	5,218	5,218	—
91	119	119	—
92	720	720	—
93	—	—	—
94	975	975	—
95	1,160	1,160	—
96	540	540	—
97	145	145	—
98	710	710	—
99	360	360	—
Totals	$155,998	$154,869	$(1,129)

The method for determining the sample size, taking the sample, and evaluating the sample results is similar to that of mean estimation. As the calculations proceed, any significant differences are noted. The first step is to estimate the standard deviation of the population of ratios (not the standard deviation of the mean). For this population of 99 accounts receivable balances, assume that the auditors use $200. The next step is to calculate the necessary sample size, the technique being the same as that for mean estimation (assume a precision of $6,000 and a standard deviation of ratios of $200).[10]

$$n = \left(\frac{\text{Stan. Dev. of Ratios} \times U_R \times N}{A} \right)^2$$

$$n = \left(\frac{200 \times 1.96 \times 99}{6,000} \right)^2$$

$$n = 42 \text{ (with replacement)}$$

$$n' = \frac{n}{1 + \dfrac{n}{N}}$$

$$n' = \frac{42}{1.42}$$

$$n' = 30 \text{ (without replacement)}$$

The next step is to take the sample, recording both the book and audit values of each sample figure. For example if Account No. 1 is chosen, the amounts are $1,020 and $1,020. If No. 7 is selected, the amounts are $5,391 and $5,418. Then the auditors will calculate the achieved precision by using methods that parallel the equivalent techniques used in mean estimation.

---

[10]The calculation of the standard deviation of ratios, even when it is based on a sample, is a very tedious operation unless a computer program can be used; $200 is used because it is the approximate standard deviation of this population of ratios. An illustration of how to calculate this standard deviation can be found in Volumes 4 and 6 of *An Auditor's Approach to Statistical Sampling* published by the AICPA.

$$A = \frac{\text{Stan. Dev. of Ratios} \times U_R \times N}{\sqrt{n}}$$

$$A = \frac{78^{11} \times 1.96 \times 99}{\sqrt{30}}$$

$A = 2,754$ (with replacement and rounded for decimal differences)

$$A' = A \times \sqrt{\frac{N - n}{N - 1}}$$

$$A' = 2,754 \times \sqrt{\frac{69}{98}}$$

$A' = 2,311$ (without replacement)

The estimate of the population total is accomplished differently in ratio estimation. A percentage relationship is established between the total book value of the sample ($53,376) and the audit amount of the sample ($53,278). The percentage is .998. Multiply .998 times the client's book amount ($155,998). The estimate of the population total is $155,712 (rounded). The auditor has 95% reliability that the true population total is within the following range.

$$\$155,712 + 2,311 = \$158,023$$
$$\$155,712 - 2,311 = \$153,401$$

Because the book population total is $155,998, the figure is well within the precision range, and as such would be acceptable to the auditors.

It certainly should not escape notice that a sample size of 30 is used in ratio estimation, regardless of the fact that no stratification is performed on a highly skewed population of 99 accounts. The difference, of course, lies in the smaller standard deviation of ratios. Also, it would not escape notice that the sample provided an unreliable estimate of the population standard deviation. For this and other reasons, firms that use ratio estimation perform some stratification anyway.

Although using ratio estimation may have some advantages, the calculations are very complicated and time-consuming. Without computer capability, the effort to derive results may not be marginally beneficial.

---

[11] Note that the standard deviation of the sample is much lower than that of the population. Here is a potential problem with ratio estimation. The sample missed the larger book/audit differences.

Difference Estimation

The objective of difference estimation is to establish the probable dollar difference between the client's book amount and the audited amount (same as ratio estimation), but the method is *slightly* different, because the auditor uses dollar rather than ratio differences.

As an example, assume that a book figure is $100,000 in a population of 1,000. A sample of $20,000 of this book amount revealed an audited amount of $21,000. The sample size is 100. The total difference in the sample is $1,000 or an average difference of $10 ($1,000 difference ÷ the sample size of 100). To estimate the population difference, the auditors multiply the average difference of $10 times the population total of 1,000. The estimated population difference is $10,000 and the estimate of the population total is $110,000.

Difference estimation should be used if the calculated audit figures are *not* proportional to the client-furnished book figures. Otherwise, the method's advantages and disadvantages are essentially the same as for ratio estimation.[12]

Summary and Critique of Variable Sampling

If the auditors decide that the use of variable sampling is beneficial in examining one or more accounts, they must convert certain criteria into numerical form. They also must obtain knowledge of certain quantitative characteristics of the population from which samples are taken. In exchange for this effort in quantifying variables, the auditors acquire statistically definitive conclusions about the accuracy and precision of their sample results.

Whether this exchange is acceptable or not depends on the auditors' view of the relative benefits to be gained from using variable sampling techniques. Though it is true that preciseness is added to the examination of certain accounts, some auditors may think that an element of risk also is added to their task. It may be easier in litigation cases, for instance, to defend less precise judgments than to justify the use of numerical reliability and precision figures.

Moreover, the auditors will consider the cost/benefit relationship of adopting variable sampling techniques. It simply is not feasible to use such methods on small populations or accounts that require special audit treatment (small populations are used in this chapter merely to facilitate discussions of theoretical concepts). In contrast, if the account(s) has a large and relatively homogeneous composition of figures, and if only routine audit procedures are required to examine the account(s), there may be good reason to use variable sampling.

What about the question of whether variable sampling saves audit time by reducing the sample size from the higher level sometimes needed for judgment

---

[12]For further comparisons of the variable sampling techniques discussed in this chapter, see John Neter and James K. Loebbecke, *Behavior of Major Statistical Estimators in Sampling Accounting Populations—Auditing Research Monograph No. 2*, published by the AICPA.

sampling? This is a debatable question and perhaps a moot one as well. Variable sampling should *optimize* the sample size so that neither under- nor over-sampling occurs.[13]

However, the most significant advantage of variable sampling is that it provides the auditors with mathematically definitive conclusions about the accuracy and precision of their sample results. It is on this basis that the justification for the use of these statistical methods should stand.

Chapter 13 References

Aly, Hamdi F., and Jack I. Duboff, "Statistical vs. Judgement Sampling: An Empirical Study of Auditing the Accounts Receivable of a Small Retail Store," *The Accounting Review,* January 1971, pp. 119–128.

Anderson, Rod, and A. D. Teitlebaum, "Dollar-Unit Sampling—A Solution to the Audit Sampling Dilemma," *CA Magazine,* April 1973, pp. 30–39.

Bedingfield, James P., "The Current State of Statistical Sampling and Auditing," *The Journal of Accountancy,* December 1975, pp. 48–55.

Boatsman, James R., and G. Michael Crooch, "An Example of Controlling the Risk of a Type II Error for Substantive Tests in Auditing," *The Accounting Review,* July 1975, pp. 610–615.

Broderick, John C., "Statistical Sampling," *The Arthur Young Journal,* Autumn/Winter 1972-73, pp. 35–44.

Deakin, Edward B., and Michael H. Granof, "Regression Analysis as a Means of Determining Audit Sample Size," *The Accounting Review,* October 1974, pp. 764–771.

Elliott, Robert K., and John R. Rogers, "Relating Statistical Sampling to Audit Objectives," *The Journal of Accountancy,* July 1972, pp. 46–55.

Goodfellow, J. L., J. K. Loebbecke, and J. Neter, "Some Perspectives on CAV Sampling Plans," (Part I), *CA Magazine,* October 1974.

---

[13]One should consider that obtaining a representative sample is not always the avowed purpose. For a discussion of different sampling objectives, see "The Four Objectives of Sampling in Auditing: Representative, Corrective, Protective and Preventive," written by Yuji Ijiri and Robert S. Kaplan, and published by *Management Accounting,* December, 1970, pp. 42–44.

Goodfellow, J. L., J. K. Loebbecke, and J. Neter, "Some Perspectives on CAV Sampling Plans," (Part II), *CA Magazine,* November 1974.

Ijiri, Yuji, and Robert S. Kaplan, "The Four Objectives of Sampling in Auditing: Representative, Corrective, Protective and Preventive," *Management Accounting,* December 1970, pp. 42–44.

Loebbecke, James K., and John Neter, "Statistical Sampling in Confirming Receivables," *The Journal of Accountancy,* June 1973, pp. 44–50.

Neter, John, and James K. Loebbecke, *Behavior of Major Statistical Estimators in Sampling Accounting Populations—Auditing Research Monograph No. 2,* AICPA, 1975.

Roberts, Donald M., "A Statistical Interpretation of *SAP* No. 54," *The Journal of Accountancy,* March 1974, pp. 47–53.

Tracy, John A., "Bayesian Statistical Confidence Intervals for Auditors," *The Journal of Accountancy,* July 1969, pp. 41–47.

Chapter 13
Questions Taken from the Chapter

13–1.   What risk do auditors encounter when they gather evidence?

13–2.   What two items of information are provided the auditors if they use variable sampling for accounts receivable confirmation requests?

13–3.   For substantive tests what is the traditional relationship between the sample size and the quality of internal control?

13–4.   How must auditors convert their judgment on the quality of internal control in order to use this judgment in deciding on the sample size for substantive tests?

13–5.   How can auditors express the persuasiveness of evidence in quantitative terms?

13–6.   Define the alpha risk. What happens to the reliability level if the alpha risk is raised?

13–7.   How can the additional cost of obtaining evidence be measured more precisely by use of variable sampling?

13–8.   How can variable sampling help auditors to express, in more definite terms, the extent to which the evidence does or does not support the client representations?

13–9.   Give the definitions of precision and reliability as presented in AU Section 320B.09 *(SAS No. 1)*.

13–10. According to AU Section 320A.10 *(SAS No. 1)*, the precision and reliability measurements can be usefully adapted to the auditor's purposes by relating precision to what audit term? Relating reliability to what audit term?

13–11. Explain with numbers the concept of letting precision represent one-half of materiality.

13–12. Define materiality.

13–13. Name three financial statement totals to which the account balance under examination can be compared for aid in determining the materiality of that account balance.

13–14. What is the beta risk?

13–15. What are two reasons why auditors might be willing to increase the beta risk?

13–16. Write down the AU Section 320B.35 *(SAS No. 1)* formula for deriving a final reliability figure. Define the parts of the formula.

13–17. What are the conditions under which a 50% beta risk might be acceptable (assuming a strong system of internal control)?

13–18. What is the possible dilemma facing auditors if they allow the alpha risk to increase because of a strong system of internal control?

13–19. Other than the quality of internal control, what is an alternate basis for judging whether the alpha risk should be varied?

13–20. Why does stratification of a numerical distribution lower the necessary sample size?

13–21. Statistically, the reliability or confidence level is the complement of what?

13–22. Normally, when does the client's customer reply to negative confirmation requests?

13–23. Under what conditions is the amount of the beta risk one-half of the amount of the alpha risk?

13–24. What happens to the beta risk if precision is raised in relation to materiality? If precision is lowered in relation to materiality?

13–25. What is the standard deviation a measure of in a numerical distribution?

13–26. Name a way to make an estimate of the population standard deviation (other than by a guess).

13–27. Name the formula for the calculation of the sample size if sampling is conducted with replacement.

13–28. Name the correction factor to determine the sample size if sampling is conducted without replacement.

13–29. What are two techniques for deriving a set of random numbers?

13–30. Give the formula for the number of sample means that can be derived from a sample of 30 from a population of 100 if sampling is done without replacement.

13–31. Give the formula for calculating the attained population precision if sampling is done with replacement.

13–32. Give the formula for the finite correction factor for attaining population precision if sampling is done without replacement.

13–33. What are the auditors' alternatives if the client's book value is not within the attained precision range?

13–34. What is the null hypothesis posed in the chapter material?

13–35. What is the alpha risk posed in the chapter material?

13–36.   What is the alternative hypothesis posed in the chapter material?

13–37.   In the chapter material, what factors are considered in setting the size of the beta risk?

13–38.   Under what conditions might it be feasible and advantageous to use ratio estimation?

13–39.   What should the relationship be between the audit and book figures in order for the use of ratio estimation to be most effective?

13–40.   Why is it advantageous to estimate a total based on ratios rather than means?

13–41.   How do the auditors use ratio estimation to make an estimate of the population total?

13–42.   What should the relationship be between audited and book figures in order to make effective use of difference estimation?

Chapter 13
Objective Questions Taken from CPA Examinations

Items 13–43 to 13–47 apply to an examination by Robert Lambert, CPA, of the financial statements of Rainbow Manufacturing Corporation for the year ended December 31, 19X9. Rainbow manufactures two products: Product A and Product B. Product A requires raw materials that have a very low per-item cost, and Product B requires raw materials that have a very high per-item cost. Raw materials for both products are stored in a single warehouse. In 19X8, Rainbow established the total value of raw materials stored in the warehouse by physically inventorying an unrestricted sample of items selected without replacement.

Mr. Lambert is evaluating the statistical validity of alternative sampling plans Rainbow is considering for 19X9. Lambert knows the size of the 19X8 sample and that Rainbow did not use stratified sampling in 19X8. Assumptions about the population, variability, specified precision (confidence interval), and specified reliability (confidence level) for a possible 19X9 sample are given in each of the following five items. You are to indicate in each case the effect upon the size of the 19X9 sample as compared to the 19X8 sample. Each of the five cases is independent of the other four and is to be considered separately. YOUR ANSWER CHOICE FOR EACH OF ITEMS 1 THROUGH 5 SHOULD BE SELECTED FROM THE FOLLOWING RESPONSES:

a. Larger than the 19X8 sample size.
b. Equal to the 19X8 sample size.
c. Smaller than the 19X8 sample size.
d. Of a size that is indeterminate based upon the information given.

13–43. Rainbow wants to use stratified sampling in 19X9 (the total population will be divided into two strata, one each for the raw materials for Product A and Product B). The population size of the raw materials inventory is approximately the same in 19X9 as in 19X8, and the variability of the items in the inventory is approximately the same. The specified precision and specified reliability are to remain the same.

Under these assumptions, the required sample size for 19X9 should be:
a. Larger than the 19X8 sample size.
b. Equal to the 19X8 sample size.
c. Smaller than the 19X8 sample size.
d. Of a size that is indeterminate based upon the information given.

13–44. Rainbow wants to use stratified sampling in 19X9. The population size of the raw materials inventory is approximately the same in 19X9 as in 19X8, and the variability of the items in the inventory is approximately the same. Rainbow specified the same precision but desires to change the specified reliability from 90% to 95%.

Under these assumptions, the required sample size for 19X9 should be
a. Larger than the 19X8 sample size.
b. Equal to the 19X8 sample size.
c. Smaller than the 19X8 sample size.
d. Of a size that is indeterminate based upon the information given.

13–45. Rainbow wants to use unrestricted random sampling without replacement in 19X9. The population size of the raw materials inventory is approximately the same in 19X9 as in 19X8, and the variability of the items in the inventory is approximately the same. Rainbow specifies the same precision but desires to change the specified reliability from 90% to 95%.

Under these assumptions, the required sample size for 19X9 should be
a. Larger than the 19X8 sample size.
b. Equal to the 19X8 sample size.
c. Smaller than the 19X8 sample size.
d. Of a size that is indeterminate based upon the information given.

13–46. Rainbow wants to use unrestricted random sampling without replacement in 19X9. The population size of the raw materials inventory for 19X9 has increased over that of 19X8, and the variability of the items in the inventory has increased. The specified precision and specified reliability are to remain the same.

Under these assumptions, the required sample size for 19X9 should be
a. Larger than the 19X8 sample size.
b. Equal to the 19X8 sample size.
c. Smaller than the 19X8 sample size.
d. Of a size that is indeterminate based upon the information given.

13–47. Rainbow wants to use unrestricted random sampling without replacement in 19X9. The population size of the raw materials inventory for 19X9 has increased over that for 19X8, but the variability of the items in the inventory has decreased. The specified precision and specified reliability are to remain the same.

Under these assumptions, the required sample size for 19X9 should be
a. Larger than the 19X8 sample size.
b. Equal to the 19X8 sample size.
c. Smaller than the 19X8 sample size.
d. Of a size that is indeterminate based upon the information given.

13–48. From prior experience, a CPA is aware of the fact that cash disbursements contain a few unusually large disbursements. In using statistical sampling, the CPA's best course of action is to
a. Eliminate any unusually large disbursements which appear in the sample.
b. Continue to draw new samples until no unusually large disbursements appear in the sample.
c. Stratify the cash-disbursements population so that the unusually large disbursements are reviewed separately.
d. Increase the sample size to lessen the effect of the unusually large disbursements.

13–49. Approximately 5% of the 10,000 homogeneous items included in Barletta's finished goods inventory are believed to be defective. The CPA examining Barletta's financial statements decides to test this estimated 5% defective rate. He learns that by sampling without replacement, a sample of 284 items from the inventory will permit specified reliability (confidence level) of 95% and specified precision (confidence interval) of $\mp.025$. If specified precision is changed to $\pm.05$, and specified reliability remains 95%, the required sample size is
a. 72.
b. 335.
c. 436.
d. 1,543.

13–50.   The "reliability" (confidence level) of an estimate made from sample data is a mathematically determined figure that expresses the expected proportion of possible samples of a specified size from a given population

a.   That will yield an interval estimate that will encompass the true population value.

b.   That will yield an interval estimate that will not encompass the true population value.

c.   For which the sample value and the population value are identical.

d.   For which the sample elements will not exceed the population elements by more than a stated amount.

13–51.   A CPA's client wishes to determine inventory shrinkage by weighing a sample of inventory items. If a stratified random sample is to be drawn, the strata should be identified in such a way that

a.   The overall population is divided into sub-populations of equal size so that each sub-population can be given equal weight when estimates are made.

b.   Each stratum differs as much as possible with respect to expected shrinkage but the shrinkages expected for items within each stratum are as close as possible.

c.   The sample mean and the standard deviation of each individual stratum will be equal to the means and standard deviations of all other strata.

d.   The items in each stratum will follow a normal distribution so that probability theory can be used in making inferences from the sample data.

13–52.   In estimating the total value of supplies on repair trucks, Baker Company draws random samples from two equal-sized strata of trucks. The mean value of the inventory stored on the larger trucks (Stratum 1) was computed as $1,500, with a standard deviation of $250. On the smaller trucks (Stratum 2), the mean value of inventory was computed as $500, with a standard deviation of $45. If Baker had drawn an unstratified sample from the entire population of trucks, the expected mean value of inventory per truck would be $1,000, and the expected standard deviation would be

a.   Exactly $147.50.

b.   Greater than $250.

c.   Less than $45.

d.   Between $45 and $250, but not $147.50.

Items 13–53 to 13–57 are based on the following information:

An audit partner is developing an office training program to familiarize his professional staff with statistical decision models applicable to the audit of dollar-value balances. He wishes to demonstrate the relationship of sample sizes to population size and variability and the auditor's specifications as to precision and confidence level. The partner prepared the following table to show comparative population characteristics and audit specifications of two populations.

	Characteristics of Population 1 Relative To Population 2		Audit Specifications as to a Sample from Population 1 Relative to a Sample from Population 2	
	Size	Variability	Specified Precision	Specified Confidence Level
Case 1	Equal	Equal	Equal	Higher
Case 2	Equal	Larger	Wider	Equal
Case 3	Larger	Equal	Tighter	Lower
Case 4	Smaller	Smaller	Equal	Lower
Case 5	Larger	Equal	Equal	Higher

In each of Items 53 through 57 you are to indicate for the specified case from the table above the required sample size to be selected from Population 1 relative to the sample from Population 2. YOUR ANSWER CHOICE SHOULD BE SELECTED FROM THE FOLLOWING RESPONSES:

a. Larger than the required sample size from Population 2.
b. Equal to the required sample size from Population 2.
c. Smaller than the required sample size from Population 2.
d. Indeterminate relative to the required sample size from Population 2.

13–53. In Case 1 the required sample size from Population 1 is

13–54. In Case 2 the required sample size from Population 1 is

13–55. In Case 3 the required sample size from Population 1 is

13–56. In Case 4 the required sample size from Population 1 is

13–57. In Case 5 the required sample size from Population 1 is

Items 13–58 to 13–62 apply to an examination by Lee Melinda, CPA, of the financial statements of Summit Appliance Repair Co. for the year ended June 30, 19X9. Summit has a large fleet of identically stocked repair trucks. It establishes the total quantities of materials and supplies stored on the delivery trucks at year-end by physically inventorying a random sample of trucks.

Mr. Melinda is evaluating the statistical validity of Summit's 19X9 sample. Assumptions about the size, variability, specified precision (confidence interval), and specified reliability (confidence level) for the 19X9 sample are given in each of the following five items. You are to indicate in each case the effect upon the size of the 19X9 sample as compared to the 19X8 sample. Each of the five cases is independent of the other four and is to be considered separately. YOUR ANSWER CHOICE SHOULD BE SELECTED FROM THE FOLLOWING RESPONSES:

a. Larger than the 19X8 sample size.
b. Equal to the 19X8 sample size.
c. Smaller than the 19X8 sample size.
d. Of a size that is indeterminate based upon the assumptions as given.

13–58. Summit has the same number of trucks in 19X9, but supplies are replenished more often, meaning that there is less variability in the quantity of supplies stored on each truck. The specified precision and specified reliability remain the same.

Under these assumptions the required sample size for 19X9 should be

13–59. Summit has the same number of trucks; supplies are replenished less often (greater variability); Summit specifies the same precision but decides to change the specified reliability from 95% to 90%.

Under these assumptions the required sample size for 19X9 should be

13–60. Summit has more trucks in 19X9. Variability and specified reliability remain the same, but with Melinda's concurrence, Summit decides upon a wider specified precision.

Under these assumptions the required sample size for 19X9 should be

13–61. The number of trucks and variability remain the same, but with Melinda's concurrence, Summit decides upon a wider specified precision and a specified reliability of 90% rather than 95%.

Under these assumptions the required sample size for 19X9 should be

13–62. The number of trucks increases, as does the variability of quantities stored on each truck. The specified reliability remains the same, but the specified precision is narrowed.

Under these assumptions the required sample size for 19X9 should be

13–63. How should an auditor determine the precision required in establishing a statistical sampling plan?
a. By the materiality of an allowable margin of error the auditor is willing to accept.

b.   By the amount of reliance the auditor will place on the results of the sample.
c.   By reliance on a table of random numbers.
d.   By the amount of risk the auditor is willing to take that material errors will occur in the accounting process.

13–64.   An auditor makes separate compliance and substantive tests in the accounts payable area which has good internal control. If the auditor uses statistical sampling for both of these tests, the confidence level established for the substantive tests is normally
a.   The same as that for tests of compliance.
b.   Greater than that for tests of compliance.
c.   Less than that for tests of compliance.
d.   Totally independent of that for tests of compliance.

13–65.   *Statement on Auditing Standards No. 1* suggests a formula for determining the reliability level for substantive tests *(S)* based upon the reliance assigned to internal accounting control and other relevant factors *(C)* and the combined reliability level desired from both internal control and the substantive tests *(R)*. This formula is

a. $S = 1 - \dfrac{C}{R}$

b. $S = R - C$

c. $S = 1 - \dfrac{(1 - R)}{(1 - C)}$

d. $S = R - \dfrac{1}{C}$

13–66.   Precision is a statistical measure of the maximum likely difference between the sample estimate and the true but unknown population total, and is directly related to
a.   Reliability of evidence.
b.   Relative risk.
c.   Materiality.
d.   Cost benefit analysis.

13–67.   An important statistic to consider when using a statistical sampling audit plan is the population variability. The population variability is measured by the
a.   Sample mean.
b.   Standard deviation.
c.   Standard error of the sample mean.
d.   Estimated population total minus the actual population total.

Chapter 13
Discussion/Case Questions and Problems

13–68. Assume that a population has the following characteristics.

Number of elements—300
Client's population total—$600,000
The estimated standard deviation of the population—$2,000
The desired reliability—96% (assume U_R of 2.00)
Needed sample size—100

Answer the following questions:
a. The desired precision is $_____.
b. If the sample mean is $1,800, the estimate of the population total is $_____.
c. The estimated standard error of the mean is $_____.
d. If ratio estimation is used and (1) the sample book figure is $200,000, (2) the sample audit figure is $210,000, the estimate of the population total is $_____.
e. If difference estimation is used and (1) the sample book figure is $200,000, (2) the sample audit figure is $195,000, the estimate of the population total is $_____.

13–69. The population consists of 1,000 accounts. The sample size is 100. The client's book total is $850,000. A 95% reliability is needed (U_R of 1.96).

The desired precision is $6,000. The audit total of the sample is $80,000. The sum of the squared deviations is $89,100.

Required:
Give computations to show whether the client book total is acceptable.

13–70. The requirements of this problem are to take the accompanying population and perform the following.
a. Estimate the standard deviation.
b. Determine the sample size. The needed reliability is 80% and the desired precision is $1,000.
c. Take, in random fashion, the sample of the size called for in b.
d. Estimate the population total based on the sample selected in c.
e. Calculate a precision range around the population estimate derived in d.

(The computer program(s) labeled RATIO and/or PSD can be used to solve this problem. The program listings are at the end of the appendix.)

Number	Amount	Number	Amount	Number	Amount
1	218	55	234	109	235
2	209	56	190	110	210
3	200	57	191	111	190
4	196	58	180	112	185
5	170	59	181	113	220
6	178	60	182	114	215
7	180	61	209	115	210
8	199	62	208	116	205
9	201	63	214	117	200
10	230	64	188	118	195
11	228	65	193	119	190
12	212	66	196	120	185
13	197	67	194	121	185
14	190	68	202	122	180
15	202	69	207	123	175
16	196	70	206	124	170
17	233	71	184	125	172
18	181	72	182	126	173
19	196	73	178	127	174
20	211	74	175	128	175
21	210	75	209	129	176
22	208	76	213	130	177
23	186	77	215	131	178
24	190	78	231	132	179
25	191	79	232	133	180
26	183	80	175	134	118
27	224	81	176	135	180
28	226	82	178	136	191
29	215	83	180	137	192
30	208	84	182	138	193
31	200	85	185	139	194
32	193	86	188	140	195
33	183	87	189	141	196
34	203	88	191	142	197
35	181	89	193	143	198
36	228	90	195	144	199
37	230	91	197	145	200
38	232	92	199	146	201
39	188	93	202	147	202
40	199	94	204	148	203
41	187	95	206	149	204
42	212	96	208	150	205
43	230	97	210	151	206
44	201	98	212	152	207
45	214	99	215	153	208
46	215	100	218	154	209
47	189	101	220	155	210
48	178	102	221	156	211
49	234	103	224	157	212
50	204	104	226	158	213
51	205	105	228	159	214
52	220	106	230	160	215
53	198	107	232	161	216
54	169	108	234	162	158

Number	Amount	Number	Amount	Number	Amount
163	171	176	196	189	178
164	240	177	102	190	196
165	235	178	198	191	206
166	234	179	195	192	216
167	190	180	206	193	226
168	182	181	206	194	236
169	203	182	205	195	196
170	171	183	205	196	186
171	220	184	174	197	185
172	198	185	174	198	184
173	233	186	180	199	183
174	202	187	179	200	186
175	197	188	188		
				Total	$39,816
				Mean	$199.08

13–71. This problem is based on the population of 200 numbers used in No. 13–70. Assume that the auditors are testing the null hypothesis that the true population total is $39,816. The test is two-tailed and the alpha risk is 10%. Assume a population standard deviation of $20 and a sample size of 36.

Required:
Take 10 random samples of sample size 36, without replacement, and make an estimate of the population total for each one. Determine which estimates result in a Type I error if the null hypothesis is true.

13–72. This problem is based on the population of 200 used in No. 13–70. Assume that the auditors are testing the alternative hypothesis that the true population total is >$39,816 with an assumed population total of $42,000. The beta risk is 5% (precision is one-half of materiality). Assume a population standard deviation of $20 and a sample size of 36.

Required:
Use the 10 random samples of sample size 36 obtained in No. 13–71, or take 10 random samples of sample size 36, without replacement. Make an estimate of the population total for each of the 10 samples and determine which estimates result in a Type II error if the alternative hypothesis is true. (*Hint:* Look in the table of areas under the normal curve.)

13–73. A variable sampling test results in a population estimate of $1,000,000 with a lower precision limit of $900,000 and an upper precision limit of $1,100,000. The client's book value is $1,200,000. The auditors agree that the client should be asked to make an adjusting entry, but there is disagreement as to the amount. Auditor A suggests that the books should be adjusted to $1,100,000, the upper precision limit. Auditor B suggests adjustment to $1,000,000, the population estimate. Auditor C suggests adjustment to $900,000, the lower precision limit.

Required:
Give reasons that might be given in support of all three proposed adjustments. Which do you support? Why?

13–74. The alpha risk sometimes is called the client's risk, and the beta risk is sometimes called the auditor's risk. Explain.

13–75. The following list of tests is typical of those usually performed by auditors in gathering evidence to ascertain the fairness of financial statement balances. For each listed test, indicate whether it might be useful to apply variable statistical sampling techniques. Support your *yes* or *no* answers with reasons.
a. Confirming accounts receivable.
b. Observing physical inventory counts.
c. Confirming insurance information with the client's insurance agent.
d. Testing the clerical accuracy of the accounts receivable aging schedule.
e. Examining invoices that support the repairs and maintenance expenses in the income statement.
f. Examining securities in the client's safety deposit box.
g. Reconciling the client's bank reconciliations.
h. Ascertaining the reasonableness of the fixed asset depreciation rates.
i. Examining the invoices that support the additions to fixed assets.

13–76. During the planning of the examination of Strong Company, the partner in charge of the audit was discussing sampling plans with a manager and a senior. The discussion centered upon the potential use of estimation sampling for variables in testing the clerical accuracy of the perpetual inventory records.

The senior maintained that the mean estimation technique was the appropriate method. Experience has shown that the dollar amounts of clerical errors in the inventory records are small, although the number of errors often is large. Because the difference in "book" and "audited" values should be small, the mean estimation technique, with proper stratification, should be used.

The manager disagreed with the suggestion to use the mean estimation technique. He contended that an increasing number of auditors are using the newer variable sampling techniques of ratio and difference estimation. These newer methods do not require as much stratification as mean estimation because the standard deviation of ratios and/or differences usually is smaller than the standard deviation of audited values.

Although, the senior agreed, ratio and difference estimation was used by some auditors, he was "afraid" of this sampling technique. He did not agree with ratio and difference estimation's implicit assumption that the ratio/differences between the audited and book values found in the sample are typical of the ratio/differences in the population. If a few large dollar errors existed in the population and none were found in the sample, the results could be misleading.

The manager believed that the potential problem presented by the senior was overrated. Experience shows that perpetual inventory records contain small dollar errors. The use of ratio and difference estimation will lower the sample size and be at least as effective as mean estimation.

At this point, the audit partner became a bit impatient. He was anxious to complete the planning of the examination and start the field work. He suggested that the group confer with a newly promoted senior who had recently taken an up-to-date auditing course and had attended a company-sponsored professional development course in statistical sampling.

Required:
Assume that you are the newly promoted senior and have been asked to suggest a sampling plan for the inventory testing.

Summarize the merits and flaws in each argument. Then, arrive at a conclusion, supporting your conclusion with reasons.

13-77.   You desire to evaluate the reasonableness of the book value of the inventory of your client, Draper, Inc. You satisfied yourself earlier as to inventory quantities. During the examination of the pricing and extension of the inventory, the following data were gathered using appropriate unrestricted random sampling with replacement procedures.

*Total items in the inventory (N)	12,700
*Total items in the sample (n)	400
*Total audited value of items in the sample	$38,400
* $\displaystyle\sum_{j=1}^{400} (Xj - \bar{X})^2$	312,816
*Formula for estimated population standard deviation	$S_{xj} = \sqrt{\displaystyle\sum_{j=1}^{j=N} \frac{(xj - \bar{x})^2}{n-1}}$
*Formula for estimated standard error of the mean	$SE = \dfrac{S_{xj}}{\sqrt{n}}$

*Confidence level coefficient
of the standard error
of the mean at a 95%
confidence (reliability) level* ± 1.96

Required:

a. Based on the sample results, what is the estimate of the total value of the inventory? Show computations in good form where appropriate.

b. What statistical conclusions can be reached regarding the estimated total inventory value calculated in a. above at the confidence level of 95%? Present computations in good form where appropriate.

c. Independent of your answers to a. and b., assume that the book value of Draper's inventory is $1,700,000, and on the basis of the sample results, the estimated total value of the inventory is $1,690,000. The auditor desires a confidence (reliability) level of 95%. Discuss the audit and statistical considerations the auditor must evaluate before deciding whether the sampling results support acceptance of the book value as a fair presentation of Draper's inventory.

(AICPA adapted)

13–78.   During the course of an audit engagement, a CPA attempts to obtain satisfaction that there are no material misstatements in the accounts receivable of a client. Statistical sampling is a tool that the auditor often uses to obtain representative evidence to achieve the desired satisfaction. On a particular engagement an auditor determined that a material misstatement in a population of accounts would be $35,000. To obtain satisfaction the auditor had to be 95% confident that the population of accounts was not in error by $35,000. The auditor decided to use unrestricted random sampling with replacement and took a preliminary random sample of 100 items (n) from a population of 1,000 items (N). The sample produced the following data:

Arithmetic mean of sample items ($\bar{x}$) $4,000

Standard deviation of sample items (SD) $ 200

The auditor also has available the following information:

Standard error of the mean (SE) $= SD \div \sqrt{n}$

Population precision (P) $= N \times R \times SE$

Partial List of Reliability Coefficients

If Reliability Coefficient (R) is	Then Reliability is
1.70	91.086%
1.75	91.988
1.80	92.814
1.85	93.568
1.90	94.256
1.95	94.882
1.96	95.000
2.00	95.450
2.05	95.964
2.10	96.428
2.15	96.844

Required:

a. Define the statistical terms "reliability" and "precision" as applied to auditing.

b. If all necessary audit work is performed on the preliminary sample items and no errors are detected,

    (1) What can the auditor say about the total amount of accounts receivable at the 95% reliability level?

    (2) At what confidence level can the auditor say that the population is not in error by $35,000?

c. Assume that the preliminary sample was sufficient.

    (1) Compute the auditor's estimate of the population total.

    (2) Indicate how the auditor should relate this estimate to the client's recorded amount.

(AICPA adapted)

13–79.   If the population total on the client's books is not within the precision range calculated by the auditor, there are several suggested alternatives that the auditor can take. Listed below are criticisms of these alternatives. In each case indicate how you would answer this criticism.

a. If the auditor takes another sample, he is guilty of hunting for acceptable results rather than being objective and accepting the results.

b. If the auditor accepts the client's book total even though it is slightly outside the calculated precision range, he is guilty of changing the statistical objectives after the statistical result is derived.

c. If the auditor asks the client to adjust the books, he is engaging in record-keeping, because he is asking for a figure *he* derived, not a figure derived by the client.

13–80. One argument that has been given against the use of statistical sampling for conducting audit tests is that auditors are "sticking their necks out" by using definite reliability and precision percentages. Why not confine judgments to general statements and avoid that problem? In the event of a lawsuit, wouldn't the auditor have a better defense by stating that judgment was used rather than having to justify certain percentages? Reply to this statement.

13–81. a. Identify and explain four areas where judgment may be exercised by a CPA in planning a statistical sampling test.

b. Assume that a CPA's sample shows an unacceptable book value. Describe the various actions that he might take based upon this finding.

(AICPA adapted)

13–82. After applying variable statistical sampling techniques to the examination of a client's accounts receivable records, a group of auditors obtained the following results.

Estimate of the population total	$215,000
Attained precision	± 30,000
Reliability	90%

The client's book value is $250,000; therefore, the figure is above the upper limit of the precision range ($245,000). Previous audits of these records has revealed numerous large dollar errors, most of them resulting in book overstatements. Therefore, there was no reason to believe that an extended sample will produce substantially different results.

One auditor proposes that the client be asked to adjust its book value from $250,000 to at least $245,000. Another auditor believes that the proposed $5,000 adjustment is too small to be concerned with and should be ignored. Another auditor contends that it is improper to ask the client to change a $250,000 figure supported by individual customer accounts. He asks what the debit and credit in the adjusting entry will be and which individual accounts will be reduced.

Required:
Comment on each of the three auditors' contentions. Would you suggest another alternative?

Appendix to Chapter 13

Variable Sampling

A LIST OF NOTATIONS AND FORMULAS

Notation	Definition	Formula
1. N	Number of items in a population	—
2. n	Number of items in a sample	—
3. μ	Population mean	$\mu = \dfrac{\Sigma X}{N}$
4. $\bar{X}$	Sample mean	$\bar{X} = \dfrac{\Sigma X}{n}$
5. X	A variable or element	
6. Σ	Summation	
7. x	Value of a deviation	$x = X - \bar{X}$ (sample) $x = X - \mu$ (population)
8. σ	Standard deviation of a population	$\sigma = \sqrt{\dfrac{\Sigma x^2}{N}}$
9. s	Standard deviation of a sample	$s = \sqrt{\dfrac{\Sigma x^2}{n^*}}$
10. v	Coefficient of variation	$v = \dfrac{s}{\bar{x}}$
11. m_1	The first moment for a distribution of values	$m_1 = \dfrac{\Sigma x}{n}$
12. m_2	The second moment for a distribution of values	$m_2 = \dfrac{\Sigma x^2}{n}$
13. m_3	The third moment for a distribution of values	$m_3 = \dfrac{\Sigma x^3}{n}$
14. m_4	The fourth moment for a distribution of values	$m_4 = \dfrac{\Sigma x^4}{n}$
15. a_3	Skewness	$a_3 = \dfrac{m_3}{s^3}$
16. a_4	Kurtosis	$a_4 = \dfrac{m_4}{s^4}$
17. $\mu_{\bar{x}}$	The mean of the sampling distribution of means	$\mu_{\bar{x}} = \mu$
18. $\sigma_{\bar{x}}$	The standard error of the mean of the population (standard deviation of the sampling distribution of means)	$\sigma_{\bar{x}} = \dfrac{\sigma}{\sqrt{n}}$ (infinite population) $\sigma_{\bar{x}} = \dfrac{\sigma}{\sqrt{n}} \sqrt{\dfrac{N-n}{N-1}}$ (finite population)
19. $s_{\bar{x}}$	The estimate of the standard error of the mean (standard deviation of the sampling distribution of means)	$s_{\bar{x}} = \dfrac{s}{\sqrt{n-1}}$ (infinite population)** $s_{\bar{x}} = \dfrac{s}{\sqrt{n-1}} \sqrt{\dfrac{N-n}{N-1}}$ (finite population)**

20. R Reliability or confidence level

21. U_R Reliability factor

22. A Precision

---

*$n - 1$ is used for an estimate of the population standard deviation.
**Sometimes n is used.

A LIST OF RELIABILITY FACTORS

Reliability	One Side or One-Tailed Factor	Two Side or Two-Tailed Factor
80	.84	1.28
85	1.04	1.44
90	1.28	1.64
95	1.64	1.96
99	2.33	2.58

Numerical Distributions

The Standard Deviation of a Distribution

The standard deviation is an important measure of dispersion in a distribution of numbers. Illustration I–1 shows how two distributions with the same mean can have different standard deviations.

ILLUSTRATION I–1
DISTRIBUTIONS WITH THE SAME MEAN
BUT DIFFERENT STANDARD DEVIATIONS

Number X	Mean $\bar{X}$	Deviation x	Deviation Squared x^2	Calculation of Standard Deviation
10	20	-10	100	$s = \sqrt{\dfrac{\Sigma x^2}{n}}$
15	20	-5	25	
15	20	-5	25	
20	20	0	0	
20	20	0	0	$s = \sqrt{\dfrac{300}{9}}$
20	20	0	0	
25	20	5	25	
25	20	5	25	
30	20	10	100	$s = 5.77$
			300	

Number X	Mean $\bar{X}$	Deviation x	Deviation Squared x^2	Calculation of Standard Deviation
5	20	−15	225	$s = \sqrt{\dfrac{\Sigma x^2}{n}}$
15	20	−5	25	
15	20	−5	25	
20	20	0	0	
20	20	0	0	
20	20	0	0	$s = \sqrt{\dfrac{550}{9}}$
25	20	5	25	
25	20	5	25	
35	20	15	225	$s = 7.82$
			550	

Extreme values in a distribution can have a significant effect on the standard deviation. Illustration I–2 contains an example of such an effect.

ILLUSTRATION I–2
STANDARD DEVIATION OF A
DISTRIBUTION WITH AND WITHOUT EXTREME VALUES

	Number X	Mean $\bar{X}$	Deviation x	Deviation Squared x^2	Calculation of Standard Deviation
	15	20	−5	25	$s = \sqrt{\dfrac{\Sigma x^2}{n}}$
	18	20	−2	4	
(No	18	20	−2	4	
Extreme	20	20	0	0	
Values)	20	20	0	0	$s = \sqrt{\dfrac{66}{9}}$
	20	20	0	0	
	22	20	2	4	
	22	20	2	4	
	25	20	5	25	$s = 2.71$
				66	
	11	23	−12	144	
	21	23	−2	4	$s = \sqrt{\dfrac{\Sigma x^2}{n}}$
	21	23	−2	4	
(Extreme	23	23	0	0	
Values)	23	23	0	0	
	23	23	0	0	
	25	23	2	4	$s = \sqrt{\dfrac{304}{9}}$
	25	23	2	4	
	35	23	12	144	$s = 5.81$
				304	

The significance of the change in the standard deviation caused by extreme values in the distribution is discussed in a later section.

The Coefficient of Variation

The standard deviation is an absolute value. This measure can be placed in relative terms by dividing it by the mean. The coefficient of variation of both distributions in Illustration I–2 is calculated as follows.

$$v = \frac{s}{\bar{x}} = \frac{2.71}{20} = .14 \text{ (Distribution with no extreme values)}$$

$$v = \frac{s}{\bar{x}} = \frac{5.81}{23} = .25 \text{ (Distribution with extreme values)}$$

It is difficult to state whether a coefficient of variation is too "high" or "low." Coefficients of two or more distributions can be compared in order to determine which distribution has the most variation.

Skewness

A distribution of numbers is symmetrical if the skewness is zero (the mean, median, and mode are the same). Some skewness exists if there are extreme values at one end of the distribution with no counterbalancing values at the other end. Illustration I–3 shows the difference between a distribution with and without skewness. Illustration I–4 shows how the skewness is calculated.

ILLUSTRATION I–3
A DISTRIBUTION WITH AND WITHOUT SKEWNESS

				20		
(No		15	20	25	
Skewness)	10	15	20	25	30

The total $(\Sigma X) = 180$
The mean $(\mu) = 20$
The median $= 20$
The mode $= 20$

Note that both the two sets of numbers, 15 and 25, are the same distance from the mean. Also the two numbers, 10 and 30, are the same distance from the mean.

			20		
(Skewness)		15	20	25	
	10	15	20	25	48

The total (ΣX) = 198
The mean (μ)   = 22
The median   = 20
The mode   = 20

Note that the number 10 is not the same distance from the mean as the number 48. Therefore, the distribution is skewed.

ILLUSTRATION I–4
CALCULATION OF SKEWNESS USING ILLUSTRATION I–3 NUMBERS

	Number X	Mean $\bar{X}$	Deviation x	Deviation Squared x^2	The 3rd Power of the Deviation x^3
	10	20	-10	100	$-1,000$
	15	20	-5	25	-125
(No	15	20	-5	25	-125
Skewness)	20	20	0	0	0
	20	20	0	0	0
	20	20	0	0	0
	25	20	$+5$	25	$+125$
	25	20	$+5$	25	$+125$
	30	20	$+10$	100	$+1,000$
				300	0

$$m^3 = \frac{\Sigma x^3}{n} = \frac{0}{9} = 0$$

$$a^3 = \frac{m_3}{s^3} = 0 \text{ (The skewness is zero.)}$$

	Number X	Mean $\bar{X}$	Deviation x	Deviation Squared x^2	The 3rd Power of the Deviation x^3
	10	22	-12	144	$-1,728$
	15	22	-7	49	-343
	15	22	-7	49	-343
	20	22	-2	4	-8
(Skewness)	20	22	-2	4	-8
	20	22	-2	4	-8
	25	22	$+3$	9	$+27$
	25	22	$+3$	9	$+27$
	48	22	$+26$	676	$+17,576$
				948	$+15,192$

$$m^3 = \frac{\Sigma x^3}{n} = \frac{15,192}{9} = 1,688 \qquad s = \sqrt{\frac{\Sigma x^2}{n}} = 10.26$$

$$s^3 = 10.26^3 = 1,080$$

$$a_3 = \frac{m_3}{s^3} = \frac{1,688}{1,080} = 1.56 \text{ (The measure of positive skewness)}$$

Kurtosis

Another characteristic of a normal distribution is that its kurtosis is 3. Kurtosis is the measure of peakness in the distribution, and a value of 3 means that particular distribution is mesokurtic. If the distribution is "flat," the kurtosis is <3; if the distribution is "narrow" or too peaked, the kurtosis is >3.

Illustration I–5 shows a distribution that is close to normal with a skewness of zero and kurtosis >3. The same illustration contains numbers with a skewness of zero but a kurtosis <3.

ILLUSTRATION I–5
CALCULATION OF KURTOSIS

Number X	Mean $\bar{X}$	Deviation x	Deviation Squared x^2	The 4th Power of the Deviation x^4
10	20	-10	100	10,000
17	20	-3	9	81
17	20	-3	9	81
20	20	0	0	0
20	20	0	0	0
20	20	0	0	0
23	20	$+3$	9	81
23	20	$+3$	9	81
30	20	$+10$	100	10,000
			236	20,324

$$m_4 = \frac{\Sigma x^4}{n} = \frac{20,324}{9} = 2,258 \qquad s = \sqrt{\frac{236}{9}} = 5.12$$

$$s^4 = 5.12^4 = 687$$

$$a_4 = \frac{m_4}{s^4} = \frac{2,258}{687} = 3.29 \text{ kurtosis (slightly peaked distribution)}$$

	Number X	Mean $\bar{X}$	Deviation x	Deviation Squared x^2	The 4th Power of the Deviation x^4
	10	20	−10	100	10,000
	12	20	−8	64	4,096
	12	20	−8	64	4,096
("Flat"	20	20	0	0	0
Distri-	20	20	0	0	0
bution)	20	20	0	0	0
	28	20	+8	64	4,096
	28	20	+8	64	4,096
	30	20	+10	100	10,000
				456	36,384

$$m_4 = \frac{\Sigma x^4}{n} = \frac{36,384}{9} = 4,043 \qquad s = \sqrt{\frac{456}{9}} = 7.12$$

$$s^4 = 7.12^4 = 2,570$$

$$a_4 = \frac{m_4}{s^4} = \frac{4,043}{2,570} = 1.57 \text{ kurtosis (flat distribution)}$$

Summary of Skewness and Kurtosis

The following guidelines can be used to describe a normal distribution, skewness, and kurtosis.

1. A distribution is normal if its skewness is zero and its kurtosis is 3.

2. A distribution is negatively skewed if there is an extreme value(s) lower than the mean and not counterbalanced by an extreme value(s) higher than the mean. The skewness factor is < 0.

3. A distribution is positively skewed if there is an extreme value(s) higher than the mean and not counterbalanced by an extreme value(s) lower than the mean. The skewness is > 0.

4. A distribution is "flat" if its kurtosis is < 3.

5. A distribution is "peaked" if its kurtosis is > 3.

Illustration I–6 contains diagrams of five distributions, each representing the characteristics listed in this summary. These diagrams, as well as those in other appendix illustrations, are constructed with discrete distributions. However, for teaching purposes, the diagrams are drawn with straight lines connecting the points. When this method is used, the diagrams resemble the continuous distributions used in statistics books.

ILLUSTRATION I–6

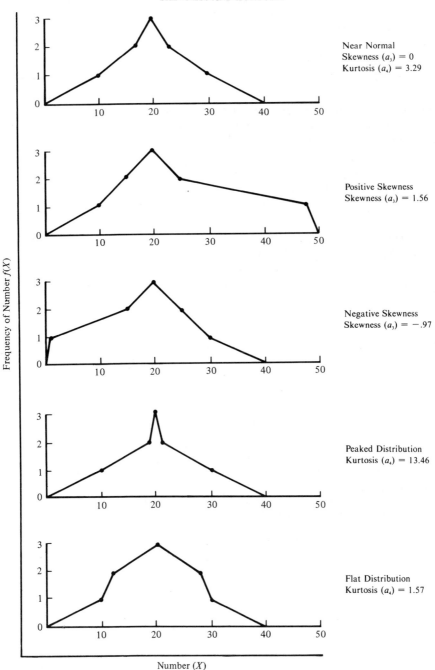

DISTRIBUTIONS SHOWING MEASURES OF
SKEWNESS AND KURTOSIS

Near Normal
Skewness $(a_3) = 0$
Kurtosis $(a_4) = 3.29$

Positive Skewness
Skewness $(a_3) = 1.56$

Negative Skewness
Skewness $(a_3) = -.97$

Peaked Distribution
Kurtosis $(a_4) = 13.46$

Flat Distribution
Kurtosis $(a_4) = 1.57$

Frequency of Number $f(X)$

Number (X)

Illustration 1–6

Sampling from Distributions

Normal Distributions and the Standard Deviation

In a normal distribution, a certain percentage of the numbers fall within a certain multiple of the standard deviation from the mean of that distribution. This is true whether the distribution is of a population or sample means. In the case of a distribution of sample means, however, the statistical measure is called the standard error of the mean.

The list of two-tailed reliability factors in the early part of the appendix contains several standard deviation multiples associated with a percentage of the numbers in a normal distribution. For example, in a normal distribution, 80% of the numbers fall within ± 1.28 standard deviations of the mean of the distribution, 90% fall within ± 1.64 standard deviations, etc.

Illustration I–7 contains a distribution of 45 numbers that is reasonably close to normal. The skewness is zero and the kurtosis is 2.91. Note that the percentages of numbers falling within certain distances of the mean are close to the reliability factors listed in the earlier section of the appendix.

ILLUSTRATION I–7
STANDARD DEVIATION MULTIPLES IN
A DISTRIBUTION CLOSE TO NORMAL

				20				
				20				
				20				
			18	20	22			
			18	20	22			
		15	18	20	22	25		
		15	18	20	22	25		
		15	18	20	22	25		
		15	18	20	22	25		
	13	15	18	20	22	25	27	
10	13	15	18	20	22	25	27	30

The mean ($\bar{X}$) is 20
The standard deviation is 4.11

Multiple of the Standard Deviation	Range Around the Standard Deviation	Numbers Included in the Range	Percentage of the Distribution Included in the Range
1.28	$20 + 1.28(4.11) = 25.26$	15,15,15,15,15, 15	86.7% (80% in the Reliability
	$20 - 1.28(4.11) = 14.74$	18,18,18,18,18, 18,18,18	Table)
		20,20,20,20,20, 20,20,20,20,20, 20	
		22,22,22,22,22, 22,22,22	
		25,25,25,25,25, 25	

Multiple of the Standard Deviation	Range Around the Standard Deviation	Numbers Included in the Range	Percentage of the Distribution Included in the Range
1.96	20 + 1.96(4.11) = 28.06 20 − 1.96(4.11) = 11.94	13,13,15,15,15, 15,15,15 18,18,18,18,18, 18,18,18 20,20,20,20,20, 20,20,20,20,20, 20 22,22,22,22,22, 22,22,22 25,25,25,25,25, 25,27,27	95.6% (95% in the Reliability Table)

In a discrete distribution, the percentage of numbers will not equate, exactly, with the reliability tables, but should be close.

Distribution of Sample Means

Out of any population (*N*), there is a given combination of sample means of a given sample size (*n*) that can be drawn. For example, in a population consisting of the numbers 10, 15, and 20, three sample means can be drawn if the sample size is two and if the sampling method is without replacement (the same number cannot be drawn twice). These sample means are:

$$\frac{10 + 15}{2} = 25 \div 2 = 12.5$$

$$\frac{10 + 20}{2} = 30 \div 2 = 15.0$$

$$\frac{15 + 20}{2} = 35 \div 2 = 17.5$$

There is a formula that can be used to calculate the number of sample means of sample size (*n*) out of a population (*N*) when the sampling method is without replacement.[1] The following is an example of the formula and its application.

[1]Throughout the appendix, sampling without replacement is assumed unless otherwise specified.

$$\frac{N!^2}{n!(N-n)!} = \frac{9!}{2!(9-2)!} = \frac{362880}{10080} = 36 \text{ sample means of sample size two from a population of nine.}$$

Illustration I–8 shows the sample means of a sample size of two out of a population of nine. Illustration I–9 shows the population distribution and the distribution of sample means in chart form.

ILLUSTRATION I–8
A DISTRIBUTION OF SAMPLE MEANS

	Population (N) = 9
Mean (μ) = 20	20
Standard Deviation (σ) = 4.62	16 20 24
Skewness = 0	12 16 20 24 28
Kurtosis = 2.25	

	Distribution of Sample Means Sample Size (n) = 2
Mean ($\bar{X}$) = 20	18 20 22
Standard Error of the Mean ($\sigma_{\bar{x}}$) = 3.06	18 20 22
Skewness = 0	18 20 22
Kurtosis = 2.39	18 20 22
	16 18 20 22 24
	16 18 20 22 24
	14 16 18 20 22 24 26
	14 16 18 20 22 24 26

The standard deviation of the distribution of sample means is called the standard error of the mean and is calculated by the following formula, when the sampling is done without replacement.

$$\sigma_{\bar{x}} = \frac{\sigma}{\sqrt{n}}\sqrt{\frac{N-n}{N-1}} \qquad \sigma_{\bar{x}} = \frac{4.62}{\sqrt{2}}\sqrt{\frac{9-2}{9-1}} = 3.06$$

---

[2]Refers to factorial, e.g., $4 \times 3 \times 2 \times 1$.

ILLUSTRATION I-9

A POPULATION AND A DISTRIBUTION OF SAMPLE MEANS

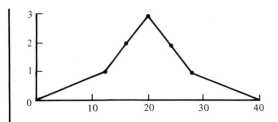

Population (N) = 9
Mean (μ) = 20
Standard Deviation (σ) = 4.62
Skewness = 0
Kurtosis = 2.25

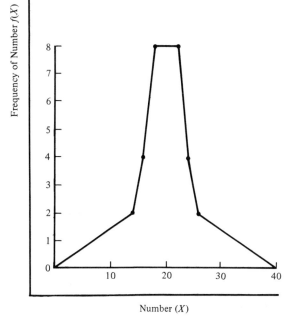

Sample Size (n) = 2
Mean ($\overline{X}$) = 20
Standard Error of the
Mean ($\sigma_{\overline{X}}$) = 3.06
Skewness = 0
Kurtosis = 2.39

Frequency of Number $f(X)$

Number (X)

Illustration 1–9

Sampling from a Skewed Distribution

Any distribution of sample means will be normally distributed if the sample is taken from a normal distribution. However, even if the distribution from which the sample is taken is skewed, the distribution of sample means will approach normality as the sample size becomes larger.

A sample size of 30 (which is considered large by statisticians) is expected to produce an approximately normal distribution of sample means even if the distribution from which the sample size is taken is skewed.

Illustration I–10 contains a skewed distribution of 9 numbers. The illustration also shows distributions of sample means of samples sizes of 2 and 3. Illustration I–11 contains the same distributions in chart form. Note that the distribution of sample means is less skewed with a larger sample size.

ILLUSTRATION I–10
DISTRIBUTIONS OF SAMPLE MEANS
FROM A SKEWED DISTRIBUTION

	Population (N) = 9
Mean (μ) = 22	20
Standard Deviation (σ) = 9.43	14 20
Skewness = .83	10 14 20 26 32 42
Kurtosis = 2.80	

Distribution of Sample
Means Sample Size (n) = 2

```
            17
            17     20      23
            17     20      23  26
      15    17     20      23  26            31
12    15    17     20      23  26  28        31
12 14 15    17  18 20  21  23  26  28  29 31 34  37
```

Mean ($\bar{X}$) = 22
Standard Error of the Mean ($\sigma_{\bar{x}}$) = 6.24
Skewness = .45
Kurtosis = 2.43

Estimation of the Population Total

Distribution of Population Estimates

Although the two distributions of sample means shown in Illustration I–11 are not normal, we will assume for purposes of discussion that the 84 sample means shown on page 607 are close enough to normal so that we may use the two-tailed reliability factors shown in the earlier part of the appendix.

ILLUSTRATION 1-10 (Cont'd)

Distribution of Sample Means Sample Size (n) = 3

12.6	14.6	16.0	16.6	18.0	18.6	20.0	20.6	22.0	22.6	23.3	24.0	25.3	26.0	27.3	28.0	29.3	31.3	33.3
								22.0										
								22.0										
								22.0										
	14.6			18.0		20.0		22.0			24.0							
	14.6		16.6	18.0	18.6	20.0		22.0			24.0	25.3		27.3		29.3		
	14.6		16.6	18.0	18.6	20.0		22.0			24.0	25.3		27.3		29.3		
	14.6	16.0	16.6	18.0	18.6	20.0	20.6	22.0			24.0	25.3	26.0	27.3		29.3	31.3	
	14.6	16.0	16.6	18.0	18.6	20.0	20.6	22.0			24.0	25.3	26.0	27.3		29.3	31.3	
	14.6	16.0	16.6	18.0	18.6	20.0	20.6	22.0			24.0	25.3	26.0	27.3		29.3	31.3	
12.6								22.6	23.3				28.0					33.3

Mean $(\bar{X})$ = 22

Standard Error of the Mean $(\sigma_{\bar{x}})$ = 4.71

Skewness = .24

Kurtosis = 2.32

ILLUSTRATION I–11

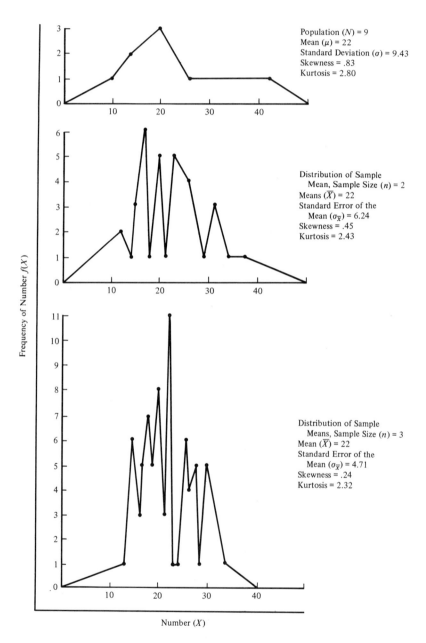

DISTRIBUTION OF SAMPLE MEANS FROM
A SKEWED DISTRIBUTION

Population (N) = 9
Mean (μ) = 22
Standard Deviation (σ) = 9.43
Skewness = .83
Kurtosis = 2.80

Distribution of Sample
    Mean, Sample Size (n) = 2
Means $(\overline{X})$ = 22
Standard Error of the
    Mean $(\sigma_{\overline{x}})$ = 6.24
Skewness = .45
Kurtosis = 2.43

Distribution of Sample
    Means, Sample Size (n) = 3
Mean $(\overline{X})$ = 22
Standard Error of the
    Mean $(\sigma_{\overline{x}})$ = 4.71
Skewness = .24
Kurtosis = 2.32

Frequency of Number $f(X)$

Number (X)

Illustration 1–11

These reliability factors are shown in terms of multiples of the standard error of the mean ($\sigma_{\bar{x}}$). For example, the $\sigma_{\bar{x}}$ on page 607 of Illustration I–10 is 4.71. The multiple for a 90% reliability is then $1.64 \times 4.71 = 7.72$. The statement can then be made that if a random sample of 3 is taken from the population of 9 shown at the top of Illustration I–10, approximately 90% of the 84 sample means will fall within $\pm$ 7.72 of 22 (which is the mean of the sampling distribution and the true mean of the population).

By multiplying each sample mean times the population size (9), a distribution of population estimates is derived. The mean of this distribution is the true population total. Illustration I–12 shows how the distribution of population estimates is derived. Illustration I–13 contains the distribution in chart form.

Reliability and Precision

The distribution of population estimates shown in Illustration I–12 can now be used as a basis for discussing the relationship between reliability (confidence level) and precision (confidence interval).

If a random sample of 3 is taken from the population of 9 shown at the top of Illustration I–10, the mean of the sample is one of the 84 numbers shown at the bottom of Illustration I–10, and the estimate of the population total is one of the 84 numbers shown in the population estimate column of Illustration I–12. The sampler does not know how far the estimate varies from the true population total (198). However, if the distribution of population estimates is approximately normal, the sampler knows at a certain reliability percentage the maximum distance between the population estimate and the true population total.

For example, if the 3 numbers of 14, 14, and 20 are drawn from the population of 9 shown at the top of Illustration I–10, the sample mean is 16 $[(14 + 14 + 20) \div 3]$, and the population estimate is 144 (16×9). Illustration I–14 on page 611 contains an array of reliability percentages and precision ranges of population estimates.

Note that in order to attain higher reliability, the sampler must accept a wider precision. This is true as long as the sample size is 3. If the sample size increases, a narrower precision range can be attained at given reliability levels, since a different distribution of sample means is produced and the standard error of the mean decreases. At *any* given sample size, however, there is an inverse relationship between the size of the reliability percentage and the narrowness of the precision.

Summary of Material to Date

1. A distribution of sample means is a list of the means of *all* possible samples of a given sample size that can be drawn from a given population. The formula for the number of possible sample means is

$$\frac{N!}{n!(N - n)!}.$$

ILLUSTRATION I–12
DISTRIBUTION OF POPULATION ESTIMATES

Sample Mean	Population Size	Population Estimate	Sample Mean	Population Size	Population Estimate
12.6	9	114	22.0	9	198
14.6	9	132	22.0	9	198
14.6	9	132	22.0	9	198
14.6	9	132	22.0	9	198
14.6	9	132	22.0	9	198
14.6	9	132	22.6	9	204
14.6	9	132	23.3	9	210
16.0	9	144	24.0	9	216
16.0	9	144	24.0	9	216
16.0	9	144	24.0	9	216
16.6	9	150	24.0	9	216
16.6	9	150	24.0	9	216
16.6	9	150	24.0	9	216
16.6	9	150	24.0	9	216
16.6	9	150	24.0	9	216
18.0	9	162	25.3	9	228
18.0	9	162	25.3	9	228
18.0	9	162	25.3	9	228
18.0	9	162	25.3	9	228
18.0	9	162	25.3	9	228
18.0	9	162	25.3	9	228
18.0	9	162	26.0	9	234
18.6	9	168	26.0	9	234
18.6	9	168	26.0	9	234
18.6	9	168	26.0	9	234
18.6	9	168	27.3	9	246
18.6	9	168	27.3	9	246
20.0	9	180	27.3	9	246
20.0	9	180	27.3	9	246
20.0	9	180	27.3	9	246
20.0	9	180	28.0	9	252
20.0	9	180	29.3	9	264
20.0	9	180	29.3	9	264
20.0	9	180	29.3	9	264
20.0	9	180	29.3	9	264
20.6	9	186	29.3	9	264
20.6	9	186	31.3	9	282
20.6	9	186	31.3	9	282
22.0	9	198	31.3	9	282
22.0	9	198	33.3	9	300
22.0	9	198			
22.0	9	198			
22.0	9	198			
22.0	9	198			

Note: If approximately 90% of the sample means fall within ± 7.72 (4.71 × 1.64) of 22 (the true mean), then approximately 90% of the true population estimates fall within ± 69.48 (7.72 × 9) of 198 (the true population total).

ILLUSTRATION I–13

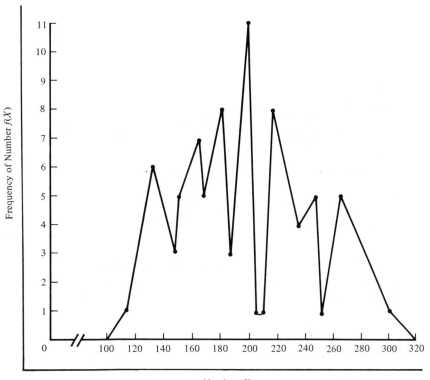

DISTIBUTION OF POPULATION ESTIMATES

Illustration 1–13

ILLUSTRATION I–14
*RELIABILITY PERCENTAGES AND PRECISION RANGES
ESTIMATION OF POPULATION TOTAL IS 144*

Reliability or Confidence Level	*Precision Range Within Which the True Population Total Falls at This Reliability*
80%	$144 + (4.71 \times 1.28 \times 9) = 198.26$
	$144 - (4.71 \times 1.28 \times 9) = 89.74$
90%	$144 + (4.71 \times 1.64 \times 9) = 213.52$
	$144 - (4.71 \times 1.64 \times 9) = 74.48$
95%	$144 + (4.71 \times 1.96 \times 9) = 227.08$
	$144 - (4.71 \times 1.96 \times 9) = 60.92$

2. The distribution of sample means of a large sample is close to normal even if the population from which the sample is taken is skewed; 30 is considered by statisticians to be a large sample.

3. In a distribution of sample means that is approximately normal, the sampler can state the percentage of sample means that fall within a certain distance of the mean of the distribution, which is also the true mean of the population.

4. The preceding statement means that a large random sample can be taken and a mean of the sample calculated. The sampler can state, at a certain reliability percentage, that the true mean of the population is within a certain distance from the sample mean. Likewise, the sampler can state, at a certain reliability percentage, that the true population total is within a certain distance from the estimate of the population total.

5. At a given sample size, there is an inverse relationship between the reliability percentage and the precision that can be attained, i.e., as the reliability percentage goes up, the precision interval widens, and vice versa.

6. As the sample size increases, a narrower precision can be obtained with the same percentage of reliability, or a higher percentage of reliability can be obtained with the same precision.

Determining Sample Size

Infinite Population

Up to this point, much of the discussion has centered on the determination of reliability and precision, given a certain sample size. However, we should now turn the known and unknown variables around and *find* the appropriate sample size for the reliability and precision that we wish to achieve.

The formula for this appropriate sample size is derived from the following.[3]

1. The formula for the standard error of the mean of an infinite population is $\sigma_{\bar{x}} = \dfrac{\sigma}{\sqrt{n}}$ where $\sigma =$ the standard deviation and $n =$ the sample size.

Using arbitrary numerical examples, $\sigma_{\bar{x}} = \dfrac{250}{\sqrt{60}} = 32.26$.

2. 32.26 is one standard error of the mean. An 80% reliability is measured by 1.28 standard errors. Therefore, $32.26 \times 1.28 = 41.29$.

---

[3] The purpose of this explanation is to encourage conceptual understanding of the sample size formula used by auditors in variable sampling.

3. To obtain a population precision, the multiple of the standard error of the mean is multiplied by the population size (assume 200). Therefore, $41.29 \times 200 = 8,258$ (rounded off).

4. The formula for population precision (A) if the sampling is done with replacement or from an infinite population is:

$$A = \frac{\sigma}{\sqrt{n}} \times u_R \times N$$

$$A = \frac{250}{\sqrt{60}} \times 1.28 \times 200$$

$$A = 8,258$$

5. By taking the formula in Item 4 and applying algebra, the following formula for sample size is determined.

$$\sqrt{n} = \frac{\sigma \times u_R \times N}{A}$$

$$n = \left(\frac{\sigma \times u_R \times N}{A}\right)^2$$

$$n = \left(\frac{250 \times 1.28 \times 200}{8,258}\right)^2$$

$$n = 60$$

Finite Population

If sampling is done without replacement (a finite population), the sample size can be smaller. A finite population correction factor is multiplied times the standard error of the mean ($\sigma_{\bar{x}}$); this lowers the value of $\sigma_{\bar{x}}$ and, thus, narrows the attained precision.

Therefore, when sampling is performed without replacement, the formula shown in Item 5 of the previous section is divided by a correction factor in order to bring the sample size down. This factor is $1 + \dfrac{n}{N}$. The additional calculation to obtain the sample size without replacement is

$$n' = \frac{n}{1 + \dfrac{n}{N}}$$

$$n' = \frac{60}{1.30}$$

$$n' = 47$$

Explanation for the Sample Size Formula

Taking the formula $n = \left(\dfrac{\sigma \times u_R \times N}{A}\right)^2$, we can easily ascertain the effect on the sample size of any increase or decrease in the value of a variable. The following is a chart of such changes.

Standard Deviation σ	Reliability Coefficient u_R	Population Size N	Precision A	Sample Size n
Increases	Same	Same	Same	Increases
Same	Increases	Same	Same	Increases
Same	Same	Increases	Same	Increases
Same	Same	Same	Widens	Decreases
Decreases	Same	Same	Same	Decreases
Same	Decreases	Same	Same	Decreases
Same	Same	Decreases	Same	Decreases
Same	Same	Same	Narrows	Increases

In describing precision, the terms wide and narrow are used rather than increase and decrease.

Estimate of the Standard Deviation

In determining the sample size, an estimate of the population standard deviation is made, since the actual figure is not known. If the sampler cannot make a reasonable estimate of the standard deviation, a random sample of 30 can be taken and the standard deviation of this sample used as the best estimate of the population figure.

In this case, the formula for the standard deviation of a sample is used. However, since a sample standard deviation is biased downward from a population standard deviation, $n - 1$ is used as a denominator rather than n. Thus the modified formula is written as follows. $s = \sqrt{\dfrac{\Sigma x^2}{n-1}}$

The estimated standard error of the mean is written in the following manner.

$$s_{\bar{x}} = \frac{s}{\sqrt{n}} \text{ (infinite population)}$$

$$s_{\bar{x}} = \frac{s}{\sqrt{n}} \sqrt{\frac{N-n}{N-1}} \text{ (finite population)}$$

Evaluating Sample Results

Point Estimate of the Population Total

Regardless of whether the sampling is done with or without replacement, the estimate of the population total is made by taking the sample, calculating the sample mean, and multiplying this sample mean times the population size.

Assume that the true, but unknown, population total is 100,000 and that the population size is 200. If a sample of 60 is taken and the sample mean is 490, the estimate of the population total is 98,000 (490 × 200).

Reliability and Precision—Infinite Population

Although the sampler may not know how close the 98,000 estimate is to the true population total, a precision range can be established around the population estimate. This precision range will include, at a certain reliability level, the true population total.

If 80% confidence is required, the precision range is calculated by using a coefficient (u_R) of 1.28. The calculation is as follows:

1. Assume that the standard deviation (s) of the sample of 60 is 240. This figure then becomes the best estimate of the standard deviation of the population, replacing the estimate of 250 used to establish the sample size.

2. The *attained* precision is

$$\frac{s}{\sqrt{n}} \times u_R \times N, \frac{240}{\sqrt{60}} \times 1.28 \times 200 = 7,928.$$

3. The statement can then be made that there is an 80% reliability that the true population total is between 105,928 (the 98,000 estimate + 7,928) and 90,072 (the 98,000 estimate − 7,928).

4. Note that the attained precision of 7,928 is narrower than the 8,258 target precision on which the sample size is calculated. These two precision figures will usually be different, since the standard deviation of the sample differs from the original estimate of the population standard deviation. A narrower attained precision is of benefit to the sampler, since the precision range is closer to the estimate of the population total.

Reliability and Precision—Finite Population

In determining precision for sampling with replacement (infinite population), the formula for the estimate of the standard error of the mean is

$$s_{\bar{x}} = \frac{s}{\sqrt{n}}.$$

However, in sampling without replacement (finite population), the same formula is

$$s_{\bar{x}} = \frac{s}{\sqrt{n}} \sqrt{\frac{N-n}{N-1}}.$$

The use of the finite correction factor, $\sqrt{\dfrac{N-n}{N-1}}$, lowers the value of $s_{\bar{x}}$, which results in a narrower (and better) precision. Therefore, when sampling is done without replacement, a correction factor is applied to the sample size that is used when sampling with replacement. This correction factor results in a lower size when sampling without replacement.

The correction factor is $1 + \dfrac{n}{N}$. If the sample size with replacement is calculated at 60, the sample size without replacement is

$$\frac{60}{1 + \dfrac{n}{N}} = \frac{60}{1 + \dfrac{60}{200}} = \frac{60}{1.30} = 47 \text{ (rounded off)}.$$

Using 47 as the sample size rather than 60, assume that the estimate of the population total is 98,000 and the standard deviation of the sample is 240 (the same as that obtained in the sample of 60 with replacement). The calculation of the attained precision is as follows:

$$\frac{s}{\sqrt{n}} \sqrt{\frac{N-n}{N-1}} \times u_R \times N$$

$$\frac{240}{\sqrt{47}} \sqrt{\frac{200-47}{200-1}} \times 1.28 \times 200 = 7{,}847 \text{ (rounded off)}.$$

Note that the sample of 47 taken without replacement achieved 7,847 precision, which is close to the 7,928 precision attained by taking a sample of 60 with replacement. Of course, the different samples of 47 and 60 would produce slightly different population and standard deviation estimates. The purpose of this illustration is to show that a smaller sample size can be taken without replacement and precision results obtained that are comparable to those achieved when a larger sample size is taken with replacement.

The Sampling Risks and Tests of Hypotheses

The Null Hypothesis and the Alpha Risk

In estimation sampling, an estimate of the population total is made and a precision range around this estimate is calculated, at a certain reliability or confidence level. In hypothesis testing, a null hypothesis is stated and this hypothesis is either accepted or rejected, at a certain level of significance.

The null hypothesis could be stated as: "The assumed mean of the population is the correct or true mean."

As the difference between the assumed mean of the population and the sample mean becomes larger, there is less likelihood that the null hypothesis is correct. However, the sample mean does not have to be the same as the assumed mean in order for the null hypothesis to be accepted. In fact, the sampler can define how much difference can exist between the assumed and sample mean without rejecting the null hypothesis. This can be done by setting a level of significance.

The level of significance is not the same as a confidence level or reliability as it is used in estimation sampling. It is the probability that the difference between the assumed mean and the sample mean is so high that the null hypothesis is rejected when, in fact, the null hypothesis is correct.

As an example, we can take the population of 9 numbers and the 36 sample means shown in Illustration I–8.[4] Both sets of numbers are listed below.

Population (N) = 9					*Distribution of Sample Means* *Sample Size (n) = 2*						
		20					18	20	22		
	16	20	24				18	20	22		
12	16	20	24	28			18	20	22		
						16	18	20	22	24	
						16	18	20	22	24	
					14	16	18	20	22	24	26
					14	16	18	20	22	24	26

Assume that the null hypothesis is that the purported mean of 20 is the true mean (in this case, the null hypothesis is correct). Also, assume that the null hypothesis is tested at a 10% level of significance.

A 10% level means that the sampler is willing to take a 10% risk that the null hypothesis will be rejected when it is true. This risk is referred to as the alpha risk, and the error of incorrectly rejecting the null hypothesis is called a Type I error.

[4] It is recognized that a sample size of 30 is desirable, but this sample is used (and assumed to be normal) so that a more workable distribution of sample means can be employed.

We can show how this error could occur by using the 36 sample means listed above. Of the sample means, 90% will fall within 1.64 $\sigma_{\bar{x}}$ of the true mean of 20, and 10% of the means will fall outside this range. If any sample is selected that is within 1.64$\sigma_{\bar{x}}$ of 20, this figure will be correctly accepted as the true mean. This acceptance will occur 90% of the time, since 90% of the sample means are within 1.64 $\sigma_{\bar{x}}$ of 20.

On the other hand, 10% of the sample means fall outside 1.64 $\sigma_{\bar{x}}$ of the true mean of 20, which means that 10% of the time the figure of 20 will be incorrectly rejected as the true mean. The sampler can set the level of significance as high as wanted. The higher the level, the greater the alpha risk and the greater the probability of making a Type I error.

Illustrations I–15 and I–16 contain explanations of the concepts described in the foregoing paragraphs.

ILLUSTRATION I–15

EXAMPLE OF THE NULL HYPOTHESIS AND ALPHA RISK OF 10%

A Distribution of Sample Means
Sample Size (n) = 2, True Mean of Population = 20

	14.98				25.02	
		18	20	22		
		18	20	22		
		18	20	22		
		18	20	22		
	16	18	20	22	24	
	16	18	20	22	24	
14	16	18	20	22	24	26
14	16	18	20	22	24	26

Region of Incorrect Rejection of Null Hypothesis that 20 is the True Mean (5%)	Region of Correct Acceptance of Null Hypothesis that 20 is the True Mean (90%)	Region of Incorrect Rejection of Null Hypothesis that 20 is the True Mean (5%)

$\sigma_{\bar{x}} = 3.06$

$(1.64)\sigma_{\bar{x}} = 5.02$

$20 + 5.02 = 25.02$

$20 - 5.02 = 14.98$

The null hypothesis is that the assumed mean of 20 is the true mean. If the sample mean selected is any of the numbers 16–24 (32 of 36 sample means, approximately 90%), the null hypothesis is correctly accepted. If either of the two sample means of 14 or the two sample means of 26 is selected, the null hypothesis is incorrectly rejected. As can be seen, the alpha risk can be increased by "drawing in the lines of acceptance" to exclude more sample means.

ILLUSTRATION I–16
EXAMPLE OF THE NULL HYPOTHESIS
AND ALPHA RISK OF APPROXIMATELY 32%

A Distribution of Sample Means,
Sample Size (n) = 2, True Mean
of Population = 20

		16.94				23.06	
		18	20	22			
		18	20	22			
		18	20	22			
		18	20	22			
	16	18	20	22		24	
	16	18	20	22		24	
14	16	18	20	22		24	26
14	16	18	20	22		24	26

Region of Incorrect Rejection of Null Hypothesis that 20 is the True Mean (about 16%)

Region of Correct Acceptance of Null Hypothesis that 20 is the True Mean (about 68%)

Region of Incorrect Rejection of Null Hypothesis that 20 is the True Mean (about 16%)

$\sigma_{\bar{x}} = 3.06$
$(1.00)\sigma_{\bar{x}} = 3.06$
$20 + 3.06 = 23.06$
$20 - 3.06 = 16.94$

The null hypothesis is that the assumed mean of 20 is the true mean. If the sample selected has a mean of 18, 20, or 22 (24 of 36 sample means, approximately 68%), the null hypothesis is correctly accepted. If the sample selected has a mean of 14, 16, 24, or 26, the null hypothesis is incorrectly rejected.

Further Discussion of the Alpha Risk

In studying Illustrations I–15 and I–16, two items should be noted.

1. The sampler can change the probability of a Type I error if such a change is desired. Low alpha risk probabilities result in larger sample sizes or wider precision intervals because the range of acceptance is wide. The opposite, of course, is also true.

2. This particular test of hypothesis is "two-tailed" in that rejection of the null hypothesis occurs if sample means are selected that are either *higher*

or *lower* than the region of acceptance. For example, Illustration I–15 contains a diagram that shows that there is a 5% probability that the null hypothesis will be rejected because the sample mean is too high (26), and a 5% probability that the null hypothesis will be rejected because the sample mean is too low (14).

The Alternate Hypothesis and the Beta Risk

The alpha risk is the risk that a correct figure will be incorrectly rejected, the null hypothesis being that the assumed figure is correct. The beta risk, on the other hand, is the risk that the null hypothesis will be incorrectly accepted; such an acceptance represents a Type II error.

In order to understand how the probability of a Type II error is constructed, we should pose an alternate hypothesis that a figure other than the assumed amount is correct. Incorrect acceptance of the null hypothesis is equivalent to incorrect rejection of the alternate hypothesis.

Illustration I–17 on page 621 contains an explanation of the beta risk. A distribution of sample means is drawn around both the assumed means of 20 and 30. The null hypothesis is that the true mean is ≤ 20. The alternate hypothesis is that the true mean is > 20, with the assumed mean at 30.

Comments on Illustration I–17

It should be noted that the examples shown in Illustration I–17 are "one-tailed" tests, since the null hypothesis is that the true mean is ≤ 20 and the alternate hypothesis is that the true mean is > 20. This means that there is a 5% alpha risk that the sample mean selected would be so high (26) as to reject the null hypothesis.

If the null hypothesis is stated that the true mean is $= 20$, there is a 10% alpha risk that a sample mean selected would be so low (14) or so high (26) as to reject the hypothesis.

Also, it is important to understand that the calculation of these alpha and beta risks in Illustration I–17 relates only to the same sample sizes and the same $\sigma_{\bar{x}}$ of 3.06. The precision around the mean at a 90% confidence level is 1.64 $\sigma_{\bar{x}}$ = 5.02.

There is a difference of 10 between the mean of 20 for the null hypothesis and the assumed mean of 30 for the alternate hypothesis. This figure of 10 is considered to be the material difference.[5] When the material difference is twice the precision (as it is in Illustration I–17), the following conditions exist.

1. The beta risk (5%) is equal to the alpha risk (5%) if a one-tailed test of the null hypothesis is conducted.

2. The beta risk (5%) is one-half the alpha risk (10%) if a two-tailed test of the null hypothesis is conducted.

---

[5]We are using an auditing term so that we can relate this material to earlier chapters.

However, the material difference can be more or less than a figure which is twice the amount of precision. Illustration I–18 on page 622 contains a difference of 8 and an alternate hypothesis that the true mean is > 20, with an assumed mean of 28. Note that as the material difference narrows, the beta risk or probability of committing a Type II error increases. The opposite, of course, is true.

ILLUSTRATION I–17

EXAMPLE OF THE ALTERNATE

HYPOTHESIS AND A BETA RISK OF 5%

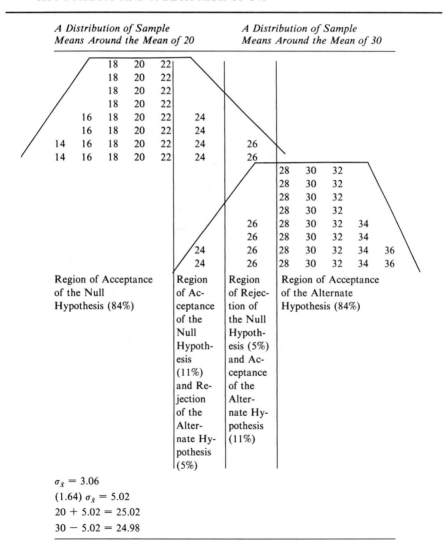

| *A Distribution of Sample* *Means Around the Mean of 20* | | *A Distribution of Sample* *Means Around the Mean of 30* |

Region of Acceptance of the Null Hypothesis (84%)

Region of Acceptance of the Null Hypothesis (11%) and Rejection of the Alternate Hypothesis (5%)

Region of Rejection of the Null Hypothesis (5%) and Acceptance of the Alternate Hypothesis (11%)

Region of Acceptance of the Alternate Hypothesis (84%)

$\sigma_{\bar{x}} = 3.06$

$(1.64)\ \sigma_{\bar{x}} = 5.02$

$20 + 5.02 = 25.02$

$30 - 5.02 = 24.98$

For the data above, the following table shows the probabilities of correct and incorrect decisions.

	The Null Hypothesis Is True and the True Mean is ≤ 20	The Alternate Hypothesis is True and the True Mean is > 20, with an Assumed Mean of 30
The Probability of Accepting the Null Hypothesis is	95% (Correct Decision)	5% (Type II Error)
The Probability of Rejecting the Null Hypothesis is	5% (Type I Error)	95% (Correct Decision)

ILLUSTRATION I–18
EXAMPLE OF THE ALTERNATE
HYPOTHESIS AND A BETA RISK OF 16%

A Distribution of Sample Means Around the Mean of 20						A Distribution of Sample Means Around the Mean of 28				
		18	20	22						
		18	20	22						
		18	20	22						
		18	20	22						
	16	18	20	22	24					
	16	18	20	22	24					
14	16	18	20	22	24	26				
14	16	18	20	22	24	26				
						26	28	30		
						26	28	30		
						26	28	30		
						26	28	30		
					24	26	28	30	32	
					24	26	28	30	32	
				22	24	26	28	30	32	34
				22	24	26	28	30	32	34

Region of Acceptance of the Null Hypothesis (62%)	Region of Acceptance of the Null Hypothesis (33%) and Rejection of the Alternate Hypothesis (16%)	Region of Rejection of the Null Hypothesis (5%) and Acceptance of the Alternate Hypothesis (22%)	Region of Acceptance of the Alternate Hypothesis (62%)

$\sigma_{\bar{x}} = 3.06$

$(1.64)\ \sigma_{\bar{x}} = 5.02$ [The one-tailed U_R for 95% (100% − 5%) is 1.64].

$20 + 5.02 = 25.02$

$28 - 3.06 = 24.94$ [The one-tailed U_R for 84% (100% − 16%) is approximately 1].

For the data above, the following table shows the probabilities of correct and incorrect decisions.

	The Null Hypothesis Is True and the True Mean is ≤ 20	The Alternate Hypothesis is True and the True Mean is > 20, with an Assumed Mean of 28
The Probability of Accepting the Null Hypothesis is	95% (Correct Decision)	16% (Type II Error)
The Probability of Rejecting the Null Hypothesis is	5% (Type I Error)	84% (Correct Decision)

Summary of Explanation of Tests of Hypotheses

1. A null hypothesis is a statement that an assumed figure is the correct amount. The hypothesis can be stated as follows: The true mean = 20, the true mean is ≤ 20, or the true mean is ≥ 20.

2. A null hypothesis is tested at a certain level of significance, e.g., the probability that a sample mean could be far enough away from the assumed mean so that the null hypothesis is rejected, although the null hypothesis is correct.

3. The alpha risk is the risk that the null hypothesis will be incorrectly rejected. Such a rejection represents a Type I error.

4. The null hypothesis may be tested on a one-tailed or two-tailed basis. If a one-tailed basis is used, the null hypothesis will be rejected only if the sample figure is too far removed in one direction from the assumed figure. If a two-tailed basis is employed, the null hypothesis will be rejected if the sample figure is too far removed in either direction from the assumed figure.

5. An alternate hypothesis is a statement that a figure other than the assumed figure is correct.

6. The difference between the assumed figure for the null hypothesis and the assumed figure for the alternate hypothesis is called the material difference.

7. The beta risk is the risk that the null hypothesis will be incorrectly accepted and that the alternate hypothesis will be incorrectly rejected. Such a rejection represents a Type II error.

8. The larger the material difference between the assumed figure for the null hypothesis and the assumed figure for the alternate hypothesis, the smaller the beta risk and the smaller the probability of a Type II error. The smaller the material difference, the larger the beta risk and the larger the probability of a Type II error.

Appendix to Chapter 13
Questions Taken from the Appendix

13A–1. What does the standard deviation measure?

13A–2. Describe how the standard deviation is calculated.

13A–3. What is the effect on the standard deviation of extreme values in the numerical distribution?

13A–4. What is the coefficient of variation?

13A–5. Under what conditions does skewness exist in a numerical distribution?

13A–6. Under what conditions is a distribution of numbers symmetrical?

13A–7. What does kurtosis measure in a distribution of numbers?

13A–8. In a normal distribution, what is the numerical value of skewness? Kurtosis?

13A–9. What condition exists in a numerical distribution when it is negatively skewed? Positively skewed?

13A–10. What is the standard deviation of a distribution of sample means?

13A–11. In a normal distribution, what relationships exist between the standard deviation and the mean?

13A–12. In a skewed population, what happens to the distribution of sample means as the sample size becomes larger?

13A–13. How can a distribution of sample means be converted to a distribution of population estimates?

13A–14. What is another name for reliability? For precision?

13A–15. At a given sample size, what is the relationship between precision and reliability?

13A–16. What is a distribution of sample means?

13A–17. As the sample size increases and precision stays the same, what happens to reliability?

13A–18. As the sample size increases and the reliability stays the same, what happens to precision?

13A–19. Name the four variables that are used to determine the sample size.

13A–20. Derive the formula for the sample size assuming an infinite population. Start with one standard error of the mean.

13A–21. What type of sampling is done when an infinite population is assumed? When a finite population is assumed?

13A–22. What correction factor is applied to the sample size determination when sampling is done without replacement?

13A–23. What happens to the sample size when all other factors remain the same and
   a.   The standard deviation increases? decreases?
   b.   The reliability coefficient increases? decreases?
   c.   The population size increases? decreases?
   d.   The precision widens? narrows?

13A–24. How is the estimate of the population total calculated?

13A–25. What is the formula for calculating the *attained* precision around the estimate of the population total when sampling from an infinite population?

13A–26. Why is the attained precision usually different from the target precision used to determine the sample size?

13A–27. What is the formula for the finite correction factor that is multiplied times the estimate of the standard error of the mean?

13A–28. Give an example of a null hypothesis.

13A–29. What happens to the probability of the null hypothesis being correct as the difference between the assumed and sample means becomes larger?

13A–30. What is the level of significance?

13A–31. What is the alpha risk?

13A–32. What is a Type I error?

13A–33. How can the alpha risk be increased?

13A–34. What is the relationship between the alpha risk percentage and the sample size?

13A–35. What is the difference between a "two-tailed" and a "one-tailed" test of the null hypothesis?

13A–36. What is the beta risk?

13A–37. What is a Type II error?

13A–38. What is an alternate hypothesis?

13A–39. What is the material difference?

13A–40. If the material difference is twice the precision in a one-tailed test of the null hypothesis, what is the relationship between the alpha risk and beta risk percentages? What if there is a two-tailed test of the null hypothesis?

13A–41. What happens to the beta risk as the material difference narrows?

Appendix to Chapter 13

Problems Based on Appendix Material

Some of the problems can be solved with the aid of one or both of the computer programs listed at the end of this appendix. The programs are written in IBM BASIC. Some of the problems can be assigned to be run on terminals. Problems that can be solved with the computer programs are labeled. The name of the program is shown after the problem listing. All of the problems can be worked by hand, since they relate to the material presented in this appendix.

Problems on Numerical Distributions

13A–42. (Ratio) Calculate the standard deviation of the following set of numbers.

$$
\begin{array}{ccc}
16 & & \\
18 & 30 & \\
10 & 20 & 32 \\
12 & 24 & 36
\end{array}
$$

13A–43. Calculate the coefficient of variation of the numbers listed in Problem 13A–42.

13A–44. (Ratio) Calculate the skewness of the numbers listed in Problem 13A–42.

13A–45. (Ratio) Calculate the kurtosis of the numbers listed in Problem 13A–42.

Problems on Sampling from Distributions

13A–46. Calculate and show in chart form a distribution of sample means of sample size 2 taken from the distribution of numbers listed in Problem 13A–42. Assume sampling without replacement.

13A–47. (Ratio) Calculate the standard deviation of the distribution of sample means derived in Problem 13A–46.

13A–48. (Ratio) Calculate the skewness of the distribution of sample means derived in Problem 13A–46.

13A–49. (Ratio) Calculate the kurtosis of the distribution of sample means derived in Problem 13A–46.

Problems on Estimation of the Population Total

13A–50. Calculate and show in chart form a distribution of population estimates based on the sample means derived in Problem 13A–46.

13A–51. (Ratio) Using the following list of 100 numbers, select without replacement a random sample of 30 and estimate the population total.

Number	Amount	Number	Amount
1	$2,525	51	$2,385
2	2,984	52	2,975
3	2,615	53	2,993
4	2,336	54	2,226
5	2,453	55	2,564
6	2,704	56	2,949
7	2,054	57	2,338
8	2,948	58	2,676
9	2,839	59	2,878
10	2,455	60	2,211
11	2,680	61	2,853
12	2,542	62	2,580
13	2,886	63	2,746
14	2,887	64	2,774
15	2,340	65	2,111
16	2,957	66	2,541
17	2,815	67	2,306
18	2,319	68	2,591
19	2,646	69	2,353
20	2,691	70	2,870
21	3,116	71	2,370
22	2,159	72	2,555
23	2,234	73	2,941
24	2,428	74	2,141
25	2,816	75	2,303
26	2,160	76	2,983
27	2,882	77	2,449
28	2,413	78	2,325
29	2,704	79	2,610
30	2,666	80	2,723
31	2,583	81	2,363
32	2,646	82	2,752
33	2,266	83	2,261
34	2,942	84	2,996
35	2,936	85	2,939
36	2,566	86	2,382
37	2,811	87	2,242
38	2,054	88	2,563
39	2,275	89	2,564
40	2,610	90	2,329
41	2,471	91	2,048
42	2,402	92	2,574
43	2,499	93	2,778
44	2,624	94	2,244
45	2,499	95	2,235
46	2,492	96	2,257
47	2,570	97	2,491
48	2,950	98	2,474
49	2,857	99	2,260
50	2,301	100	2,127

Problems on Determining Sample Size

13A–52. (Ratio or PSD) Using the list of numbers in Problem 13A–51, select, without replacement, a random sample of 30 and calculate the standard deviation of this sample. This calculation will serve as an estimate of the standard deviation of the 100 numbers.

13A–53. (PSD) Take the standard deviation calculated in Problem 13A–52. Assume a desired reliability or confidence level of 90%. Assume a desired precision of 6,000. Assume a population size of 100. Calculate the necessary sample size if (1) sampling with replacement is used, (2) sampling without replacement is used.

13A–54. (PSD) As a continuation of Problem 13A–53, use the same standard deviation and population size but different reliability and precision amounts. Calculate six more sample sizes if (1) sampling with replacement is used, (2) sampling without replacement is used.

Problems on Evaluating Sample Results

13A–55. (PSD) Take the sample size calculated in Problem 13A–53 (1) (sampling with replacement) and select the sample. Perform the following calculations: (1) the estimate of the population total based on the sample mean, (2) the achieved precision and the precision range around the estimate of the population total.

13A–56. (PSD) As a continuation of Problem 13A–55, use the same sample and estimate of the population total. Calculate three more precision amounts and precision ranges using different reliability levels.

13A–57. (PSD) Take the sample size calculated in Problem 13A–53 (2) (sampling without replacement) and select the sample. Perform the following calculations: (1) the estimate of the population total based on the sample mean, (2) the achieved precision and precision range around the estimate of the population total.

13A–58. (PSD) As a continuation of Problem 13A–57, use the same sample and estimate of the population total. Calculate three more precision amounts and precision ranges using different reliability levels.

Problems on Sampling Risks

13A–59. Take the distribution of sample means calculated in Problem 13A–46 and the standard deviation of this distribution calculated in Problem 13A–47. Assuming a 10% level of significance and conducting a two-tailed test of the null hypothesis that the true mean = 22, determine which sample means, if selected, would constitute a Type I error.

13A–60. Take the distribution of sample means calculated in Problem 13A–46 and the standard deviation of this distribution calculated in Problem 13A–47. Calculate the precision of this distribution at a 90% reliability. Construct another distribution of sample means with the same standard deviation. The numerical difference between the means of the two distributions should be twice the amount of the precision. Using a level of significance of 5% and a one-tailed test of the null hypothesis, determine which sample means, if selected, constitute (1) a Type I error, (2) a Type II error.

Appendix to Chapter 13
Listings of Computer Programs and Examples of Input and Output

---

Program Listing (Ratio)

```
1 DIM T(200),B(200),A(200),S(200),C(200),P(200),Q(200),F(200),G(200),L(200)
2 LET W=0
3 PRINT 'IF SAMPLES OF MEANS ARE NEEDED, INPUT 1. OTHERWISE, INPUT 0.'
4 INPUT M
5 IF M≠1 THEN 26
6 PRINT
7 PRINT 'INPUT TOTAL NO. OF MEANS NEEDED.'
8 INPUT N
9 PRINT
10 PRINT 'INPUT THE NUMBER SIZE FOR EACH GROUP FOR WHICH A MEAN WILL BE CALCULATED.'
11 INPUT Y
12 FOR I=1 TO N
13 PRINT
14 PRINT 'INPUT INDIVIDUAL NUMBERS.'
15 FOR Z=1 TO Y
16 INPUT L
17 LET W=W+L
18 NEXT Z
19 LET W1=W/Y
20 LET T(I)=W1
21 PRINT
22 PRINT 'THE MEAN OF THE INDIVIDUAL NUMBERS IS', W1, 'THE TOTAL OF THE NUMBERS IS', W
23 LET W=0
```

```
24 NEXT I
25 GO TO 48
26 PRINT
27 PRINT 'INPUT THE SAMPLE SIZE NEEDED.'
28 INPUT N
29 PRINT 'INPUT THE INDIVIDUAL SAMPLES.'
30 FOR I=1 TO N
35 INPUT T(I)
47 NEXT I
48 FOR I=1 TO N
55 LET C(I)=T(I)
60 NEXT I
70 FOR I=1 TO N
80 LET H=C(1)
90 LET H1=1
100 FOR J=2 TO N
110 IF H≥C(J) THEN 140
120 LET H=C(J)
130 LET H1=J
140 NEXT J
150 LET S (H1)=I
155 LET P(I)=H
157 LET Q(I)=I
160 LET C(H1)=-1.0
170 NEXT I
172 PRINT 'IF YOU WISH A LIST OF THE NUMBERS IN DESCENDING ORDER, INPUT 1, OTHERWISE INPUT
    ANOTHER NUMBER.'
173 INPUT M1
```

```
174 IF M1≠1 THEN 197
175 PRINT
180 PRINT 'ALL NUMBERS', 'RANK'
185 FOR I=1 TO N
187 PRINT
190 PRINT P(I), Q(I)
191 PRINT
195 NEXT I
197 PRINT
200 LET R1=0
201 LET R2=0
202 LET R3=0
205 FOR I=1 TO N
210 LET R1=R1+T(I)
211 LET R2=T(I)×T(I)
212 LET R3=R3+R2
213 NEXT I
214 LET R4=R1/N
215 LET R5=R4×R4
216 LET R6=R5×N
217 LET R7=R3−R6
218 LET R8=R7/N
220 LET S=SQR(R8)
222 LET R=R1/N
225 PRINT
260 PRINT 'THE STAN. DEV. IS', S
262 PRINT
264 PRINT 'THE TOTAL IS', R1
```

```
266 PRINT
268 PRINT 'THE MEAN IS', R4
270 PRINT
290 FOR I=1 TO N
295 LET D2=T(I)−R4
300 LET D3=(D2×D2)×D2
302 LET D4=D3×D2
304 LET T3=T3+D3
306 LET T4=T4+D4
308 NEXT I
310 LET Q3=T3/N
315 LET S3=(S×S)×S
320 LET M3=Q3/S3
325 PRINT 'THE SKEWNESS IS', M3
330 PRINT
335 LET Q4=T4/N
340 LET S4=S3×S
345 LET M4=Q4/S4
350 PRINT 'THE KURTOSIS IS', M4
370 END
```

Example of Input (Ratio)

RUN

RATIO

IF SAMPLES OF MEANS ARE NEEDED, INPUT 1. OTHERWISE, INPUT 0.
?0

INPUT THE SAMPLE SIZE NEEDED.
?100

INPUT THE INDIVIDUAL SAMPLES.

?2525	?2882	?2975	?2325
?2984	?2413	?2993	?2610
?2615	?2704	?2226	?2723
?2336	?2666	?2564	?2363
?2453	?2583	?2949	?2752
?2704	?2646	?2338	?2261
?2054	?2266	?2676	?2996
?2948	?2942	?2878	?2939
?2819	?2936	?2211	?2382
?2455	?2556	?2853	?2242
?2680	?2811	?2580	?2563
?2542	?2054	?2774	?2564
?2886	?2275	?2111	?2329
?2887	?2610	?2541	?2048
?2340	?2471	?2306	?2574
?2957	?2402	?2591	?2778
?2815	?2499	?2353	?2244
?2319	?2624	?2870	?2235
?2646	?2499	?2370	?2257
?2691	?2492	?2555	?2491
?3116	?2570	?2941	?2474
?2159	?2950	?2141	?2260
?2234	?2857	?2303	?2127
?2428	?2301	?2983	?2746
?2816	?2385	?2449	
?2160			

Example of Output (Ratio)

THE STAN. DEV. IS 268.985

THE TOTAL IS 255777

THE MEAN IS 2557.77

THE SKEWNESS IS 6.09411E-02

THE KURTOSIS IS 1.9815

Program Listing (PSD)

```
10 REM THIS PROGRAM IS FOR AUDITING STUDENTS. IT CAN BE USED FOR VARIABLE SAMPLING
   EXERCISES.
15 REM THE PROGRAM PERFORMS THE FOLLOWING FUNCTIONS:
20 REM CALCULATION OF THE ESTIMATED STANDARD DEVIATION OF A POPULATION BASED ON A
   PRELIMINARY SAMPLE OF 30.
25 REM CALCULATION OF THE PRECISION OF THE ESTIMATE, THE ESTIMATE OF THE POPULATION
   TOTAL, AND THE RANGE INTO
30 REM WHICH THE TRUE POPULATION TOTAL SHOULD FALL, AT A GIVEN LEVEL OF CONFIDENCE.
130 LET C=0
132 PRINT 'IF EST. OF STAN. DEV. IS KNOWN, INPUT 1 AND CONTINUE WITH PROGRAM.'
133 PRINT 'IF SAMPLE OF 30 IS NEEDED, INPUT 0.'
134 INPUT X
136 IF X=1 THEN 218
138 PRINT 'INPUT THE NUMBER 30, FOLLOWED BY THE 30 ELEMENTS WHEN THE NEXT THIRTY QUESTION
    MARKS COME UP.'
140 INPUT K
142 LET E1=0
```

```
144 LET E3=0
150 FOR J=1 TO K
160 INPUT E
165 LET E1=E1+E
170 LET E2=E×E
180 LET E3=E3+E2
182 NEXT J
184 LET E4=E1/K
186 LET E5=E4×E4
188 LET E6=E5×K
190 LET E7=E3−E6
200 LET E8=E7/(K−1)
210 LET S=SQR (E8)
211 IF K>30 THEN 268
212 GO TO 220
218 PRINT 'INPUT THE EST. STAN. DEV.'
219 INPUT S
220 PRINT
225 PRINT 'BASED ON THE SAMPLE OR ANOTHER EST., THE EST. STAN. DEV. OF THE POPULATION IS', S
227 PRINT 'THE MEAN OF THE SAMPLE IS', E4
228 LET C=C+1
229 IF C=2 THEN 268
230 PRINT
231 PRINT 'INPUT RELIAB. %, RELIAB. COEFFIC., NO. OF ELEMENTS IN POP., AND PRECISION. SEPARATE BY
    COMMAS.'
235 INPUT C1,R,N,P
240 LET S1=((S×R×N)/P) × ((S×R×N)/P)
242 PRINT
```

```
243 PRINT 'IF SAMPLING WITH REPLACEMENT, INPUT 1. IF SAMPLING WITHOUT REPLACEMENT, INPUT
    ANOTHER NUMBER.'
244 INPUT R1
245 IF R1 = 1 THEN 250
246 LET S1 = S1/(1+(S1/N))
250 PRINT 'SAMPLE SIZE IS', S1
252 PRINT 'IF YOU WISH ANOTHER SAMPLE SIZE OF ANOTHER RELIAB. AND/OR ANOTHER PRECISION,
    INPUT 1, OTHERWISE INPUT 0'
253 INPUT A1
254 IF A1 = 1 THEN 231
255 PRINT
256 PRINT 'IF NO SAMPLE OF 30 WAS PREVIOUSLY INPUT, PUT IN THE FINAL SAMPLE SIZE FOLLOWED BY
    NUMBERS. INPUT SIZE AGAIN.'
257 IF X = 1 THEN 140
258 PRINT 'IF A SAMPLE OF 30 WAS PREVIOUSLY INPUT AND FINAL SAMPLE SIZE IS >30, PUT IN FINAL
    SIZE AND THEN NUMBERS.'
260 IF S1>30 THEN 140
264 PRINT
266 PRINT 'INPUT 30 EVEN THOUGH THE FINAL SAMPLE SIZE IS LESS THAN 30.'
267 GO TO 270
268 PRINT
269 PRINT 'INPUT THE SAMPLE SIZE AGAIN.'
270 INPUT F
272 LET S2 = S/SQR(F)
274 IF R1 = 1 THEN 280
276 LET S2 = S2×SQR(1−(F/N))
280 LET P1 = S2×R×N
290 LET T = E4×N
```

```
300 LET T1=T+P1
310 LET T2=T−P1
312 PRINT 'THE ESTIMATED STAN. DEV. BASED ON THE FINAL SAMPLE IS', S
314 PRINT 'THE ESTIMATED STAN. ERROR OF THE MEAN BASED ON THE FINAL SAMPLE IS', S2
320 PRINT 'THE ESTIMATED TOTAL BASED ON THE FINAL SAMPLE MEAN IS', T
330 PRINT 'THE PRECISION IS', P1, 'AT A RELIABILITY OR CONFIDENCE LEVEL OF', C1
340 PRINT 'THE RANGE IS', T2, T1
342 PRINT
343 PRINT 'IF PRECISION AND RANGE ARE WANTED ON OTHER RELIAB. %, INPUT 1, PUSH RETURN AND
    INPUT RELIAB. % AND COEFFIC.'
344 PRINT 'IF NO OTHER PRECISION AND RANGE AMOUNTS ARE WANTED, INPUT ANY NUMBER
    EXCEPT 1.'
345 INPUT M
346 IF M≠1 THEN 350
347 INPUT C1,R
348 GO TO 280
350 END
```

Example of Input and Output (PSD)

```
RUN

PSD

IF EST. OF STAN. DEV. IS KNOWN, INPUT 1 AND CONTINUE WITH PROGRAM
IF SAMPLE OF 30 IS NEEDED, INPUT 0.
?1
INPUT THE EST. STAN. DEV.
?280
```

BASED ON THE SAMPLE OR OTHER EST., THE EST. STAN. DEV. OF THE POPULATION IS 280
THE MEAN OF THE SAMPLE IS 0

INPUT RELIAB. %, RELIAB. COEFFIC., NO. OF ELEMENTS IN POP., AND PRECISION. SEPARATE BY COMMAS.
?95,1.96,100,7000

IF SAMPLING WITH REPLACEMENT, INPUT 1. IF SAMPLING WITHOUT REPLACEMENT, INPUT ANOTHER NUMBER.
?0

SAMPLE SIZE IS 38.0673

IF YOU WISH ANOTHER SAMPLE SIZE OF ANOTHER RELIAB. AND/OR ANOTHER PRECISION, INPUT 1, OTHERWISE INPUT 0
?0

IF NO SAMPLE OF 30 WAS PREVIOUSLY INPUT, PUT IN THE FINAL SAMPLE SIZE FOLLOWED BY NUMBERS. INPUT SIZE AGAIN.
?39
?22984
?22615
?22336
?22948
?22886
?22887
?22815
?23116
?22159
?22428
?22704
?22646
?22936

```
?22566
?22275
?2471
?22624
?22570
?22950
?22857
?22975
?22226
?22564
?22676
?22306
?2591
?22870
?22941
?22303
?22363
?22752
?22261
?22242
?22048
?22778
?22244
?22474
?22260
?22127
INPUT THE SAMPLE SIZE AGAIN.
?39
```

THE ESTIMATED STAN. DEV. BASED ON THE FINAL SAMPLE IS 292.924
THE ESTIMATED STAN. DEV. ERROR OF THE MEAN BASED ON THE FINAL SAMPLE IS 36.6343
THE ESTIMATED TOTAL BASED ON THE FINAL SAMPLE MEAN IS 258395.
THE PRECISION IS 7180.31 AT A RELIABILITY OR CONFIDENCE LEVEL OF 95
THE RANGE IS 251214. 265575.
IF PRECISION AND RANGE ARE WANTED ON OTHER RELIAB. %, INPUT 1, PUSH RETURN AND INPUT RELIAB. % AND COEFFIC.
IF NO OTHER PRECISION AND RANGE AMOUNTS ARE WANTED, INPUT ANY NUMBER EXCEPT 1.
?1
?90,1.64
THE ESTIMATED STAN. DEV. BASED ON THE FINAL SAMPLE IS 292.924
THE ESTIMATED STAN. DEV. ERROR OF THE MEAN BASED ON THE FINAL SAMPLE IS 36.6343
THE ESTIMATED TOTAL BASED ON THE FINAL SAMPLE MEAN IS 258395.
THE PRECISION IS 6008.02 AT A RELIABILITY OR CONFIDENCE LEVEL OF 90
THE RANGE IS 252387. 264403.
IF PRECISION AND RANGE ARE WANTED ON OTHER RELIAB. %, INPUT 1, PUSH RETURN AND INPUT RELIAB. % AND COEFFIC.
IF NO OTHER PRECISION AND RANGE AMOUNTS ARE WANTED, INPUT ANY NUMBER EXCEPT 1.
?1
?80,1.28
THE ESTIMATED STAN. DEV. BASED ON THE FINAL SAMPLE IS 292.924
THE ESTIMATED STAN. DEV. ERROR OF THE MEAN BASED ON THE FINAL SAMPLE IS 36.6343
THE ESTIMATED TOTAL BASED ON THE FINAL SAMPLE MEAN IS 258395.
THE PRECISION IS 4689.18 AT A RELIABILITY OR CONFIDENCE LEVEL OF 80
THE RANGE IS 253706. 263084.
IF PRECISION AND RANGE ARE WANTED ON OTHER RELIAB. %, INPUT 1, PUSH RETURN AND INPUT RELIAB. % AND COEFFIC.
IF NO OTHER PRECISION AND RANGE AMOUNTS ARE WANTED, INPUT ANY NUMBER EXCEPT 1.
?0

The Use of EDP to Gather Audit Evidence

Learning Objectives *After reading and studying the material in this chapter, the student should*

Know the difference between the use of the computer to aid in gathering evidence and testing the computer system (as discussed in Chapter 7).

Understand the purposes and uses of a generalized computer audit program software package.

Know the audit evidence acquired whenever a generalized computer audit program is used for a given account in the financial statements.

Understand the limits of a generalized computer audit program in gathering audit evidence.

In the other chapters of this section of the book, the general nature of evidence-gathering, working-paper techniques, and the use of certain statistical sampling methods have been discussed. In this chapter, then, we shall explore the many ways in which computers can be used to expedite the evidence-gathering phase of the examination and enhance its quality.

In the last decade, CPA firms have greatly increased the use of their own computer programs to extract various types of evidence from and make analyses of data stored on clients' computer files. As more and more audited companies make use of EDP capabilities and continue to acquire sophisticated equipment, employment of generalized computer audit programs may become a necessity rather than a desirable option.

Therefore, we will describe the various techniques that CPAs *actually use* to gather evidence by employing their own computer software packages on the client's EDP data files. To integrate the material in this chapter with other discussions in the text, we have chosen to extend the illustrations started in Chapter 5 and continued through Chapters 6–8 and 13. This chapter explains the types of information that users of a generalized computer audit program might extract from the accounts receivable records utilized in this text.

Although it is a bit of an oversimplification, we might describe the auditors' relationship with EDP as falling into one or both of the following categories:

1. The study and testing of controls maintained in an EDP system.

2. The use of computer hardware and software to audit records kept on computer files.

The material in Chapter 7 contains descriptions of this first relationship in which the auditors study and evaluate the system of internal control in an EDP environment. As pointed out, the auditors' programs can be utilized to test the client's computerized records, although the use of test data is also an option.

In regard to the second relationship, the auditors can use a series of software packages to perform various data-processing operations on the client's computerized files, which will result in output that can be used in the examination of the account(s).

Functions of Generalized Computer Audit Programs[1]

Many functions can be performed by a generalized computer audit program. Here is a fairly comprehensive, but not exhaustive, list of these program capabilities.

1. Scanning or examining computer records for exceptional or unusual characteristics and obtaining a list of these characteristics. Some examples are:
   a. Accounts receivable balances over a certain amount, over the credit limit, or containing a credit balance.
   b. Unusually large inventory balances.
   c. Unusual payroll situations such as terminated employees or excessive overtime.

2. Making or checking computations and obtaining for the auditors' review computations that are incorrect. Here are some examples:

[1]This term is a broad description of the many software packages used by CPA firms. Each company has its individual name. For more detail on the subject, see W. Thomas Porter, "Generalized Computer-Audit Programs," *The Journal of Accountancy,* January, 1969, pp. 54–62.

a. Payroll calculations.
b. Interest calculations.
c. Depreciation calculations.
d. Calculations of various prepaid items.

3. Comparing data on different records or files and listing unusual or irregular results. For instance:
    a. Comparing accounts receivable master file balances between two dates with accounts receivable debits and credits contained in detail transaction files between the same dates.
    b. Comparing items on the master payroll files with items on personnel records.

4. Selecting and obtaining various types of samples. Included among these might be:
    a. Accounts receivable and accounts payable confirmations.
    b. Fixed asset changes.
    c. Inventory items.
    d. Selected items from an income statement account.

5. Preparing various analyses and listings that facilitate the examination of certain accounts. Some of these are:
    a. Accounts receivable aging schedules.
    b. Analysis of inventory by date of purchase.
    c. Statistics relating to various accounts, such as accounts receivable and inventory turnover.
    d. Trial balances.

All of these listed capabilities of generalized computer audit programs have one common thread. The auditors employ a developed series of computer programs to acquire various lists, analyses, etc. from data maintained on the client's computerized files. Access to this information is considered essential, or certainly desirable, in order to perform the audit. If a manual system were employed by the client, some of this same information would be gained through "hand" procedures similar to those discussed in Chapters 9–12. If an EDP system were used and no generalized computer audit programs were available, printouts of "raw data" would probably be obtained from the computer files. Manual analyses would then be made from such data.

Description of Generalized Computer Audit Programs

Generalized computer audit programs should not be thought of as mysterious and difficult to understand. Although training is necessary to apply the programs in actual audit practice, the concepts can be illustrated in chart and narrative form. For example, the following diagram shows the *general* method that is employed, although individual systems differ from one CPA firm to the next.

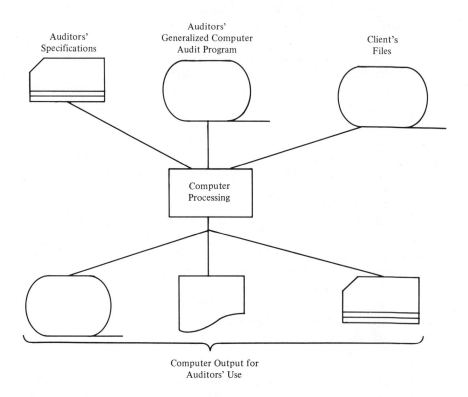

Computer Output for
Auditors' Use

The auditors' specifications are the descriptions of the examination objectives and the types of files to be used in computer processing. The generalized computer audit program itself is written by or for the individuals performing the audit. It is not necessary for the auditors to understand the minute features of the programming techniques, but they need to be knowledgeable of the audit objectives and the characteristics of the client's system from which the information is taken.

Although the client's files are assumed to be on magnetic tape, they could be on magnetic disk, punched cards, or other types of storage devices. The computer output to be used by the auditors in meeting examination objectives might be tape, printed material, cards, or other types of output. Whatever the form, the auditors have a tool that is very usable, one that becomes accessible in a manual system only as a result of a time-consuming effort.

An added feature utilized by at least one large CPA firm is a special software processor that converts the auditors' specification data into a problem-oriented or third-level language program. This program, in turn, processes the client's files, producing the desired information to be used in the examination of the appropriate account(s). A technique such as this reduces the auditors' need for first-hand knowledge of the programming language. It does, however, sharpen the necessity of understanding the audit objectives and the manner in which the data are stored in the client's computer files.

Audit Objectives and EDP Capabilities

One should not be left with the impression that a generalized computer audit program produces all of the necessary evidence in the examination of a company's financial statements. While it is true that arithmetic, comparisions, and other clerical functions may be made through this computerized process, it also remains for the auditors to take much of the output and utilize it by conducting the *same* procedures that they would have if the information had been acquired through manual methods.

To illustrate this point, let us relist some of the output obtained from the processing of the client's files by a generalized computer audit program. To the right of each listing are descriptions of appropriate audit steps to be taken *with* the information produced by computer processing.

Output Produced by the Generalized Audit Program	*Audit Procedures Performed with This Output*
1. Listing of accounts receivable balances that exceed the credit limit.	1. Discussions with client personnel on causes and possible corrective measures.
2. Listing of accounts receivable credit balances.	2. Inquiries concerning the question of reclassifying such balances as current liabilities.
3. Listing of unusually large inventory balances.	3. Investigation for possible obsolescence; observation of client's inventory count.
4. Listing of recently terminated employees.	4. Investigation of the disposition of unclaimed payroll checks.
5. Accounts receivable and accounts payable confirmations.	5. A tally of confirmation results and followup of customer's exceptions.
6. Listing of fixed asset changes.	6. Examination and analysis of source documents supporting such changes.

Certainly, this tabulation demonstrates that generalized computer audit programs produce necessary information for the conduct of an examination, but in many instances additional work must be performed. Such programs provide the most aid in phases of the audit that call for calculations, listings, clerical comparisons, etc. On the other hand, while computer printouts furnish information that can be *used* to make oral inquiries and conduct observations, this computer output is no *substitute* for these audit procedures themselves.

Gathering Information from an Accounts Receivable Master File

Description of the System

The basic functions of generalized computer audit programs, along with explanations of their advantages and limitations have been discussed. The next step is to provide a relevant example of an application. Although many illustrations could be chosen, we have selected one that is both easily identifiable and integrated with material in previous chapters.

Assume that the client processes accounts receivable charges and credits on a computerized batch-processing system, which includes magnetic tape as the storage device for the master files. As charge sales, cash receipts, and non-cash credits are read into computer memory, the data on the master tape are also put in. One of the outputs of the computer processing is a new master tape containing updated data on each customer's account.[2]

This data-processing function can best be illustrated by referring to examples from the accounts receivable records shown in Chapters 5 and 7. We will use three customer's accounts, Block Wholesale Co., Xanthos Bros. Furn. Co., and Urban Home Furnish. Their manually determined ledger accounts balances and their records on master tape file are on page 649.

Before an application of the generalized computer audit program is made, the objectives of the computer processing will be defined. For purposes of illustration, we can list some objectives that are similar to, if not the same as, those sought in an actual engagement.

1. The auditors wish to select for positive confirmation purposes accounts receivable balances that have certain characteristics.

    a. One such characteristic is the dollar size. Positive confirmations are circularized on large account balances for the reasons enumerated in Chapter 9, i.e., better audit evidence is obtained on a significant percentage of the total dollar amount of accounts receivable. We will assume that the computer program contains instructions to select for positive confirmation circularization any balance in excess of $5,000.

[2]For an overview of the entire process, see the systems flow chart in Chapter 7.

Customer Number	Name	Balance Notation	Date	Posting Source	Debit	Credit	Balance
0001	Block Wholesale Co.	Bal. 9-30-X9			$	$	$ 940.00
			10-10-X9	S-02117	1,020.00		1,960.00
			10-12-X9	R-02375		940.00	1,020.00
0037	Xanthos Bros. Furn. Co.		8-18-X9	S	1,190.00		1,190.00
			9-15-X9	R		200.00	990.00
			10-22-X9	CM-00375		145.00	845.00
0091	Urban Home Furnish.		5-15-X9	S	1,010.00		1,010.00
			6-20-X9	S	1,010.00		2,020.00
			10-10-X9	S-02114	3,198.00		5,218.00

If these same balances were kept on magnetic tape, they might appear as follows.

RECORDS ON MASTER TAPE FILE

Customer Account No.	Name	Credit Status*	Current Charges	Current Cash Receipts	Current Sales Receipts	Current Sales Returns	Total Balance	Current Balance	Balance 31-60 Days	Balance 61-90 Days	Balance Over 90 Days
000001	Block Wholesale Co.	1	0102000	0094000	0000000	0000000	0102000	0102000	0000000	0000000	0000000
030037	Xanthos Bros. Furn. Co.	2	0000000	0000000	0014500	0000000	0084500	0000000	0000000	0084500	0000000
000091	Urban Home Furnish.	3	0319800	0000000	0000000	0000000	0521800	0319800	0000000	0000000	0202000

*1—Unlimited credit
2—$500 or more balance over 60 days—credit limit of $2,000
3—$1,000 over 90 days—no additional credit given

The credit status exists after the current month's processing. Customer No. 37's balance does not exceed the credit limit, but it would have if current charges had exceeded $1,155.00 ($2,000.00 − $845.00). Customer No. 91's credit limit has been exceeded since there is a $2,020.00 balance in the over-90-day category, and current charges were made to the account.

b. Another characteristic is the age of the account. The auditors wish to establish, as firmly as possible, the authenticity of accounts with older balances. Therefore, we assume that another audit procedure is to send positive confirmations to all customers whose balances, or any portion thereof, are in excess of 90 days.

2. The auditors wish to select for negative confirmation requests a systematic sample of accounts other than those selected for positive requests. We will assume that every third account is selected starting with the first customer number. If an account is already chosen for a positive request, it will be skipped for a negative request.

3. The third and last auditors' objective is to produce a report of special account characteristics for further analysis or investigation. Within an accounts receivable master file, there are many information items that auditors might find useful. We will confine our output to the following:

a. A list of accounts with balances that exceed the credit limit. This list will be discussed with the credit manager or other appropriate personnel.

b. A list of accounts with all or any portion of the balance in the 61–90-day category and in which current charges were made. Accounts such as this are not necessarily doubtful of collection but may furnish the auditors with information on the company's handling of slow-paying accounts. This list may also be discussed with the client.

c. A list of accounts that have had credit balances for a period in excess of 30 days. If the total of these accounts is considered material, the auditors may wish to propose an adjusting entry reclassifying such balances as a current liability.

d. A list of accounts with all or any portion of the balance in excess of 90 days. The auditors will probably discuss the collectibility of such accounts with appropriate personnel. Ultimately, this type of information may be used as partial support for a proposed adjusting entry to the allowance for doubtful accounts.

It should also be pointed out that the accounts receivable aging schedule could be printed by the computer on which the generalized computer audit program is run. Whether this would be a part of the auditors' objective or not depends on the client's data-processing output. If such a schedule is furnished by the client, the auditors may wish to have it test-checked as a part of their procedures.

A Program Flow Chart of the Auditors' Objectives

Any reasonably complicated computer program is supported by a flow chart detailing the logic operations of the routine being processed. By way of review we might remember that a systems flow chart shows the overview of the data movement from one location or machine to another. Conversely, a program flow chart (exhibited here) contains descriptions of the computer-processing logic and is used to code the computer program itself as well as to provide needed documentation.

On page 652 is a program flow chart for this hypothetical application of a generalized computer audit program.

Input and Output of the Generalized Computer Audit Program

On page 653 is a tabulation of selected accounts receivable customer data.

The numbers in the columns are arranged on the tabulation as they might be if the data were stored on magnetic tape. These particular customers are selected because their file contains all of the characteristics asked for in the list of auditors' objectives (balance over 90 days, etc.).

We will assume that the master record data are stored on magnetic tape and are read into the main memory of the computer as the client's input file.

On pages 654–655 is an example of the output that would be generated by the application of the generalized computer audit program to a selected group of accounts.

Examples from Audit Practice

The large CPA firms have developed their own packages of generalized computer audit programs. The names of the software programs differ, of course (Strata, Auditape, Auditpak, Audex, etc.), but most of them have common attributes.

1. The ability to read client data stored on computer files and perform certain audit procedures such as footings, extensions, etc.

2. The ability to select computerized client data with certain characteristics and print the data for the auditor's scrutiny and review.

3. The ability to select samples based on a predesigned plan and perform certain statistical sampling calculations.

PROGRAM FLOW CHART FOR THE
PROGRAM TO READ MASTER FILES

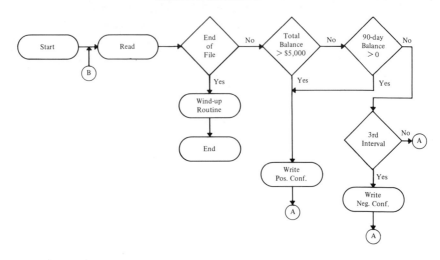

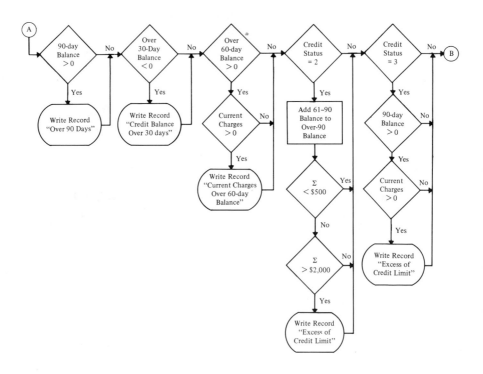

MASTER RECORD DATA FOR AUDITOR'S PROGRAM

Customer Account Number	Credit Status	Current Charges	Current Cash Receipts	Current Sales Returns	Current Write-Offs	Total Balance	Current Balance	Balance 31–60 Days	Balance 61–90 Days	Balance Over 90 Days
0001(3)	1	0102000	0094000	0000000	0000000	0102000	0102000	0000000	0000000	0000000
C003	1	0020800	0000000	0000000	0000000	0020800	0020800	0000000	0000000	0000000
C004(3)	1	0174700	0092000	0000000	0000000	0184700	0174700	0010000	0000000	0000000
0005	1	0000000	0000000	0014500	0000000	0000000	0000000	0000000	0000000	0000000
0007(1)	1	0541800	0325000	0000000	0000000	0541800	0541800	0000000	0000000	0000000
0009(1)(5)	1	0516000	0000000	0000000	0000000	0619800	0516000	0094800	0009000	0000000
0015	1	0133500	0102500	0000000	0000000	0133500	0133500	0000000	0000000	0000000
0016(2)(5)(4)	2	0271500	0010000	0000000	0000000	0324000	0271500	0000000	0032000	0020500
0025(3)	1	0038500	0000000	0000000	0000000	0077000	0038500	0038500	0000000	0000000
0027(1)(5)	1	0470900	0000000	0000000	0000000	0704900	0470900	0204000	0030000	0000000
0030(6)	1	0000000	0000000	0000000	0000000	− 092000	0000000	− 092000	0000000	0000000
0044(1)(2)(4)	2	0505100	0020000	0000000	0000000	0555100	0505100	0000000	0000000	0050000
0090	1	0051600	0000000	0000000	0000000	0087300	0051600	0035700	0000000	0000000
0091(2)(4)(1)	3	0319800	0000000	0000000	0000000	0521800	0319800	0000000	0000000	0202000

Characteristics that the generalized computer audit program will search for and output to meet the auditors' objectives

(1) Over $5,000—for positive confirmation.
(2) Over 90 days—for positive confirmation and an exception report.
(3) Every 3rd account—for negative confirmation.
(4) Over the credit limit—for an exception report.
(5) Balance 61–90 days with current charges—for an exception report.
(6) Credit balance over 30 days—for an exception report.

The following is an example of the output, given the listed input on page 653.

ACCOUNTS FOR POSITIVE CONFIRMATION
CRITERIA: TOTAL BALANCE IN EXCESS OF $5,000.00, OR ANY PART IN EXCESS OF 90 DAYS

Cus Num	Credit Status	Curr Char	Current Cash Rcts	Current Sales Retns	Write-Offs	Tot Bal	Curr Bal	31–60 Bal	61–90 Bal	Over 90
7	1	$5,418.00	$3,250.00	0.0	0.0	$5,418.00	$5,418.00	0.0	0.0	0.0
9	1	5,160.00	0.0	0.0	0.0	6,108.00	5,160.00	948.00	90.00	0.0
16	2	2,715.00	100.00	0.0	0.0	3,240.00	2,715.00	0.0	320.00	205.00
27	1	4709.00	0.0	0.0	0.0	7,049.00	4,709.00	2,040.00	300.00	0.0
44	2	5,151.00	200.00	0.0	0.0	5,551.00	5,051.00	0.0	0.0	500.00
91	3	3,198.00	0.0	0.0	0.0	5,218.00	3,198.00	0.0	0.0	2,020.00

ACCOUNTS FOR NEGATIVE CONFIRMATION
*CRITERIA: EVERY THIRD ACCOUNT, EXCLUDING ACCOUNTS FOR POSITIVE CONFIRMATION**

Cus Num	Credit Status	Curr Char	Current Cash Rcts	Current Sales Retns	Write-Offs	Tot Bal	Curr Bal	31–60 Bal	61–90 Bal	Over 90
1	1	$1,020.00	$ 940.00	0.0	0.0	$1,020.00	$1,020.00	0.0	0.0	0.0
5	1	0.0	0.0	145.00	0.0	0.0	0.0	0.0	0.0	0.0
15	1	1,335.00	1,025.00	0.0	0.0	1,335.00	1,335.00	0.0	0.0	0.0
90	1	516.00	0.0	0.0	0.0	873.00	516.00	357.00	0.0	0.0

*If all 100 accounts were processed, the listed accounts would be different. Only 14 accounts are illustrated. Every third account is based on these 14 accounts.

EXCEPTION LISTINGS

Cus Num	Credit Status	Curr Char	Current Cash Rcts	Current Sales Retns	Write-Offs	Tot Bal	Curr Bal	31–60 Bal	61–90 Bal	Over 90
9	1	$5,160.00	0.0	0.0	0.0	$6,108.00	$5,160.00	$ 948.00	$ 90.00	0.0
9		61–90 Balance with current charges								
16	2	2,715.00	100.00	0.0	0.0	3,240.00	2,715.00	0.0	320.00	205.00
16		Balance exceeds credit limit								
16	2	2,715.00	100.00	0.0	0.0	3,240.00	2,715.00	0.0	320.00	205.00
16		61–90 Balance with current charges								
16	2	2,715.00	100.00	0.0	0.0	3,240.00	2,715.00	0.0	320.00	205.00
16		Balance exceeds ninety days								
27	1	4,709.00	0.0	0.0	0.0	7,049.00	4,709.00	2,040.00	300.00	0.0
27		61–90 Balance with current charges								
30	1	0.0	0.0	0.0	0.0	–920.00	0.0	–920.00	0.0	0.0
30		Credit balance exceeds thirty days								
44	2	5,051.000	200.00	0.0	0.0	5,551.00	5,051.00	0.0	0.0	500.00
44		Balance exceeds credit limit								
44	2	5,051.00	200.00	0.0	0.0	5,551.00	5,051.00	0.0	0.0	500.00
44		Balance exceeds ninety days								
91	3	3,198.00	0.0	0.0	0.0	5,218.00	3,198.00	0.0	0.0	2,020.00
91		Balance exceeds credit limit								
91	3	3,198.00	0.0	0.0	0.0	5,218.00	3,198.00	0.0	0.0	2,020.00
91		Balance exceeds ninety days								

Auditpak II[3]

One such package of programs developed by Coopers & Lybrand is called Auditpak II. Although COBOL is used to process client's records with Auditpak II, the auditor need not be familiar with this programming technique. His part in running the program consists of completing a questionnaire that sets out the audit objectives in the English language. The questionnaire information is placed on punched cards (or on another machine-readable form). Then, a preprocessor computer program produces a COBOL program that is used to extract the desired information from the client's files and/or perform the desired audit procedures. Printed reports are produced for the auditor's review or other actions.

For example, some of the audit objectives for inventory might be:

1. To compare physical counts with perpetual records.

2. To print a list of inventory items selected for test counts.

3. To compare inventory prices with master price lists.

Some of the audit objectives for accounts receivable and sales might be:

1. To prepare an aged schedule of accounts receivable.

2. To test footings of sales and cash receipts journals.

3. To select accounts receivable accounts for confirmation.

4. To identify account receivable balances in excess of the credit limit.

Some of the audit objectives for fixed assets might be:

1. To select certain fixed asset additions or disposals for review.

2. To test the computation of depreciation expense.

As can be seen from the foregoing lists, the audit objectives described on an Auditpak II questionnaire and designed for computer usage are basically the same as the objectives if the client's records are in manual form. The computer is used as a tool to expedite the audit process and reduce the chances of clerical mistakes in the audit procedures. In all likelihood, more use of these techniques will be made by auditors in the future.

The following charts contain illustrations of the Auditpak II process, starting with the auditor's questionnaire and ending with the printed reports.

---

[3]This particular package of programs was selected for illustration and is used with permission of Coopers & Lybrand.

the auditpak II loop

how to get results quickly

1. Define your objectives and complete the AUDITPAK II questionnaire.

2. Keypunch the questionnaire and run AUDIT-PAK II preprocessor to generate a COBOL program tailored to your client's exact needs.

3. Prepare Job Control Statements and compile and execute an "error-free" COBOL program.

4. Review your output and take action accordingly.

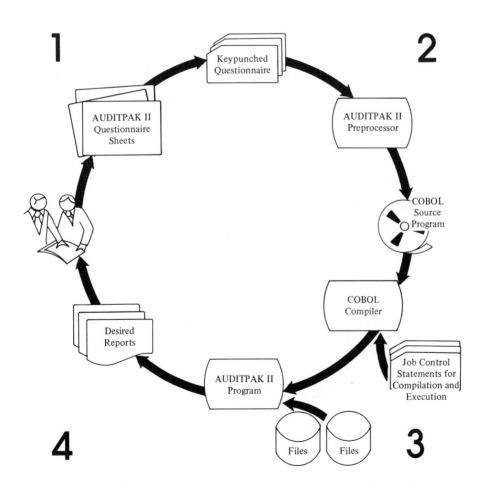

the auditpak II
reports:

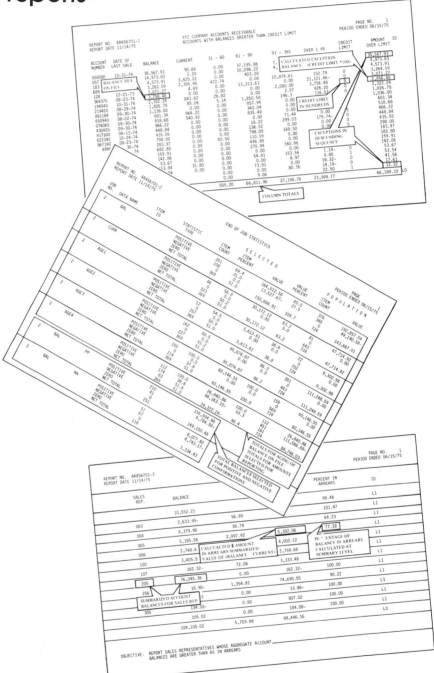

The AY Stratified Sampling System

A package of software programs developed by Arthur Young & Company has a special capability. The package is called the AY Stratified Sampling System. The system actually is a series of computer programs designed to allow auditors to use statistical sampling techniques on clients' computerized files (such as accounts receivable, inventory, etc.). The statistical sampling method used in these programs is estimation sampling for variables, with stratification.[4] Applications can be made on a wide variety of computer equipment common to the business world.

The AY Stratified Sampling System is divided into five components, each making use of a separate program.

1. The first computer program reads the client data file and, in accord with certain program specifications, prints a histogram. The histogram contains a breakdown of the client's account balances according to selected statistical characteristics, such as the number of accounts within certain dollar balance intervals, the mean and standard deviation of these accounts, etc.

2. The second computer program reads the histogram, and designs and prints sampling plans. The plans include strata boundaries, population sizes, total book values, standard deviations of the book values, and the calculated sample sizes. The plans are based on specified confidence levels, precision, and criteria pertaining to the number of strata and their dollar boundaries.[5]

3. The third computer program reads the pre-stratified client data file, reads the auditor's sampling plan, and selects the sample items using the appropriate stratification. At this stage of the computer processing, the sample items represent the account balances on the client's book.

4. The fourth computer program reads the selected sample plans, reads the audited amounts of the sample items (obtained by applying the appropriate audit procedures), and edits the audited sample items.

5. The fifth computer program reads the audited sample items, reads the selected sampling plan, and calculates the estimated population total, precision limits, etc. using the criteria of the selected sampling plan. The program prints these calculated amounts. Pages 660–661 contains a flow chart of all the described computer processing and the histogram.

---

[4]This technique estimates the population total of financial statement figures by dividing the population into subgroups. See Chapter 13 for more details.

[5]Precision is the range within which the sample result is expected to be accurate. Confidence levels, for purposes of this software package, are defined as the probability that the population value will be within the confidence interval, as determined by the point estimate and achieved precision.

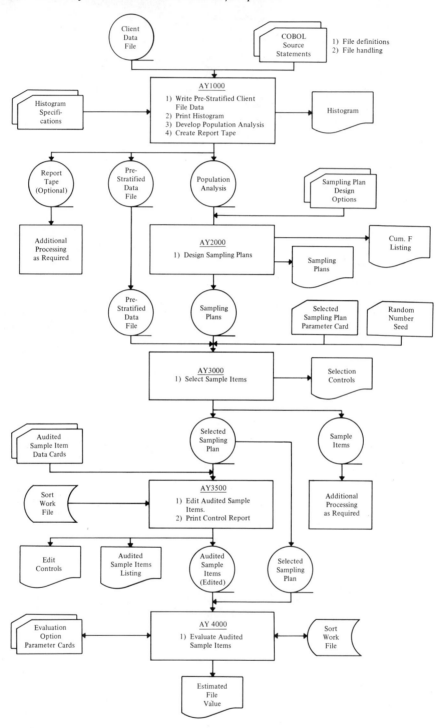

```
                    S T R A T I F I E D   S A M P L I N G   S Y S T E M                    VERSION 3.0
                              PRE-STRATIFICATION HISTOGRAM                        REPORT DATE - 04/05
                                    LOGARITHMIC METHOD

FILE 1 L12                                                                   STANDARD   DISTRIBUTION
SEQ                                                          CUM                        DEVIATION  SCALED 1 FOR
 #    INTERVAL----BOUNDARIES      # ITEMS      VALUE           F        MEAN
```

SEQ #	INTERVAL----BOUNDARIES	# ITEMS	VALUE	CUM F	MEAN	STANDARD DEVIATION	DISTRIBUTION SCALED 1 FOR
001	-35.01 TO -30.00	2	67.71-	3.17	33.85-	1.16	*
002	-29.99 TO -20.00	21	480.40-	17.66	22.88-	3.05	*
003	-19.99 TO -10.00	73	949.35-	44.68	13.00-	3.17	*
004	-9.99 TO -9.00	12	114.89-	48.14	9.57-	.27	*
005	-8.99 TO -8.00	11	93.55-	51.46	8.50-	.25	*
006	-7.99 TO -7.00	7	52.52-	54.10	7.50-	.27	*
007	-6.99 TO -6.00	9	58.52-	57.10	6.50-	.35	*
008	-5.99 TO -5.00	26	135.61-	62.20	5.22-	.29	*
009	-4.99 TO -4.00	26	115.77-	67.30	4.45-	.33	*
010	-3.99 TO -3.00	12	43.18-	70.77	3.60-	.27	*
011	-2.99 TO -2.00	24	53.55-	75.67	2.23-	.23	*
012	-1.99 TO -1.00	46	57.89-	82.45	1.26-	.30	*
013	-0.99 TO -.01	68	21.13-	90.65	.31-	.29	*
014	.00 TO .00		.00	90.65	.00	.00	*
015	.01 TO .99	157	86.58	103.12	.55	.28	*
016	1.00 TO 1.99	36	54.24	109.12	1.51	.31	*
017	2.00 TO 2.99	25	61.24	114.12	2.45	.26	*
018	3.00 TO 3.99	31	103.29	119.69	3.33	.30	*
019	4.00 TO 4.99	38	169.66	125.85	4.46	.27	*
020	5.00 TO 5.99	44	238.86	132.49	5.43	.28	*
021	6.00 TO 6.99	55	353.78	139.90	6.43	.30	*
022	7.00 TO 7.99	52	385.00	147.11	7.40	.25	*
023	8.00 TO 8.99	42	355.84	153.59	8.47	.28	*
024	9.00 TO 9.99	29	274.83	158.98	9.48	.28	*
025	10.00 TO 19.99	412	6,150.31	223.17	14.93	2.89	***
026	20.00 TO 29.99	399	9,953.73	286.33	24.95	2.97	***
027	30.00 TO 39.99	409	14,140.11	350.29	34.57	2.86	***
028	40.00 TO 49.99	358	16,052.43	410.12	44.84	2.96	***
029	50.00 TO 59.99	354	19,362.00	469.62	54.69	3.02	***
030	60.00 TO 69.99	370	24,056.62	530.44	65.02	2.89	***
031	70.00 TO 79.99	326	24,356.62	587.54	74.71	2.84	***
032	80.00 TO 89.99	298	25,292.93	642.13	84.88	2.92	***
033	90.00 TO 99.99	290	27,498.92	695.98	94.82	3.00	***
034	100.00 TO 199.99	2,455	367,712.30	1,191.46	149.78	29.22	***********************
035	200.00 TO 299.99	2,259	567,834.30	1,666.75	251.37	29.13	**********
036	300.00 TO 399.99	1,611	558,787.29	2,068.12	346.33	29.06	********
037	400.00 TO 499.99	885	395,502.09	2,365.61	446.33	28.53	******
038	500.00 TO 599.99	469	254,540.53	2,582.18	542.73	28.35	****
039	600.00 TO 699.99	218	140,273.40	2,729.83	643.46	29.03	**
040	700.00 TO 799.99	110	81,781.89	2,834.71	743.47	27.44	*
041	800.00 TO 899.99	61	51,472.38	2,912.81	843.81	28.87	*
042	900.00 TO 999.99	34	32,024.68	2,971.12	941.90	29.35	*
043	1,000.00 TO 1,999.99	37	45,008.19	3,163.47	1,216.44	200.61	*
044	2,000.00 TO 2,361.09	1	2,361.09	3,182.47	2,361.09	.00	*

```
** T O T A L S **                    12,202    2,663,500.49                2,361.09

HIGHEST VALUE ON FILE----------------                          2,361.09
LOWEST VALUE ON FILE-----------------                            35.01-
```

As one can see, the AY Stratified Sampling System makes use of computer programs to gather evidence on account balances in the client's financial statements. The evidence-gathering technique is estimation sampling for variables, using stratification. One should take note of the fact that manual methods could be used to gather the same evidence with the same statistical sampling techniques. The process would be burdensome and costly, however, and might discourage use of an important quantitative tool. An audit software package of the type illustrated here expedites evidence-gathering and encourages the employment of sophisticated audit techniques.

Critique of Generalized Computer Audit Programs

The Auditors' EDP Proficiency

Now that conceptual discussions and illustrations of generalized computer audit programs have been presented, we turn to broad considerations of the role of EDP training for the auditors. One question that should be immediately asked is how much computer programming proficiency, if any, auditors need to possess in order to use the tools illustrated in this chapter. The answer generally given by CPA firms is that none is necessary, although some knowledge may be desirable.

It seems logical, however, to assume that the most efficient use of these software packages occurs when the users comprehend the computer system in use. To possess this understanding, auditors should have a workable EDP "vocabulary"[6] and should be familiar with the manner in which data are processed on machine-readable records and stored in computerized files.

Auditors also need an understanding of flow charting, and should possess the ability to read various types of documentation, e.g., record layouts. Perhaps of primary importance is the desirability for auditors to visualize their objectives, the file systems used by the client, and the output required from application of the generalized computer audit program to these files. Beyond this, an ability to code the computer languages is helpful but not absolutely necessary.[7]

Fraud Detection

There seems to be little question but that a proper application of a generalized computer audit program will enhance the quality of an examination. However, can the use of these software systems increase the probability of detecting fraud? This would be a difficult question to answer even if actual case studies were available. Fraud detection is not the major purpose of the ordinary examination, and generalized computer audit programs were not designed to accomplish this goal.

[6]See terminology in Chapter 7.

[7]We see no inconsistency in requiring auditing students to perform programming exercises that are relevant to auditing applications, for the same reason that we see no inconsistency in requiring students in financial accounting to solve problems in order to enhance their ability to understand statements and reports.

At the same time, though, it should be acknowledged that the use of auditors' software provides some advantages over manual auditing should certain types of fraud exist in the client's organization. If computers are being employed to perpetrate irregularities, an "around-the-computer" approach is less likely to provide evidence of these occurrences.

One of the most publicized frauds in recent years is the *Equity Funding* case. As indicated in Chapter 7, this case is not technically a computer fraud but an irregularity aided and partially concealed through the *use* of EDP equipment. In this case the company maintained a set of fictitious insurance policies on machine-readable files (accessible for manual scrutiny only through printouts).

Through the use of some clever programming techniques, Equity Funding restricted the policy information shown on the printouts, so that all such policies had the appearance of legitimacy.[8] Also, other measures were taken by the client to hamper the auditors' evidence-gathering. Germane to this discussion, however, is the point that client computer printouts of data stored in the files did not provide sufficient evidence to detect the irregularity.

It is mere conjecture to speculate as to whether the correct application of a generalized computer audit program would have caught or prevented the fraud. Whenever data are stored on magnetized "invisible" devices, the application of a CPA firm's own program would be expected to produce more reliable results. An important question is whether the cost of developing and applying such methods is worth the benefit to be derived.

At the present time, only large multi-office CPA firms have the resources and the expertise to develop and use generalized computer audit programs. Supervisory responsibility for program operation is placed on specially trained individuals who acquire the necessary computer knowledge either through in-house seminars or prior academic background. In the not too distant future, however, the option of using such programs may evolve into a necessity as more companies develop computerized systems. At that time, understanding of generalized computer audit programs should become more commonplace.

Chapter 14 References

Adams, Donald L., and John F. Mullarkey, "A Survey of Audit Software," *The Journal of Accountancy,* September 1972, pp. 39–66.

Andrews, Frederick, "Why Didn't Auditors Find Something Wrong with Equity Funding?" *The Wall Street Journal,* May 4, 1973, pp. 1, 16.

[8]There are many sources of information on the *Equity Funding* fraud. One fairly extensive study was issued by the AICPA in 1975 entitled *Report of the Special Committee on Equity Funding.* A capsule version of the computer aspects of the fraud appeared in *The Wall Street Journal,* May 4, 1973, pp. 1, 16: Frederick Andrews, "Why Didn't Auditors Find Something Wrong with Equity Funding?"

Auditing EDP Systems Committee, AICPA, "Technical Proficiency for Auditing Computer Processed Accounting Records," *The Journal of Accountancy,* October 1971, pp. 74–82.

Porter, W. Thomas, "Generalized Computer-Audit Programs," *The Journal of Accountancy,* January 1969, pp. 54–62.

Report of the Special Committee on Equity Funding, AICPA, 1975.

Chapter 14
Questions Taken from the Chapter

14–1.   Name two categories into which auditor/EDP relationships fall.

14–2.   Give five examples of functions that can be performed by a generalized computer audit program.

14–3.   What is the common thread or trait of all the generalized computer audit program capabilities listed in the chapter?

14–4.   Reproduce and label the flow chart or diagram that shows the general method used to operate a generalized computer audit program.

14–5.   Name three types of storage devices that can be used for client files that are processed with generalized computer audit programs.

14–6.   Describe the special software processor used by a large CPA firm.

14–7.   Take the following examples of output produced by generalized computer audit programs and list a possible audit procedure that would be performed with each item of output.
   a.   Listing of accounts receivable balances that exceed the credit limit.
   b.   Listing of unusually large inventory balances.
   c.   Listing of fixed asset changes.

14–8.   Indicate the phases of an audit in which generalized computer audit programs provide the most aid.

14–9.   Name three items of knowledge that auditors should possess in order to comprehend the computer system used by the client.

14–10.   In the *Equity Funding* case, why were the client computer printouts of file data inadequate to provide the auditors with sufficient evidence?

Chapter 14
Discussion/Case Questions

14–11.   A CPA's client, Boos & Baumkirchner, Inc., is a medium-sized manufacturer of products for the leisure-time activities market (camping equipment, scuba gear, bows and arrows, etc.). During the past year, a computer system was installed, and inventory records of finished goods and parts were converted to computer processing. The inventory master file is maintained on a disk. Each record of the file contains the following information:

Item or part number
Description
Size
Unit of measure code
Quantity on hand
Cost per unit
Total value of inventory on hand at cost
Date of last sale or usage
Quantity used or sold this year
Economic order quantity
Code number of major vendor
Code number of secondary vendor

In preparation for year-end inventory, the client has two identical sets of preprinted inventory count cards prepared. One set is for the client's inventory counts and the other is for the CPA's use to make audit test counts. The following information has been keypunched into the cards and interpreted on their face:

*Item or part number
*Description
*Size
*Unit of measure code

In taking the year-end inventory, the client's personnel will write the actual counted quantity on the face of each card. When all counts are complete, the counted quantity will be keypunched into the cards. The cards will be processed against the disk file, and quantity-on-hand figures will be adjusted to reflect the actual count. A computer listing will be prepared to show any missing inventory count cards and all quantity adjustments of more than $100 in value. These items will be investigated by client personnel, and all required adjustments will be made. When adjustments have been completed, the final year-end balances will be computed and posted to the general ledger.

The CPA has available a general-purpose computer audit software package that will run on the client's computer and can process both cards and disk files.

Required:
a. In general and without regard to the facts above, discuss the nature of general-purpose computer audit software packages and list the various types and uses of such packages.
b. List and describe at least five ways a general-purpose computer audit software package can be used to assist in all aspects of the audit of the inventory of Boos & Baumkirchner, Inc. (For example, the package can be used to read the disk inventory master file and list items and parts with a high unit cost or total value. Such items can be included in the test counts to increase the dollar coverage of the audit verification.)

(AICPA adapted)

14–12.   The following is a list of tasks performed by auditors in gathering evidence to support their opinion. For each task, indicate whether a generalized computer audit program could be used. Indicate the reasons for your answer.
a. Selecting accounts receivable confirmations.
b. Comparing subsidiary amounts to general ledger control amounts.
c. Investigating accounts receivable confirmation responses.
d. Inquiries on the collectibility of customer accounts.
e. Extracting important items from the minutes of meetings of the board of directors.
f. Preparing a list of fixed asset additions in excess of a certain dollar amount.
g. Checking capital stock issuances with legal authorizations.
h. Determining the adequacy of insurance coverage.
i. Test-checking the accounts receivable aging schedule with supporting documents.

14–13.   In many cases the output of a generalized computer audit program is used by the auditor for the same purpose it would be if the output were obtained through manual means. For each of the following examples of generalized computer audit program output, indicate the use that the auditor would make of the output in conducting the examination. For example, a computer-generated list of unusually large inventory balances might be used to check for possible inventory obsolescence.
a. Bank reconciliations for all the company's cash accounts.

   b. An aged accounts receivable schedule.
   c. A list of unusually large inventory balances (list a procedure other than checking for possible inventory obsolescence).
   d. A list of fixed asset additions and retirements.
   e. A list of vendors from whom large dollar purchases were made during the year.
   f. A list of employees who have received overtime pay in excess of a certain amount.
   g. A set of comparative financial statements for the two previous years.

14-14. Refer in the body of the chapter to the program flow chart to read accounts receivable master files. The following is an explanation of the flow chart logic with certain steps left blank. Fill in the blank steps.
   a. Start.
   b. Read a record.
   c. If it is the end of the file, end the process. If not, go to the next step.
   d. _____

   e. If all or part of the account has a balance > 90 days old, write a positive confirmation and go to the next step Ⓐ. If not, go to the next step Ⓐ.
   f. If this is the 3rd interval of accounts, write a negative confirmation and go to the next step Ⓐ. If not, go to the next step Ⓐ.
   g. _____

   h. If there is a credit balance in the over-30-day category, write a record and go to the next step. If not, go to the next step.
   i. _____

   j. _____

   k. If the credit status is 2, add the 61–90-day balances to the over-90-day balances. If not, go to the step that tests for credit status of 3.
   l. _____

   m. If the sum of the 61–90-day balances plus the over-90-day balances > $2,000, write a record and go the next step. If not, go to the next step.
   n. _____

   o. If the 90-day balance is > 0, test for current charges. If not, read another record.
   p. _____

14–15.   Refer in the body of the chapter to the master record data for auditor's program. Assume the following information about three customer accounts.
   a.   Customer No. 500 has an $8,000 balance, of which $6,000 is current, $1,000 is 31–60 days old, and $1,000 is 61–90 days old.
   b.   Customer No. 501 has a $4,000 balance, of which $3,500 is current and $500 is over 90 days old.
   c.   Customer No. 502 has a $3,000 balance, all of which is current.

Required:
Using the criteria shown in the program flow chart to read accounts receivable master files, indicate which accounts would have positive confirmations sent by the auditors. Also indicate the error messages, if any, that would be sent on these three accounts.

14–16.   An auditor obtains a magnetic tape that contains the dollar amounts of all client inventory items by style number. The information in the tape is in no particular sequence. By use of a generalized computer program, how can the auditor best ascertain that no consigned merchandise is included on the tape?

(AICPA adapted)

14–17.   An auditor's client has a magnetic tape that contains the detail of their customers' insurance policies by policy number. Unknown to the auditor is the fact that many of the policies are for non-existent customers. To prevent these "non-existent" policies from being printed and checked out by the auditors, a special code was placed on the magnetic tape. When the computer read this code, the policy data associated with this code were not printed.

Required:
Indicate how the auditor could have used a generalized computer audit program to detect this coverup.

14–18.   Roger Peters, CPA, has examined the financial statements of the Solt Manufacturing Company for several years and is making preliminary plans for the audit for the year ended June 30, 19X9. During this examination, Mr. Peters plans to use a set of generalized computer audit programs. Solt's EDP manager has agreed to prepare special tapes of data from company records for the CPA's use with the generalized programs.

The following information is applicable to Mr. Peters' examination of Solt's accounts payable and related procedures:

1. The formats of pertinent tapes are shown on page 670.

2. The following monthly runs are prepared:
   a.   Cash disbursements by check number.
   b.   Outstanding payables.
   c.   Purchase journals arranged (1) by account charged, and (2) by vendor.

3. Vouchers and supporting invoices, receiving reports, and purchase order copies are filed by vendor code. Purchase orders and checks are filed numerically.

4. Company records are maintained on magnetic tapes. All tapes are stored in a restricted area within the computer room. A grandfather-father-son policy is followed for retaining and safeguarding tape files.

Required:

a. Explain the grandfather-father-son policy. Describe how files could be reconstructed when this policy is used.
b. Discuss whether company policies for retaining and safeguarding the tape files provide adequate protection against losses of data.
c. Describe the controls that the CPA should maintain over:
   1. Preparing the special tape.
   2. Processing the special tape with the generalized computer audit programs.
d. Prepare a schedule for the EDP manager outlining the data that should be included on the special tape for the CPA's examination of accounts payable and related procedures. This schedule should show the:
   1. Client tape from which the item should be extracted.
   2. Name of the item of data.

(AICPA adapted)

MASTER FILE—VENDOR NAME

Rec Type | Vendor Code | Space | Blank | Vendor Name | Blank | Card Code 100

MASTER FILE—VENDOR ADDRESS

Rec Type | Vendor Code | Space | Blank | Address—Line 1 | Address—Line 2 | Address—Line 3 | Blank | Card Code 120

TRANSACTION FILE—EXPENSE DETAIL

Rec Type | Vendor Code | Blank | Voucher Number | Batch | Voucher Number | Voucher Date | Vendor Code | Invoice Date | Due Date | Invoice Number | Purchase Order Number | Debit Account | Prod Type Code | Product Code | Blank | Amount | Quantity | Card Code 160

TRANSACTION FILE—PAYMENT DETAIL

Rec Type | Vendor Code | Voucher Number | Batch | Voucher Number | Voucher Date | Vendor Code | Invoice Date | Due Date | Invoice Number | Purchase Order Number | Check Number | Check Date | Blank | Amount | Blank | Card Code 170

Chapter 15

The General Nature of Reports

Learning Objectives   *After reading and studying the material in this chapter, the student should*

Know the four standards of reporting.

Be able to relate the standards of reporting to the standard short-form report.

Understand the significance of each key phrase in the auditor's report.

Understand auditors' reporting obligations as they relate to prior-year and consolidated financial statements and references to other auditors.

In the first four chapters of this text, the organizational structures under which auditors operate, and the legal and ethical framework in which the audit function is performed were discussed, and an overview of the audit process was provided. In the next four chapters, we explored some of the methods that auditors use in their study and evaluation of the company's system of internal control. In the succeeding four chapters, we discussed the gathering of evidence to provide a basis for the auditor's opinion on the fairness of the financial statements.

Therefore, the next section of this text covers the reporting methods that auditors use to communicate their opinions to the users of their clients' financial statements. It is the purpose of this chapter to lay a general framework for reporting, with particular emphasis on the standard short-form report.

The Standards of Reporting

The guidelines used to write the audit report are the standards of reporting established by the AICPA. Of the ten generally accepted auditing standards contained in AU Section 150.02 of *AICPA Professional Standards, Volume 1 (SAS No. 1),* four relate to the subject of how various types of reports should be written.

These four standards of reporting are:

1. The report shall state whether the financial statements are presented in accordance with generally accepted accounting principles.

2. The report shall state whether such principles have been consistently observed in the current period in relation to the preceding period.

3. Informative disclosures in the financial statements are to be regarded as reasonably adequate unless otherwise stated in the report.

4. The report shall either contain an expression of opinion regarding the financial statements, taken as a whole, or an assertion to the effect that an opinion cannot be expressed. When an overall opinion cannot be expressed, the reasons therefor should be stated. In all cases where an auditor's name is associated with financial statements, the report should contain a clear-cut indication of the character of the auditor's examination, if any, and the degree of responsibility he is taking.

The Standard Short-Form Report

Prior to 1934, the opinion paragraph of the standard audit report began with the phrase "we certify that in our opinion . . ." which lead to the reports being referred to as "auditor's certificates." Although the term certificate is still found occasionally, "auditor's report" (with such variations as "report of independent auditor" and "independent auditor's report") is now the most prevalent phrase and is the wording used by the AICPA Auditing Standards Executive Committee.

An Examination of the Report

As explained in Chapter 3, the standard short-form report consists of a scope paragraph (a description of the financial statements the auditors examined and the standards that they used to perform this examination) and an opinion paragraph (the auditors' opinion as to the fairness of the financial statements based on their audit). The report form currently suggested in AU Section 509.07 *(SAS No. 2)* is reproduced below.

(Report Date)

(Scope Paragraph)

We have examined the balance sheet of X Company as of (at) December 31, 19xx, and the related statements of income, retained earnings and changes in financial position for the year then ended. Our examination was made in accordance with generally accepted auditing standards and accordingly, included such tests of the accounting records and such other auditing procedures as we considered necessary in the circumstances.

(Opinion Paragraph)

In our opinion, the financial statements referred to above present fairly the financial position of X Company as of (at) December 31, 19xx, and the results of its operations and the changes in its financial position for the year then ended, in conformity with generally accepted accounting principles applied on a basis consistent with that of the preceding year.

Although changes to the wording above have been discussed on numerous occasions by the senior auditing committee of the AICPA, the report style has remained almost intact for thirty years, and it represents, at the present time, the message that auditors wish to convey to users of financial statements concerning the scope of their responsibilities and the results of their examination.[1] Departures from the opinion paragraph illustrated in this standard report form, with the exception of reference to other auditors (which is discussed later in this chapter), constitute qualifications. These are discussed in the next chapter. Let us review, then, the important aspects of the report and relate them to the standards of reporting and to the material covered in previous chapters of the text, where appropriate.

Addressing the Report

A survey of auditor's reports included in annual reports of public companies would find some addressed to the stockholders, some addressed to the board of directors, and some addressed to both. AU Section 509.08 *(SAS No. 2)* contains a statement that the audit report "may be addressed to the company whose financial statements are being examined or to its board of directors or stockholders," except where the company whose financial statements the auditor is engaged to examine is not his client; in those cases, it is customary to address the report to the auditor's client. In the increasing number of cases where the auditor

---

[1]For other opinions on the auditor's intended message, see H. M. Anderson and J. W. Giese, "The Auditor's Belief and His Opinion—The Need for Consistency," *The CPA Journal,* January, 1973, pp. 49–54, and *The Commission on Auditors' Responsibilities: Report, Conclusions, and Recommendations,* Chapter 7.

is elected or ratified by the stockholders, the auditor should acknowledge this by addressing his report to them. In other cases, the inclusion of the stockholders in the address seems desirable, although not mandatory.

Examination of Financial Statements

The first significant phrase in the report is that the auditors have examined the various financial statements (which are specifically named). Although the distinction between examination of accounting records and financial statements may appear academic to some report-readers, it is an important difference to the auditors. Reference to financial statements has a broad connotation in that it implies the gathering of evidence beyond that produced by the client's accounting records.

For instance, it is explained in Chapter 12 that auditors request the client's legal counsel to confirm to them in writing the status of any pending litigation as of the balance sheet date. A possible result of these attorneys' letters is a footnote disclosure in the financial statements or an adjusting entry recording a liability. An audit of the records alone might not disclose this situation.

The confirmation of accounts payable is another example of the examination of financial statements rather than records. Auditors often send letters to some of the client's creditors asking for the detail of the amounts owed by the client at the balance sheet date. Generally, no figure is supplied to the creditor in the anticipation that if an accounts payable is omitted on the client's records, the creditor's independently derived figure will disclose the omission.

Since the scope of the audit generally includes many procedures similar to the illustrations in the two preceding paragraphs, auditors consider it proper to report that they have examined the *financial statements* rather than the *accounting records*. In addition, if the examination was related to the accounting records, some limiting phrase would be required in recognition of the fact that the sampling and testing procedures employed would probably result in not all records being examined.

Use of Generally Accepted Auditing Standards

The next important phrase is reference to generally accepted auditing standards. During the course of an audit, dozens, perhaps hundreds, of procedures are used to acquire the evidence on which the opinion is based. It would be ponderous and repetitious to reference each of these procedures, particularly since they are acts performed under the general guidelines of the ten auditing standards. Therefore, the report merely contains a statement that generally accepted auditing standards were followed and that all necessary tests and audit procedures were performed.[2] Furthermore, there is some belief that modification

---

[2]Note that it is not necessary for an auditor to perform all *customary* audit procedures. If, for example, the auditor did not confirm accounts receivable or observe inventories because it was impracticable or impossible to do so, but satisfied himself by means of alternative auditing procedures, he need not describe the circumstances or alternative procedures employed. What is required is that he perform all audit procedures *necessary* for him to form an opinion regarding the financial statements.

of the scope paragraph might be misunderstood by the readers. As an example, an auditor who uses the work of a specialist (such as an actuary, appraiser, attorney, or engineer) in performing an audit is prohibited from referring to the specialist in his audit report (AU Section 336), because such reference might be considered (1) a qualification of the report, (2) a division of responsibility, or (3) an indication that a more thorough audit had been made than would appear if no such reference had been made.

Although the three general standards are discussed throughout the text, one, in particular, warrants special attention—the standard on independence. It would be improper for the auditors to issue a standard short-form report with an unqualified opinion if they did not adhere to the AICPA rule on independence. A special type of disclaimer, shown in Chapter 17, is applicable to the situation in which independence is missing.

The second standard of field work is another example of a required auditing standard for the issuance of a typical short-form report. Although there is no mention of the study and evaluation of internal control in the scope paragraph of the standard report (many years ago such mention was made), there is an assumption that such a study and evaluation were conducted and that the second standard of field work was followed if the examination was made in accordance with generally accepted auditing standards.

The concepts of materiality and relative risk as set forth in AU Section 150.03–.05 *(SAS No. 1)* apply to all generally accepted auditing standards, including the standards of reporting.

The Auditor's Opinion

The second paragraph of the standard short-form report begins with the words "in our opinion." This phrase is an expression of the fact that, although auditors have a special expertise in accounting and auditing, no guarantee or factual statement regarding accuracy or even fairness can be made to readers of the report.

We have demonstrated in several places throughout the text that evidence-gathering techniques designed to determine financial statement fairness are imprecise. In Chapters 5 and 7 we explained that the evaluation of the strengths and weaknesses of internal control is subjective, regardless of whether a manual or computer system is employed. In Chapters 5, 6, and 9–14, we have shown that auditors use samples to draw conclusions about the validity of the client's records, a practice that leaves open the possibility that these conclusions may be incorrect. Closely related to this is the fact that auditors conduct their examination within the framework of a cost and time constraint, i.e., the benefit to be gained from further inquiry or additional testing should be worth the effort in order to justify its being done.

Finally, it was pointed out (particularly in Chapter 1) that the financial statements themselves are imprecise measurements of a company's operations over a period of time and its financial status at a point in time. Since the client exercises judgment in deriving financial statement amounts, it is reasonable to assume that auditors must do the same in their examination of these statements.

Nevertheless, the readers and users of financial statements have a right to expect that the phrase "in our opinion" represents the sound judgment of professional experts, and this is the message that auditors intend to convey. Such an expression of opinion is required to meet the fourth standard of reporting.

A complete set of financial statements normally consists of a balance sheet (also referred to as a statement of financial position or condition), a statement of income (or statement of earnings), a statement of retained earnings (sometimes combined with the statement of income), and a statement of changes in financial position; although other statements may be needed in certain circumstances. The auditor must understand which phrase in his report applies to each financial statement.

The phrase "financial position" in the opinion paragraph refers to the balance sheet. Accordingly, if an auditor were expressing an opinion on a balance sheet only, an example of the opinion paragraph would be:

> In our opinion, the *balance sheet* of X Company presents fairly its *financial position* as of December 31, 19xx, in conformity with. . . .

Note that the example above makes no reference to results of operations or changes in financial position. It would be inappropriate to do so, since the balance sheet does not purport to present either of these.

A statement of income and a statement of retained earnings (or a combined statement of income and retained earnings) are ordinarily considered essential for a fair presentation of "results of operations." Therefore, an audit report covering these two statements would include the following wording in the opinion paragraph:

> In our opinion, the *statements of income and retained earnings* of X Company present fairly its *results of operations* for the year ended December 31, 19xx, in conformity with. . . .

No reference is made to financial position or changes in financial position in the example above, since the financial statements needed to express an opinion as to those attributes of a company (the balance sheet and statement of changes in financial position) have not been presented.

The importance of an understanding of the relation between the financial statements and the opinion being expressed will become apparent when qualifications (sometimes applying to only one of several financial statements) of the audit report are discussed in the next chapter.

Fair Presentation

The middle part of the second paragraph of the standard short-form report contains the phrase "present fairly the financial position of X Company as of (at)

December 31, 19xx, and the results of its operations and the changes in its financial position for the year then ended, in conformity with generally accepted accounting principles . . ."

It was pointed out in Chapter 3 that the current report style, which includes the phrase above, evolved from earlier versions that contained no reference to fairness or to generally accepted accounting principles and used such words as "true" and "correct." Gradually, the terms "true" and "correct" gave way to "fairly," a word the accounting profession felt better expressed the subjective nature of financial statements.

However, for many years there was no clear agreement within the accounting profession as to the proper relationship between the word "fairly" and the phrase "generally accepted accounting principles." Some believed that there was a general presumption that statements were fair *if* they were presented in accordance with generally accepted accounting principles, while others believed that two separate opinions were needed; one that the financial statements were fairly presented, and the other that they were prepared in accordance with generally accepted accounting principles.

Influential groups outside of the accounting profession have been unwilling to accept the former definition. One of the first real evidences of this reluctance appeared in the *Continental Vending* case, summarized in Chapter 3. A statement was made by the judiciary to the effect that adherence to generally accepted accounting principles was only *one* determinant of fairness and was not a complete defense for the auditors.

Since then, the question of the relationship between fairness and accounting principles has gained increasing attention, especially in the accounting literature. As a result of this attention, the Auditing Standards Executive Committee issued *SAS No. 5,* entitled *The Meaning of "Present Fairly in Conformity with Generally Accepted Accounting Principles" in the Independent Auditor's Report.* One conclusion of the statement is that:

> The independent auditor's judgment concerning the "fairness" of the overall presentation of financial statements should be applied within the framework of generally accepted accounting principles. Without that framework the auditor would have no uniform standard for judging the presentation of financial position, results of operations, and changes in financial position in financial statements.[3]

Coupled with the preceding statement are further comments and discussions which seem to indicate that the concepts of fairness and accounting principles are broadening somewhat. For example, AU Section 411.07 *(SAS No. 5)* contains these additional comments:

> Generally accepted accounting principles recognize the importance of recording transactions in accordance with their substance. The auditor should consider whether the substance of transactions differs materially from their form.

[3]AU Section 411.03 *(SAS No. 5).*

AU Section 411.04 *(SAS No. 5)* goes on to specify that the auditor must form a judgment as to whether

> . . . the information presented in the financial statements is classified and summarized in a reasonable manner . . . the financial statements reflect the underlying events and transactions in a manner that presents the financial position, results of operations, and changes in financial position stated within a range of acceptable limits, that is, limits that are reasonable and practicable to attain in financial statements.

In view of this statement, it seems reasonably clear that when auditors use the term "present fairly," they are implying a broader responsibility to statement-users than was previously thought to be the case. Auditors are now rendering their opinion that financial statements are informative of matters that may affect their use, understanding, and interpretation, and that, in their opinion, the statements are not misleading.

Generally Accepted Accounting Principles

The term "generally accepted accounting principles," as it is understood in the accounting profession, is set forth in AU Section 411.02 *(SAS No. 5)* as follows:

> The first standard of reporting requires an auditor who has examined financial statements in accordance with generally accepted auditing standards to state in his report whether the statements are presented in accordance with generally accepted accounting principles. The phrase "generally accepted accounting principles" is a technical accounting term which encompasses the conventions, rules, and procedures necessary to define accepted accounting practice at a particular time. It includes not only broad guidelines of general application, but also detailed practices and procedures. . . .

The reader may remember from Chapter 2 that Rule 203 of the AICPA *Code of Professional Ethics* prohibits a member of the AICPA from expressing an opinion that financial statements are presented in accordance with generally accepted accounting principles if they contain any departure from an accounting principle established by a body designated to issue such principles (at present the Committee on Accounting Principles, the Accounting Principles Board, and the Financial Accounting Standards Board), unless it can be demonstrated that owing to unusual circumstances, the principle would cause the financial statements to be misleading. AU Section 411 *(SAS No. 5)* acknowledges this source of accounting principles and goes on to suggest other possible sources, including AICPA accounting interpretations and industry audit guides, pronouncements of other professional associations and regulatory agencies, and accounting textbooks and articles.

AU Section 411.09 *(SAS No. 5)* concludes with the following paragraph:

Specifying the circumstances in which one accounting principle should be selected from among alternative principles is the function of bodies having authority to establish accounting principles. When criteria for selection among alternative accounting principles have not been established to relate accounting methods to circumstances, the auditor may conclude that more than one accounting principle is appropriate in the circumstances. The auditor should recognize, however, that there may be unusual circumstances in which the selection and application of specific accounting principles from among alternative principles may make the financial statements taken as a whole misleading.

Thus, where two or more accounting principles are applicable in a particular circumstance, and all of them are generally accepted, the management of the company being audited, as well as the auditor, must determine the *most appropriate* principle to be used. As noted in Chapter 3, this was the essence of the courts' positions in the *Barchris Construction Company* case (in which the court decided that a gain on a sale/lease-back should have been deferred and amortized over the life of the related lease rather than being recognized as income in one year, although both practices were in existence at the time), and *Continental Vending* case (in which the court ruled that compliance with generally accepted accounting principles was not a complete defense if the financial statements were not fairly presented).[4]

Although it is included as a separate standard of reporting, adequate disclosure in financial statements is often considered to be encompassed within generally accepted accounting principles. Whether adequate disclosure is considered as a separate requirement or as part of generally accepted accounting principles, there is no question but that it is essential to the fair presentation of financial statements. Examples of such disclosures include important subsequent events, restrictions on payment of dividends, guarantees of debt, commitments and contingencies, and depreciation and inventory methods.

Consistent Application of Accounting Principles

The last phrase of the standard report states that the accounting principles are applied on a basis consistent with that of the preceding year, and is responsive to the second standard of reporting. It is important that this phrase be placed in the standard short-form report so that readers will know that there is comparability of accounting principles between periods. It is implicit in the standard that the accounting principles have been consistently applied within each period.

The desirability of providing the foregoing information stems from the fact that a variety of generally accepted accounting principles can be used in financial statement reporting. As an illustration, assume that from one year to the next a company's net income increased by $2.00 a share. Readers of the financial statements, unless informed otherwise, have a right to assume that this $2.00

---

[4]For more opinions on fairness and generally accepted accounting principles see Geraldine F. Dominiak and Joseph G. Louderback, "Present Fairly and Generally Accepted Accounting Principles," *The CPA Journal,* January, 1972, pp. 45–49.

increase resulted from events other than changes in accounting principles, such as a change in the method of inventory valuation.

If the financial statements of only one period are being reported on, as in the example on page 673, the consistency reference relates to the period preceding the period being audited, and the appropriate wording is "applied on a basis consistent with that of the preceding year." If the auditor's report relates to two or more years, the consistency reference relates to consistency among the periods being reported on and appropriate wording is "applied on a consistent basis." The report relates also to the consistency of those years with the prior year *if* the prior-year financial statements are presented. The report should then include such words as "consistently applied during the periods and on a basis consistent with that of the preceding year."

It must be noted that not all factors affecting comparability of financial statements between years result in an exception as to consistency in the auditor's report. AU Section 420.06–.11 *(SAS No. 1)* lists the following factors that affect consistency:

1. A change in accounting principle (for example, a change from the straight-line method to the declining balance method of depreciation)

2. A change in the reporting entity (for example, changing specific subsidiaries comprising the group of companies for which consolidated financial statements are presented)

3. A correction of an error in principle (for example, a change from an accounting principle that is not generally accepted to one that is generally accepted)

4. A change in principle inseparable from a change in an estimate (for example, a change from amortizing certain costs to direct charge-off based on a new estimate that the costs no longer have value)

AU Section 420.12–.18 *(SAS No. 1)* also lists the following factors, which would normally not affect the standard consistency reference in the auditor's report:

1. A change in accounting estimate (for example, the estimated lives of depreciable assets).

2. An error correction not involving an accounting principle (for example, the correction of mathematical mistakes).

3. A change in classification or reclassification (for example, a reclassification made in previously issued financial statements to enhance comparability with current financial statements).

4. Variations in format and presentation of the statement of changes in financial position (for example, changes to or from a balanced form, but changing from a cash to a working capital presentation may affect consistency).

5. A substantially different transaction or event (for example, a change in the type of business conducted by a company).

6. A change expected to have a material future effect (for example, a change having no material effect in the current year, but expected to have a material future effect).

While the factors listed above do not require a consistency exception in the auditor's report, adequate disclosure should be made of them in the financial statements.

However, it should be pointed out that it is not the task of the auditors to apprise report readers of all the reasons why company operations differ from one year to another. Auditors do not serve the function of financial analysts; rather, they give to statement-users an opinion relating to generally accepted accounting principles and their consistent application.

Report Date

The date of an auditor's report carries with it a significance not fully understood by the general public. This is the date to which the auditor assumes responsibility for detection of subsequent events that might have a material effect on the audited financial statements. This is normally the date on which the auditor completes his work in the client's office (often referred to as the end of field work).

Consider the following facts. An audit is being performed as of December 31, 1979. The auditors must allow the company time to close and balance its financial records before work on the audit can be started, so it is January 17, 1980, before the audit work commences in the client's office. By February 3, 1980, the audit staff has substantially completed its work, and the audit partner arrives at the client's office to review the audit work and clear any problems with the client. This review is completed on February 4, 1980, and that evening the audit staff return to their own office to have a quality-control review made of the audit report and audit working papers and to have the audit report typed and duplicated. Because February is a very busy time of the year, the quality-control reviews and report duplication are not completed until February 8, 1980. The completed and signed report is delivered to the client on February 9, 1980. What should be the date of the auditor's report?

The audit report should be dated as of the last day of field work, which in this case would be February 4, 1980. The audit staff is in the client's office through this date and has access to financial records and management personnel necessary to perform a review of transactions and events subsequent to the audit date that might affect the audited financial statements. After the audit staff leave the client's office, this information is no longer readily available, and the auditor has no responsibility to make any inquiry or to perform any auditing procedure subsequent to this date (except with respect to filings under the Securities Act of 1933—see Chapter 3).

Although the auditor has no responsibility to search for subsequent events affecting the financial statements after the date of his report, any event coming to his attention between that date and the date the financial statements are issued should be reflected in the financial statements, or the auditor should qualify his report. In either event, the dating of the auditor's report will be affected. The auditor would have the option of dual dating his report, that is, using the original date for the overall report and a subsequent date for the subsequent event or transaction (for example, February 4, 1980, except for Note 8, as to which the date is February 7, 1980), or using only the later date. In the latter case, it would be necessary for the audit staff to return to the client's office and perform a subsequent review from the original date to the later date.

Report Coverage of Prior-Year Financial Statements

The example of the audit report shown on page 673 of this chapter is applicable to financial statements covering only one year. It has become common (and is required in many SEC filings) for companies to present financial statements for two or more years. In such cases, the language of the standard short-form report must be modified. *SAS No. 15* requires that an auditor update his report (re-express a previous opinion or, depending on the circumstances, express a different opinion from that previously expressed) on the prior-year financial statements if he has audited them. The updating is performed by covering both the current and prior year(s) financial statements in his report. The following form is suggested in *SAS No. 15:*

> We have examined the balance sheets of ABC Company as of (at) December 31, 19X2 and 19X1, and the related statements of income, retained earnings, and changes in financial position for the years then ended. Our examinations were made in accordance with generally accepted auditing standards and, accordingly, included such tests of the accounting records and such other auditing procedures as we considered necessary in the circumstances.
>
> In our opinion, the financial statements referred to above present fairly the financial position of ABC Company as of (at) December 31, 19X2 and 19X1, and the results of its operations and the changes in its financial position for the years then ended, in conformity with generally accepted accounting principles applied on a consistent basis.

The reissued report will normally be dated as of the end of field work of the most recent audit. Reports with differing opinions and the modification of previously issued reports are discussed in Chapter 16. If the prior-year financial statements were unaudited or audited by other auditors, this fact must be disclosed in the current auditor's report, as indicated in the next section.

Prior-Year Financial Statements Not Audited

Where the prior-year financial statements have not been audited, notations to this effect should be placed above the appropriate column headings in the financial statements and a sentence similar to the following placed at the end of the scope paragraph:

> We have not examined the financial statements as of December 31, 19X8 (end of prior year) or for the year then ended and we express no opinion on them.

SAS No. 15 provides that the disclaimer of opinion may be omitted in certain SEC filings (although the applicable financial statements must still be labeled "unaudited"). There is no justification for this exception in *SAS No. 15,* and the authors believe it is unwarranted.

Prior-Year Financial Statements Audited by Predecessor Auditor

If prior-year financial statements are presented that were examined by a predecessor auditor, the predecessor auditor may reissue his report. However, before doing so he should consider whether his original opinion is still applicable. Accordingly, he should (a) read the financial statements of the current period, (b) compare the prior-period financial statements that he reported on with the financial statements to be presented for comparative purposes, and (c) obtain a letter of representation from the successor auditor regarding matters that might have a material effect on, or require disclosure in, the financial statements reported on by the predecessor auditor. The date of a reissued report should be the same date as that used in the original report to avoid the impression that the predecessor auditor's examination extended beyond the original date. If the predecessor auditor's report or the applicable financial statements require revision as a result of the procedures set forth above, dual dating should be used.

If prior-year financial statements are presented, and they were audited by a predecessor auditor whose report is not presented, this fact should be disclosed in the successor auditor's report. The following sentence to be added to the scope paragraph is an example of such disclosure:

> The financial statements for 19X8 (the prior year) were examined by other auditors whose report dated March 1, 19X9, expressed an unqualified opinion on those statements.

If prior-year financial statements are presented and the predecessor auditor's report on those statements was qualified, this fact should also be disclosed in the successor auditor's report.

Reference to Other Auditors'
Participation in an Examination

Occasionally, more than one auditor may be involved in an examination of a company's financial statements, particularly where numerous subsidiaries, divisions, or investments are involved. In this case, the principal or lead auditor (normally the one examining the parent company and at least one-half of the total enterprise in terms of revenues and total assets) must determine if he has performed sufficient substantive work to serve as the principal auditor. (See Rule 401 of the AICPA Rules of Conduct where an auditor may insist on auditing any component of a business which in his judgment is necessary to warrant the expression of his opinion.) If he has, he must decide upon the manner in which he will utilize the reports of the other auditors. He may decide either to assume responsibility for the other auditors' work, in which case he will make no reference to them in his report, or he may not assume responsibility for their work and state the division of responsibility in his audit report. AU Section 543 *(SAS No. 1)* includes a discussion of factors to consider in making the decision whether or not to make reference to other auditors and inquiries to make regarding their professional reputation and independence.

If a principal auditor decides to refer to another auditor in his report, the standard short-form report would be modified along the following lines:

(Scope Paragraph)

... necessary in the circumstances. We did not examine the financial statements of B Company, a consolidated subsidiary, which financial statements reflect total assets and revenues constituting 20 percent and 22 percent, respectively, of the related consolidated totals. These statements were examined by other auditors whose report thereon has been furnished to us, and our opinion expressed herein, insofar as it relates to the amounts included for B Company, is based solely upon the report of the other auditors.

(Opinion Paragraph)

In our opinion, based upon our examination and the report of the other auditors referred to above, the accompanying. . . . [5]

The other auditor is normally not named, but he may be, provided he consents and his report is included with the report of the principal auditor.

Reports on Consolidated Statements

To this point we have assumed that the financial statements being reported on were those of a single entity. In practice, the auditor often reports on consoli-

[5] AU Section 543.09 *(SAS No. 1)*.

dated (or combined) financial statements, as well as consolidating financial statements.

No particular problem is encountered when the standard short-form report is applied to consolidated financial statements. The words "consolidated" and "subsidiaries" are inserted at the appropriate places to produce the following partial example:

> We have examined the *consolidated* balance sheet of X Company *and subsidiaries* as of (at) December 31, 19xx, and the related *consolidated* statements of income, retained earnings, and changes in financial position for the year then ended. Our examination was made. . . .

A more complex situation arises when consolidating financial statements are presented in columnar form, and the auditor either has not audited, or is expressing a qualified opinion regarding, one or more of the consolidating companies, the effects of which may or may not carry through to the consolidated financial statements. It is important that the auditor recognize that the scope of the work necessary to express an opinion on the individual companies in a consolidating financial statement may be significantly greater than that necessary to express an overall opinion on the consolidated financial statements.

Annual Reports and Other Documents Containing Audited Financial Statements

Financial statements, together with the related auditor's report, are often included in annual reports to stockholders, reports to regulatory agencies, and other documents. Although the auditor's responsibility does not extend beyond the financial statements in these cases, he should read the documents containing the financial statements and determine that the information included in them is not materially inconsistent with the information contained in the financial statements. For example, if a company's normal operations for a year produced a net loss, but an extraordinary gain more than offset the loss and resulted in net income, the text of an annual report which discussed net income for the year without stating that it resulted from an extraordinary gain could be materially inconsistent with the financial statements.

AU Section 550.04 *(SAS No. 8)* provides that in the event of such a material inconsistency, the auditor must consider an explanatory paragraph in his report, the withholding of his report, or withdrawal from the engagement. In the event of a material misstatement in the text, as opposed to a material inconsistency between the text and the financial statements, AU Section 550.05 *(SAS No. 8)* contains a suggestion that legal counsel be consulted.

Where audited financial statements are included in a registration statement filed pursuant to the Securities Act of 1933, the auditor may write a letter to underwriters whereby he takes additional responsibility for certain information contained in the text of the registration statement. These letters are discussed in Chapter 17.

Summary

Although the short-form audit report is highly standardized, the auditor must exercise care in specifying the periods he desires to cover (when more than one period is presented), the entities he desires to cover (when consolidating or other multi-company financial statements are presented), and the report date. He must also be aware of the meanings of the important phrases in the report, which have been discussed in this chapter, so that they can be used as a standard against which to evaluate his work and the financial statements upon which he intends to express an opinion. It is important that this evaluation be consciously made at the conclusion of each audit, and that the auditor avoid the mistake of utilizing the standard short-form report without giving sufficient thought to the opinion being expressed.

Chapter 15 References

Anderson, H. M., and J. W. Giese, "The Auditor's Belief and His Opinion—The Need for Consistency," *The CPA Journal,* January, 1973, pp. 49–54.

Dominiak, Geraldine F. and Joseph G. Louderback, "Present Fairly and Generally Accepted Accounting Principles," *The CPA Journal,* January, 1972, pp. 45–49.

Lahey, James M., "Toward a More Understandable Auditor's Report," *The Journal of Accountancy,* April, 1972, pp. 48–53.

Rosenfield, Paul, and Leonard Lorensen, "Auditors' Responsibilities and the Audit Report," *The Journal of Accountancy,* September, 1974, pp. 73–83.

Chapter 15
Questions Taken from the Chapter

15–1.   How do auditors communicate with the users of financial statements?

15–2.   What guidelines must the auditor use in preparing audit reports?

15–3.   List the important concepts of the four standards of reporting.

15–4.   The standard short-form audit report consists of two paragraphs. State the name of each paragraph and the information each conveys.

15–5.   What is the difference between an audit certificate and an audit report?

15–6.   To whom should the audit report normally be addressed?

15–7.   Since much of an auditor's work involves an examination of various client financial records, why does the auditor's report state that he has examined the financial statements rather than the financial records?

15–8.   How can a reader of an audit report determine what audit steps were performed if they are not listed in the audit report?

15–9.   What is the significance of the phrase "in our opinion" in the audit report?

15–10.   Define the word "fairly" as used within the context of the auditor's report.

15–11.   Can financial statements not be fair and yet be in accordance with generally accepted accounting principles?

15–12.   List the conditions specified in AU Section 411 *(SAS No. 5)* for financial statements to "present fairly."

15–13.   What are the rule-making bodies whose pronouncements have been designated as generally accepted accounting principles by the AICPA?

15–14.   Name five sources of generally accepted accounting principles other than the pronouncements of rule-making bodies referred to in Question 15–13.

15–15.   What responsibility does the auditor have where either of two alternative generally accepted accounting principles could be applied by his client?

15–16.   Explain the significance of consistent application of accounting principles.

15–17. List four factors that affect the comparability of financial statements between periods and the consistency reference in the auditor's report. Give an example of each.

15–18. List six factors that affect the comparability of financial statements between periods but do not affect the consistency reference in the auditor's report. Give an example of each.

15–19. List the individual financial statements normally included in a complete set and indicate which phrase of the opinion paragraph applies to each.

15–20. What is the significance of the date of the auditor's report?

15–21. How does the auditor determine the date of the audit report?

15–22. What is the auditor's responsibility for events during the period between the date of his report and the date the report is issued?

15–23. What is meant by the phrase "dual dating of a report," and when would an auditor use this practice?

15–24. What reporting responsibility does an auditor have for prior-year financial statements presented with current-year statements he has audited if (1) he audited the prior year, (2) the prior year was unaudited, and (3) the prior year was audited by another auditor?

15–25. When more than one auditor is involved in an examination of a company's financial statements, what two important decisions regarding reporting must the principal auditor make?

15–26. What disclosure is made in a principal auditor's report if he decides to assume responsibility for another auditor's work? If he decides not to assume responsibility for another auditor's work?

15–27. Is a principal auditor ethically justified in making an inquiry into the professional reputation and independence of another auditor involved in work for the same client?

15–28. What two indicators of the size of the portion of an enterprise being examined by other auditors are required to be disclosed in the principal auditor's report?

15–29.　How is the standard short-form report modified to apply to consolidated financial statements?

15–30.　What is the auditor's responsibility for the text portion of annual reports and other documents containing audited financial statements?

15–31.　What actions should an auditor consider if he notes a material inconsistency between information in the text of an annual report and the audited financial statements?

Chapter 15
Objective Questions Taken from CPA Examinations

15–32.　The consistency standard does *not* apply to an accounting change that results from a change in
a. Accounting principle.
b. Accounting estimate.
c. Reporting entity.
d. Accounting principle inseparable from a change in accounting estimate.

15–33.　What is the objective of the reporting standard relating to consistency?
a. To give assurance that adequate disclosure will be made so that there will be comparability of financial statements between companies in the same industry.
b. To give assurance that the comparability of financial statements between periods has not been materially affected by changes in accounting principles.
c. To give assurance that the comparability of financial statements between periods has not been materially affected by any change.
d. To give assurance only that the same accounting principles have been applied to all similar transactions within each period presented.

15–34.　What recognition, if any, should be given to an accounting change which has *no* material effect on the financial statements in the current year but the change is reasonably certain to have substantial effect in later years?
a. There is no need for recognition because when the change was made, it had no material effect.
b. The change should be disclosed in the notes to the financial statements whenever the statements of the year of change are presented, but the CPA 'need not recognize the change in the CPA's opinion as to consistency.

    c. The change should be disclosed in the notes to the financial statements whenever the statements of the year of change are presented, and the CPA must recognize the change in the CPA's opinion as to consistency.

    d. The change should be recognized in the CPA's opinion only when it is not disclosed in notes to the financial statements.

15–35. The auditor's formal review of subsequent events normally should be extended through the date of the

    a. Auditor's report.

    b. Next formal interim financial statements.

    c. Delivery of the audit report to the client.

    d. Mailing of the financial statements to the stockholders.

15–36. On August 5, 1975, a CPA completed field work on a client's financial statements for the year ended June 30, 1975. On August 20, 1975, the anticipated date for delivery of the report, the CPA read in the paper that one of the client's plants burned to the ground. The CPA calls the client to confirm the information and tells the client this subsequent event should be disclosed in the financial statements. The client agrees to add a footnote but still needs the report today. The CPA quickly revises the statements by adding a "Note G" and delivers the report on time. What date should the report bear?

    a. August 5, 1975.

    b. August 5, 1975, except for Note G which is August 20, 1975.

    c. August 20, 1975.

    d. June 30, 1975.

15–37. In forming an opinion on the consolidated financial statements of Albom Corp., a CPA relies upon another auditor's examination of the financial statements of Henig Company, a wholly owned subsidiary. Henig's auditor expressed an unqualified opinion. In the report on Albom's consolidated financial statements, the CPA expresses an unqualified opinion and refers to the other auditor's examination. This indicates that

    a. The CPA concludes, on the basis of his review of the other auditor's work, that the same responsibility can be assumed as though the CPA had audited Henig.

    b. The CPA is satisfied with the other auditor's work in general but has some reservation relating to that auditor's independence or professional standing.

    c. The CPA's review of the subsidiary's financial statements indicates a problem that may have escaped the other auditor.

    d. The CPA is satisfied with the other auditor's independence and professional standing but is unwilling to take responsibility for his work.

15–38. A principal auditor decides to assume responsibility for the work of another CPA insofar as the other CPA's work relates to the principal auditor's expression of an opinion on the financial statements taken as a whole. The opinion of the other CPA is qualified, but the subject of the qualification is *not* material in relation to the financial statements taken as a whole. In discharging the reporting obligation, the principal auditor

a. Must qualify his opinion.

b. Must make reference to the other CPA's qualified opinion and state that it is not material to the financial statements taken as a whole.

c. Need not make reference in the report to the qualification, but must disclose, in a note to the financial statements, the subject of the qualification and its effect on financial position, results of operations, and changes in financial position.

d. Need not make reference in the report to the qualification.

15–39. Which of the following should *not* be treated as an accounting change that affects the standard of reporting relating to consistency?

a. Change in the service lives of all assets.

b. Change from cash basis to accrual method of accounting.

c. Change from FIFO to LIFO inventory method.

d. Change from straight-line to declining-balance method of depreciation.

15–40. The standard short-form auditor's report is generally considered to have a scope paragraph and an opinion paragraph. In the report the auditor refers to both generally accepted accounting principles (GAAP) and generally accepted auditing standards (GAAS). In which of the paragraphs are these terms used?

a. GAAP in the scope paragraph and GAAS in the opinion paragraph.

b. GAAS in the scope paragraph and GAAP in the opinion paragraph.

c. GAAS in both paragraphs and GAAP in the scope paragraph.

d. GAAP in both paragraphs and GAAS in the opinion paragraph.

15–41. A CPA has completed the initial audit of a new client which has *not* previously been audited. The client has prepared an annual report with comparative financial statements of the prior year which have been marked "unaudited." What comment should the CPA make to his client or in his report relative to the prior year's financial statements?

a. The CPA should tell his client that if the client wants to include the CPA's report on this year's financial statements in the annual report the client cannot include the comparative statements because the CPA has not audited them.

b. The CPA must add a disclaimer of opinion on the prior year's financial statements to his report.

c. The CPA must qualify his report to the effect that comparison with the prior-year figures may not be valid because they are unaudited.

d. The CPA need do nothing.

15–42. Which of the following would be an inappropriate addressee for an auditor's report?

a. The corporation whose financial statements were examined.

b. A third party, even if the third party is a client who engaged the auditor for examination of a non-client corporation.

c. The president of the corporation whose financial statements were examined.

d. The stockholders of the corporation whose financial statements were examined.

15–43. If the basic financial statements are accompanied by a separate statement of changes in stockholders' equity, this statement

a. Should not be identified in the scope paragraph but should be reported on separately in the opinion paragraph.

b. Should be excluded from both the scope and opinion paragraphs.

c. Should be identified in the scope paragraph of the report but need not be reported on separately in the opinion paragraph.

d. Should be identified in the scope paragraph of the report and must be reported on separately in the opinion paragraph.

Chapter 15
Discussion/Case Questions

15–44. You recently hired an accounting major, a graduate of a local university, to assist you in your CPA practice. He has been assisting you in the audit of the balance sheet of a small client. When you were called out of town for a day near the end of the audit, you asked him to draft the audit report for your review when you returned. Upon your return you found the following:

I have audited the books and accounts of March Company as of October 31, 19X9 and I certify that, in my opinion, the Balance Sheet correctly reflects the financial condition of March Company as of that date.

List and discuss the departures from generally accepted auditing standards of reporting that you would call to your assistant's attention.

15–45. You were recently appointed to a committee formed by the local CPA Society to investigate possible substandard reporting practices. The following report was submitted to the committee and, as the newest member of the committee, you were asked to make the initial review of the report and to give the committee your recommendations.

We have examined the balance sheet and related statements of income and expense and surplus of Holly Corp. as at March 31, 19X9. Our examination was made in accordance with the generally accepted accounting principles and accordingly, included such tests of the accounting records and such other auditing procedures as we considered necessary in the circumstances.

In our opinion, the accompanying balance sheet and related statements of income and expense and surplus presents fairly the financial position of Holly Corp. as at March 31, 19X9, in conformity with generally accepted accounting principles applied on a basis consistent with that of the preceding year.

State whether or not you would recommend to the committee that the audit report be found to be in violation of generally accepted auditing standards of reporting, and if so, why.

15–46. A number of years ago a large public accounting firm used the following form of opinion paragraph.

In our opinion, the financial statements referred to above present fairly the financial position of X Company as of December 31, 19xx, and the results of its operations and the changes in its financial position for the year then ended, and were prepared in conformity with generally accepted accounting principles applied on a basis consistent with that of the preceding year.

How does this opinion differ from the one recommended in *SAS No. 1?*

What additional responsibility, if any, is the auditor assuming in the opinion paragraph above?

How does the concept of fairness in the opinion above compare with the concept in *SAS No. 5?*

15–47. The following form of audit report was recommended by a special commission on auditors' responsibilities:

The accompanying consolidated balance sheet of XYZ Company as of December 31, 1976, and the related statements of consolidated income and changes in consolidated financial position for the year then ended, including the notes, were prepared by XYZ Company's management, as explained in the report by management.

In our opinion, those financial statements in all material respects present the financial position of XYZ Company at December 31, 1976, and the results of its operations and changes in financial position for the year then ended in conformity with generally accepted accounting principles appropriate in the circumstances.

We audited the financial statements and the accounting records supporting them in accordance with generally accepted auditing standards. Our audit included a study and evaluation of the company's accounting system and the controls over it. We obtained sufficient evidence through a sample of the transactions and other events reflected in the financial statement amounts and an analytical review of the information presented in the statements. We believe our auditing procedures were adequate in the circumstances to support our opinion.

Based on our study and evaluation of the accounting system and the controls over it, we concur with the description of the system and controls in the report by management. Nevertheless, in the performance of most control procedures, errors can result from personal factors. Also, control procedures can be circumvented by collusion or overridden. Furthermore, projection of any evaluation of internal accounting control to future periods is subject to the risk that changes in conditions may cause procedures to become inadequate and the degree of compliance with them to deteriorate.

We reviewed the information appearing in the annual report other than the financial statements, compared it to the statements, and found no material disagreement between them.

We reviewed the process used by the company to prepare the quarterly information released during the year. Our reviews were conducted each quarter. Our reviews which consisted primarily of inquiries of management, analysis of financial information and comparisons of that information to information and knowledge about the company obtained during our audits, were based on our reliance on the company's internal accounting control system. Any adjustments or additional disclosures we recommended have been reflected in the information.

We reviewed the company's policy statement on employee conduct, described in the report by management, and reviewed and tested the related controls and internal audit procedures. While no controls or procedures can prevent or detect all individual misconduct, we believe the controls and internal audit procedures have been appropriately designed and applied during the year.

We met with the audit committee of XYZ Company sufficiently often to inform it of the scope of our audit and to discuss any significant accounting or auditing problems encountered and any other services provided to the company.

    a. Discuss the substantive differences between this report and the standard short-form report.
    b. Identify any representations made in this report that are not implied in the standard short-form report.
    c. State the advantages and disadvantages of this form of report.

15–48.   Upon completion of all field work on September 23, 1975, the following "short-form" report was rendered by Timothy Ross to the directors of The Rancho Corporation.

To the Directors of
The Rancho Corporation:

We have examined the balance sheet and the related statement of income and retained earnings of The Rancho Corporation as of July 31, 1975. In accordance with your instructions, a complete audit was conducted.

In many respects, this was an unusual year for The Rancho Corporation. The weakening of the economy in the early part of the year and the strike of plant employees in the summer of 1975 led to a decline in sales and net income. After making several tests of sales records, nothing came to our attention that would indicate that sales have not been properly recorded.

In our opinion, with the explanation given above, and with the exception of some minor errors that are considered immaterial, the aforementioned financial statements present fairly the financial position of The Rancho Corporation at July 31, 1975, and the results of its operations for the year then ended, in conformity with pronouncements of the Accounting Principles Board and the Financial Accounting Standards Board applied consistently throughout the period.

Timothy Ross, CPA
September 23, 1975

Required:
List and explain deficiencies and omissions in the auditor's report. The type of opinion (unqualified, qualified, adverse, or disclaimer) is of no consequence and need not be discussed.

Organize your answer sheet by paragraph (scope, explanatory, and opinion) of the auditor's report.

(AICPA adapted)

15–49. Various types of "accounting changes" can affect the second reporting standard of the generally accepted auditing standards. This standard reads, "The report shall state whether such principles have been consistently observed in the current period in relation to the preceding period."

Assume that the following list describes changes that have a material effect on a client's financial statements for the current year.

1. A change from the completed-contract method to the percentage-of-completion method of accounting for long-term construction-type contracts.
2. A change in the estimated useful life of previously recorded fixed assets based on newly acquired information.
3. Correction of a mathematical error in inventory pricing made in a prior period.
4. A change from prime costing to full absorption costing for inventory valuation.

5. A change from presentation of statements of individual companies to presentation of consolidated statements.
6. A change from deferring and amortizing pre-production costs to recording such costs as an expense when incurred because future benefits of the costs have become doubtful. The new accounting method was adopted in recognition of the change in estimated future benefits.
7. A change to including the employer share of FICA taxes as "Retirement benefits" on the income statement from including it with "Other taxes."
8. A change from the FIFO method of inventory pricing to the LIFO method of inventory pricing.

Required:
Identify the type of change described in each item above, state whether any modification is required in the auditor's report *as it relates to the second standard of reporting,* and state whether the prior year's financial statements should be restated when presented in comparative form with the current year's statements. Organize your answer sheet as shown below.

Item No.	Type of Change	Should Auditor's Report Be Modified?	Should Prior Year's Statements Be Restated?

(AICPA adapted)

15–50. Richard Adkerson, CPA, was reviewing with the president of Central Pond Shipping Co. the standard short-form audit report he intended to issue to this new client. The president had the following comments about phrases in the report that he did not understand.

a. How could Adkerson have examined the financial statements when they were not prepared until the audit was almost complete? The president suggested changing the wording to "examined the accounting records."
b. The president knew Adkerson spent considerable time examining canceled checks and suggested that this fact be included in the scope paragraph.
c. The president asked for a slightly stronger statement than "in my opinion" and suggested "in my informed opinion."
d. The president requested that some wording be included stating that the accounting records were correct, since he questioned the competency of his bookkeeper.

e. The president did not completely trust his brother, who was also associated with the company. He asked Adkerson to include in his report a statement that he had not detected any material fraud.

f. The president asked how Adkerson could comment on consistency when there had been numerous management changes during the year.

How should Adkerson reply to each of these comments?

15–51. The management of Onbeach Supply Co. desires to include four years' financial statements in this year's annual report. R. Vaughn, CPA, has audited the financial statements of the current and immediately preceding year; however, another CPA audited the financial statements for the year prior to that, and the earliest year was unaudited.

Discuss or draft the audit report that Vaughn should issue, assuming that unqualified opinions were given in all of the audited years.

15–52. In the current year the operating revenue of State Mountain Gas Co. declined to $1,700,000 from $2,200,000 in the prior year. Owing to a change in accounting principle, however, net income increased by $500,000 in the current year.

The financial statements in the annual report were properly prepared to report the effect of the change in accounting principle as a separate item before net income. The president's letter, however, discussed only the increase in net income.

What responsibility does the auditor of the financial statements have for the president's letter? What action, if any, should be taken in this case?

15–53. In certain filings with the Securities and Exchange Commission, audited financial statements for both (1) the parent company, and (2) the parent company and consolidated subsidiaries, are required.

Discuss or draft a single standard short-form audit report for XYZ Company that covers both parent company financial statements for three years ended December 31, 19X9 and consolidated financial statements for five years ended December 31, 19X9.

Chapter 16

Departures From the Standard Audit Report

Learning Objectives *After reading and studying the material in this chapter, the student should*

Be able to relate the fourth standard of reporting to the various types of audit reports.

Know the circumstances that require a departure from an unqualified opinion.

Be able to paraphrase a report form for all types of departures from an unqualified opinion.

Although many audit reports are of the standard short form discussed in Chapter 15, circumstances may arise which make it necessary for the auditor to depart from this standard form. Such departure is provided for in the fourth standard of reporting, which requires that the auditor either express an opinion regarding the financial statements or state that an opinion cannot be expressed, in which case the reasons therefor are to be stated.[1]

AU Section 509.09 of *AICPA Professional Standards,* Volume 1 *(SAS No. 2)* lists the following circumstances that can result in a departure from the standard audit report.

1. The auditor's opinion is based in part on the report of another auditor.

---

[1]For an in-depth discussion of the fourth standard of reporting, see D. R. Carmichael, *The Auditor's Reporting Obligation, Auditing Research Monograph No. 1,* AICPA, 1972.

2. The auditor wishes to emphasize a matter pertaining to the financial statements.

3. The financial statements are affected by a departure from an accounting principle promulgated by a body designated by the AICPA Council to establish such principles.

4. The scope of the auditor's examination is limited with respect to one or more audit procedures considered necessary in the circumstances.

5. The financial statements are affected by a departure from a generally accepted accounting principle.

6. The financial statements are affected by uncertainties concerning future events, the outcome of which cannot be estimated reasonably at the date of the auditor's report.

7. Accounting principles have not been consistently applied.

Departures Other Than Qualifications, Disclaimers, or Adverse Opinions

The first three circumstances listed result in an unqualified opinion even though the report is a departure from the standard audit report.

The situation in which an auditor's opinion is based in part on the report of another auditor was discussed in Chapter 15, and an example was provided. The purpose of referring to another auditor is to set forth clearly the division of responsibility for the performance of the examination. Such reference is not considered a qualification of the audit report. However, if the report of the other auditor is qualified, the principal auditor must determine whether the effect of the qualification on the financial statements he is reporting on is material; if so, he should qualify his report. If the effect of the other auditor's qualification is judged not to be material, then the principal auditor need make no reference in his report to the qualification.

AU Section 509.27 *(SAS No. 2)* permits, but does not require, an auditor to emphasize a matter pertaining to the financial statements by using a separate paragraph in the audit report. Some auditors are reluctant to use this device, because they are required to select the disclosures to be emphasized and there are few guidelines for such selection. Other auditors have found the separate paragraph useful to emphasize related-party transactions, changes in accounting estimates, and changes in operating conditions.

The following example is a separate paragraph used to emphasize a change in an accounting estimate.

(Middle Paragraph)

Effective January 1, 19XX, the Company revised its estimates of the residual values of its equipment and the obsolescence rates of inventories of material and supplies, as discussed in Note 1. This revision reflects primarily a change in conditions and not a change in accounting principles or practices. As a result of this revision, with which we concur, net income for the year ended December 31, 19XX, was reduced by $500,000.

Because the use of a separate paragraph under these circumstances is for the purpose of emphasis rather than disclosure, there is no intention to qualify the auditor's report; accordingly, no reference should be made to the separate paragraph in the opinion paragraph. The use of a phrase such as "with the foregoing explanation" in the opinion paragraph is not appropriate, because it suggests that a qualification may be intended.

Chapter 2 includes a discussion of ET Section 203.01 (Rule 203) of the AICPA *Code of Professional Ethics* (see page 49). This rule severely restricts the conditions under which an auditor may issue an unqualified audit report on financial statements containing a departure from an accounting principle published by a body designated by the AICPA Council to establish such principles. The rule requires the auditor to demonstrate that, because of unusual circumstances, the financial statements would be misleading if the pronouncement of a designated body were followed. AU 509.19 *(SAS No. 2)* requires the use of a separate paragraph to set forth the information needed to comply with Rule 203. Circumstances involving this aspect of reporting are relatively rare. The following illustration demonstrates one such situation. Only the middle paragraph is reproduced.

In October, 1973, the Company extinguished a substantial amount of debt through a direct exchange of new equity securities. Application of *Opinion No. 26* of the Accounting Principles Board to this exchange requires that the excess of the debt extinguished over the present value of the new securities should be recognized as a gain in the period in which the extinguishment occurred. While it is not practicable to determine the present value of the new securities issued, such value is at least $2,000,000 less than the face amount of the debt extinguished. It is the opinion of the Company's Management, an opinion with which we agree, that no realization of a gain occurred in this exchange (Note 1), and therefore, no recognition of the excess of the debt extinguished over the present value of the new securities has been made in these financial statements.[2]

Note 1 states that the terms and conditions of the new equity securities are substantially similar to those of the debt securities extinguished, both on the basis of the company's continuing operations and in the event of liquidation, and that, in the opinion of management, no gain was realized as a result of the exchange.

If departures from the prescribed principles can be justified, as in the foregoing case, it is appropriate for the auditor to express an unqualified opinion on the financial statements.

[2]Hortense Goodman and Leonard Lorensen, *Illustrations of Departures from the Auditor's Standard Report, Financial Report Survey 7,* American Institute of Certified Public Accountants, p. 97.

Departures Resulting in
Qualifications, Disclaimers, or Adverse Opinions

The last four departure circumstances listed at the beginning of this chapter require qualifications, disclaimers, or adverse opinions if the effects are material. These circumstances can be summarized as:

1. Scope limitations.

2. Departures from generally accepted accounting principles, which include inadequate disclosures.

3. Uncertainties.

4. Accounting principles not consistently applied.

Before these circumstances are discussed, the forms of qualifications, disclaimers, and adverse opinions are covered to provide an understanding of the conclusions reached by the auditor and conveyed by him to the readers of his report in each of the circumstances.

When an auditor expresses a qualified opinion, disclaimer of opinion, or an adverse opinion, he must disclose in an explanatory paragraph in his report all substantive reasons for such an opinion, as well as the effect of the matter on the financial statements, if reasonably determinable. If the effect is not determinable, this fact should be stated. The one exception to the requirement for an explanatory paragraph is when the opinion paragraph is modified only because of an inconsistency in the application of accounting principles. Even then, an explanatory paragraph may be, and often is, used.

It is very important that an auditor clearly state in this paragraph any reservations he has about the financial statements, including differences of opinion with his client regarding accounting principles and the possible adverse effects of uncertainties. He must avoid any temptation to use vague wording that might lessen the impact of the qualification for his client. Though it is permissible and often desirable to refer to a footnote to the financial statements for additional details and explanations of the circumstances involved, it is not adequate to refer to a footnote explanation in lieu of an explanatory paragraph.

Qualified Reports

The Use of "Except For" and "Subject To"—The phrases "except for" and "subject to" are used to indicate a qualification of an auditor's opinion.

The phrase "except for" means what the words imply—exception, objection, demurral. The auditor takes exception to, or objects to, some aspects of the financial statements. This qualifying phrase is used if the financial statements contain a departure from generally accepted accounting principles, because the auditor would take exception to such departure. It is also used if accounting principles have not been applied consistently between periods, because the auditor must indicate in his report that there has been an exception to the consistent application of such principles. Finally, the phrase "except for" is used if the scope of an auditor's work has been limited in some manner; the auditor would be objecting to or taking exception to the lack of evidential matter, or to restrictions imposed on the amount of evidential matter he has gathered.

The phrase "subject to" means contingent upon, or with provision or condition. The auditor believes the financial statements are fairly presented, contingent upon the occurrence or non-occurrence of some future and currently undeterminable event. This type of qualification is used if an uncertainty affecting the financial statements cannot be resolved.

The auditor would issue a qualified report only if, in his judgment, the subject of the qualification does or could have a material effect on the financial statements, and if the subject of the qualification does not require a disclaimer or an adverse opinion. Thus the qualifying phrases can be summarized as follows:

Condition

Effect	Scope Limitation	Departure From GAAP	Uncertainty	Inconsistency
Material	Except for	Except for	Subject to	Except for

Apply the foregoing discussion to the following example. An auditor learns that the Internal Revenue Service recently examined the income tax returns of a client and has proposed a deficiency of $200,000 for additional income taxes. Management of the client states that they intend to protest the proposed deficiency and that, though they are willing to make appropriate footnote disclosure of the matter, they are unwilling to record a liability for the proposed deficiency at the present time. The auditor considers the $200,000 to be material in relation to the financial statements. After reviewing the report of the Internal Revenue agent and management's protest to the deficiency, the auditor concludes that the Internal Revenue Service is correct, and that the proposed deficiency will be upheld. What type of qualification, if any, should be included in the auditor's report? (Ignore any other possible implications of this example.)

In this case, a "subject to" qualification would not be appropriate. Sufficient information has been obtained to conclude that an additional liability for income taxes exists and that, without the recording of this liability, the financial statements are not presented fairly in accordance with generally accepted accounting principles. Therefore, an "except for" qualification would be required. Note that footnote disclosure of the matter is not adequate if the financial statements require adjustment.

Assume in the foregoing example that review of the Internal Revenue agent's report and management's proposed protest to the deficiency leaves uncertainty as to the final outcome of the matter. In this case, the auditor is faced with an uncertainty that is not susceptible to reasonable estimation. He is unsure whether or not the financial statements are presented fairly in accordance with generally accepted accounting principles. Consequently, a "subject to" opinion would be appropriate.

Assume further that in the foregoing example the review of the Internal Revenue agent's report and management's proposed protest to the deficiency convinces the auditor that the client would incur little, if any, additional income tax as a result of the examination. This conclusion may be difficult to reach in practice, but if it could be reached, the auditor could issue an unqualified report because his opinion would be that the financial statements are presented fairly in accordance with generally accepted accounting principles.

The Location of the Qualifying Phrase—The location of the qualifying phrase depends upon which financial statement or statements are affected by the qualification. In many cases a qualification will affect all of the basic financial statements, but in other situations only individual statements are affected.

If there has been a departure from generally accepted accounting principles that affects all of the basic financial statements (for example, the failure to record an adequate provision for income taxes in the current year), the qualifying phrase "except for" should be placed at the beginning of the opinion paragraph, immediately after the phrase "In our opinion." In this location, it qualifies the fair presentation in accordance with generally accepted accounting principles of (1) financial position (the balance sheet), (2) results of operations (statements of income and retained earnings), and (3) changes in financial position (statement of changes in financial position).

Assume that the client has improperly charged to retained earnings a loss on sale of securities that did not meet the criteria for a prior-period adjustment set forth in *FASB No. 16*. Obviously, net income for the year is overstated by the amount of the loss, and the financial statements presenting net income (statements of income, retained earnings, and changes in financial position) contain a departure from generally accepted accounting principles. But what is the effect on the balance sheet? The amount shown in the balance sheet as retained earnings is the same whether the loss is charged directly to retained earnings or indirectly through the income statement. If the auditor decides that an "except for" qualification is appropriate (rather than an adverse opinion), he must place the qualifying phrase in the opinion paragraph so that it applies only to results of operations and changes in financial position. An example of such wording follows.

> In our opinion, the accompanying balance sheet presents fairly the financial position of X Company as of December 31, 19XX, and except for the effect of not recording a loss on sale of securities as a reduction of net income as discussed in the preceding paragraph, the accompanying statements of income, retained earnings, and changes in financial position present fairly the results of its operations and the

changes in its financial position for the year ended December 31, 19XX, in conformity with generally accepted accounting principles applied on a basis consistent with that of the preceding year.

As an exercise, draft the opinion paragraph of an audit report for the following conditions. While performing a subsequent review in connection with an audit of Y Company for the year ended December 31, 19XX, the auditor finds that a contractor who is constructing a new administration building for Y Company was late in submitting invoices to Y Company for work completed through the end of the year. As a result, Y Company did not record the liability for this work and refuses to adjust the financial statements at this late date, although it acknowledges a liability. The building is incomplete at the end of the year. What is the effect of this matter on the balance sheet? Is there any effect on the statements of income, retained earnings, or changes in financial position? Draft an adjusting journal entry to record the transaction if this will aid in determining the overall effect on the financial statements. Be sure not to qualify any financial statements that are presented fairly.

Disclaimers of Opinion

A disclaimer of opinion states that the auditor does not express an opinion on the financial statements. The auditor uses disclaimers if (1) there have been limitations on the scope of his examination, or (2) there are uncertainties regarding the financial statements. Disclaimers also are used in connection with unaudited financial statements (covered in the next chapter).

Disclaimer Due to Scope Limitations[3]—In the previous section it was stated that a qualified opinion ("except for") would be used by the auditor in the event of a scope limitation that did not require a disclaimer of opinion. How does the auditor determine whether a disclaimer of opinion or a qualification is appropriate when he encounters a scope limitation? The answer turns on the materiality of the scope limitation and the potential effect on the financial statements. This is another area in which the auditor must apply his judgment. Few guidelines are given in AU Section 509.45–.47 *(SAS No. 2)* and other official pronouncements as to how material an item must be to require a disclaimer rather than a qualification. Probably the best guidance can be provided by an illustration of the two possibilities—one situation requiring a qualification and one requiring a disclaimer. Between these two situations are less clear cases in which the auditor must rely on judgment, intuition, and experience.

Suppose an auditor is engaged to perform an audit of a company after the end of its fiscal year and therefore was not on hand to observe the taking of the year-end physical inventory. Also, the inventory records are not adequate to allow a retroactive verification of the year-end quantities. If the amount of

---

[3] Auditors have encountered legal trouble because of a disclaimer. See David B. Isbell and D. R. Carmichael, "Disclaimers and Liability—The Rhode Island Trust Case," *The Journal of Accountancy,* April, 1973, pp. 37–42.

inventories shown in the balance sheet is $100,000, compared with total current assets of $500,000, total assets of $900,000, total equity of $600,000, and net income for the year of $200,000 (ignore for this purpose the problem of auditing beginning inventories), the maximum potential misstatement on the downside is 20% of current assets, 11% of total assets, 17% of total equity, and 25% of net income for the year. Though the auditor would consider many other factors in evaluating materiality, such as the trend in earnings, the effect on the current ratio, and the possibility of understatement, solely on the basis of the percentages shown he might conclude that the effect would be material enough to require qualification but not so material to the overall financial statements as to require a disclaimer of opinion. In contrast, if the amount of inventories shown in the balance sheet is $450,000 instead of $100,000, the maximum potential effect for misstatement on the downside is 90% of current assets, 50% of total assets, 75% of total equity, and 112% of net income for the year. On the basis of these percentages, the auditor might conclude that the potential for misstatement pervades all of the financial statements to the extent that he is unable to express an opinion. Thus, it would be appropriate to issue a disclaimer of opinion in this case.[4]

Disclaimer Due to Uncertainties—When the Auditing Standards Executive Committee issued an initial exposure draft of *SAS No. 2,* they decided to eliminate disclaimers of opinion for uncertainties. When the official pronouncement was released, however, the use of disclaimers for uncertainties was retained, although it was relegated to a footnote in AU Section 509.25 *(SAS No. 2)*. It is fairly clear that the Committee preferred a qualified opinion and regarded a qualification as being sufficiently informative to users of the financial statements.

For what reason would auditors issue a disclaimer of opinion rather than a "subject to" opinion? The answer, as in the case of scope exceptions, relates to materiality. Some uncertainties are more material than others in that their final resolutions are likely to have a greater impact on the financial statements. For example, an event that threatens to impair a firm's entire asset base is more serious than a possible loss on disposition of an investment that constitutes a relatively small portion of the assets. Though there are no definite guidelines on whether a disclaimer or a qualified opinion should be issued, it is unlikely that some auditors would issue a disclaimer of opinion unless the future existence of the firm were in such jeopardy that the going-concern basis for the recording of the assets was no longer applicable.

Such an event might occur if a law were passed rescinding or greatly modifying the manner in which a company is allowed to operate. Legislation

---

[4]The percentages used in these illustrations are for teaching purposes only and should not be considered applicable in all circumstances in the determination of materiality. See *The Auditor's Reporting Obligation, Auditing Research Monograph No. 1,* written by D. R. Carmichael and published by the AICPA, 1972, pp. 53–66, for a discussion of factors affecting the degree of materiality. Included in this discussion are relative magnitude, probability (including uncertainty of outcome and likelihood of error), and utility (nature of item, pervasiveness, and expertise of auditor in the area).

forbidding racetrack betting certainly would hamper a firm operating such a business. Also, a company's survival prospects might be extremely poor if the market for its products appeared to be diminishing. Auditors might question whether a going-concern assumption is valid for clients in this type of financial condition.

Form of a Disclaimer—The opinion paragraph of a report containing a disclaimer may take the following general format:

> Because of the significance of the [scope limitation or uncertainty] discussed in the preceding paragraph, we are unable to express, and we do not express, an opinion on the accompanying financial statements.

It is important that every report containing a disclaimer of opinion include an explanatory paragraph describing clearly and precisely all significant conditions that gave rise to the disclaimer. The auditor also must disclose any reservations or exceptions with regard to fairness of presentation or consistency in the application of accounting principles. In other words, an auditor may not hide behind a disclaimer of opinion if he is aware of some deficiency in the financial statements resulting from improper application or inconsistent use of accounting principles.

In some cases it may be appropriate to disclaim an opinion on one or more financial statements and to express an unqualified or qualified opinion on others where scope limitations are involved. This approach would be inappropriate, however, in the case of an uncertainty, because the basis for such a disclaimer is that some future effect may be so material as to prevent an expression of opinion on the financial statements taken as a whole.

A form of report referred to as a "piecemeal opinion" (expression of opinion on specific financial statement items such as cash, accounts receivable, or accounts payable following a disclaimer of opinion on the financial statements taken as a whole or an adverse opinion) was permissible in certain circumstances prior to the effectiveness of AU Section 509 *(SAS No. 2)*. Because a piecemeal opinion tended to offset or soften the effect of a disclaimer or adverse opinion, its use is now prohibited.

Adverse Opinions

An adverse opinion is an opinion that the financial statements do not present fairly a company's financial position, results of operations, or changes in financial position in conformity with generally accepted accounting principles. The adverse opinion is used if the financial statements being reported on contain a departure from generally accepted accounting principles so material that it permeates the financial statements taken as a whole. Thus, the auditor's decision of whether an adverse opinion or "except for" qualification is appropriate in the case of a departure from a generally accepted accounting principle rests on the materiality of the amounts involved. This decision is similar to the choice between a disclaimer of opinion and an "except for" qualification in the case of a scope limitation (discussed on page 704).

Every adverse opinion requires an explanatory paragraph that clearly sets forth the subject of or reason for the adverse opinion and the amounts involved or estimated effect on the financial statements, if reasonably determinable. Because an adverse opinion states that the financial statements are not in accordance with generally accepted accounting principles, it is meaningless to refer to consistency, although any inconsistencies should be pointed out in the explanatory paragraph of the audit report.

The opinion paragraph of a report containing an adverse opinion may take the following form.

> In our opinion, because of the effects of [the departure from generally accepted accounting principles] discussed in the preceding paragraph, the financial statements referred to above do not present fairly, in conformity with generally accepted accounting principles, the financial position of X Company as of December 31, 19XX, or the results of its operations and changes in its financial position for the year then ended.

It is obvious from reading the paragraph that an adverse opinion is likely to have a very negative effect on the readers of the opinion and the related financial statements; therefore, such opinions are issued only after all attempts to persuade the client to adjust the financial statements have failed. The only other option available to the auditor in this situation is withdrawal from the engagement. If an auditor's relationship with the client is such that he is unable to persuade the client not to issue financial statements that contain a departure from generally accepted accounting principles so significant that an adverse opinion is required, then a severance of the auditor-client relationship should be considered.

Summary of the Forms of Qualifications, Disclaimers, and Adverse Opinions

The summary on page 702 of the types of qualifications required for material scope limitations, departures from generally accepted accounting principles, uncertainties, and inconsistencies now can be expanded to include circumstances so material as to require a disclaimer or adverse opinion. Although such circumstances have no specific designation in AU Section 509 *(SAS No. 2)* or other official pronouncements, they are referred to as "supermaterial" for teaching purposes in this text.

Condition

Effect	*Scope Limitation*	*Departure From GAAP*	*Uncertainty*	*Inconsistency*
Material	Except for Qualification	Except for Qualification	Subject to Qualification	Except for Qualification
Supermaterial	Disclaimer	Adverse	Subject to Qualification or Disclaimer	Not Applicable

It can be seen from the summary that only one type of inconsistency in the application of accounting principles is recognized and that the "except for" type of qualification is applicable in all inconsistency cases.

Scope Limitations

An auditor should perform all of the auditing procedures he considers necessary to express an unqualified opinion on the financial statements being examined, unless it is impracticable to perform certain procedures or he is instructed specifically by the client to omit or limit certain procedures. Any limitation on the scope of an auditor's work, whether client-imposed or otherwise, constitutes a scope limitation unless the auditor is able to satisfy himself by other means. If, for example, an auditor is engaged by a client after the end of the client's fiscal year and does not observe the year-end physical inventory, he may be able to observe an inventory at a later date and reconcile to the audit date if there are reliable perpetual inventory records and strong internal controls. If the auditor is able to satisfy himself by this means that the inventory as of the end of the year is stated properly, he need not consider the scope of his work to be limited.

In the previous section it was stated that a scope limitation would require either an "except for" qualification or a disclaimer of opinion, depending on the materiality of the potential misstatement. Obviously, a scope limitation also involves a modification of the scope paragraph to qualify the phrase "included such tests of the accounting records and such other auditing procedures as we considered necessary in the circumstances." If there has been a limitation on the scope of an auditor's work, clearly he has not performed the auditing procedures he considered necessary in the circumstances. Therefore, it is necessary to indicate an exception in both the scope and opinion paragraphs for a scope limitation. An example of a modification of the scope paragraph is shown below.

> ... included such tests of the accounting records and such other auditing procedures as we considered necessary in the circumstances, except as explained in the following paragraph.

It would be improper to discuss a scope limitation in a footnote to the financial statements because the limitation applies to the work of the auditor, and the financial statements constitute the representations of management.

A scope limitation may result from the specific request of a client. An example of an audit report covering this circumstance follows (assume that the potential misstatement is supermaterial).

> ... necessary in the circumstances, except as explained in the following paragraph.
>
> As instructed, we did not request confirmation of accounts receivable balances directly from the company's customers as of December 31, 19XX, and we were unable to satisfy ourselves as to the balances by means of other auditing procedures.
>
> Because of the significance of the matter discussed in the preceding paragraph, the scope of our work was not sufficient to enable us to express, and we do not express, an opinion on the accompanying financial statements.

In the example, the words "As instructed" in the explanatory paragraph make clear that the limitation was imposed by the client and was not of the auditor's choosing. Needless to say, an auditor should consider carefully whether he should perform an audit if significant scope limitations have been imposed by a client.

Inadequate accounting records also may give rise to a scope limitation. The following example illustrates a report qualified (assume that a disclaimer is unnecessary) because of inadequate accounting records.

> ... for the year then ended. Except as explained in the following paragraph, our examination was made in accordance with generally accepted auditing standards and, accordingly, included such tests of the accounting records and such other auditing procedures as we considered necessary in the circumstances.
>
> Because of the inadequacy of prior-year records, we were unable to obtain sufficient evidence to form an opinion as to whether at December 31, 19XX, manufacturing equipment ($500,000) is stated at cost and as to the adequacy of the related allowance for depreciation ($150,000) and depreciation expense for the year ($40,000).
>
> In our opinion, except for the effect of such adjustments, if any, as might have been disclosed with respect to manufacturing equipment and the related allowance for depreciation and depreciation expense had prior-year records been adequate, the accompanying financial statements present fairly the. ...

Note that the qualification is placed so that it applies to the statement of income as well as to the balance sheet and statement of changes in financial position. A misstatement of manufacturing equipment also results in a misstatement of depreciation expense for the year, so the income statement should be, and is, properly covered by the qualification.

A scope limitation also may arise in initial audits. In many cases, an auditor who is engaged during a year will be unable to satisfy himself with regard to the quantities or consistent determination of beginning inventory because he was not present to observe it. If he is unable to satisfy himself by other means, and if the potential effect is material, he will have a scope exception because of the effect of beginning inventory on the determination of cost of sales, net income for the year, and consistency. An example of report wording covering this situation follows (assume that the potential effect is supermaterial).

> ...as we considered necessary in the circumstances, except as explained in the following paragraph.
>
> We did not observe the physical inventory ($1,700,000) taken as of December 31, 19X1, because this date was prior to our initial engagement as auditors for the company. Because of inadequacies in the company's accounting records for the previous year, it was not practicable to extend our auditing procedures sufficiently to permit us to make adequate retroactive tests of inventory quantities and pricing or to express an opinion on the consistency of application of accounting principles with that of the preceding year.
>
> In our opinion, the accompanying balance sheet presents fairly the financial position of X Company as of December 31, 19X2, in conformity with generally accepted accounting principles. Because of the significance of the matters discussed in the preceding paragraph, the scope of our work was not sufficient to enable us to express, and we do not express, an opinion on the statements of income and changes in financial position for the year ended December 31, 19X2.

Note that the report is worded so that the scope limitation applies to the statements of income and changes in financial position and consistency; an unqualified opinion is expressed on the balance sheet, because the beginning inventory has no effect on this statement.

As brought out in Chapter 12, letters from the client's legal counsel can be a valuable source of evidence in determining the status of contingent liabilities affecting the financial statements. If the auditors are successful in obtaining an appropriate response from the client's legal counsel, and if all necessary disclosures and adjustments have been made in the financial statements, an unqualified opinion usually can be issued. Unfortunately, the appropriate replies regarding contingent liabilities are not always forthcoming. The legal counsel may not furnish the desired reply about potential liabilities arising from claims or lawsuits, because (1) they do not believe it is proper to divulge information on claims or litigation, or (2) the outcome of the claim or litigation is so uncertain that the legal counsel cannot make an accurate evaluation.

Refusal by an attorney to furnish the information requested from him (see page 529 for an example of an audit inquiry letter to legal counsel) normally would constitute a limitation on the scope of an auditor's examination and would preclude issuance of an unqualified report. A scope limitation has occurred, because the requested information is available, but is being withheld by the attorney. In contrast, the inability of an attorney to respond because the outcome of claims or litigation is uncertain and not subject to reasonable estimation is not a scope limitation, because the information requested (for example, the outcome of a lawsuit) is not available. The latter case is an example of an uncertainty and requires a different type of qualification. Uncertainties are discussed in another section.

Departures From Generally Accepted Accounting Principles

Departures from generally accepted accounting principles require a qualified ("except for") opinion if the effect is material, and an adverse opinion if the effect is supermaterial. Inadequate disclosure is considered a departure from generally accepted accounting principles in AU Section 509.17 *(SAS No. 2)*. The meaning of the term "generally accepted accounting principles" within the context of the auditor's report is discussed in Chapter 15. The reader should refer to that discussion and AU Section 411 *(SAS No. 5)* if the meaning of the term is not understood.[5]

An audit report on financial statements involving a departure from generally accepted accounting principles (assume that an adverse opinion is not required) might be worded as follows.

. . . considered necessary in the circumstances.

As explained more fully in Note 1, the provision for pension expense for the year ($200,000) was less than the minimum amount required by *Opinion No. 8* of the Accounting Principles Board. Had the minimum provision been made, pension expense would have been increased by approximately $150,000 and net income would have been decreased by approximately $75,000 ($.16 per share).

In our opinion, except for the effect of the underprovision for pension expense as described in the preceding paragraph, the accompanying financial statements present fairly. . . .

In the example, it must be assumed that the effect of the unrecorded pension liability is material in relation to the balance sheet, because the qualification applies to it as well as to the statements of income and changes in financial position.

An illustration of an auditor's report containing an adverse opinion as a result of a departure from generally accepted accounting principles is shown below.

. . . considered necessary in the circumstances.

[5]See also, Marshall S. Armstrong, "Some Thoughts on Substantial Authoritative Support," *The Journal of Accountancy,* April, 1969, pp. 44–50.

As set forth in Note 2, land owned by the company is stated in the accompanying balance sheet at appraised value, which is $700,000 in excess of cost. Had this land been stated at cost, in accordance with generally accepted accounting principles, property and equipment and stockholders' equity would be reduced by this amount as of December 31, 19XX. The recording of appraised value has no effect on the statements of income and changes in financial position.

In our opinion, because of the significant effect of recording appraised value as discussed in the preceding paragraph, the accompanying balance sheet does not present fairly the financial position of X Company as of December 31, 19XX, in conformity with generally accepted accounting principles. However, in our opinion, the statements of income and retained earnings and changes in financial position present fairly the results of operations and changes in financial position for the year ended December 31, 19XX, in conformity with generally accepted accounting principles applied on a basis consistent with that of the preceding year.

This example illustrates how a departure from generally accepted accounting principles can affect one rather than all of the basic financial statements.

The requirement for adequate disclosure is stated specifically in the third standard of reporting; however, the auditor must not confuse long and rambling footnotes with adequate disclosure (they actually may make it more difficult for the reader to understand the matter involved). It is very important that the auditor also understand that disclosure is not and cannot be used as a substitute for proper accounting. For example, the failure to record a significant and known liability cannot be remedied by disclosure of the unrecorded liability in a footnote; the financial statements can be corrected only by recording the proper liability.

The topic of auditor's liability included a discussion in Chapter 3 of the *Continental Vending Machine Corporation (Continental)* case. There were several aspects to this case. One of the most important involved the adequacy of the disclosure of balances and transactions between Continental, its president, and an affiliated company. A comparison of the footnote to the financial statements describing these balances and transactions with the statement the prosecution successfully contended should have been made illustrates the necessity for clear and adequate disclosure. Repeated hereafter is the footnote as it appeared in the financial statements.

The amount receivable from Valley Commercial Corporation (an affiliated company of which Mr. Harold Roth is an officer, director and stockholder) bears interest at 12% a year. Such amount, less the balance of the notes payable to that company, is secured by the assignment to the Company of Valley's equity in certain marketable securities. As of February 15, 1963, the amount of such equity at current market quotations exceeded the net amount receivable.[6]

The government contended that the footnote should have read as follows.

[6]Denzil Y. Causey, Jr., *Duties and Liabilities of the CPA,* Bureau of Business Research, The University of Texas at Austin, 1973, p. 239.

The amount receivable from Valley Commercial Corporation (an affiliated company of which Mr. Harold Roth is an officer, director and stockholder), which bears interest at 12% a year, was uncollectible at September 30, 1962, since Valley had loaned approximately the same amount to Mr. Roth, who was unable to pay. Since that date, Mr. Roth and others have pledged as security for the repayment of his obligation to Valley and its obligation to Continental (now $3,900,000, against which Continental's liability to Valley cannot be offset) securities which, as of February 15, 1963, had a market value of $2,978,000. Approximately 80% of such securities are stock and convertible debentures of the Company.[7]

An auditor must avoid any inclination to temper a necessarily harsh disclosure in order to pacify a client.

If the disclosures necessary for a fair presentation are omitted from the financial statements, the auditors' report might be worded in the following manner (assume that an adverse opinion is not required):

> . . . considered necessary in the circumstances.
>
> The Company's pension plan is available to employees over the age of 25. Total pension expense for the year ended September 30, 19XX, was approximately $500,000. The Company's policy is to fund pension cost accrued, including amortization of estimated prior service cost of approximately $300,000 over a 40-year period. The value of vested benefits as of September 30, 19XX, exceeded the total of the pension fund by approximately $1,000,000.
>
> In our opinion, except for the omission of the information in the preceding paragraph, the accompanying financial statements present fairly. . . .

In practice, qualifications and adverse opinions necessitated by inadequate disclosure are rare. Because the auditor must disclose the omitted information in an explanatory paragraph of his report, the client normally prefers to make the disclosure in the financial statements to avoid receiving an audit report that contains the omitted disclosure and takes exception to the fair presentation of the financial statements.

Uncertainties

The term "uncertainties," as used here, refers to some future events or occurrences that are not susceptible to reasonable estimation or that depend upon decisions of parties other than client management. Examples include outcomes of Internal Revenue Service examinations, proceedings of regulatory commissions, and lawsuits, as well as estimates of recoverability of asset values, losses on discontinued operations, and the ability of a company to continue as a going concern. An uncertainty normally requires a qualification ("subject to"), unless

[7]*Ibid.,* p. 240.

it is so significant that it negates an opinion on the financial statements taken as a whole, in which case a disclaimer of opinion may be appropriate.

A careful distinction must be made between an uncertainty and a departure from generally accepted accounting principles. If the value of an asset is indeterminable, as may be the case for a long-term investment security that is not traded publicly and for which no market value can be ascertained, then an uncertainty exists. However, although the amount to be realized ultimately from a marketable security may not be determinable, an excess of carrying value over market value would represent a departure from generally accepted accounting principles.

An example of report wording for an uncertainty regarding the outcome of a tax examination follows.

> . . . considered necessary in the circumstances.
>
> As more fully discussed in Note 1, the Internal Revenue Service has reviewed the Company's federal income tax returns for the years 19XX to 19XX and has proposed the assessment of additional income taxes totaling $1,000,000. The principal item being disputed is the taxability of earnings of certain foreign operations. The proposed assessment is being protested by the Company and no provision has been made for any additional federal income taxes which may arise from this matter.
>
> In our opinion, subject to the effect of the income tax matter referred to in the preceding paragraph, the accompanying financial statements present fairly. . . .

If the uncertainty is so significant as to require a disclaimer of opinion, the following wording could be used.

> . . . considered necessary in the circumstances.
>
> The company incurred a loss of $1,000,000 for the year ended June 30, 19XX, and it discontinued operations of plants that accounted for $5,000,000 of the total sales of $10,000,000 during the year. Management has indicated that there is a serious question concerning the ability of the company to continue operations unless substantial relief can be obtained through deferred payment agreements with creditors. If the company were required to liquidate its assets, it may be unable to realize its investment in inventories, property and equipment, and deferred charges.
>
> Because of the significance of the possible losses on realization of the investments in the assets noted in the preceding paragraph, we are unable to express, and we do not express, an opinion on the accompanying financial statements of X Company.

The foregoing example illustrates the uncertainty of the ability of a company to recover the cost of its assets if it is not operating at least on a break-even basis, or if it may be forced into liquidation.

Accounting Principles Not Consistently Applied

A change in accounting principle, if material, requires a modification of the auditor's report concerning consistency. Although the absence of a qualification or adverse opinion regarding a departure from generally accepted accounting principles implies that a newly adopted accounting principle is generally accepted, concurrence with the change must be stated explicitly by insertion of the expression "with which we concur" after the description of the newly adopted principle. Also, the auditor must consider compliance with AC Section 1051.16 *(APB Opinion No. 20)*, which requires that a change be justified on the basis that it implements a preferable principle.

AC Section 1051.19 *(APB Opinion No. 20)* also prescribes whether or not a change in accounting principle is required or permitted to be applied retroactively by restatement of prior financial statements. If a change is reflected retroactively, there is no inconsistency between the accounting principles used in the financial statements of the year of the change and financial statements of prior years after they have been restated. In this case, the modification of the consistency reference in the opinion paragraph is for the purpose of disclosure of the change rather than for the purpose of expressing an exception. An example of a consistency reference after a change in accounting principle that was applied retroactively would be as follows.

> In accordance with *Statement No. 2* of the Financial Accounting Standards Board, all research and development costs have been expensed as incurred and prior-year results have been restated to expense such costs that were deferred in prior years.
>
> In our opinion, the accompanying financial statements present fairly the financial position of X Company as of December 31, 19X1 and 19X0, and its results of operations and changes in financial position for the years then ended, in conformity with generally accepted accounting principles applied on a consistent basis, after restatement for the change, with which we concur, in the method of accounting for research and development costs as described in the preceding paragraph.[8]

If a change in accounting principle is reflected retroactively, as in the example, the reference in the auditor's report to the change is required only in the year of the change.

If a change in accounting principle is not applied retroactively, there is an inconsistency that requires an exception in the auditor's report. Such an exception might be worded in the following manner (assume that the auditor is reporting only on the year during which the change was made).

> ... and the results of its operations for the year then ended, in conformity with generally accepted accounting principles which, except for the change (with which

---

[8]As discussed in a previous section, the use of an explanatory paragraph in the report is optional for exceptions as to consistency; reference to a footnote explanation is sufficient.

we concur) to the last-in, first-out method of determining inventory costs as indicated in Note 1, were applied on a basis consistent with that of the preceding year.

If an auditor's report covers two or more years and a change in accounting principle occurred in one of these years (and the financial statements were not restated), the reference to the inconsistency must be continued as long as the year of change is included in the years being reported on. If the year of change is the earliest year being reported on, then the reference to the inconsistency is changed to indicate that the inconsistency is not within the periods covered by the report, but with the preceding period. An illustration of such wording would be as follows:

> ... and the results of its operations and the changes in its financial position for the years then ended, in conformity with generally accepted accounting principles consistently applied during the periods subsequent to the change made in 19XX, with which we concur, in the method of accounting for the conversion privilege of its subordinated debentures as described in Note 1.

The following example summarizes the discussion in this section. A change is made in an accounting principle in year 19X3, and the auditor's report covers three years.

Year	Years Covered By Auditor's Report	Reference to Inconsistency Assuming: Restatement	No Restatement
19X3	19X1–X3	... applied on a consistent basis during the periods after restatement for the change. ...	... applied on a consistent basis during the periods, except for the change. ...
19X4	19X2–X4	No reference	... applied on a consistent basis during the periods, except for the change. ...
19X5	19X3–X5	No reference	... applied on a consistent basis during the periods subsequent to the change made in 19X3. ...
19X6	19X4–X6	No reference	No reference

If an auditor has not examined the financial statements of a company for the year preceding the first year covered by his report, he must perform sufficient procedures for the preceding year to satisfy himself as to the consistency of accounting principles between the years.

Reports With Differing Opinions

If an auditor is reporting on financial statements covering two or more years, he may wish to express different opinions in the same report. For example, if a significant uncertainty such as a lawsuit arose in the latest year being reported on, the auditor might express a qualified opinion on the latest year and an unqualified opinion on the previous year.

An example of the reporting format for such a report, taken from *SAS No. 15,* is shown below (explanatory and opinion paragraphs only):

> As discussed in Note X, during 19X2 the company became a defendant in a lawsuit relating to the sale in 19X2 of a wholly owned subsidiary. The ultimate outcome of the lawsuit cannot be determined, and no provision for any liability that may result has been made in the 19X2 financial statements.
>
> In our opinion, subject to the effects on the 19X2 financial statements of such adjustments, if any, as might have been required had the outcome of the uncertainty referred to in the preceding paragraph been known, the financial statements referred to above present fairly the financial position of ABC Company as of December 31, 19X2 and 19X1, and the results of its operations and the changes in its financial position for the years then ended, in conformity with generally accepted accounting principles applied on a consistent basis.

Changes in Updated Opinions

The requirements for updating previously issued opinions on prior-year comparative financial statements was discussed in Chapter 15. Circumstances such as the subsequent resolution of an uncertainty, discovery of an uncertainty in a subsequent period and the subsequent restatement of prior-year financial statements may cause the auditor to issue an updated opinion that differs from the original opinion expressed. In these situations, *SAS No. 15* states that the auditor should express the appropriate opinion and include an explanatory paragraph disclosing (a) the date of the auditor's previous report, (b) the type of opinion previously expressed, (c) the circumstances that caused the auditor to express a different opinion, and (d) that the auditor's updated opinion on the financial statements of the prior period is different from his previous opinion on those statements.

An example of an explanatory paragraph describing a change in an updated opinion (from *SAS No. 15*) is shown hereafter:

> In our report dated March 1, 19X2, our opinion on the 19X1 financial statements was qualified as being subject to the effects on the 19X1 financial statements of such adjustments, if any, as might have been required had the outcome of certain litigation been known. As explained in Note X, the litigation was settled as of

November 1, 19X2, at no material cost to the Company. Accordingly, our present opinion on the 19X1 financial statements, as presented herein, is different from that expressed in our previous report.

Commentary

The significant effect on a company of receiving an audit report containing a qualification, disclaimer, or adverse opinion can be illustrated by reference to the practices of the Securities and Exchange Commission (SEC). Generally, the SEC will not accept for filing under the Securities Act of 1933 or the Securities Exchange Act of 1934 financial statements for which the related auditor's report is modified because of a scope limitation or a departure from a generally accepted accounting principle.[9] The SEC generally will accept modifications of the auditor's report due to inconsistent application of accounting principles and uncertainties; however, there are exceptions to this policy, as specified in *Accounting Series Release No. 115,* which states, ". . . a registration statement under the 1933 Act will be considered defective because the certificate does not meet the requirements of Rule 2-02 of Regulation S–X when the accountant qualifies his opinion because of doubt as to whether the company will continue as a going concern."

The failure of the management of a company to make the required filings under the securities acts may result in the suspension of trading in the company's stock and possible civil and criminal penalties. Thus, the SEC views qualifications, disclaimers, and adverse opinions as serious matters, and it has provided the auditor of companies subject to SEC jurisdiction with a powerful tool to prevent limitations on the scope of his work and departures from generally accepted accounting principles. Most other users of financial statements such as bankers, other credit grantors, and stockholders share the SEC's view as to the serious nature of modifications of the auditor's report.

Chapter 16 References

Armstrong, Marshall S., "Some Thoughts on Substantial Authoritative Support," *The Journal of Accountancy,* April, 1969, pp. 44–50.

Carmichael, D. R., "Auditors' Reports—A Search for Criteria," *The Journal of Accountancy,* September, 1972, pp. 67–74.

Carmichael, D. R., *The Auditor's Reporting Obligation, Auditing Research Monograph No. 1,* AICPA, 1972.

[9]See Rule 2-02 of Regulation S–X and *Accounting Series Release No. 4.*

Epaves, Richard A., Laurence R. Paquette, and Michael A. Pearson, "A Flow Chart Conceptualization of Auditors' Reports on Financial Statements," *The Accounting Review,* October, 1976, pp. 913–916.

Goodman, Hortense, and Leonard Lorensen, *Illustrations of Departures from the Auditor's Standard Report, Financial Report Survey 7,* AICPA.

Isbell, David B., "The Continental Vending Case: Lessons for the Profession," *The Journal of Accountancy,* August, 1970, pp. 33–40.

Isbell, David B., and D. R. Carmichael, "Disclaimers and Liability—The Rhode Island Trust Case," *The Journal of Accountancy,* April, 1973, pp. 37–42.

"Nothing to Hide," *Forbes*, October 1, 1973, pp. 61–62.

Smith, E. Wakefield, and John Shellenberger, "Some Unpublished Audit Reports," *The Journal of Accountancy,* November, 1970, pp. 45–52.

"Well Done," *Forbes,* June 1, 1973, pp. 32, 34.

"When Companies and Auditors Fight—and Switch," *Business Week,* March 2, 1974.

Chapter 16
Questions Taken from the Chapter

16–1. What are the requirements of the fourth standard of reporting?

16–2. State the circumstances listed in AU Section 509.09 *(SAS No. 2)* that can result in a departure from the standard audit report.

16–3. List the departures from the standard audit report that do not result in a qualification, disclaimer, or adverse opinion.

16–4. What is the purpose of referring to another auditor in the audit report?

16–5. Discuss the reasons for and against the use of a separate paragraph in the audit report to emphasize a matter regarding the financial statements.

16–6. Give three examples of matters that might be emphasized in a separate paragraph of the audit report.

16–7. Under what conditions should the phrase "with the foregoing explanation" be used in the opinion paragraph to refer to a matter emphasized in a separate paragraph?

16–8. Under what condition may an auditor issue an unqualified opinion on financial statements containing a departure from an accounting principle published by a body designated by the AICPA Council to establish accounting principles? If this condition is met, what form will the auditor's report take?

16–9. List the four circumstances given in AU Section 509.09 *(SAS No. 2)* that result in a qualification, disclaimer, or adverse opinion.

16–10. Which of the following types of reports require an explanatory paragraph: "except for" qualification, "subject to" qualification, disclaimer of opinion, or adverse opinion? Discuss the information that normally would be included in an explanatory paragraph.

16–11. Explain the meaning and use of the "except for" qualification.

16–12. Explain the meaning and use of the "subject to" qualification.

16–13. Where should a qualifying phrase be placed if it is the auditor's intention to qualify all of the basic financial statements?

16–14. Give an example of a situation in which only the balance sheet and statement of changes in financial position should be qualified, and draft an opinion paragraph with the qualifying phrase properly located to qualify only these two statements.

16–15. What is the purpose of a disclaimer of opinion and when is it used?

16–16. Under what conditions would an auditor issue a disclaimer of opinion rather than an "except for" qualification?

16–17. Discuss the circumstances under which an auditor would issue a disclaimer of opinion rather than a "subject to" qualification.

16–18. Draft the opinion paragraph for a disclaimer of opinion.

16–19.   Discuss the auditor's reporting obligation if the scope of his work was so limited that a disclaimer of opinion was required, but he had reservations about the fairness of presentation of the financial statements.

16–20.   In what cases may it be appropriate for an auditor to issue a partial disclaimer of opinion? Give an example.

16–21.   What is a piecemeal opinion and under what conditions is it appropriate for the auditor to use it?

16–22.   What does an adverse opinion state and when is it used?

16–23.   How does an auditor decide whether to use an "except for" qualification or an adverse opinion?

16–24.   Draft the opinion paragraph for an adverse opinion.

16–25.   Discuss the significance of an adverse opinion to the auditor, his client, and a reader of the audit report.

16–26.   List three conditions or reasons for limitations on the scope of an auditor's work.

16–27.   Discuss the forms an auditor's report may take as the result of a scope limitation.

16–28.   What modification is required of the scope paragraph of an auditor's report if his scope has been limited? Why is this modification necessary?

16–29.   Why would it be improper to discuss a scope limitation in a footnote?

16–30.   Discuss how failure of an auditor to observe the physical inventory at the beginning of a year may give rise to a scope exception.

16–31.   What forms of audit report are required to describe departures from generally accepted accounting principles and when is each used?

16–32.   Why are qualifications and adverse opinions because of inadequate disclosure rare in actual practice?

16–33.   What kind of uncertainties result in a modification of the auditor's report? Give three examples.

16–34.   What forms of audit report are required to describe uncertainties, and when is each used?

16–35.   How does an auditor indicate in his report that a client's newly adopted accounting principle is generally accepted?

16–36.   Discuss the effect on the auditor's reference to consistency of a change in accounting principles that is reflected retroactively and of one that is not.

16–37.   For what periods of time must a reference to inconsistent use of accounting principles be continued in an auditor's report?

16–38.   What is the general policy of the SEC with regard to qualifications, disclaimers, and adverse opinions on financial statements filed under the federal securities acts?

Chapter 16
Objective Questions Taken from CPA Examinations

16–39.   Because an expression of opinion as to certain identified items in financial statements tends to overshadow or contradict a disclaimer of opinion or adverse opinion, it is inappropriate for an auditor to issue
a. A piecemeal opinion.
b. An unqualified opinion.
c. An "except for" opinion.
d. A "subject to" opinion.

16–40.   When expressing a qualified opinion, the auditor generally should include a separate explanatory paragraph describing the effects of the qualification. The requirement for a separate explanatory paragraph does *not* apply when the opinion paragraph has been modified because of
a. A change in accounting principle.
b. Inability to apply necessary auditing procedures.
c. Reclassification of an expense account.
d. Uncertainties.

16–41. Keller, CPA, was about to issue an unqualified opinion on the audit of Lupton Television Broadcasting Company when he received a letter from Lupton's independent counsel. The letter stated that the Federal Communications Commission had notified Lupton that its broadcasting license will *not* be renewed because of some alleged irregularities in its broadcasting practices. Lupton cannot continue to operate without this license. Keller has also learned that Lupton and independent counsel plan to take all necessary legal action to retain the license. The letter from independent counsel, however, states that a favorable outcome of any legal action is highly uncertain. Based on this information, what action should Keller take?

   a. Issue a qualified opinion, subject to the outcome of the license dispute, with disclosure of the substantive reasons for the qualification in a separate explanatory paragraph of his report.

   b. Issue an unqualified opinion if full disclosure is made of the license dispute in a footnote to the financial statements.

   c. Issue an adverse opinion on the financial statements and disclose all reasons therefor.

   d. Issue a piecemeal opinion with full disclosure made of the license dispute in a footnote to the financial statements.

16–42. McPherson Corp. does *not* make an annual physical count of year-end inventories, but instead makes test counts on the basis of a statistical plan. During the year, Mullins, CPA, observes such counts as he deems necessary and is able to satisfy himself as to the reliability of the client's procedures. In reporting on the results of his examination, Mullins

   a. Can issue an unqualified opinion without disclosing that he did not observe year-end inventories.

   b. Must comment in the scope paragraph as to his inability to observe year-end inventories, but can nevertheless issue an unqualified opinion.

   c. Is required, if the inventories were material, to disclaim an opinion on the financial statements taken as a whole.

   d. Must, if the inventories were material, qualify his opinion.

16–43. Trotman, Inc., a manufacturing company, has engaged a CPA to examine its financial statements for the year ended June 30, 19X3. The CPA observed the physical inventory count at June 30, 19X3, but there was no physical inventory taken at June 30, 19X2. The CPA has *not* been able to satisfy himself as to the value of the inventory at June 30, 19X2. Assuming that there were *no* other exceptions in the financial statements, the CPA should

   a. Issue an unqualified opinion on the financial statements taken as a whole but clearly indicate in the scope paragraph of his auditor's report the limitations on his work.

   b. Issue an unqualified opinion on the financial statements taken as a whole; no mention of the scope limitation is necessary.

    c. Disclaim an opinion on the statements of income and retained earnings but issue an unqualified opinion on the balance sheet and the statement of changes in financial position.

    d. Disclaim an opinion on the statements of income, retained earnings, and changes in financial position but issue an unqualified opinion on the balance sheet.

16–44. An auditor's "except for" report is a type of
    a. Adverse opinion.
    b. "Subject to" opinion.
    c. Qualified opinion.
    d. Disclaimer of opinion.

16–45. A CPA will issue an adverse auditor's opinion if
    a. The scope of his examination is limited by the client.
    b. His exception to the fairness of presentation is so material that an "except for" opinion is not justified.
    c. He did not perform sufficient auditing procedures to form an opinion on the financial statements taken as a whole.
    d. Such major uncertainties exist concerning the company's future that a "subject to" opinion is not justified.

16–46. An auditor's opinion exception arising from a limitation on the scope of his examination should be explained in
    a. A footnote to the financial statements.
    b. The auditor's report.
    c. Both a footnote to the financial statements and the auditor's report.
    d. Both the financial statements (immediately after the caption of the item or items which could not be verified) and the auditor's report.

16–47. It is *less* likely that a disclaimer of opinion would be issued when the auditor has reservations arising from
    a. Inability to apply necessary auditing procedures.
    b. Uncertainties.
    c. Inadequate internal control.
    d. Lack of independence.

16–48. On January 2, 19X4, the Retail Auto Parts Co. received a notice from its primary suppliers that, effective immediately, all wholesale prices would be increased 10%. On the basis of this notice, Retail Auto Parts Co. revalued its December 31, 19X3, inventory to reflect the higher costs. The inventory constituted a material portion of total assets; however, the effect of the revaluation was material to current assets but *not* to total assets or net income.

In reporting on the company's financial statements for the year ended December 31, 19X3, in which inventory is valued at the adjusted amounts, the auditor should

a. Issue an unqualified opinion provided the nature of the adjustment and the amounts involved are disclosed in footnotes.

b. Issue a qualified opinion.

c. Disclaim an opinion.

d. Issue an adverse opinion.

16–49. The opinion paragraph of a CPA's report begins: "In our opinion, based upon our examination and the report of the other auditors, the accompanying consolidated balance sheet and consolidated statements of income and retained earnings and of changes in financial position present fairly. . ." This is

a. A partial disclaimer of opinion.

b. An unqualified opinion.

c. An "except for" opinion.

d. A qualified opinion.

16–50. An auditor will express an "except for" opinion if

a. The client refuses to provide for a probable federal income tax deficiency that is material.

b. The degree of uncertainty associated with the client company's future makes a "subject to" opinion inappropriate.

c. He did not perform procedures sufficient to form an opinion on the consistency of application of generally accepted accounting principles.

d. He is basing his opinion in part upon work done by another auditor.

16–51. In determining the type of opinion to express, an auditor assesses the nature of the reporting qualifications and the materiality of their effects. Materiality will be the primary factor considered in the choice between

a. An "except for" opinion and an adverse opinion.

b. An "except for" opinion and a "subject to" opinion.

c. An adverse opinion and a disclaimer of opinion.

d. A "subject to" opinion and a piecemeal opinion.

16–52. Your independent examination of the Day Company reveals that the firm's poor financial condition makes it unlikely that it will survive as a going concern. Assuming that the financial statements have otherwise been prepared in accordance with generally accepted accounting principles, what disclosure should you make of the company's precarious financial position?

a. You should issue an unqualified opinion, but in a paragraph between the scope and opinion paragraphs of your report direct the reader's attention to the poor financial condition of the company.

b. You should insist that a note to the financial statements clearly indicate that the company appears to be on the verge of bankruptcy.

c. You need not insist on any particular disclosure, since the company's poor financial condition is clearly indicated by the financial statements themselves.

d. You should provide adequate disclosure and appropriately modify your opinion because the company does not appear to be a going concern.

16–53. Approximately 90% of Helena Holding Company's assets consist of investments in wholly owned subsidiary companies. The CPA examining Helena's financial statements has satisfied himself that changes in underlying equity in these investments have been properly computed based upon the subsidiaries' unaudited financial statements, but he has not examined the subsidiaries' financial statements. The auditor's report should include

a. An adverse opinion.

b. An "except for" opinion.

c. A "subject to" opinion.

d. A disclaimer of opinion.

16–54. When a client declines to disclose essential data in the financial statements or to incorporate such data in the footnotes, the independent auditor should

a. Provide the necessary supplemental information in the auditor's report and appropriately qualify the opinion.

b. Explain to the client that an adverse opinion must be issued.

c. Issue an unqualified report and inform the stockholders of the improper disclosure in a separate report.

d. Issue an opinion "subject to" the client's inclusion of the essential information in future financial statements.

Chapter 16
Discussion/Case Questions

16–55. You are the auditor of X Co. and during your audit for the year ended September 30, 19X7, you note that approximately 72% of the company's sales were to one customer. In the prior year, no single customer accounted for more than 28%. You believe disclosure of this fact is significant because the loss of this customer could seriously affect future operations of X Co. Management of X Co. refuses to include this information in the financial statements, maintaining that this type of information is not often found in financial statements and that it could have a negative effect on the negotiation of the renewal of a bank loan now in progress.

a. How would you reply to X Co. management with regard to this disclosure?

b. Assume that you are unable to convince X Co. management to make the disclosure that you consider appropriate and draft the audit report that you would issue.

16–56. Y Co., which never has been audited, is to be acquired by one of your present clients and combined on a pooling-of-interest basis. Your client includes 10 years of financial statements in its annual report and has instructed you to audit Y Co. for the previous 10 years so that 10 years of restated financial statements can be included in the annual report again this year. During this audit, you note that Y Co. had substantial gains from sales of long-term investment securities in 19X7 and 19X8. However, the gains in 19X7 were presented as extraordinary items, whereas the 19X8 gains were reported as part of ordinary operations. You find that the different reporting was caused by a change in the criteria for extraordinary items contained in *APB Opinion No. 30,* which became effective in late 19X7. Thus, both transactions were reported properly in conformity with the accounting principles in effect during the respective years, and the inconsistency arose from action taken by the APB, not Y Co.
   a. Will you include a consistency exception in your audit report? Explain your reasoning.
   b. Draft the audit report you will issue, assuming that you have no other reservations regarding the financial statements.

16–57. In connection with your audit of Z Co. for the year ended December 31, 19XX, you find that as a result of an improper cutoff of inventory shipments at the end of the year, approximately $6,000 of sales applicable to the subsequent year was recorded in the current year. During the current year, sales totaled $5,000,000, and net income was $850,000. Management of Z Co. refuses to adjust the financial statements for the $6,000 error.
   a. What position would you take with the management of Z Co. regarding the correction of this error? How would you explain your position?
   b. Draft the audit report you will issue, assuming that you are satisfied with the financial statements otherwise.

16–58. You have been engaged to audit W Co., a new company formed to explore for oil and gas. During the year, W Co. acquired several large undeveloped lease holdings for $1,000,000 and drilled several exploratory wells, all of which were dry. At December 31, 19XX, the financial statements showed total assets of $1,100,000 (including the cost of the undeveloped leases) and net worth of $400,000. In your discussion with the management of W Co. regarding the recoverability of the cost of the leases, they maintain that, despite the unsuccessful efforts to date, the cost of the leases could be recovered, because the geologic features underlying certain of the leases are favorable for the formation of hydrocarbons and a productive oil well recently had been drilled near their leases. They are able to supply you with an opinion of an outside geologist and other documentary support for their position.

a. What type of audit report would you issue to W Co.? Explain your reasons fully.

b. Draft the audit report you will issue, assuming that you are otherwise satisfied with the financial statements.

c. Assume that management believes that the leases probably will not be productive of oil or gas, but insists that they be shown in the financial statements at cost, as this presentation would be in accordance with generally accepted accounting principles. Would this change the type of audit report you would issue? Explain why or why not.

d. Draft the audit report you would issue on the basis of the assumption in c, if it would differ from the report you drafted in b.

16–59. During the current year, your client, U Co., acquired B Co. in a business combination accounted for as a pooling-of-interest. B Co. previously had reported on a fiscal year ending on June 30, whereas U Co. used a year ending December 31. Considerable time and cost would be involved in restating and auditing the prior-year financial statements of B Co. on a December 31 year-end for combination with U Co. on a pooling-of-interest basis. The management of U Co. decides to combine B Co. for the current year, which they believe is most important to readers of the financial statements, but not for prior years, because they do not consider the extra cost justified. You think the financial statements of B Co. are significant in relation to those of U Co. in the current and prior years, and that failure to combine them would make the financial statements misleading when taken as a whole. Current and one-year-prior financial statements are to be included in the annual report.

a. Discuss the propriety of management's position regarding the restatement of current and prior-year financial statements.

b. Explain the reporting problems, if any, in the case described.

c. Draft an appropriate audit report.

16–60. As a normal procedure in all of your audits, you request management of your clients to furnish you a management representation letter including, among other things, representations that the financial statements are presented fairly, that there are no material unrecorded liabilities, etc. One of your clients, T Co., has had a recent change in management near the end of the company's fiscal year, December 31. As you are concluding your audit for the year, you ask the new management to sign the normal management representation letter. They respond that they are unwilling to do so because they do not have full knowledge of the company's affairs for a significant portion of the year being audited.

a. Discuss the reasonableness of the new management's position.

b. Under these particular circumstances, would failure to obtain a management representation letter be considered a scope limitation?

c. Draft the audit report you would issue.

16–61. Your client, S Co., is a major manufacturer of widgets. During the current year, several class-action suits were filed against the company claiming damages as a result of alleged price fixing within the widget industry during the previous three years. The company is contesting the suits, and in connection with the current-year audit, legal counsel for S Co. has furnished you a letter stating that they are optimistic, but still uncertain, as to the final outcome of these suits.

a. Discuss the reporting problem involved in the case described.

b. Draft an audit report covering two years based on the information provided.

16–62. During 19X7, R Co., your client, changed its method, effective January 1, 19X7, of computing inventory cost from the average cost method to the last-in, first-out method. This change had the effect of reducing net income for the year by $300,000 ($.26 per share), a material amount.

a. How can the company justify the change in inventory pricing as required by *APB Opinion No. 20?*

b. Draft an audit report covering two years that you will issue (use an explanatory paragraph), assuming that you are otherwise satisfied with the financial statements.

16–63. Q Co. is a gas distribution company that is regulated by the state public service commission. The commission must approve all rate changes made by Q Co., but Q Co. is allowed to collect higher rates subject to refund in the event they are disapproved subsequently by the commission. During the year ended September 30, 19XX, Q Co. increased its rates to its customers and collected an additional $200,000. You have completed your audit of Q Co. in November, and the state public service commission has not yet approved or disapproved the rate increase.

a. Explain the reporting problem involved in the case.

b. Draft an audit report that you will issue covering two years.

16–64. You recently were engaged to audit the financial statements of P Co., a company that previously has not been audited. During your work, you learn that three years ago a serious fire destroyed most of the company's accounting records, including those substantiating the cost of its property and equipment. You physically observe the major items of property and equipment included in the financial statements, but you are unable to determine when and at what price much of it had been acquired. Property and equipment is a material item in the balance sheet.

a. Describe the reporting problem involved in the case.

b. Draft an audit report covering the matter, assuming that you have no other reservations about the financial statements.

    c. If your report is qualified, how long do you anticipate the same qualification will continue?

16–65.   Roscoe, CPA, has completed the examination of the financial statements of Excelsior Corporation as of and for the year ended December 31, 19X8. Roscoe also examined and reported on the Excelsior financial statements for the prior year. Roscoe drafted the following report for 19X8.

March 15, 19X9

We have examined the balance sheet and statements of income and retained earnings of Excelsior Corporation as of December 31, 19X8. Our examination was made in accordance with generally accepted accounting standards and accordingly included such tests of the accounting records as we considered necessary in the circumstances.

In our opinion, the above mentioned financial statements are accurately prepared and fairly presented in accordance with generally accepted accounting principles in effect at December 31, 19X8.

                        Roscoe, CPA
                        (Signed)

Other Information:

Excelsior is presenting comparative financial statements.

Excelsior does not wish to present a statement of changes in financial position for either year.

During 19X8, Excelsior changed its method of accounting for long-term construction contracts and properly reflected the effect of the change in the current year's financial statements and restated the prior year's statements. Roscoe is satisfied with Excelsior's justification for making the change. The change is discussed in Footnote No. 12.

Roscoe was unable to perform normal accounts receivable confirmation procedures, but alternate procedures were used to satisfy Roscoe as to the validity of the receivables.

Excelsior Corporation is the defendant in a litigation, the outcome of which is highly uncertain. If the case is settled in favor of the plaintiff, Excelsior will be required to pay a substantial amount of cash, which might require the sale of certain fixed assets. The litigation and the possible effects have been properly disclosed in Footnote No. 11.

Excelsior issued debentures on January 31, 19X7, in the amount of $10,000,000. The funds obtained from the issuance were used to finance the expansion of plant facilities. The debenture agreement restricts the payment of future cash dividends to earnings after December 31, 19Y3. Excelsior declined to disclose this essential data in the footnotes to the financial statements.

Required:

Consider all facts given and rewrite the auditor's report in acceptable and complete format, incorporating any necessary departures from the standard (short-form) report.

Do not discuss the draft of Roscoe's report but identify and explain any items included in "Other Information" that need not be part of the auditor's report.

(AICPA adapted)

16–66.   Charles Burke, CPA, has completed field work for his examination of the Willingham Corporation for the year ended December 31, 19X8, and now is in the process of determining whether to modify his report. Presented below are two independent, unrelated situations which have arisen.

Situation I

In September, 19X8, a lawsuit was filed against Willingham to have the court order it to install pollution-control equipment in one of its older plants. Willingham's legal counsel has informed Burke that it is not possible to forecast the outcome of this litigation; however, Willingham's management has informed Burke that the cost of the pollution-control equipment is not economically feasible and that the plant will be closed if the case is lost. In addition, Burke has been told by management that the plant and its production equipment would have only minimal resale values and that the production that would be lost could not be recovered at other plants.

Situation II

During 19X8, Willingham purchased a franchise amounting to 20% of its assets for the exclusive right to produce and sell a newly patented product in the northeastern United States. There has been no production in marketable quantities of the product anywhere to date. Neither the franchisor nor any franchisee has conducted any market research with respect to the product.

Required:

In deciding the type-of-report modification, if any, Burke should take into account such considerations as follows:
—Relative magnitude
—Uncertainty of outcome
—Likelihood of error
—Expertise of the auditor
—Pervasive impact on the financial statements
—Inherent importance of the item

Discuss Burke's type-of-report decision for each situation in terms of the foregoing and other appropriate considerations. Assume that each situation is adequately disclosed in the notes to the financial statements. Each situation should be considered independently. In discussing each situation, ignore the others. *It is not necessary for you to decide the type of report that should be issued.*

(AICPA adapted)

16–67. The M & N Manufacturing Company, Inc., Nashville, Tennessee, produces various parts for use in automobiles. It has net sales of about $10,000,000 annually. Net earnings after federal income taxes are approximately 5% of net sales.

The condensed balance sheets of the Company at December 31, 19X1, and December 31, 19X0, are as follows:

Condensed Balance Sheets
M & N Manufacturing Company, Inc.

December 31

Assets	19X1	19X0
Cash	$1,030,000	$ 820,000
Trade accounts receivable	450,000	350,000
Inventories	250,000	175,000
Prepaid expenses	20,000	25,000
Accounts receivable—officers	20,000	5,000
Property, plant, and equipment—net	890,000	800,000
Patents and goodwill	110,000	100,000
	$2,770,000	$2,275,000
Liabilities		
Notes payable to banks	$ 50,000	$ 150,000
Trade accounts payable	100,000	150,000
Accrued expenses	20,000	25,000
Federal income taxes	450,000	450,000
Long-term debt	200,000	—
Common stock	525,000	450,000
Capital in excess of par value of shares	75,000	50,000
Retained earnings	1,350,000	1,000,000
	$2,770,000	$2,275,000

You previously examined the financial statements of this Company for the year ended December 31, 19X0. The Company uses generally accepted accounting principles.

Certain additional information about the Company for 19X1 follows:

1. Provision for depreciation for the year was $100,000. This is significantly higher than last year, because the Company has changed from the straight-line method to an accelerated method for both book and tax purposes on the current year's additions.

2. The net income after income taxes was $500,000.

3. Cash dividends paid during the year were $150,000.

4. An additional 7,500 shares of common stock, par value $10, were sold for $100,000.

5. There were certain notes to the financial statements.

Required:
Prepare the accountants' report for this Company for the year ended December 31, 19X1. Assume that the financial statements covered by the report are a balance sheet and statements of income and retained earnings, other changes in stockholders' equity, and changes in financial position. You finish the field work on February 7, 19X2.

(Used with permission of Ernst & Ernst)

16–68. The field work on Triple Steel Corporation has been completed. You were responsible for preparing the preliminary draft of the financial statements and accountants' report; however, a staff accountant assisted you. You had the draft statements and report typed for presentation to the supervisor, Bob Fisher. You have worked for Bob before and know he demands the very best work.

You are now reviewing the draft to make sure it is final in all respects.

Required:
a. Review the following financial statements and accountants' report. Note any changes that will have to be made to them before you submit them to the supervisor. Changes may be required because of omissions (e.g., inadequate disclosure) or carelessness in drafting the report (e.g., different amounts appearing for the same item in different places). It will not be necessary to foot the statements.
b. List any matters that you want to be sure have been investigated and discussed with management.

(Used with permission of Ernst & Ernst)

ACCOUNTANTS' REPORT

Board of Directors
Triple Steel Corporation
Detroit, Michigan

We have examined the financial statements of Triple Steel Corporation for the year ended June 30, 19X2. Our examination was made in accordance with generally accepted auditing standards, and accordingly included such tests of the accounting records and such other auditing procedures as we considered necessary in the circumstances. We previously made a similar examination of the financial statements for the preceding year.

In our opinion, the accompanying balance sheet and statements of operations and retained earnings present fairly the financial position of Triple Steel

Corporation at June 30, 19X2, and the results of its operations for the year then ended, in accordance with generally accepted accounting principles.

Detroit, Michigan
August 10, 19X2

BALANCE SHEET
TRIPLE STEEL CORPORATION

	JUNE 30	
	19X2	*19X1*
ASSETS		
CURRENT ASSETS		
Cash and certificates of deposit	$ 3,548,583	$ 3,447,058
Trade receivables, less allowance		
of $30,000	1,465,410	2,326,510
Inventories—Notes A and C:		
Finished products	1,671,230	1,146,555
Work in process	1,068,312	1,038,238
Materials and supplies	914,173	879,668
	3,653,715	3,064,461
Prepaid taxes and insurance	340,290	420,086
TOTAL CURRENT ASSETS	9,007,998	9,258,115
OTHER ASSETS		
Cash value of life insurance	136,298	127,618
Miscellaneous deposits and accounts	203,342	133,060
	339,640	260,678
PROPERTIES—on the basis of cost—Note B		
Land	815,130	721,161
Buildings	3,886,339	3,802,183
Machinery and equipment	20,965,042	19,710,505
	25,666,511	24,233,849
Less allowances for depreciation and		
amortization	12,595,431	11,645,132
	13,071,080	12,588,717
	$22,418,718	$22,107,510

BALANCE SHEET
TRIPLE STEEL CORPORATION

	JUNE 30	
	19X2	*19X1*
LIABILITIES AND STOCKHOLDERS' EQUITY		
CURRENT LIABILITIES		
Accounts payable	$ 1,278,981	$ 1,103,641
Payrolls and amounts withheld therefrom	229,594	256,010
Taxes, other than income taxes	342,333	241,339
Pension plan contributions and other current liabilities	210,939	222,635
Federal income taxes, less U.S. Government securities of $495,801 in 19X1	130,488	185,050
Current portion of long-term debt	400,000	400,000
TOTAL CURRENT LIABILITIES	2,592,335	2,408,675
LONG-TERM DEBT		
5½% First Mortgage Sinking Fund Bonds—Note C	4,000,000	4,400,000
DEFERRED FEDERAL INCOME TAXES	32,000	–0–
STOCKHOLDERS' EQUITY—Notes D, E, and F		
Common stock, $2.50 par value:		
Authorized 500,000 shares		
Issued and outstanding 357,419 shares		
(1971—357,024)	893,548	892,560
Additional paid-in capital	2,842,772	2,835,016
Retained earnings	12,058,063	11,571,259
	15,794,383	15,298,835
	$22,418,718	$22,107,510

See notes to financial statements.

STATEMENT OF OPERATIONS AND RETAINED EARNINGS
TRIPLE STEEL CORPORATION

	YEAR ENDED JUNE 30	
	19X2	*19X1*
Net sales	$21,643,276	$24,019,044
Interest and other income	151,213	203,998
	21,794,489	24,223,042
Costs and expenses:		
Cost of products sold, exclusive of depreciation	17,467,856	18,324,089
Provision for depreciation of properties	1,397,773	1,274,509
Selling, administrative, and general expenses	777,850	763,246
Interest on long-term debt	264,000	286,000
Federal income taxes	900,000	1,780,000
	20,807,479	22,427,844
NET INCOME FOR THE YEAR	987,010	1,795,198
Retained earnings at July 1	11,571,260	10,275,441
	12,558,270	12,070,639
Cash dividends paid—$1.40 a share	500,206	499,380
RETAINED EARNINGS AT END OF YEAR	$12,058,064	$11,571,259

STATEMENT OF SOURCE AND APPLICATION OF FUNDS
TRIPLE STEEL COMPANY
June 30, 19X2

SOURCE OF FUNDS	
From operations:	
Net income	$ 987,010
Provision for depreciation	1,397,773
Increase in deferred income taxes	32,000
TOTAL FROM OPERATIONS	2,416,783
Proceeds from sale of common stock	8,744
Decrease in net current assets	433,777
TOTAL	$2,859,304
APPLICATION OF FUNDS	
Additions to property, plant, and equipment	1,880,136
Increase in other assets	78,962
Cash dividend	500,206
Reduction in long-term debt	400,000
TOTAL	$2,859,304

See notes to financial statements.

NOTES TO FINANCIAL STATEMENTS
TRIPLE STEEL CORPORATION
June 30, 19X2

Note A—Inventory. Inventories have been stated at the lower of cost or market prices. Cost as to raw material content of certain inventories, in the amount of $2,568,882 at June 30, 19X2, has been determined by the last-in, first-out method and, as to the remainder of the inventories, has been determined by methods that are substantially equivalent to the first-in, first-out method.

Note B—Pensions. The Company has pension plans covering substantially all employees. The total pension expense for 19X2 and 19X1 was $110,000 and $115,000, respectively, including amortization of prior service cost over a period of 40 years. The Company's policy is to fund pension cost accrued. Unfunded prior service cost under the plans was approximately $350,000 at June 30, 19X2.

Note C—First Mortgage Bonds. The Corporation is required to pay $400,000 annually on the First Mortgage Bonds until retirement on June 30, 19X3.

All properties (with minor exceptions), life insurance policies, and certain inventories ($2,687,320) are pledged in connection with the Bonds.

Note D—Common Stock Optioned to Officers and Employees. Certain officers and key employees have been granted options entitling them to purchase shares of common stock at $19.50 and $29.875 per share, which represented the fair market value on the dates of grant. At July 1, 19X1, options were outstanding for 941 shares. During 19X2, options for 393 shares were exercised for an aggregate option price of $8,744. At June 30, 19X2, options for 546 shares were outstanding and exercisable for an aggregate option price of $13,639. No shares were available for additional options at the beginning or end of the year.

The excess of proceeds over par value of the Common Stock sold during 19X2, amounting to $7,755, has been credited to additional capital and comprised the only change in that account.

Note F—Restrictions on Dividends. The indenture relating to the First Mortgage Bonds requires the Company to maintain net current assets, as defined, of not less than $3,000,000 and otherwise restricts the declaration of dividends, other than stock dividends. At June 30, 19X2, retained earnings of $1,549,391 were free of such restrictions.

Chapter 17

Special Types of Reports

Learning Objectives *After reading and studying the material in this chapter, the student should*

Know the types of reports that qualify as special reports.

Be able to paraphrase a report for each type of special report.

Understand the responsibility assumed by the CPA when unaudited financial statements are issued.

Understand the purpose and major sections of comfort letters, long-form reports, and interim reports.

In the two preceding chapters the form of audit report applicable to general-purpose financial statements is discussed. This form, generally recognized by readers of published annual reports, applies to all financial statements that purport to show financial position, results of operations, and changes in financial position. There are, however, other types of audit reports less familiar to the general public, some of which are referred to as special reports or special purpose reports.

SAS No. 14 defines special reports or special-purpose reports as those issued in connection with (a) financial statements that are prepared in accordance with a comprehensive basis of accounting other than generally accepted accounting principles, (b) specified elements, accounts, or items of a financial statement, (c) compliance with aspects of contractual agreements or regulatory requirements related to audited financial statements, and (d) financial statements presented in prescribed form, or schedules that require a prescribed form of auditor's report. This chapter contains discussions of some of these types of reports, as well as long-form reports, letters for underwriters, and reports applicable to unaudited financial statements, forecasts, and interim statements.

Unaudited Financial Statements

Many small companies need assistance in the preparation of financial statements. CPAs provide a useful service by rendering such assistance. They must be certain, however, that the client fully understands the limitations involved in the preparation of unaudited financial statements. AU Section 516.02 *(SAS No. 1),* as amended by *SAS No. 2,* includes the following comments regarding unaudited financial statements:

> Financial statements are unaudited if the certified public accountant (a) has not applied any auditing procedures to them or (b) has not applied auditing procedures which are sufficient to permit him to express an opinion concerning them as described in Section 509. The certified public accountant has no responsibility to apply any auditing procedures to unaudited financial statements.

Section (a) of the foregoing description of unaudited financial statements is self-evident, but Section (b) requires an exercise of judgment. Remember from the previous chapter that an audit can be performed even though it is impracticable for the auditor to perform certain audit procedures, or he is instructed specifically by the client to omit or limit certain procedures. Such circumstances normally result in a qualified (except for) opinion or disclaimer of opinion. At what point in the performance of selected audit procedures does an auditor's work move from preparation of unaudited financial statements to an audit engagement? A primary factor is the purpose or intent of the engagement. If an expression of the auditor's opinion on the financial statements is contemplated, then an audit would normally be presumed. On the other hand, assistance in the preparation of financial statements would generally indicate that no audit is to be performed. The auditor must bear in mind that his original intent may be modified by circumstances encountered or procedures performed during the engagement.[1]

Arrangement letters specifying the services to be performed are indispensable if the preparation of unaudited financial statements is to be undertaken. The lack of such a letter was one of several issues involved in the *1136 Tenants' Corp.* case discussed in Chapter 3. Another issue that had serious implications for CPAs preparing unaudited financial statements was the holding by the court that the CPAs had a duty to investigate "suspicious" circumstances noted during their work, even though an audit was not being performed. This holding conflicts directly with the statement in AU Section 516.02 *(SAS No. 1)* that a CPA has no responsibility to perform any auditing procedures on unaudited financial state-

---

[1]Alan J. Winters summarizes the state of this service in practice in "Unaudited Statements: Review Procedures and Disclosures," *The Journal of Accountancy,* July, 1976, pp. 52–59.

ments. There are few, if any, engagements in which a CPA does not encounter circumstances that could be considered, at least by hindsight, unusual, questionable, or suspicious. Therefore, a significant expansion of the CPA's work is necessary to meet the guidelines of the court.[2] For this and other reasons (low fees, desire of the staff for more challenging work, etc.), some accounting firms discourage clients seeking preparation of unaudited financial statements. Others have found this service to be an important element in the development of their professional practice.

When the preparation of unaudited financial statements is undertaken, the CPA is well advised to develop a reasonable understanding of the business operations of his client. This knowledge will allow him to make overall evaluations as to the reasonableness of the financial data he prepares. Such evaluations cannot assure that the financial statements are presented fairly; only a complete audit could do that. But they are more desirable than merely copying numbers from a general ledger to a financial statement without an understanding of what they mean and what relationships should reasonably be expected.

A CPA must attach a disclaimer of opinion to any unaudited financial statements with which he is associated. Association is defined broadly to include the preparation of, or assistance in the preparation of, any such statements, whether or not the CPA's name appears with the statements. In other words, a CPA cannot avoid attaching a disclaimer to unaudited financial statements which he prepares by omitting any mention of his name and instructing the client not to say that he prepared them. This requirement protects the client from an untenable position if he is questioned about the preparation of the statements.

An example follows of the form that a disclaimer of opinion on unaudited financial statements might take.[3]

The accompanying balance sheet of X Co. as of December 31, 19XX, and the related statements of income, retained earnings, and changes in financial position for the year then ended have been prepared from the accounts of the Company. We have not audited these financial statements and, therefore, express no opinion on them.

In addition to attaching a disclaimer to the financial statements, the CPA should make sure that each page of the financial statements is marked clearly as unaudited.

AU Section 516.05 *(SAS No. 1)* also provides for preparation of financial statements for a client's internal use. Such statements may omit certain footnotes and disclosures that otherwise would be considered necessary. However, the

---

[2]For more explanation of the auditors' expanded responsibility in this area, see Emanuel Saxe, "Accountants' Responsibility for Unaudited Financial Statements," *The New York Certified Public Accountant* (now *The CPA Journal*), June, 1971, pp. 419–423.

[3]It should be noted that the wording of this type of disclaimer is different from that used in the case of a scope limitation on an audit. See Chapter 16.

CPA must add wording to his disclaimer restricting the use of the financial statements to those within the company and stating that all necessary disclosures have not been made.

In some cases, for his own satisfaction, a CPA may wish to perform certain steps in connection with the preparation of unaudited financial statements that might be considered auditing procedures. For example, he may wish to review bank reconciliations, tax returns, and other documents supporting amounts shown in the financial statements. However, he should make no reference to these steps in his report. To describe any procedures performed might cause the reader to believe that an audit was made.

A CPA may detect errors in the data used to prepare financial statements even though he does not perform an audit. For example, he may note that incorrect prices were used to compute inventory or that certain necessary accrued liabilities for such items as payroll or income taxes have been omitted. In this situation, he should insist that the accounting records and financial statements be corrected. If the client refuses to make the necessary corrections, the CPA must either state his reservations regarding the financial statements in his disclaimer or withdraw from the engagement. An example of the wording that might be added to a disclaimer is shown below.

> The accompanying financial statements do not include $25,000 of income tax expense and liability in the statement of income and balance sheet, respectively, as required by generally accepted accounting principles.

It would not be appropriate to issue a qualified or adverse opinion, because no audit has been performed and without an audit, no opinion—unqualified, qualified or adverse—can be expressed.

Closely related to the situation presented by unaudited financial statements is the situation in which a CPA is not independent. A CPA is not prohibited from being an officer, director, or stockholder of a company, although many accounting firms place restrictions on such relationships. As pointed out in Chapter 2, however, any of these relationships, plus several others, result in a loss of independence, and without independence there can be no audit. No matter how extensive the procedures a CPA performs, they cannot be in accordance with generally accepted auditing standards if the CPA lacks independence. Accordingly, if a CPA who is not independent is associated (as previously defined) with financial statements, he must issue a disclaimer and specifically state that he is not independent (although reasons for lack of independence are not to be given). The form recommended by AU Section 517.03 *(SAS No. 1)* follows.

> We are not independent with respect to XYZ Company, and the accompanying balance sheet as of December 31, 19XX, and the related statements of income and retained earnings and changes in financial position for the year then ended were not audited by us; accordingly, we do not express an opinion on them.

When a CPA lacks independence, the requirements to state any reservations about the financial statements, to omit a description of any work performed, and to mark all pages as unaudited are the same as those discussed for unaudited financial statements.

Reports on Financial Statements Prepared in Accordance with a Comprehensive Basis of Accounting Other than Generally Accepted Accounting Principles

A comprehensive basis of accounting other than generally accepted accounting principles is defined in AU Section 621 *(SAS No. 14)* as one with at least one of the following characteristics:

    a.   A basis of accounting required to comply with the regulations of a government regulatory agency having jurisdiction over an entity.

    b.   A basis of accounting used for income tax reporting purposes.

    c.   The cash basis of accounting.

    d.   A basis of accounting having "substantial support" [the term "substantial support" was not defined in *SAS No. 14* and this resulted in disagreement among certain members of the AudSEC Committee] such as the price-level basis of accounting described in APB Statement No. 3.

If the financial statements being reported on have at least one of the aforementioned characteristics, the scope paragraph should identify the statements and state whether or not generally accepted auditing standards were applied. Also, an explanatory paragraph should state or refer to a note to the financial statements that describes the basis of accounting employed and how it differs from generally accepted accounting principles (the monetary effect need not be stated). Furthermore, a statement must be made that the financial statements are not intended to be presented in accordance with generally accepted accounting principles. Finally, the opinion paragraph should state the auditor's opinion as to whether or not the financial statements are presented fairly in conformity with the basis of accounting described in the explanatory paragraph or footnote, on a basis consistent with that used in the prior period.

The cash basis of accounting will be considered as an example. Cash-basis financial statements are considered a form of presentation as outlined in the preceding paragraph. These statements are prepared on the basis of cash receipts and disbursements without adjustment for uncollected assets (e.g., accounts and interest receivable) or unpaid liabilities (e.g., accounts and accrued interest payable). If the omitted assets or liabilities are material, financial statements prepared on a cash basis would not be in accordance with generally accepted accounting principles. In certain situations, however, cash-basis financial statements are necessary, such as when a company uses the cash basis for income tax purposes or when a partnership is required by its articles of partnership to compute and distribute income to its partners on a cash basis.

When a CPA audits cash-basis financial statements, the form of his report depends on whether or not the statements purport to represent general-purpose financial statements (i.e., present financial position, results of operations, and changes in financial position). Financial statements not considered to be of general purpose are referred to as statements of assets and liabilities arising from cash transactions (rather than balance sheets) and statements of cash receipts and disbursements (rather than statements of income).

If cash-basis statements purport to represent general purpose financial statements, the reporting requirements discussed in Chapters 15 and 16 are applicable, i.e., (1) an unqualified opinion may be expressed (with disclosure of the basis of reporting) if the cash basis is not materially different from the accrual basis, (2) a qualified opinion ("except for" because of the departure from generally accepted accounting principles) should be expressed if the difference is material, and (3) an adverse opinion should be expressed if the difference is supermaterial.

If the cash-basis statements do not purport to represent general-purpose financial statements, they are deemed to be special reports. Disclosure should be made of the basis on which the financial statements were prepared and the nature of omitted assets and liabilities. A report on such statements might be worded as follows.

> We have examined the statement of assets, liabilities and capital arising from cash transactions of A Co. as of December 31, 19XX, and the related statements of cash revenues received and expenses paid and changes in partners' capital accounts for the year then ended. Our examination was made in accordance with generally accepted auditing standards and, accordingly, included such tests of the accounting records and such other auditing procedures as we considered necessary in the circumstances.
>
> As further described in Note 1 and as provided in the partnership agreement, revenue and expenses are recognized only when cash is received or paid, and receivables, payables, and inventories are not included in the accompanying financial statements. Accordingly, these statements do not purport to present financial position and results of operations in conformity with generally accepted accounting principles.
>
> In our opinion, the accompanying financial statements present fairly the assets, liabilities, and capital arising from cash transactions of A Co. as of December 31, 19XX, and the cash revenues received and expenses paid and changes in the partners' capital accounts for the year then ended, on the basis of accounting described above, which basis has been applied in a manner consistent with that of the preceding year.

The general standards and standards of field work are applicable to audits of cash-basis financial statements.

Reports on Specified Elements, Accounts, or Items of a Financial Statement

The need for an audit report on specific elements of a financial statement could arise from provisions of franchise, lease, and royalty agreements (where payments are based on sales, production, etc.), or provisions of agreements for business combinations (where the value of the acquired company may be based in part on its net current assets).

The auditor must approach such engagements with care, because materiality must be considered within the context of the individual element or account being reported on and not the financial statements as a whole. If the engagement is not performed in connection with an examination of the financial statements, the auditor should also be aware that it may be necessary to examine several or even most of the accounts in order to express an opinion on only one. For example, a report on inventories might require the auditor to examine cash, cost of sales, accounts payable, and perhaps other accounts. He must also be certain that he is not placed in the position of interpreting rather than reporting compliance with legal agreements or contracts (e.g., a royalty agreement may not specify whether sales are to be adjusted for returns and allowances, bad debts, other income, etc.). An opinion of legal counsel should be requested for any ambiguous provisions.

An example of such a report, taken from AU Section 621 *(SAS No. 14)*, relating to the amount of sales for purposes of computing rental is as follows:

We have examined the schedule of gross sales (as defined in the lease agreement dated March 4, 19XX, between ABC Company, as lessor, and XYZ Stores Corporation, as lessee) of XYZ Stores Corporation at its Main Street store, [City], [State], for the year ended December 31, 19XX. Our examination was made in accordance with generally accepted auditing standards and, accordingly, included such tests of the accounting records and such other auditing procedures as we considered necessary in the circumstances.

In our opinion, the schedule of gross sales referred to above presents fairly the gross sales of XYZ Stores Corporation at its Main Street store, [City], [State], for the year ended December 31, 19XX, on the basis specified in the lease agreement referred to above.

The report above identifies the account examined, states that the examination was made in accordance with generally accepted auditing standards, identifies the basis on which the information is presented (in this case, the lease agreement), and expresses an opinion that the account is fairly presented on the basis indicated.

Long-Form Reports

Long-form audit reports contain not only the standard scope and opinion paragraphs, but also details of amounts shown in the financial statements and explanatory comments regarding these amounts. For example, such a report might contain a section on accounts receivable including (1) analyses by type, age, and major customers compared by amount and percentage with those of the prior year; (2) an analysis of the allowance for doubtful accounts for the year with ratio analyses of total amount written off to total revenues, total accounts receivable balance more than 90 days old to balance in the allowance account, etc., and (3) a description of the more significant audit procedures performed in the accounts receivable area. Similar comments and analyses would be included for all significant balance sheet and income statement accounts.

The inclusion of such details opens the possibility that a reader will misunderstand the responsibility the auditor is assuming for the data in the long-form report. Remember that the auditor's opinion relates to the financial statements taken as a whole and not to individual accounts. Therefore, in commenting on individual accounts, he must be careful that his wording does not convey more assurance than he is prepared to give.

The long-form report normally begins with the scope and opinion paragraphs of the short-form report to establish the auditor's responsibility for the basic financial statements. In addition, the following paragraph, which disclaims responsibility for the supplemental data, may be included. (The CPA also may accept some responsibility for the supplemental data.)

> Our examination has been made primarily for the purpose of forming the opinion stated in the preceding paragraph. The data in Schedules 1 through 6 are not considered necessary for a fair presentation of financial position, results of operations, and changes in financial position in accordance with generally accepted accounting principles, but are presented solely as supplemental information. The data in those schedules have been summarized from the company's records and have not been subjected to the audit procedures applied in the examination of the basic financial statements. Accordingly, we express no opinion on such data.

Long-form reports are not nearly as common today as they were several years ago. At one time, the annual long-form auditor's report was the only reliable financial information available to many managers and stockholders, because internal financial statements were not prepared. Today, however, few companies attempt to operate without periodic, timely, and reliable financial information. For such companies, the long-form report would be a needless duplication of effort. Also, because of the continuously rising cost of professional auditing services, a long-form report would be much more expensive to prepare today than it was several years ago. The combination of reduced need and increased expense has diminished the use of long-form reports significantly.

Negative Assurance

When an auditor issues a standard short-form report, he expresses the type of opinion that constitutes positive assurance with regard to the related financial statements. In contrast, negative assurance involves double negatives to state *"nothing* came to our attention to indicate that the financial statements are *not* fairly presented."* The accounting profession limits the use of negative assurance to letters for underwriters in connection with SEC registrations and certain special reports. Many stockholders and others might not understand the significant difference between positive and negative assurance if both were used to report on general-purpose financial statements. To appreciate the implications of a negative assurance, one must remember that the less work an auditor performs, the stronger negative assurance (that he knows of nothing wrong with the financial statements) he could give. In other words, the more ignorant he is about the financial statements, the surer he is that he knows of no reason why the statements are not presented fairly.

Despite the problems involved with negative assurance, in certain situations (1) it is the only practicable form of reporting to use (letters for underwriters), or (2) it serves a useful function (special reports). Both of these situations are discussed in the following sections.

Letters for Underwriters

Before considering the form of letters for underwriters (also commonly referred to as "comfort letters"), one must understand their purpose. They are *not* required by any of the federal securities acts, and copies are *not* filed with the SEC. They may be, and usually are, requested by underwriters (normally, investment bankers assisting a company in the sale of securities either by purchase for subsequent resale or on a commission basis) as one means of meeting their responsibilities under the Securities Act of 1933. In general, this act provides that certain persons (including underwriters) may limit or avoid liability under the act if, *after reasonable investigation,* they had *reasonable grounds to believe* and did believe that the registration statement (a document filed with the SEC containing a detailed description of a company and its operations, including financial statements) was true and no material facts were omitted. A comfort letter is considered one element of a reasonable investigation by the underwriters.

Comfort letters have been a source of conflict between auditors and underwriters and their attorneys for many years. For their own protection, underwriters would like to expand the data covered in comfort letters, whereas auditors have tried to restrict the data covered by such letters. *SAP No. 48* (later codified into *SAS No. 1*) was issued in 1971 in an attempt to resolve the conflicts. One of its important provisions is a clear statement that the underwriters are responsible for the sufficiency of any procedures performed by the auditor to give negative assurance in the comfort letter.

The conflicts, however, have continued. For example, although auditors attempt to restrict data included in the letter to that of a financial nature, the definition of financial data is not always clear. Consider backlog information as an example. This information is expressed in dollars and often is thought of as financial data, but an auditor might have difficulty commenting on it, because it is not normally a part of the basic accounting records that is subjected to audit. There may be legal questions as to whether certain orders are firm or contingent, and records may be inadequate to determine the portions of contracts that have and have not been completed and included in revenue. Thus, the determination of the data to be covered in a comfort letter is often a matter of negotiation between the auditor and the underwriter.

A comfort letter may cover all or some of the following topics.

1. Independence of the auditor.

2. Compliance of the audited financial statements included in the registration statement with SEC accounting requirements.

3. Negative assurance regarding unaudited financial statements included in the registration statement.

4. Negative assurance regarding decreases in certain financial statement items subsequent to the date of the latest financial statements included in the registration statement.

5. Negative assurance regarding financial data included in the text section (as opposed to the financial statement section) of the registration statement.

The topic of independence is not controversial. The auditor must be able to state that he is independent in order to have audited the necessary financial statements for inclusion in the registration statement. The representation concerning independence is a statement of fact rather than of opinion.

In contrast, the representation regarding compliance with SEC accounting requirements is the auditor's opinion and is presented as such. The SEC accounting requirements include those within the 1933 Act, the form on which the registration statement is being prepared, Regulation S-X, and the *Accounting Series Releases.*

If the audited financial statements included in the registration statement are not current (the determination of what is current under various conditions is defined by rules of the SEC and policies of underwriters), unaudited financial statements (often referred to as "stub period numbers") as of a current date must be included. It is on these unaudited financial statements that the auditor may give negative assurance. In most cases, it is impractical, because of time limitations, to perform an audit as of a current date, yet the investing public, through the underwriters as their representatives, needs some assurances about the unaudited statements. The accounting profession concluded that the public interest would be served best if negative assurance on general-purpose financial statements were allowed in this situation. The auditor must include in the letter the specific steps performed as a basis for his negative assurance. In this and other sections of the comfort letter, the auditor should avoid general terms (checked, tested, etc.) having uncertain meanings in describing his work. Preciseness is very important throughout the letter.

The fourth topic included in comfort letters is negative assurance regarding decreases in certain financial statement items subsequent to the date of the latest financial statements included in the registration statement. The purpose of this section is to alert the underwriter at the latest practicable time (usually within five days of the effective date of the registration statement) to decreases in important financial statement items (such as declines in sales, net income, net assets, etc.) that the underwriters may consider adverse.[4] Thus the underwriters have an opportunity to cancel the offering or at least delay it until all material adverse conditions are properly disclosed in the registration statement. The auditor searches for decreases in important financial statement items by reviewing minutes and any financial statements prepared after those included in the registration statement, and by inquiring of certain company officials as to significant transactions during the period subsequent to the date of the latest financial statements. As a result of this work, the auditor may give negative assurance that there have been no decreases in important financial statement items. If decreases have occurred, he must disclose them in his letter.

The final comfort letter topic is negative assurance regarding financial data included in the text section of the registration statement. This topic has been the source of most of the controversy between auditors and underwriters. The controversy arises because of requests by underwriters that audit procedures be performed which the auditor considers (1) beyond his competence (such as procedures related to non-financial or subjective data), or (2) lacking in substance (such as comparison of amounts in the registration statement with amounts shown on a client-prepared worksheet). AU Section 630.37 *(SAS No. 1)* recommends that the auditor limit his procedures and comments to information that is expressed in dollars and has been obtained from accounting records subject to internal controls of the company's accounting system or has been computed from such records. The data to which the procedures were applied, the procedures themselves, and the auditor's findings must be enumerated clearly.

Because of the negative assurances contained in comfort letters, each letter must contain a statement restricting its use to the underwriter. Comfort letters are never distributed beyond the client and the underwriter, and the general public generally is unaware that they exist.

AU Section 630 *(SAS No. 1)* contains examples of comfort letters that may be appropriate in various situations. As in the case of the standard short-form

---

[4]Prior to issuance of *SAP No. 48,* the auditor was required to report any material adverse change in the financial statement items, but *SAP No. 48* shifted the responsibility for determining what constitutes a material adverse change to the underwriter.

report, the auditor should use the standard wording of a comfort letter when it is appropriate, but he must consider carefully whether the letter accurately communicates the work he has done and his findings for each engagement.

Other Forms of Negative Assurance

Other forms of negative assurance are limited to certain special reports. In these reports, as in comfort letters, the scope of the work should be described specifically and the use of the report should be restricted.

In practice, the auditor sometimes encounters special provisions in clients' loan agreements. An example is the requirement that a letter be prepared by the auditor stating that as a result of his audit he has no knowledge of defaults under the loan agreement. The auditor's letter normally takes the form of negative assurance and might be worded as follows:

> We have examined the financial statements of X Company as of December 31, 19XX, and for the year then ended and have issued our report thereon dated February 5, 19XX. Our examination was made in accordance with generally accepted auditing standards, and accordingly included such tests of the accounting records and such other auditing procedures as we considered necessary in the circumstances. In this connection we read the loan agreement between X Company and B Bank dated June 21, 19XX, particularly paragraphs 10 and 11 containing the financial ratios to be maintained by X Company.
>
> In the examination referred to above, nothing came to our attention that caused us to believe that there were any events of default under the terms of the loan agreement insofar as they pertain to accounting matters.

The foregoing example is an appropriate use of negative assurance, because it supplements rather than replaces an examination of the basic financial statements made in accordance with generally accepted auditing standards.

Reports on Projections and Forecasts

There is considerable controversy within the accounting profession about the desirability of CPAs' public association with forecasts. Chapter 2 includes a discussion of the ethical considerations of such association and generally concludes that a CPA may be associated with a forecast in a manner that does not lead to the belief that he vouches for its achievability. Even though some CPAs object to public association with forecasts, most of them will assist their clients in the preparation of forecasts for internal use. If this assistance is substantial, the CPA normally issues a report to accompany the forecast disclaiming any responsibility for its achievability. The CPA also may wish to disclaim responsibility for the assumptions on which the forecast is based. An example of a form which might be used follows.

We have reviewed the accounting principles and recomputed the calculations in the accompanying projected statements of income of X Company for the three years ended December 31, 19XX. The assumptions used as a basis for the forecasts are set forth in Note 1 and were made by the management of X Company.

Our work indicated that the accompanying projections were (1) prepared on the basis of accounting principles consistent with those used in the historical financial statements dated December 31, 19XX, (2) clerically accurate, and (3) compiled in accordance with the assumptions in Note 1 (which were made by management). Because the projections relate to the future, we cannot and do not express an opinion as to the achievability of the forecast or the assumptions on which it is based.

This letter is solely for use of the management of X Company and is not to be disclosed to any other person, nor are any statements to be made implying a greater responsibility for the accompanying projection than stated in this letter.

The final paragraph would be omitted if the CPA were associated publicly with the projection.

Reports on Interim Statements

The role of the CPA was extended significantly with the issuance of *SAS No. 10*[5] and *No. 13*[6] authorizing limited reviews of interim financial information. Although CPAs often had consulted with management of their clients about the proper presentation of interim financial data, the form and extent of their work were flexible, and written reports seldom were issued.

The initial impetus for CPA association with interim financial data was provided by the SEC with the issuance of *Accounting Series Release No. 177*. This release required disclosure in a note to the annual financial statements of unaudited quarterly data, and a review of this data (procedures for the review were established by the SEC) by an independent public accountant. The SEC encouraged, but did not require, companies to have their unaudited statements reviewed quarterly by their auditor, rather than at the end of the year. In the release, the SEC stated its preference for the accounting profession to determine the auditing standards and procedures underlying an auditor's report, and urged the Auditing Standards Executive Committee of the AICPA to adopt its own standards and procedures to be followed in a review of the unaudited data. The SEC indicated that it would withdraw its review procedures if a satisfactory statement were issued by the Committee. *SAS No. 10* was issued approximately three months later, and the SEC followed through with the withdrawal.

AU Section 720 *(SAS No. 10)* reaffirms the responsibility of management and the board of directors for the preparation of interim statements and defines a

[5]Now incorporated as AU Section 720.
[6]Now incorporated as AU Section 519.

role for the CPA of performing a "limited review" of the interim statements to assist the board of directors in meeting their responsibility for a fair presentation of such data. The auditor's report on the limited review is restricted to the board of directors, and, therefore, a form of negative assurance is used. The term "limited review" now has a precise meaning to auditors including the procedures to be employed (inquiries regarding the interim statements, the accounting system, and internal control; an analytical review of interim financial information; reading of minutes, etc.). AU Section 720.22 *(SAS No. 10)* includes an example of a report on a limited review, part of which is reproduced below.

We have performed a limited review (describe the data or statements subjected to such review) of ABC Company and consolidated subsidiaries for the three-month and nine-month periods ended September 30, 19X1.

Our limited review was performed in accordance with standards for such reviews promulgated by the American Institute of Certified Public Accountants and, accordingly, consisted principally of obtaining an understanding, by inquiries, of the accounting system for preparation of interim financial information; making an analytical review of pertinent financial data; and making inquiries of and evaluating responses from certain officials of the Company who have responsibility for financial and accounting matters. . . .

Because our limited review did not constitute an examination made in accordance with generally accepted auditing standards, we express no opinion on (describe the data or statements subjected to the limited review).

In connection with our limited review . . . the following (no) matters came to our attention that we believe should be reported to you:

(Describe the matters that should be brought to the attention of the Board of Directors.)

Had we performed additional procedures or had we made an examination . . . in accordance with generally accepted auditing standards, (other) matters might have come to our attention that would have been reported to you.

This report is solely for the information of the Board of Directors and Management and is not to be quoted in documents setting forth the unaudited interim financial information or in any other document available to the public.

AU Section 519 *(SAS No. 13)* authorizes *public* reports on limited reviews of interim financial information. To avoid negative assurance in public reports, the AudSEC approved the following specific form of report in AU Section 519.05 *(SAS No. 13)*.

We have made a limited review, in accordance with standards established by the American Institute of Certified Public Accountants, of (describe the information or statements subjected to such review) of ABC Company and consolidated subsidiaries as of September 30, 19X1 and for the three-month and nine-month periods then ended. Since we did not make an audit, we express no opinion on the (information or statements) referred to above.

This form of report would be modified to describe a change in accounting principle that is not in conformity with generally accepted accounting principles and material adjustments or disclosures proposed to the client that were not made. However, no modification of the report would be required owing to an uncertainty or a lack of consistency in the application of accounting principles.

Just before *SAS No. 13* was issued, the SEC notified the AudSEC that the statement was not acceptable, because it did not comment on the representations made by management. This objection caused a last-minute revision of the statement to include a provision that the following sentence could be included in reports on interim financial information submitted to the SEC on Form 10-Q (quarterly reports).

To comply with the requirements of the Securities and Exchange Commission, we confirm the Company's representations concerning proposed adjustments and disclosures included in the accompanying Form 10-Q for the period ended September 30, 19X1, in accordance with the related instruction K.

This additional sentence provides negative assurance because the Company's representations are that no known adjustments or disclosures have been omitted. Form 10-Q and the information therein are a matter of public record. Therefore, AU Section 519 *(SAS No. 13)* incongruously provides for the filing of negative assurance reports as public documents, but forbids their distribution to interested stockholders or others.

Despite the pressure brought to bear on the accounting profession by the SEC in this matter, a question has been raised by some members of the profession as to whether assistance to management in the presentation of interim financial data is a proper function of an independent auditor.

Commentary

The work that results in reports of the types discussed in this chapter historically has been a relatively small part of most CPAs' practices. However, the movement of the profession toward more acceptance (some voluntary and some not) of negative assurance reports in response to public demand (i.e., forecasts, interim statements, etc.) may increase the importance of these reports in the future. The cost-benefit ratio to the public is yet to be determined.

Chapter 17 References

Carmichael, D. R., "An Engagement Letter for Unaudited Statements," *The Journal of Accountancy,* March, 1971, pp. 69–70.

Chang, Davis L. S., and Shu S. Liao, "Measuring and Disclosing Forecast Reliability," *The Journal of Accountancy,* May, 1977, pp. 76–87.

Chazen, Charles, and Kenneth I. Solomon, "The 'Unaudited' State of Affairs," *The Journal of Accountancy,* December, 1972, pp. 41–45.

Saxe, Emanuel, "Accountants' Responsibility for Unaudited Financial Statements," *The New York Certified Public Accountant* (now *The CPA Journal*), June, 1971, pp. 419–423.

Winters, Alan J., "Unaudited Statements: Review Procedures and Disclosures," *The Journal of Accountancy,* July, 1976, pp. 52–59.

Chapter 17
Questions Taken from the Chapter

17–1.   Define special reports according to *SAS No. 14.*

17–2.   What are some examples of special-purpose reports discussed in *SAS No. 14?*

17–3.   Why are arrangement letters important for work involving preparation of unaudited financial statements?

17–4.   List two implications of the *1136 Tenants' Corp.* case for CPAs preparing unaudited financial statements.

17–5.   What are some reasons for and against the preparation of unaudited financial statements by CPAs?

17–6.   How will a reasonable understanding of the business and operations of a client assist a CPA in preparing unaudited financial statements?

17–7.   Explain how a CPA may be associated with unaudited financial statements.

17–8. What type of report should an auditor issue with unaudited financial statements? Draft an example of such a report.

17–9. Explain the content of a CPA's report on unaudited financial statements prepared for the internal use of the client that omit footnotes and other required disclosures.

17–10. To what extent should a CPA disclose any audit procedures performed on unaudited financial statements?

17–11. If a CPA discovers errors in unaudited financial statements with which he is associated, what action should he take?

17–12. What type of report should a CPA issue on financial statements if he has performed all necessary audit procedures, but he is not independent? Draft an example of such a report.

17–13. What action should a CPA take if he has reservations about financial statements with respect to which he is not independent?

17–14. Describe cash-basis financial statements.

17–15. Give two examples of permissible use of cash-basis financial statements.

17–16. What are the reporting requirements when cash-basis financial statements purport to represent general-purpose financial statements?

17–17. If cash-basis financial statements are deemed to be special reports, discuss the applicability of each of the four standards of reporting. Draft an example of such a report.

17–18. How do long-form audit reports differ from the standard short-form report?

17–19. Give an example of what the accounts receivable section of a long-form report might contain.

17–20. How does a CPA indicate the responsibility he is assuming for information in a long-form report?

17–21.   Why are long-form reports not as common today as they were several years ago?

17–22.   What is negative assurance within the context of the auditor's report?

17–23.   When may CPAs give negative assurance, and why is its use limited?

17–24.   What is the purpose of comfort letters?

17–25.   Why have auditors and underwriters disagreed over the content of comfort letters?

17–26.   Who is responsible for the sufficiency of any procedures performed by the auditor in order to give negative assurance in the comfort letter?

17–27.   List the five topics that may be covered in a comfort letter.

17–28.   Indicate where one can find the SEC accounting requirements for a registration statement filed on Form S-1 under the 1933 Act.

17–29.   Why are unaudited financial statements included in a registration statement?

17–30.   Why is negative assurance allowed in a comfort letter on unaudited general-purpose financial statements?

17–31.   Why do comfort letters include negative assurance regarding decreases in certain financial statement items subsequent to the date of the latest financial statements included in the registration statement?

17–32.   Why has negative assurance regarding financial data included in the text section of the registration statement been a source of controversy between auditors and underwriters?

17–33.   What distribution is made of comfort letters?

17–34.   Where, other than in comfort letters, may an auditor give negative assurance?

17–35.   What are the important aspects of a CPA's report on a forecast?

17–36. Discuss the development of the CPA's association with interim financial information.

17–37. What is the purpose of a "limited review" of interim financial statements, and what does it consist of?

17–38. List the important elements of an auditor's report based on a limited review.

Chapter 17
Objective Questions Taken from CPA Examinations

17–39. Loeb, CPA, has completed a review of the Bloto Company's unaudited financial statements and has prepared the following report to accompany them:

The accompanying balance sheet of the Bloto Company as of August 31, 19X5, and the related statements of income, retained earnings and changes in financial position for the year then ended were not examined by us and accordingly we do not express an opinion on them.

The financial statements fail to disclose that the debentures issued on July 15, 19X2, limit the payment of cash dividends to the amount of earnings after August 31, 19X3. The company's statements of income for the years 19X4 and 19X5, both of which are unaudited, show this amount to be $18,900. Generally accepted accounting principles require disclosure of matters of this nature.

Which of the following comments *best* describes the appropriateness of this report?
a. The report is satisfactory.
b. The report is deficient because Loeb does not describe the scope of the review.
c. The report is deficient because the second paragraph gives the impression that some audit work was done.
d. The report is deficient because the explanatory comment in the second paragraph should precede the opinion paragraph.

17–40. Which of the following best describes the responsibility of the CPA when he prepares unaudited financial statements for his client?
a. He should make a proper study and evaluation of the existing internal control as a basis for reliance thereon.
b. He is relieved of any responsibility to third parties.

    c.   He does not have responsibility to apply auditing procedures to financial statements.

    d.   He has only to satisfy himself that the financial statements were prepared in conformity with generally accepted accounting principles.

17–41.   Wald, CPA, is preparing unaudited financial statements for Zakin Company. During the engagement, Wald becomes aware of the fact that the statements are misleading. Wald should

    a.   Disclaim an opinion.

    b.   Insist that the statements be corrected.

    c.   Issue an adverse opinion.

    d.   Insist that the statements be audited.

17–42.   John Greenbaum, CPA, provides bookkeeping services to Santa Fe Products Co. He also is a director of Santa Fe and performs limited auditing procedures in connection with his preparation of Santa Fe's financial statements. Greenbaum's report accompanying these financial statements should include a

    a.   Detailed description of the limited auditing procedures performed.

    b.   Complete description of the relationships with Santa Fe that imperil Greenbaum's independence.

    c.   Disclaimer of opinion and statement that financial statements are un-audited on each page of the financial statements.

    d.   Qualified opinion because of his lack of independence, together with such assurance as his limited auditing procedures can provide.

17–43.   Jeffries, CPA, had prepared unaudited financial statements for a client, the Gold Company. Since the statements were only to be used internally by the client, Jeffries did *not* include any footnotes and so stated in the accompanying disclaimer of opinion. Three months after the statements were issued, the Gold Company asked Jeffries if it would be all right to give a copy of the statements to its banker, who had requested financial statements. How should Jeffries respond?

    a.   Jeffries should revise the statements to include appropriate footnotes and attach a revised disclaimer of opinion before they are released to the banker.

    b.   Gold may give the statements to the banker as long as Jeffries' disclaimer of opinion accompanies the statements.

    c.   Gold should retype the statements on plain paper and send them to the banker without Jeffries' report.

    d.   Gold may let the banker review the statements and take notes, but should not give the banker a copy of the statements.

17–44.   A CPA has been engaged to prepare unaudited financial statements for his client. Which of the following statements best describes this engagement?

    a.   The CPA must perform the basic accepted auditing standards necessary to determine that the statements are in conformity with generally accepted accounting principles.

    b.   The CPA is performing an accounting service rather than an examination of the financial statements.

    c.   The financial statements are representations of both management and the CPA.

    d.   The CPA may prepare the statements from the books, but may not assist in adjusting and closing the books.

17–45.   Which of the generally accepted auditing standards of reporting would *not* normally apply to special reports such as cash-basis statements?
    a.   First standard.
    b.   Second standard.
    c.   Third standard.
    d.   Fourth standard.

17–46.   An auditor is reporting on cash-basis financial statements. These statements are best referred to in his opinion by which one of the following descriptions?
    a.   Cash receipts and disbursements and the assets and liabilities arising from cash transactions.
    b.   Financial position and results of operations arising from cash transactions.
    c.   Balance sheet and income statement resulting from cash transactions.
    d.   Cash-balance sheet and the source and application of funds.

17–47.   Which of the following best describes the difference between a long-form auditor's report and the standard short-form report?
    a.   The long-form report may contain a more detailed description of the scope of the auditor's examination.
    b.   The long-form report's use permits the auditor to explain exceptions or reservations in a way that does not require an opinion qualification.
    c.   The auditor may make factual representations with a degree of certainty that would not be appropriate in a short-form report.
    d.   The long-form report's use is limited to special situations such as cash-basis statements, modified accrual basis statements, or not-for-profit organization statements.

17–48.   Ansman, CPA, has been requested by a client, Rainco Corp., to prepare a "long-form report" for this year's audit engagement. Which of the following is the *best* reason for Rainco's requesting a long-form report?
    a.   To provide for a piecemeal opinion because certain items are not in accordance with generally accepted accounting principles.
    b.   To provide Rainco's creditors a greater degree of assurance as to the financial soundness of the company.

    c.   To provide Rainco's management with information to supplement and analyze the basic financial statements.

    d.   To provide the documentation required by the Securities and Exchange Commission in anticipation of a public offering of Rainco's stock.

17–49.   Ansman, CPA, has been requested by a client, Rainco Corp., to prepare a "long-form report" for this year's audit engagement. In issuing a long-form report, Ansman must be certain to

    a.   Issue a standard short-form report on the same engagement.

    b.   Include a description of the scope of the examination in more detail than the description in the usual short-form report.

    c.   State the source of any statistical data and that such data have not been subjected to the same auditing procedures as the basic financial statements.

    d.   Maintain a clear-cut distinction between the management's representations and the auditor's representations.

17–50.   A CPA should *not* normally refer to which one of the following subjects in a "comfort letter" to underwriters?

    a.   The independence of the CPA.

    b.   Changes in financial statement items during a period subsequent to the date and period of the latest financial statements in the registration statement.

    c.   Unaudited financial statements and schedules in the registration statement.

    d.   Management's determination of line of business classifications.

17–51.   In a "comfort letter" to underwriters, the CPA should normally avoid using which of the following terms to describe the work performed?

    a.   Examined.

    b.   Read.

    c.   Made inquiries.

    d.   Made a limited review.

17–52.   A CPA's report accompanying a cash forecast or other type of projection should

    a.   Not be issued in any form because it would be in violation of the AICPA *Code of Professional Ethics.*

    b.   Disclaim any opinion as to the forecast's achievability.

    c.   Be prepared only if the client is a not-for-profit organization.

    d.   Be a qualified short-form audit report if the business concern is operated for a profit.

17–53.   In the course of an engagement to prepare unaudited financial statements, the client requests that the CPA perform normal accounts receivable audit confirmation procedures. The CPA agrees and performs such procedures. The confirmation procedures

    a.   Are part of an auditing service that changes the scope of the engagement to that of an audit in accordance with generally accepted auditing standards.

    b.   Are part of an accounting service and are not performed for the purpose of conducting an audit in accordance with generally accepted auditing standards.

    c.   Are not permitted when the purpose of the engagement is to prepare unaudited financial statements and the work to be performed is not in accordance with generally accepted auditing standards.

    d.   Would require the CPA to render a report that indicates that the examination was conducted in accordance with generally accepted auditing standards but was limited in scope.

17–54.   A CPA would be considered "not associated" with unaudited financial statements when

    a.   The CPA performed a limited review of a publicly traded company's unaudited financial statements which are presented in a quarterly report to the stockholders.

    b.   The CPA assisted in the preparation of the unaudited financial statements.

    c.   The CPA completed an audit and rendered a report on the financial statements which without the CPA's consent were part of a prospectus which included unaudited financial statements.

    d.   The CPA received all input data from the client, reviewed it, and returned it to the client for processing by an independent computer service company.

17–55.   A long-form report generally includes the basic financial statements but would not include

    a.   Exceptions or reservations to the standard (short-form) report.

    b.   Details of items in basic financial statements.

    c.   Statistical data.

    d.   Explanatory comments.

17–56.   When engaged to prepare unaudited financial statements, the CPA's responsibility to detect fraud

    a.   Is limited to informing the client of any matters that come to the auditor's attention which cause the auditor to believe that an irregularity exists.

    b.   Is the same as the responsibility that exists when the CPA is engaged to perform an audit of financial statements in accordance with generally accepted auditing standards.

    c.   Arises out of the CPA's obligation to apply procedures which are designed to bring to light indications that a fraud or defalcation may have occurred.

    d.   Does *not* exist unless an arrangement letter is prepared.

Chapter 17
Discussion/Case Questions

17–57. The limitations of the CPA's professional responsibilities when he is associated with unaudited financial statements often are misunderstood. Listed below are eight situations or contentions the CPA may encounter in his association with and preparation of unaudited financial statements. Briefly discuss the extent of the CPA's responsibilities and, if appropriate, the actions he should take to minimize any misunderstandings. Indicate which situations are similar to those involved in the *1136 Tenants' Corporation* case discussed in Chapter 3.

a. A CPA was engaged by telephone to perform write-up work including the preparation of financial statements. His client believes that the CPA has been engaged to audit the financial statements and to examine the records accordingly.

b. A group of businessmen who own a farm which is managed by an independent agent engage a CPA to prepare quarterly unaudited financial statements for them. The CPA prepares the financial statements from information given to him by the independent agent. Subsequently, the businessmen find that the statements are inaccurate because their independent agent was embezzling funds. The businessmen refuse to pay the CPA's fee and blame him for allowing the situation to go undetected, contending that he should not have relied on representations from the independent agent.

c. In comparing the trial balance with the general ledger, the CPA finds an account labeled "audit fees" in which the client has accumulated the CPA's quarterly billings for accounting services, including the preparation of quarterly unaudited financial statements.

d. Unaudited financial statements are accompanied by the following letter of transmittal from the CPA:

We are enclosing your company's balance sheet as of June 30, 19X7, and the related statements of income and retained earnings and changes in financial position for the six months then ended which we have reviewed.

e. To determine appropriate account classification, the CPA reviewed several of the client's invoices. He noted in his working papers that some invoices were missing, but did nothing further because he believed that they did not affect the unaudited financial statements he was preparing. When the client subsequently discovered that invoices were missing, he contended that the CPA should not have ignored the missing invoices when preparing the financial statements and had a responsibility at least to inform him that they were missing.

f. The CPA has prepared a draft of unaudited financial statements from the client's records. While reviewing this draft with his client, the CPA learns that the land and building were recorded at appraisal value.

g. Financial statements marked "unaudited" were accompanied by the following letter from the CPA:

Although I performed all of the audit procedures I considered necessary in accordance with generally accepted auditing standards, I am not independent with respect to X Co. because I own stock in the Company. Therefore, the accompanying balance sheet as of December 31, 19X7, and the related statements of income and retained earnings and changes in financial position for the year then ended were not audited by me; accordingly, I do not express an opinion on them.

h. The CPA is engaged to review without audit the financial statements prepared by the client's controller. During this review, the CPA learns of several items which by generally accepted accounting principles would require adjustment of the statements and footnote disclosure. The controller agrees to make the recommended adjustments to the statements, but says that he is not going to add the footnotes because the statements are unaudited.

(AICPA adapted)

17–58. You have been engaged by the trustees of Roger Trust to audit the trust as of December 31, 19XX. The audit is for the purpose of assuring the trustees and beneficiaries that trust assets, distributions, and income and expenses have been properly handled and accounted for in accordance with the terms of the trust instrument. The trust instrument provides that accounting records are to be maintained on the cash basis.

a. State the type of audit report you consider most appropriate in this circumstance and explain why.

b. Draft the report you would issue if, after your examination, you had no reservations about the financial statements.

17–59. May Co. is a small refinery owned by several individuals who are not active in its management. To supply them with reliable information about the company's operations for the year, you are requested to prepare a long-form report in connection with this year's audit. You are asked specifically to include the following items in your report.

a. An aged trial balance of accounts receivable as of the end of the year.

b. An analysis of the allowance for doubtful accounts for the year and your conclusion as to its adequacy.

c. An analysis of insurance coverage at the end of the year and your recommendation for any changes.

d. An analysis of the major additions to property and equipment for the year and a description of the procedures you used to audit the additions.

e. An analysis of accrued liabilities at the end of the year and a comment regarding the years remaining open subject to review by the IRS.

  f. A schedule showing the percentage of the total outstanding common stock owned by each stockholder of record as shown in the stockbook.

  g. Your comments as to why revenues increased during the year.

Discuss the appropriateness of including each of these matters in your long-form report.

17–60. As a condition for a public offering of securities by your largest and most important audit client, the underwriter requires a comfort letter containing the following items.

  a. An unequivocal statement that you are independent with respect to your client within the meaning of the Securities Act of 1933 (you believe that you are independent, but you have not read the entire Act and related rules and regulations to be positive).

  b. Negative assurance regarding the audited financial statements included in the registration statement.

  c. Negative assurance regarding decreases in certain financial statement items during a period subsequent to the date of the latest financial statements in the registration statement.

  d. Negative assurance regarding certain financial information included in the text section of the registration statement, including the percentage of capacity at which the client's plant operated during the year.

A meeting is to be held for the underwriter, your client, and you to review the procedures for the public offering. Before the meeting gets underway, the underwriter shows you a copy of a comfort letter issued by another office of your firm in connection with a previous public offering that exactly conforms with the underwriter's requirements, and the client states that the comfort letter is so standard that it is not worth discussing at this meeting. What matters, if any, would you wish to discuss concerning the comfort letter?

17–61. You are engaged by Mr. Rouge to assist him in projecting the income and expenses for the next five years if he purchases and operates an offshore oil and gas drilling rig. Although Mr. Rouge is a very competent drilling rig operator, he is not familiar with the financial aspects of drilling operations. Most of your discussions on the assumptions underlying the projections end with Mr. Rouge asking, "What do you think would be reasonable?"

After the projections are completed, you have them typed and attach a transmittal letter disclaiming an opinion on the achievability of the forecast and restricting its use to Mr. Rouge. You later receive a call from a very annoyed Mr. Rouge, who asks why you have to be so negative and how you expect him to obtain financing for the drilling rig if he cannot show the projections to anyone. He adds that if he is not allowed to use the projections to raise some funds, he will be unable to pay you for your work on the projections.

What do you tell Mr. Rouge? With the benefit of hindsight, what problems can you identify in this engagement, and what would you have done differently?

17–62. Tab Co., a public company, changed auditors and selected your firm after the predecessor auditors completed their audit for the prior year. At the end of the first quarter of the current year, before you have performed any work, you are asked to perform a limited review of the first quarter interim financial information on a timely basis.

What, if any, difficulties would you expect to encounter?

17–63. A CPA performs accounting services for a company from which he is not independent. Unaudited financial statements will be issued. Would it be acceptable for the CPA to issue a disclaimer on unaudited statements, or is it necessary for him to issue a disclaimer because of lack of independence?

17–64. To issue an audit report on a specific account, it is often necessary to perform audit procedures not only on such account, but on related accounts and controls as well.

If an auditor were to issue a separate report on each of the following accounts, which related accounts or controls might be examined in each case? Indicate your reasons for each related account or control listed.
a. Property and equipment.
b. Sales.
c. Income tax liabilities.
d. Net current assets.

17–65. Nut Tree Co. has a series of bonds outstanding under an indenture that requires an annual audit by a CPA and a supplemental report stating that, as a result of the audit, the CPA did not become aware of any events of default under the indenture. The events of default in the indenture were (1) payment of dividends in any year in excess of 50% of net income for that year, (2) expenditures for property and equipment additions in excess of $500,000 in any year, and (3) failure to maintain a ratio of current assets to current liabilities of at least 2 to 1.

As a result of the current-year audit, the CPA found the company to be in compliance with the provisions of the indenture based on the amounts shown in the audited financial statements. However, the CPA's audit report was qualified because of an uncertainty arising from anti-trust action instituted against the company. The amount, if any, of the ultimate liability from this uncertainty cannot be determined.

Discuss what effect, if any, this matter would have on the auditor's special report on compliance with the indenture.

Draft the form of special report you would issue, or discuss why you would refuse to issue such a report.

17-66. Steam Clean, Inc. produces unique washing machines under a license agreement with Tom Nessinger, who invented the unique process. Nessinger has engaged you to audit the sales of Steam Clean, Inc., which are the basis for royalty payments to him (he is to be paid $10 for each washing machine sold).

During your examination, you find that the sales amount upon which Nessinger was paid excludes (1) 15 units for which the sales prices were never collected (written off as bad debts), (2) 17 units that were returned because of defects, (3) 21 units sold at a reduced price as demonstrators, (4) 106 units manufactured and sold by a foreign subsidiary, and (5) 57 units that have been modified slightly from the original design. Nessinger disputes the exclusion of all of these units from sales on which royalty payments were made.

Discuss or draft the form of special audit report you would issue in this case.

Appendix I to the Text
Systems Evaluation Approach

Within the broad framework of an audit, as discussed in this text, each CPA firm formulates its own methods for satisfying the objective of expressing an opinion on the fairness of their clients' financial statements. To accomplish this objective in the most efficient manner, an increasing number of firms have adopted a comprehensive approach to the planning and conduct of their audits.

One CPA firm, Peat, Marwick, Mitchell & Co. (PMM&Co.), refers to their approach as the Systems Evaluation Approach (SEA). PMM&Co. describes this approach as follows:

> SEA provides a logical framework within which audits can be performed at the highest level of professionalism. This logical framework is valid for all audit engagements regardless of size, complexity or nature of business. Standards for documentation of audit activities in areas such as planning, internal control evaluation and programming are specified. Emphasis is placed on the choice of documentation techniques which are available to the auditor, depending on his decisions relative to the degree of reliance he chooses to place on internal controls. In all cases, the auditor is encouraged to select those techniques which are most efficient and that have the optimum cost-benefit relationship with his audit objectives.

> The essential ingredient of SEA is objective-oriented auditing. The auditor establishes objectives to be attained and then develops auditing procedures which enable him to achieve these objectives in an efficient and economical manner. Procedures which are not productive in attaining audit objectives should be avoided. For example, systems should be tested and verified only to the extent that the auditor intends to rely on such systems. Testing of systems upon which little reliance is intended to be placed is not productive and is, therefore, discouraged. Unless the timing, nature and extent of substantive tests of significant balances and transactions can be limited by reliance on an effectively functioning system of internal control, detailed testing of the functioning of the system achieves little in the way of attaining valid audit objectives.

In the review and evaluation of internal control, SEA emphasizes the review of general controls and the review of controls in specific transaction cycles (which may affect many accounts) rather than the review of controls in specific accounts. Examples of transaction cycles are (1) the purchases-cash disbursements cycle, (2) the revenues-cash receipts cycle, and (3) the production-payroll cycle.

The flowcharts and explanations on the following pages contain a partial description of the three phases of SEA: (1) planning, (2) interim, and (3) final

FLOW CHART OF SYSTEMS EVALUATION APPROACH

DESCRIPTION	PLANNING ENGAGEMENT CONSIDERATIONS NOTE

Initial Planning

1. Establish overall audit objectives and strategy.

2. Determine effect of current economic conditions on the audit.

3. Understand the industry and accounting and auditing problems peculiar to the industry.

4. Understand client's business, products, goals, plans, etc., and administrative and accounting control environment of client.

5. Review pertinent financial data to identify any unusual or unexpected relationships.

6. Perform initial planning procedures, confer with client and have planning meetings with appropriate personnel.

7. Perform preliminary evaluation of internal controls. The preliminary evaluation is concerned primarily with "general" controls.

8. Determine whether to rely on internal controls. If not, determine if entity is auditable. If entity is not auditable, consider alternatives, such as proceeding with the engagement with the expectation of issuing a disclaimer of opinion, or withdrawing from the engagement.

9. Prepare memoranda of discussions, inquiries, evaluations, etc., obtained during planning, including discussion of identified critical audit areas and unusual accounting matters.

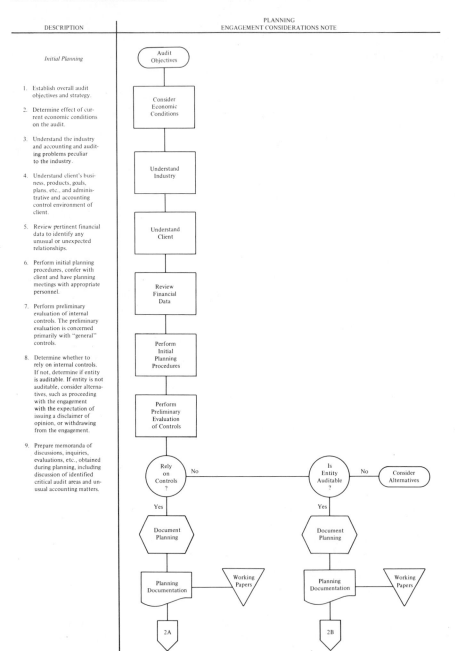

Initial Planning

The first standard of fieldwork requires that the audit engagement be adequately planned. The procedures required to effectively plan an engagement generally will be the same for all audit engagements, although the effort involved in completing the planning phase will vary significantly depending on the size of the client, complexity of operations, timing considerations due to the overall quality of internal control, and other considerations which are unique to the particular client.

1. Audit Objectives
   The auditor should develop an overall audit strategy which will allow him to accomplish his objectives with respect to each segment of his examination and his overall objective of reporting on the client's financial statements.

2. Consider Economic Conditions
   The auditor should be sensitive to the overall state of the world, U.S. and local economies, and their impact on the client's industry and, more particularly, on his client. For example, foreign exchange rates and unstable political situations in foreign governments are among those factors which may have a significant impact on international companies' present and future operations.

3. Understand Industry
   The auditor's understanding of the client's industry should encompass an understanding of accounting and auditing problems unique to that industry. Considerations such as stability of product demand, correlation of demand with consumer disposable income, major competition within the industry and technological developments may have a significant effect on the client's business and may influence audit strategy.

4. Understand Client
   The auditor should obtain an understanding of the operations of the business being examined, as well as the overall environment in which it operates. Information developed may include descriptions of the nature of the business; major products or services and relative sales volume of each; corporate structure and ownership; capitalization and long-term and short-term financing; management's long-term and short-term objectives; location of owned and leased facilities; nature of manufacturing or processing operations and level of activity for each significant plant, etc.; accounting, budgeting and reporting systems; marketing philosophy and methods of distribution and remuneration of sales force; purchasing volume; key management personnel; total number of corporate employees by function; attorneys and consultants and regulatory requirements.

5. Review Financial Data

The purpose of this review is to identify unusual relationships of financial data that may have audit significance to assist in determining the scope and relative emphasis of the audit work.

The primary emphasis is on comparison of current and historical financial and operating information. It may also be appropriate to extend the comparison to budgeted figures and average ratios for the industry when such information is available.

On many smaller engagements, interim financial information is often based on a series of estimates with no attempt to achieve accurate cut-offs, etc. In these situations, the auditor may choose not to perform a complete initial analytical review since such information may not have a reasonable degree of reliability.

6. Perform Initial Planning Procedures

The initial planning procedures are designed to enable the auditor to develop an effective audit strategy. The planning process requires involvement of the client and the auditor, and includes meetings of the audit team to discuss the engagement strategy, problem areas, areas of audit emphasis and other significant matters. In addition to procedures 1–5 discussed above, the initial planning procedures should be designed to provide the auditor with a current understanding of any unusual operating problems; critical auditing areas and unusual accounting matters; timing and report considerations; and staffing requirements, including participation by industry specialists, computer audit specialists, MC EDP consultants, statistical audit specialists, tax department representatives and management consultants. Logistical considerations, such as arrangements for adequate working space, should also be addressed during initial planning.

The extent of participation in the audit by other offices should be determined, and interoffice instructions should be sent to participating offices.

7. Perform Preliminary Evaluation of Internal Controls

The purpose of the preliminary evaluation of internal controls is to develop an understanding and perform an evaluation of the overall environment in which data is produced, processed, reviewed and accumulated within the organization. These environmental controls, classified as "general controls" are indicative of the overall control consciousness of an organization's management and staff. They have, therefore, an indirect impact on the validity of data processed by the individual systems and transaction cycles.

The general controls usually considered in the preliminary evaluation are:

a. General organizational structure,
b. Internal audit activities,
c. Administration of the accounting function, and
d. Protection of physical assets.

General controls are frequently not documented controls. They typically do not lend themselves to flowcharting techniques. The audit tests usually applied to such controls are inquiry of appropriate personnel and personal observation of the controls.

8. Rely on Controls?
Based on his evaluation of the general controls, the auditor should make a preliminary determination as to whether significant reliance can be placed on the client's internal controls in achieving his audit objectives and whether he chooses to place reliance.

9. Document Planning
SAS No. 1 indicates that working papers should include evidence of engagement planning. The written planning memorandum summarizing the findings, comments and conclusions reached during the performance of planning procedures should be included in the general section of the working papers as part of the documentation of the audit plan. Working papers documenting the analytical review of financial and operating data should be included in the current working papers or permanent file.

Interim

The procedures required to perform the interim examination will vary considerably, depending on the findings and conclusions reached during the planning phase. These procedures, discussed below, should be considered and evaluated with respect to each system or transaction cycle being audited.

10. Document System
The method of documenting a system or transaction cycle is dependent on:

   a. Whether the system or transaction cycle represents a significant audit area, and
   b. Whether the auditor's review of the general controls and other inquiries lead him to believe that significant reliance can be placed on internal controls.

If both of these conditions are met, the system is classified as a "reliance" area, and Firm policy requires that the system or transaction cycle be flowcharted, except in those instances where existing narrative documentation is current and of such clarity and quality that it permits an understanding of that system or transaction cycle. Flowcharting should be in sufficient detail to identify the major internal control strengths and weaknesses having an impact on the development of an appropriate audit program.

For "non-reliance" areas (including areas where the auditor has determined that the audit effort required to test compliance with system controls would be greater than the effort required to expand substantive tests of balances

FLOW CHART OF SYSTEMS EVALUATION APPROACH

INTERIM
SYSTEM < TRANSACTION CYCLE > CONSIDERATIONS

DESCRIPTION	RELIANCE PLACED ON INTERNAL CONTROLS	NO SIGNIFICANT RELIANCE PLACED ON INTERNAL CONTROLS

Interim

10. Prepare flow charts of major transaction cycles. Narrative write-ups should be prepared for areas not susceptible to flow-charting. When no significant reliance is placed on internal controls, documentation may be limited, but should indicate auditor's understanding of system.

11. "Walk" selected transactions through the system to test understanding of the system.

12. Identify pertinent internal control strengths and weaknesses of the system related to internal control objectives. Also determine potential errors or irregularities which could occur and are of vital concern. When no significant reliance is placed on the system, identify resulting audit concerns.

13. Prepare a workpaper which describes the control strengths and weaknesses in the system. When no significant reliance is placed on the system, document basis of determing scope and extent of substantive tests.

14. Perform an overall evaluation of internal controls identified. < This evaluation is made for each system or transaction cycle. The auditor may conclude that one transaction cycle is strong and another is weak and, therefore, approach the audit of the data produced by these systems differently >

15. Determine if we plan to rely on internal controls.

16. Design audit program and identify pertinent evidential matter available. < Modification to initial program may be required after completion of compliance tests >. Document relationship of substantive audit tests to evaluation of controls.

17. Perform compliance testing.

18. Evaluate internal control strengths when compliance testing is completed.

19. Determine if we can rely on internal controls. If not, determine if other compensating internal control strengths exist. If compensating controls do not exist, adjust audit procedures.

20. Perform any interim substantive testing.

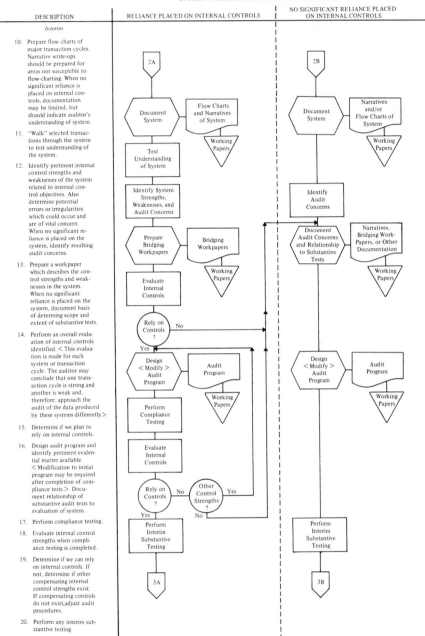

and or transactions), the auditor should document the system to the extent required to provide a level of understanding necessary to design an audit program which concentrates on substantive testing.

11. Test Understanding of System

The auditor should test the accuracy of the flow chart or narrative write-up describing the system by "walking" one or more transactions (or batches) through the transaction cycle.

12. Identify System Strengths, Weaknesses, and Audit Concerns

For reliance areas, significant internal control strengths, weaknesses and audit concerns related to the audit objectives should be identified. The possibility of undetected errors or irregularities should also be determined.

For non-reliance areas, audit concerns and objectives should be identified.

13. Prepare Bridging Workpaper

For reliance areas, a bridging workpaper is prepared to summarize and relate audit objectives with identified control strengths on which the auditor intends to rely in restricting substantive auditing procedures. Control weaknesses are also identified together with mitigating control strengths or, in their absence, compensating substantive auditing procedures. The bridging workpaper describes the audit implication of each identified control strength or weakness, and indicates (and cross-references to the audit program) the appropriate auditing procedures designed to compliance test the strengths or the substantive tests designed to compensate for the weaknesses.

For non-reliance areas, the auditor should document his audit objectives and concerns, usually in narrative form. A bridging workpaper may be prepared to facilitate performance of this procedure, but is not required.

14. Evaluate Internal Controls

Where the auditor intends to place significant reliance on the system, an evaluation of the system is made based on the identified (not tested) control strengths and identified weaknesses. In making this evaluation, prime consideration should be given to the overall strength or weakness of all related control procedures considered together, rather than to individual strengths and weaknesses.

15. Rely on Controls?

If the related controls considered together do not provide reasonable assurance against the undetected occurrence of errors and irregularities, little reliance should be placed on any of the individual controls to justify restriction of substantive testing. The auditor should consequently view the area as a non-reliance area. If the overall evaluation indicates that the system should provide reasonable assurance against undetected occurrence of errors and irregularities, the auditor should continue to audit the area as a reliance area.

16. Design Audit Program
Compliance tests should be designed to test the functioning of system strengths on which the auditor intends to place reliance. These procedures should be indicated on the bridging workpaper and cross-referenced to the audit program.

For both reliance and non-reliance areas, the auditor should document the rationale used in designing substantive auditing procedures by relating the extent of reliance that he intends to place on the system to the scope and extent of the substantive procedures. This documentation should normally be in the form of a separate memorandum, or should be indicated in narrative form on the audit program. Any significant subsequent changes to the program should be documented in a similar manner.

If, after the system has been tested and evaluated (Steps 17 and 18), the auditor determines that the system is not functioning as effectively as anticipated, the originally designed substantive procedures should be appropriately modified.

17. Perform Compliance Testing
The auditing procedures designed to test the functioning of the system strengths are performed.

18. Evaluate Internal Controls
The system is again evaluated, this time in light of the results of the compliance tests performed.

19. Rely on Controls?
If the auditor concludes that the system provides reasonable assurance against the undetected occurrence of errors and irregularities, reliance may be placed on the system.

If, however, the auditor concludes that significant reliance should not be placed on certain components of the system, he should ascertain whether there are other system strengths on which reliance may be placed. If so, these strengths should be compliance tested to verify their effectiveness. If there are no other controls on which reliance may be placed, the auditor should proceed to audit this area as a non-reliance area.

The audit program designed in Step 16 should be reevaluated in light of the results of the compliance testing performed and, where appropriate, modifications to the program should be made. The rationale for any significant changes to the audit program should be documented in the same manner as the rationale for the original program, as discussed under Step 16.

20. Perform Interim Substantive Testing

    Frequently, it is possible for the auditor to complete a substantial amount of substantive test work during interim. The amount of interim substantive testing which can be performed is dependent on several factors, including the auditor's evaluation of the system and his conclusions as to the reliance that can be placed thereon. Accounts that are often susceptible to interim substantive testing include accounts with stable and predictable periodic balances; low levels of activity; declining balances subject to systematic amortization or depreciation; increasing balances subject to systematic accrual or deferral; and non-current balance sheet accounts and non-operating income statement classifications. In addition to substantive testing of individual accounts, procedures relative to such areas as stock options and pension plans, report format and notes, and unusual accounting problems can be performed at this time.

Final

All procedures required to complete the examination are performed during the final phase. This will normally include the procedures discussed below.

21. Review Financial Data

    The review of financial data during the final phase of the engagement consists of an analytical review, using the year-end financial data, and takes place as soon as the year-end information becomes available. The purpose of this review is similar to the initial review performed during the planning phase—to identify unusual relationships that may have audit significance and thereby assist the auditor in determining whether any modification to the audit program is warranted by changes in trends and conditions subsequent to the interim audit phase of the engagement.

22. Update Review of Internal Control

    For reliance areas where the auditor has performed compliance tests up to an interim date, a decision should be made as to whether the tests should be extended to the remaining portion of the period being examined. Section 320.61 of *SAS No. 1* sets forth certain factors that should be considered in making this decision. In instances when the auditor determines that it is not necessary to extend compliance tests to the remaining period, the working papers for each system or transaction cycle should document that this judgment was made after giving proper consideration to these factors.

23. Revise Audit Program?

    Based on the results of the update review of internal control and year-end review of financial data, the auditor should make appropriate modifications to the audit program, if required.

FLOW CHART OF SYSTEMS EVALUATION APPROACH

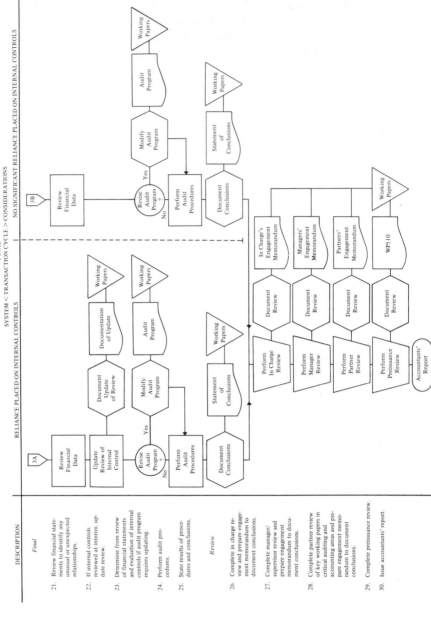

FINAL

SYSTEM < TRANSACTION CYCLE > CONSIDERATIONS

RELIANCE PLACED ON INTERNAL CONTROLS

NO SIGNIFICANT RELIANCE PLACED ON INTERNAL CONTROLS

DESCRIPTION

Final

21. Review financial statements to identify any unusual or unexpected relationships.

22. If internal controls reviewed at interim, update review.

23. Determine from review of financial statements and evaluation of internal controls if audit program requires updating.

24. Perform audit procedures.

25. State results of procedures and conclusions.

Review

26. Complete in charge review and prepare engagement memorandum to document conclusions.

27. Complete manager/supervisor review and prepare engagement memorandum to document conclusions.

28. Complete partner review of key working papers in critical auditing and accounting areas and prepare engagement memorandum to document conclusions.

29. Complete preissuance review.

30. Issue accountants' report.

24. Perform Audit Procedures

Any additional procedures considered necessary to test compliance, as well as the final substantive audit procedures designed to provide reasonable assurance of the validity of information produced by the accounting system (i.e., the individual account balances) not otherwise performed during interim, should be performed at this point.

25. State Results of Procedures and Conclusions

A conclusion statement, setting forth the auditor's conclusions with respect to the work performed, should be included in the working papers covering each section of the audit.

26.–29. Perform Review

The various members of the audit team, including the preissuance reviewer, complete their reviews. . . .

30. Issue Accountants' Report

The appropriate accountants' report will be issued at the conclusion of the audit engagement based on the results of the examination.

Appendix II to the Text
Developments in the Auditing Profession

Throughout the text, auditing is shown to be a dynamic profession in constant search for ways to meet the growing public demand for quality services. Two developments vividly exemplify this trend. One is an extensive study of the responsibilities of independent auditors conducted within the private sector of the auditing profession itself. The other is series of studies of accounting principles and auditing services conducted by committees of the U.S. Congress.

In the two following sections, each of these studies and their impact on auditing are discussed.

The Commission on Auditors' Responsibilities

In the early 1970s a gap became apparent between the public's and the auditors' perceptions of audit functions and responsibilities. A need developed to investigate the reality of such a gap and to suggest ways to close it if it did indeed exist. To accomplish this task, the Board of Directors of the AICPA appointed an independent group in 1974 to examine the problems and issues facing CPAs in the role of independent auditors. This group is called The Commission on Auditors' Responsibilities (also known as the Cohen Commission). Its duties are described in the following quotation taken from its Statement of Issues published in 1975.

> In the broadest sense, the function of independent auditors is to enhance the reliability of information used in financial decisions of a wide range of individuals and organizations. This role is an important aspect of the process of efficient allocation of resources in the economy. Therefore, it is vital to the economy that users of information have confidence in auditors. Such confidence is dependent on a mutual understanding as to the appropriate responsibilities of auditors and a belief by users that such responsibilities are being fulfilled.

> In view of the growing demands by investors, creditors, management, government, and the general public for auditors to assume a wider scope of responsibility, the American Institute of Certified Public Accountants has concluded that a full-scale study should be made of the future function of independent auditors.

> The main purpose of the study is to develop conclusions and recommendations regarding the appropriate responsibilities of independent auditors. It should consider whether a gap may exist between what the public expects or needs and what auditors can and should reasonably expect to accomplish. If such a gap does exist, it needs to be explored to determine how the disparity can be resolved.

After three years of study, the Commission issued a set of recommendations covering a wide variety of auditing issues. These recommendations are designed to clarify auditors' responsibilities in several critical areas and to restore any perceived or actual loss of confidence by the public in auditing services.

Some of the recommendations of the Commission are briefly summarized below.

1. The auditor should evaluate the choices made by management among alternative accounting principles. Traditionally, many auditors have accepted any alternative used by their clients.

2. The word "fairly" should be removed from the auditors' report because the term generally is not understood by the public. The Commission suggested a more flexible format of reporting opinions.

3. A financial statement footnote which identifies material uncertainties for users should be required. The "subject to" qualification in the auditors' report should be eliminated because the term is redundant and calls upon the auditor to predict the future financial effects of uncertainties.

4. The auditor should search for material fraud, and audit procedures should be designed to provide reasonable assurance that material fraud does not affect financial statements. In the past the auditing profession has issued pronouncements which have been interpreted by some members of the public as a denial of responsibility for fraud detection.

5. The auditor should search for illegal acts based on proper conduct policy statements issued by management. Auditors have traditionally taken the position that illegal acts would affect the scope of their engagement only if the fairness of the financial statements appeared to be affected.

6. The auditors' evaluation of internal control should be incorporated in the audit report. In the past a separate letter has been issued to management and the board of directors. Public exposure of this letter has been management's option.

7. The auditor's report should be revised further to consist of a series of paragraphs, each describing a major element of the audit function.

8. The Auditing Standards Executive Committee should be replaced by a full time board.

Some of the Commission's suggested changes are being incorporated and some changes will occur in the future. There appears to be little question, however, that the late 1970s spawned a changing public attitude toward auditors' responsibilities and that the auditing profession is responding to this change.

Investigations by the U.S. Congress

The public's "expectation gap" caused the auditing profession to suggest changes in its perception of auditors' responsibilities, and the same gap resulted in investigations by committees of the U.S. Congress. These investigations were directed at accounting practices *and* auditing services, and were conducted by committees of both houses of Congress. The findings contained in the *Report of the Subcommittee on Reports, Accounting and Management of the Committee on Governmental Affairs of the U.S. Senate* (also known as the Metcalf Report) are the basis for this subcommittee's suggestions on the improvement of auditing services offered to the public.

In an earlier subcommittee staff study certain recommendations were made which many members of the auditing profession found objectionable. Among the recommendations were the following.

1. Congress should pass legislation overturning the decision of the U.S. Supreme Court in the Hohfelder case. As indicated in Chapter 3, this Court decision disallowed damages against a CPA firm for negligence under the 1934 Securities Exchange Act.

2. There should be a mandatory change of auditors of public corporations after a given period of years. This step was suggested as a means of increasing competition among the CPA firms and insuring auditor independence.

3. To restore public confidence in published corporate reports, auditing standards should be established by some branch of the federal government.

4. To restore public confidence in the independence of auditors, the joint rendering of audit and management advisory services should be disallowed.

After a series of public hearings had been held on the staff study recommendations, a final report of the subcommittee was issued in 1977. In general the subcommittee endorsed the recommendations of the Cohen Commission and their final report appeared to accept the general propositions that the setting of auditing standards should remain in the private sector. However, the report contained certain comments that might result in changes in auditors' responsibilities.

1. The subcommittee believed that nonaccounting management services such as executive recruitment are incompatible with the function of independent auditing.

2. The subcommittee believed that the auditing profession's prohibitions against (1) advertising and (2) discussing employment with another CPA

firm's employee without notifying that CPA firm should be dropped. The subcommittee believed that such prohibitions create artificial restrictions on auditor/client communications and also lessen competition.*

3. The subcommittee believed that independent auditors of publicly owned companies should be liable for negligence to private individuals who suffer damage from such negligence.

4. The subcommittee recognized the important role of the SEC in influencing the performance of independent auditors.

Although no one can foresee all the changes that will occur in the auditing function within coming years, it is obvious that many will be made. Public expectations of auditors have grown and the profession is responding through changes of its own.

---

*The 1978 revision of the Rules of Conduct of the AICPA *Code of Professional Ethics* has modified or dropped these prohibitions. See Chapter 2.

Index